ADVERSE EFFECTS OF ENGINEERED NANOMATERIALS

EXPOSURE, TOXICOLOGY, AND IMPACT ON HUMAN HEALTH

ADVERSE EFFECTS OF ENGINEERED NANOMATERIALS

EXPOSURE, TOXICOLOGY, AND IMPACT ON HUMAN HEALTH

Edited by

BENGT FADEEL
Division of Molecular Toxicology
Institute of Environmental Medicine, Karolinska Institutet
Stockholm, Sweden

ANTONIO PIETROIUSTI
Department of Biopathology
Tor Vergata University
Rome, Italy

ANNA A. SHVEDOVA
Pathology and Physiology Research Branch
Health Effects Laboratory Division
National Institute for Occupational Safety and Health
Morgantown, WV, USA

and

Department of Physiology and Pharmacology
School of Medicine
West Virginia University
Morgantown, WV, USA

ELSEVIER

AMSTERDAM • BOSTON • HEIDELBERG • LONDON
NEW YORK • OXFORD • PARIS • SAN DIEGO
SAN FRANCISCO • SINGAPORE • SYDNEY • TOKYO
Academic Press is an imprint of Elsevier

Academic Press is an imprint of Elsevier
32 Jamestown Road, London NW1 7BY, UK
225 Wyman Street, Waltham, MA 02451, USA
525 B Street, Suite 1800, San Diego, CA 92101-4495, USA

First edition 2012

British Library Cataloguing-in-Publication Data
A catalogue record for this book is available from the British Library

Library of Congress Cataloging-in-Publication Data
A catalog record for this book is available from the Library of Congress

ISBN : 978-0-12-386940-1

For information on all Academic Press publications
visit our website at elsevierdirect.com

Typeset by TNQ Books and Journals Pvt Ltd.
www.tnq.co.in

Printed and bound in United States of America

12 13 14 15 16 10 9 8 7 6 5 4 3 2 1

Contents

C

BIOMEDICAL APPLICATIONS OF ENGINEERED NANOMATERIALS

About the Editors

BENGT FADEEL is Professor of Medical Inflammation Research at the Institute of Environmental Medicine, Karolinska Institutet, Stockholm, Sweden, and Adjunct Professor of Environmental and Occupational Health, University of Pittsburgh, Pittsburgh, PA. Dr. Fadeel received his M.D. and Ph.D. degrees from Karolinska Institutet and is the current Head of the Division of Molecular Toxicology, and Vice Chairman of the Institute of Environmental Medicine. He is a member or coordinator of several Seventh Framework Programme projects on nanosafety, member of the Working Group on Hazard in the nanoFUTURES platform of the European Commission, and member of the nanomedicines expert group at the European Medicines Agency. He is also a member of the Editorial Board of *Apoptosis* and *Nanomedicine* and author or co-author of more than 130 scientific articles. Dr. Fadeel was awarded the national Environmental Medicine Prize (2011) by the Swedish Cancer and Allergy Foundation for his research on the opportunities and risks of the emerging nanotechnologies.

ANTONIO PIETROIUSTI is Assistant Professor of Occupational Medicine at the Faculty of Medicine, Department of Biopathology, Tor Vergata University, Rome, Italy. He is also Professor of Occupational

Medicine at the Faculty of Odontostomatology of Tor Vergata University, Rome, Italy, and at the Faculties of Medicine and Odontostomatology of the University "Nostra Signora del Buon Consiglio" of Tirana, Albania. He is teacher of Occupational Medicine at the Specialization School of Internal Medicine of Tor Vergata University, Rome, Italy. Dr. Pietroiusti received his M.D. degree and specialization in Internal Medicine from Tor Vergata University. He is a member of one Seventh Framework Programme projects on nanosafety, and has been partner of one Italian Relevant National Interest Project (PRIN) on the same subject. He has been editor of one special issue of the *Italian Journal of Occupational Medicine and Industrial Hygiene* and author or co-author of more than 100 scientific articles.

ANNA A. SHVEDOVA is Lead Research Physiologist at National Institute for Occupational Safety and Health (NIOSH) at the Center for Disease Control and Prevention (CDC) and Adjunct Professor of the Department of Physiology and Pharmacology, Medical School of West Virginia University, WV. Dr. Shvedova received her Ph.D. and D.Sc. degrees from Moscow University, Russia and is currently Principal Investigator of a number of NORA/National Occupational Research Agenda and NTRC/Nanotechnology Research Center projects at the Pathology and Physiology Research Branch/NIOSH/CDC. Dr. Shvedova was founder and first President of the Dermal Toxicology Specialty Section of the Society of Toxicology (SOT). She is member of advisory board committees for US Army, Air Force and NASA projects and workpackage leader for *in vivo* studies in the FP7-NANOMMUNE project (2008-2011). Dr. Shvedova has been honored with the SOT Public Communication Award in 2001; Alice Hamilton NIOSH Award for paper of the year in Occupational Safety and Health in 2006, 2009; Bullard-Sherwood Award: Research for Practice in 2011, and Women in Toxicology SOT Award in 2007. Dr. Shvedova is board member of the working group on Skin Notation at the National Institute of Occupational Safety and Health/CDC, Associate Editor of *Toxicology & Applied Pharmacology*, and Editorial Board Member of *Regulatory Toxicology & Pharmacology*. Dr. Shvedova is author or co-author of more than 120 scientific articles and book chapters.

Preface

Should one ban aspirin? Most would say no—aspirin has helped to stave off disease in innumerable patients. Yet, this drug has also caused thousands of deaths due to gastrointestinal bleeding and severe disabilities due to hemorrhagic stroke. However, its benefits largely outweigh the costs of these adverse effects. In some way, this reasoning is even more true for engineered nanomaterials (ENM), if we consider that the enormous benefits to humans of their industrial and biomedical applications are certain, whereas their expected adverse effects are probably largely preventable. We have the duty, however, to develop a careful understanding of these effects in order to minimize them as much as possible. This book, focusing essentially on the disease potential of ENM, by examining systematically their possible effects on different organs and systems, on the basis of the experimental evidence and of plausible pathogenetic pathways, may provide a useful guide in this endeavor.

The chapters, written by acknowledged experts in the field, attempt a translation from experimental data to human disease in order to address the question: what nanomaterial-induced diseases might potentially occur in humans? The book provides up-to-date information on occupational exposure and adverse health effects of ENM, and also discusses their biomedical applications, the other side of the coin.

The first part of the book includes six chapters, aimed at giving the reader the possibility of a correct interpretation of the reported data. More specifically, the first chapter of the book gives a general overview of the interaction of ENM with the human body, describing basic phenomena such as corona formation, oxidative stress and inflammation, which may lead to adverse outcomes in different organs and conditions (from cancer, to atherosclerosis, to fetal maldevelopment). The following two chapters describe the assessment of ENM in the occupational setting and how ENM can be managed. Hence, Chapter 2 discusses exposure to ENM in different settings, methods of exposure evaluation, and available protective measures, and Chapter 3 the existence and development of reliable—and possibly specific—biological markers for health surveillance. Chapters 4 and 5 deal with the complex case of *in vivo* and *in vitro* assessments of ENM toxicity and they describe two possible solutions to the problem: on the one hand, the selection of the most appropriate *in vitro* (cell culture-based) and *in vivo* (live animal) tests and the correct interpretation of the findings (Chapter 4), and on the other, the development of predictive computer-generated models (Chapter 5). The two approaches are not mutually exclusive, since the performance of correctly designed experiments is the prerequisite for the development of reliable computational models.

The first section of the book ends with the exhaustive description of the competences and decisions of the most important EU and US regulatory and governmental agencies, aimed at the legislative safeguarding of workers, consumers, and patients from the potential risk of inadvertent or deliberate exposure to ENM.

The second part includes a series of eight chapters addressing the pathogenic potential of ENM for specific organs and tissues. Chapter 7 describes the possible adverse effects on the respiratory system, one of the most important routes of entry of ENM, whereas other organ system-specific chapters are devoted to ENM effects on the cardiovascular system (Chapter 8), neurological system (Chapter 9), skin (Chapter 11) and the gastrointestinal tract (Chapter 12). The body of evidence for the disease potential of ENM is substantial: the possibility of pulmonary fibrosis, malignant mesothelioma and atherosclerotic disease should be considered, at least for some ENM. In other cases, the link with human diseases is much more speculative, such as for neurodegenerative disorders or with Crohn's disease, an inflammatory disease of the intestines. It should be considered, however, that the amount of research on the possible pulmonary and cardiovascular effects is much higher than on neurological and gastrointestinal adverse effects. For example, nothing is known on the process of corona formation and transformation, and of the biological consequences of this process, for ENM traveling through the various compartments of the gastrointestinal lumen. The skin seems perhaps the most insensitive organ to the possible injurious action of ENM, although they might cause damage at distal sites in other, more susceptible organs if they cross the skin barrier. Chapters 10 and 14 deal with the possible impact of ENM on the immune system and on our genetic material, respectively. Both report plausible evidence of damage, a finding which may have profound implications for human health, since alterations of the immunologic and genetic integrity underpin numerous disease processes, including the development of cancer. Chapter 13 focuses on a very recently reported possible adverse effect of ENM, i.e. reproductive toxicology. This is of particular concern, for obvious reasons.

In the third and final section of the book, two chapters focusing on diagnostic (Chapter 15) and therapeutic (Chapter 16) applications of ENM are included. Although the book is oriented towards adverse effects, the positive health implications of ENM should not be ignored. The chapter on diagnostics is focused on cancer, since this represents the most relevant application and because other uses of ENM as diagnostic tools are discussed in the context of some of the other chapters (in particular, Chapters 4 and 8). Chapter 16 provides a comprehensive overview of nanomaterials for drug delivery including a critical discussion of various classes of nano-scale drug carriers that are already in clinical use or are currently in clinical trials or show promise for therapeutic and/or theranostic applications.

The annex provides information on various routes of synthesis of various classes of nanomaterials. This is not an esoteric topic of interest only to material scientists; a detailed understanding of the synthesis of ENM is needed as this may impact on biological interactions and, hence, on disease processes. For instance, the presence of contaminants can sometimes explain cytotoxic effects. Finally, the glossary at the end of the book provides an explanation of the most relevant terms used in the book.

This book is a state-of-the-art presentation of relevant research on adverse effects of engineered nanomaterials on human health. We are indebted to all the colleagues who have participated in this project. We hope that this volume will serve as a useful guide for students and researchers in the field and for clinicians, policymakers, and regulators with an interest in nanosafety and human health.

Bengt Fadeel

Karolinska Institutet

Antonio Pietroiusti,

University of Rome "Tor Vergata"

Anna A. Shvedova,

National Institute for Occupational Safety and Health and West Virginia University

List of Contributors

Christoph Alexiou Department of Oto-Rhino-Laryngology, Head and Neck Surgery, Section for Experimental Oncology and Nanomedicine, Erlangen, Germany

Giuseppe Bardi Center for Nanotechnology Innovation @ NEST, Istituto Italiano di Tecnologia, Pisa, Italy

Antonio Bergamaschi Institute of Occupational Medicine, Università Cattolica del Sacro Cuore Roma, Rome, Italy

Enrico Bergamaschi Laboratory of Industrial Toxicology, University of Parma Medical School, Parma, Italy

Diana Boraschi Institute of Biomedical Technologies of the National Research Council, Pisa, Italy

Cyrill Bussy Nanomedicine Laboratory, Center for Drug Delivery Research, The School of Pharmacy, University of London, London, UK

Luisa Campagnolo Department of Public Health and Cell Biology, University of Tor Vergata, Rome, Italy

Sandra Ceccatelli Department of Neuroscience, Karolinska Institutet, Stockholm, Sweden

Chunying Chen National Centre for Nanoscience and Technology, Beijing, China

Kai Cheng Molecular Imaging Program at Stanford (MIPS), Canary Center at Stanford for Cancer Early Detection, Department of Radiology and Bio-X Program, School of Medicine, Stanford University, Stanford, CA, USA

Zhen Cheng Molecular Imaging Program at Stanford (MIPS), Canary Center at Stanford for Cancer Early Detection, Department of Radiology and Bio-X Program, School of Medicine, Stanford University, Stanford, CA, USA

Renato Colognato Clinical Pharmacology Unit, "L. Sacco" University Hospital, Milano, Italy

Sara Correia Carreira BIRCH, Biophysics Research Unit, Department of Medical Physics & Bioengineering, University Hospitals Bristol NHS Foundation Trust, St Michael's Hospital, Bristol, UK

Wim H. De Jong National Institute for Public Health and the Environment, Laboratory for Health Protection Research, Bilthoven, The Netherlands

Shareen H. Doak Institute of Life Science, College of Medicine, Swansea University, Swansea, Wales, UK

Ken Donaldson Queen's Medical Research Institute, University of Edinburgh, Edinburgh, UK

Rodger Duffin Centre for Inflammation Research, Edinburgh University, Edinburgh, UK

Albert Duschl University of Salzburg, Department of Molecular Biology, Salzburg, Austria

Maria Dusinska NILU-Norwegian Institute for Air Research, Health Effects Laboratory, Departments of CEE and Environmental Chemistry, Kjeller, Norway

Vidana Epa CSIRO Materials Science & Engineering, Parkville, Australia

Maureen R. Gwinn US Environmental Protection Agency, Washington, DC, USA

Karin Sørig Hougaard National Research Center for the Working Environment, Copenhagen, Denmark

Ivo Iavicoli Institute of Occupational Medicine, Università Cattolica del Sacro Cuore Roma, Rome, Italy

Mark A. Jepson School of Biochemistry, University of Bristol, Bristol, UK

Vadim G. Kessler Department of Chemistry, SLU BioCenter, Uppsala, Sweden

Kostas Kostarelos Nanomedicine Laboratory, Center for Drug Delivery Research, The School of Pharmacy, University of London, London, UK

Francesca Larese Filon Unit of Occupational Medicine, University of Trieste, Trieste, Italy

Ying Liu National Centre for Nanoscience and Technology, Beijing, China

Andrea Magrini Department of Biopathology and Imaging, University of Rome "Tor Vergata", Rome, Italy

Laura Methven Nanomedicine Laboratory, Center for Drug Delivery Research, The School of Pharmacy, University of London, London, UK

Nicholas L. Mills Centre for Cardiovascular Science, Edinburgh University, Edinburgh, UK

Nancy A. Monteiro-Riviere North Carolina State University, Center for Chemical Toxicology Research and Pharmacokinetics, Raleigh, NC, USA

Andreas M. Nyström The Swedish Medical Nanoscience Center, and the Department of Neuroscience Karolinska Institutet, Stockholm, Sweden

Margriet V.D.Z. Park Maastricht University, Department of Toxicogenomics, Maastricht; National Institute for Public Health and the Environment, Laboratory for Health Protection Research, Bilthoven, The Netherlands

Robby A. Petros Department of Chemistry, University of North Texas, Denton, TX, USA

Craig A. Poland Institute of Occupational Medicine, Edinburgh, UK

Jennifer B. Raftis Centre for Inflammation Research, Edinburgh University, Edinburgh, UK

Elise Rundén-Pran NILU-Norwegian Institute for Air Research, Health Effects Laboratory, Departments of CEE and Environmental Chemistry, Kjeller, Norway

Margaret Saunders BIRCH, Biophysics Research Unit, Department of Medical Physics & Bioengineering, University Hospitals Bristol NHS Foundation Trust, St Michael's Hospital, Bristol, UK

Kai Savolainen Nanosafety Research Centre, Finnish Institute of Occupational Health, Helsinki, Finland

Gulaim A. Seisenbaeva Department of Chemistry, SLU BioCenter, Uppsala, Sweden

Birgit Sokull-Klüttgen European Commission, Joint Research Centre, Institute for Health and Consumer Protection, Unit IO5 Nanobiosciences, Ispra, Italy

Muhammet S. Toprak KTH Royal Institute of Technology, Functional Materials Division, Kista-Stockholm, Sweden

Lang Tran Institute of Occupational Medicine, Research Avenue North, Edinburgh, UK

Peter Wick Empa, Swiss Federal Laboratories for Materials Science and Technology, Laboratory for Materials-Biology Interactions, St Gallen, Switzerland

Dave Winkler CSIRO Materials Science & Engineering, Clayton, Australia; Monash Institute of Pharmaceutical Science, Parkville, Australia

SECTION A

ENGINEERED NANOMATERIALS: HAZARD, EXPOSURE AND SAFETY ASSESSMENT

CHAPTER

1

Interactions with the Human Body

Renato Colognato[1], *Margriet V.D.Z. Park*[2,3], *Peter Wick*[4], *Wim H. De Jong*[3]

[1]Clinical Pharmacology Unit, "L. Sacco" University Hospital, Milano, Italy

[2]Maastricht University, Department of Toxicogenomics, Maastricht, The Netherlands

[3]National Institute for Public Health and the Environment, Laboratory for Health Protection Research, Bilthoven, The Netherlands

[4]Empa, Swiss Federal Laboratories for Materials Science and Technology, Laboratory for Materials-Biology Interactions, St Gallen, Switzerland

OUTLINE

Adverse Effects of Engineered Nanomaterials
DOI: 10.1016/B978-0-12-386940-1.00001-5

INTRODUCTION

The Nanotechnology Era

December 29th 1959, when Professor Richard Feynman gave his talk *"There's Plenty of Room at the Bottom"* (http://www.zyvex.com/nanotech/feynman.html) at the annual meeting of the American Physical Society at the California Institute of Technology (Caltech), can probably be considered the official opening day of a new scientific era. More than 50 years have passed since that 29th December and many new events have strengthened Professor Feynman's talk and hypothesis on this so-called "Plenty of Room", which has now been partially discovered, or better to say "filled", with scientific information. The bottom has now been assessed to be everything $<10^{-9}$ m unit, and what Professor Feynman was calling the "Bottom" can now be defined as *nanotechnology*: the technology of the 10^{-9} m materials; the *nanomaterials* or *nanoparticles*.

With the rise of nanotechnology, there was a dramatic increase in pharmaceutical and drug discoveries and within this research field the delayed and controlled release of drugs was a major task. One of the first scientists integrating the pharmacological approach to materials at the nanolevel, was Professor Peter Paul Speiser at the ETH (Swiss Federal Institute of Technology) in Zürich. His research group investigated polyacrylic beads, microcapsules, for oral administration of different drug formulations [1]. In parallel, another type of nanoparticle (NP) was developed at the Department of Radiological Science at the Johns Hopkins Medical Institutions in Baltimore. This group implemented a process in which albumin was dissolved in water, followed by emulsifying this solution in cottonseed oil [2,3]. By heating to 175–185°C, the protein was denatured, resulting in NP of sizes between 300 and 1000 nm. These particles were labeled with 99mTc and were investigated for drug delivery purposes using the cytostatic mercaptopurine [4].

Some years later in Japan, Sugibayashi [5], building upon the work performed by Scheffel [3], bound 5-fluorouracil to albumin NP. After intravenous injection into Ehrlich Ascites Carcinoma-bearing mice, he found a 20% increase in life span. With these experiments NP began to be of interest for biomedical applications and conducting studies with NP had passed the point of no return.

Among all nanomaterials (NM) that have been brought to the attention of the public, this book will take into consideration engineered nanomaterials (ENM). Due to their specific design, ENM possess physicochemical properties (PCPs) that distinguish them from other well-known man-induced (welding fume, diesel exhaust) and/or environmental particulates (e.g. controlled size, mass, chemical composition, active surface, charge, agglomeration and aggregation state, etc.). The emphasis on ENM can be justified by the fact that, nowadays, ENM or the use of nanotechnology can be found in a wide variety of products like electronics, automotives, health and fitness, food and beverages, goods for children and many other product categories as indicated by the commercialization of 1317 products, with a five-fold increase since March 2006 (www.nanotechproject.org). Considering this latter issue and the wide coverage of industries in which ENM are in the production chain, it is self-evident that the interaction of ENM with biological systems should be investigated. In view of the range of different PCPs that ENM can express, this should be approached in an interdisciplinary manner, putting together biology, medicine, physics, engineering and chemistry.

The uncertainty of possible adverse effects of ENM, for which a lot of information is still incomplete or missing altogether, has been vigorously taken into consideration by many institutional organizations: the Environmental Protection Agency (EPA), the World Health Organization (WHO), the National Institute for Occupational Safety and Health (NIOSH), the European Commission (EU) and the Organization for Economic and Commercial Development (OECD). Many official documents have been published addressing the need of dedicated research in order to ensure citizen protection from adverse environmental and health risks potentially caused by ENM. In this context, the European Commission, through the Communication "Towards a European Strategy for Nanotechnology" (http://cordis.europa.eu/nanotechnology/actionplan.htm) suggests a safe and responsible strategy calling for research on appropriate methodological assays for assessing ENM toxicity. At the same time, NIOSH, in collaboration with the Center for Disease Control, through its "Approaches to Safe Nanotechnology" (http://www.cdc.gov/niosh/docs/2009-125/), recommends that specific precautions should be taken to protect workers who might be exposed to any level of ENM or ENM-containing materials. Also within the context of ISO (ISO/TC 229: Nanotechnologies, International Organization for Standardization, Geneva, Switzerland), documents are drafted for the safe and responsible production and use of ENM.

Engineered Nanomaterials (ENM): Definition and Nomenclature

In most documents and (working) definitions, the nanoscale describing the size of ENM is defined as the size range between (approximately) 1 nm and 100 nm [6–10]. Although they did not provide a regulatory definition, the Food and Drug Administration (FDA) has also recently published a draft guidance document in which the size of approximately 1 nm–100 nm is referred to as a first reference point to consider whether an FDA-regulated product contains nanomaterials, or otherwise involves application of nanotechnology [11]. In the various documents (standards and technical reports) of the ISO, the term ENM covers all materials and constructs at the nanoscale [6]. In addition to ISO/TS 80004-1:2010, there will be a whole series of documents on the vocabulary to be used for the description of specific ENM. Individual primary structures are designated ENM that can have one, two or three dimensions in the nanoscale. Agglomerates and aggregates of single primary nanomaterials are considered to be nanostructured materials and are included in the definition of ENM. It will be obvious that when a material is specifically designed to have a size below 100 nm that this is an ENM. However, although the scientific community agrees on the use of the term ENM, it becomes less clear when the definition of an ENM is discussed from a regulatory point of view. So far, regulatory definitions of ENM are mainly limited to descriptions in governmental statements or working definitions [8–11] and in one official regulation (EU Cosmetics Directive) [12]. An ENM can generally be described by its nominal size, i.e. the mean particle size. However, any size of an ENM has a size distribution, i.e. not all particles have the same size. The size distribution is one of the characteristics that describe a particular ENM and gives rise to the question of when one should consider a material to be an ENM when only part of the material's size distribution is below the 100 nm threshold. In addition, there is debate on whether the size distribution should be expressed as part of the mass or as a number size distribution. The size of the particles by number clearly indicates how many of the particles are below 100 nm and thus may pose a size-related risk.

Another issue is the change in PCPs. In some proposals for regulation this is included [8–10], as ENM may be specifically attractive to use, because they obtain specific characteristics with diminishing size. Although for some specific types of ENM, for example inorganic metal and metal oxides, a threshold of approximately 20–30 nm could be distinguished for a change in PCPs [13], for other materials this threshold was much higher, even 200–300 nm, or absent altogether [7]. There is no universal size for all ENM at which nanospecific properties are beginning to emerge.

Physical and Chemical Properties of ENM

The reason for ENM to be in the "spotlight" of both scientific research and regulators is indeed due, mainly, to the same physical and chemical properties that confer to them great potential for industrial and biomedical applications [14], such as size, shape, surface charge, aggregation/agglomeration state and solubility (soluble ions released from ENM) (Figure 1.1). Buzea and collaborators [15] state: "An important lesson we are in the process of learning from nanoscience is that simple classifications of physical behavior (and, therefore, toxicity) are overly limiting and that we must study toxicology of each material and each morphology, in addition to particle aging, to obtain accurate information to inform policy and regulatory processes". This phrase perfectly highlights the experimental approach to proceed for a better understanding of cellular responses to ENM. It is, in this context, almost impossible to describe in detail all ENM PCPs that are able to trigger biological responses, so we will only give a brief overview of those which, more than others, have been related to cellular effects.

One of the main distinguishing characteristics, always used to compare bulk versus ENM, and that can justify some biological effects in terms of nano–bio interactions, is the size, or, to be more precise, the surface to volume ratio. In fact, ENM possess a large surface area compared to the respective bulk materials and this latter issue makes ENM considerably more biologically reactive, e.g. a 300 nm NP has 5% of its atoms on the surface while the remaining atoms reside inside. In comparison, a 30 nm NP has 50% of its atoms on the surface. The atoms on the surface, which are not bonded on one side, are indeed far more active than the atoms residing inside. An increase in surface area therefore leads to an increase in surface reactivity. For some specific types of ENM, i.e. inorganic metal and metal oxides, a threshold of approximately 20–30 nm could be distinguished at which PCPs changed significantly [13] and several examples exist that show that reduction in size results in enhanced particle toxicity [16]. Whether this increase in surface area results in an increase of biological (toxicological) reactivity for ENM in general is currently under investigation.

With regard to ENM shape, there is a wide range of information available and, in general, ENM can be roughly distinguished into high- and low-aspect-ratio nanoparticles, HARN and LARN respectively. High-aspect ratio ENM include nanotubes and nanowires, with various shapes, such as helices, zigzags, belts, and may vary in length. LARN morphologies include spherical, oval, cubic, prism, helical, or pillar. A shape-dependent biological response was reported by Tarantola and co-workers [17] who demonstrated that coated spherical gold nanoparticles (AuNP) with identical surface functionalization were generally more efficiently ingested and more toxic than rod-shaped particles.

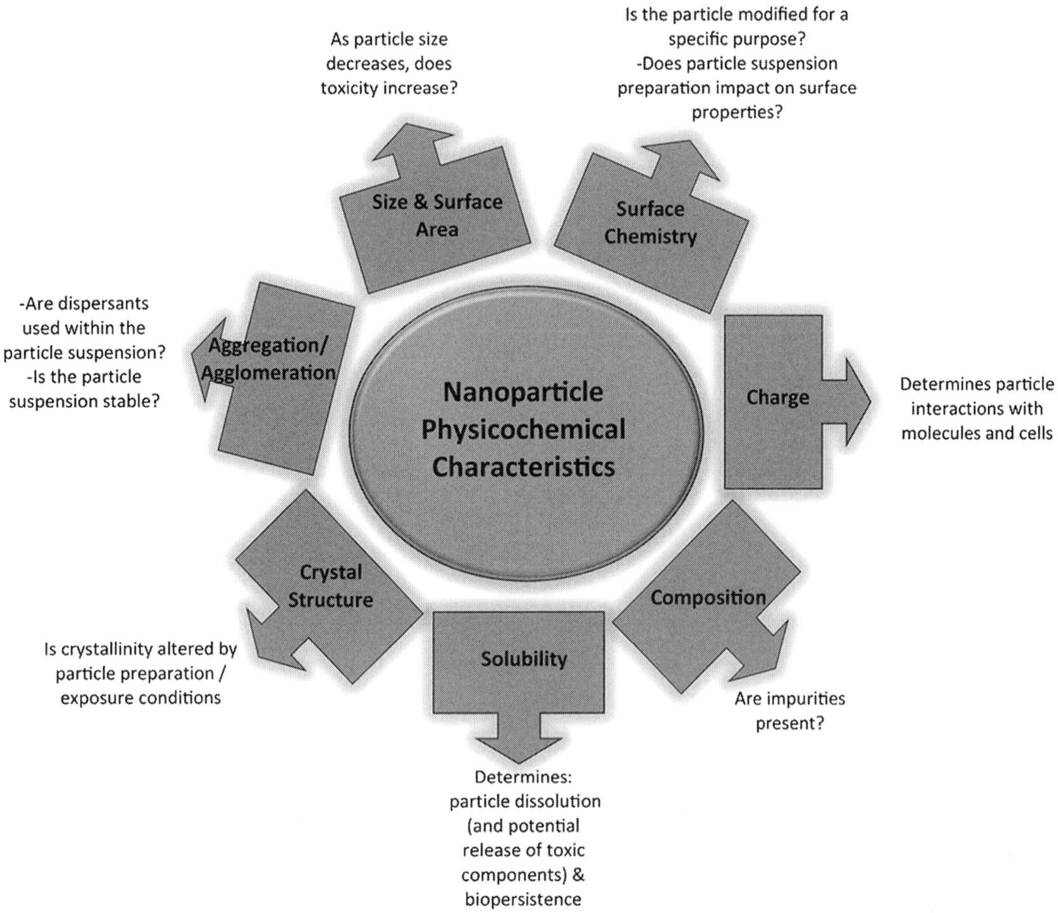

FIGURE 1.1 Engineered nanomaterials physicochemical properties. Physicochemical properties of nanomaterials that should be considered with high priority in toxicological assessments are outlined. Determining the size, surface area, aggregation/agglomeration tendency, composition, surface chemistry, charge, crystal structure, and solubility of the nanomaterial under investigation is therefore critical, as these attributes have been consistently demonstrated to influence the toxicity of nanomaterials. *(Reproduced from [26] with permission from Oxford Journal.)*

Other PCPs contributing to the behavior and biological responses of ENM are the stability and agglomeration state (additional information on the relationship between stability and biological effects can be found in Chapter 9). In aerosols and in colloidal suspensions, ENM can exist in a well-dispersed or in an agglomerated state, depending on the chemistry and electromagnetic properties of both the ENM and their surroundings. For example, magnetic ENM tend to cluster, forming an agglomerated state, unless their surfaces are coated with a non-magnetic "shell". Aggregation or agglomeration state could influence the deposition, the toxicokinetics, and the local toxicity of ENM, due to significant variations in the aerodynamic radius (important for aggregates) and the total surface area of the ENM that tends to be reduced [18–20]. Last, but not least, surface charge also has an outstanding role in the induction of biological effects. For three different AuNPs, positively, neutral and negatively charged, cell morphology was disrupted by all three NPs with a dose-dependent toxicity [21]. Nevertheless, the positively and negatively charged AuNPs displayed toxicity at concentrations at least 2.5-fold lower than neutral NPs [21]. Even though research has been unclear with regard to the toxicity mechanisms of AgNPs, research has shown a toxicity dependence on factors such as particle size, shape and surface properties [22–25]. Fullerenes (C_{60}) and carbon nanotubes are examples that clearly demonstrate the strong relationship between ENM physicochemical properties and biological effects [26,27].

ENGINEERED NANOMATERIAL–BIO INTERACTIONS

Biochemical and Biological Fluid Interactions

When materials come into contact with biological systems, a nano–bio interface is generated. This interface comprises the dynamic physicochemical interactions, kinetics and thermodynamic exchanges between the surfaces of ENM and biological components, such as proteins and phospholipids [28,29]. The solid–liquid interface is a key issue in ENM–cell interactions, but an understanding of this interface is often lacking. It is the amount and presentation of the biomolecules on the surface rather than the particles themselves that is suggested to be the cause of numerous biological responses [28,29]. In other words, it is the outer layer of biomolecules that is seen by the cells, and leads to the interaction of the particles with the cell.

The most abundant biomolecules that interact with ENM in biological environments are proteins [30–33]. There are some studies that have characterized and identified the adsorbed proteins—the corona—on different ENM or carbon nanotubes. A major outcome is that the NP size and surface properties determine the protein composition of the corona and by that influence the biological activity and molecular targeting of ENM [34,35]. However, protein binding to nanoparticles is not a static process. There is a dynamic exchange of the adsorbed proteins with the biological environment of the ENM. Another powerful tool in addition to biochemical analysis is the modeling of such binding processes. These calculations may offer a more detailed understanding of the solid–liquid interface [36,37].

Recently, it was shown that ENM have the potential to enhance protein assembly into amyloid fibrils by decreasing the lag time for nucleation in a cell-free experimental setup. In a follow-up study, prolongation of the fibrillation time could also be observed [38]. Depending on the ENM composition, the outcome of the interaction with proteins can vary. Especially, the phenomenon of increased protein assembly suggests a mechanism of surface-assisted nucleation that may increase the risk for toxic clusters and amyloid formation leading to neuronal diseases such as Alzheimer or Creutzfeld-Jacob diseases [39].

The possible portals where ENM may enter the human body are the lungs via inhalation, the gastrointestinal tract via digestion, the skin, and the blood vessels via intravenous injection [40]. However, the most critical (human) exposure route for ENM is via the pulmonary system. Upon deposition, ENMs come in initial contact with the pulmonary surfactant, which is located at the air–liquid interface. Surfactant contains 85–90% phospholipids and ≈10% surfactant proteins [41]. In a recent study with functionalized multiwalled carbon nanotubes (MWCNT), it was shown that pulmonary phospholipid coating has the potential to alter the plasma protein coating of the tubes. This change in the pattern of adsorbed proteins is hypothesized to influence significantly the interaction and subsequent effects of the MWCNT on biological systems [41,42]. The composition of the protein corona can be seen as a fingerprint of ENM in a certain environment, which might be used as a tool for early diagnostics. For example, the use of functionalized superparamagnetic nanoparticles (SPIONs) enables the identification of their specific interaction partners in a physiological environment, such as the entire intact cell or blood samples (further discussion on SPIONs can be found in Chapter 15). Proteomic analysis of the adsorbed molecules would allow the tracking of specific biomarkers related to diseases [43,44]. It is essential therefore, that further investigations be performed in order to understand this highly dynamic and complex interface.

The characterization of the solid–liquid interface is an interdisciplinary challenge. The interface is influenced by several different factors (Table 1.1). In a given medium, the most important ENM characteristics that determine the surface properties are the material's chemical composition, shape and curvature, surface functionalization, porosity and surface crystallinity and hydrophobicity or hydrophilicity [14,16]. Understanding the time evolution of a nano–bio interface in a given biological environment and, in particular, the understanding of the dynamic exchanges of biomolecules would enable the prediction of the particle's fate, biological activity or even its life time in an organism.

Transport Principle of Engineered Nanomaterials (Trojan Horse)

The basic features of compound uptake have been well described already by previous particle toxicity studies that recognized the process of phagocytosis to initiate potential adverse effects [45,46]. For ENM, these findings may not be applicable directly because phagocytosis is not the only relevant process of NP cell entry. Other mechanisms, too, are responsible for the uptake of metals, metal oxides, or other particulate nano-sized systems by the cell [47] and for the different biological reactions that may follow. Although particles

TABLE 1.1 Factors Influencing the Interface Between ENM and Biological Systems

Bio-physicochemical Actors	Main Forces Governing the Interfacial Interaction
Nanoparticles	Hydrodynamic interactions
Suspending media	Electrodynamic/electrostatic interactions
Solid–liquid interface	Solvent interactions
Nano–bio interface	Steric interactions
	Polymer bridging interactions

Adapted from Nel et al., 2009 [28].

with diameters below 100 nm are capable of entering a cell by almost any vesicle transport pathway [48–50], further options are to be considered, for example transport of ENM into cells due to receptor binding [51,52] or even "diffusion" through the plasma membranes, which is referred to as an adhesive interaction [48,53]. It was demonstrated with erythrocytes that ENM could advance into the cell interior, whereas larger particles were unsuccessful [53]. This is surprising, as the erythrocytes lack the conventional uptake mechanisms. No matter how the ENM gain entry into a cell, the process of infiltration is indeed reminiscent of a Trojan horse invasion because a veritable material package is delivered by introduction of only one particle into the cell (Figure 1.2) [40]. The effects observed are influenced by the different uptake processes. In the case of uptake by vesicular transport, particles are sheathed by membranes. Free transport through the membrane, however, would be assumed to be more critical, as it allows particles to achieve direct contact with the plasma proteins and with other molecules of the cell. Dependent on the material properties of the ENM, they may have fatal consequences for the cells. Soluble ENM such as ZnO may, once taken up, dissolute and generate a hot spot of free zinc at a much higher level than the surrounding average concentration would be [54–56]. Particles that do not dissolve but remain catalytically "active" have the potential to stress the cell or tissue as well [47]. Even inert particles may affect cell physiology by introducing xenobiotics adsorbed on the particle surface, which detach after cellular uptake. This piggy-back transportation is well known for combustion-derived particles where different types of polycyclic aromatic hydocarbons (potential mutagenic compounds) were identified on these particle surfaces [57]. The uptake of certain non-reactive materials may be particularly critical when they are nano-sized: particle

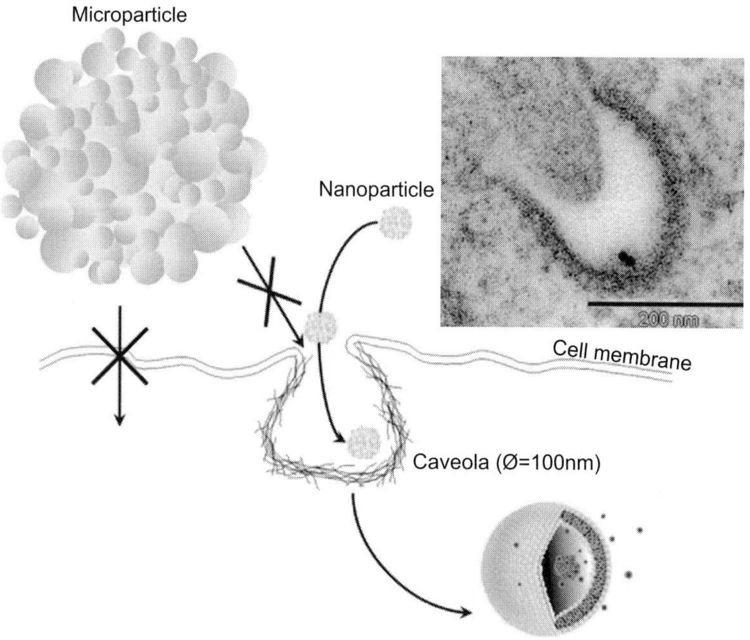

FIGURE 1.2 Comparison of nanoparticles and microparticles due to their possible uptake into cells via vesicular pathways (caveolae). Only small particles with a diameter of less than 100 nm fit into vesicular structures such as caveolae with which they will be transported into the plasma. Within the cells, these vesicles may fuse to build up a lysosome with an acidic interior, facilitating the dissolution of materials such as ZnO. The ions can move relatively freely inside the cells (dots). The TEM image shows such a situation where two nanoparticles (22 nm) are located within a caveolae of a lung epithelial cell (A549) in culture. *(From [40] with permission of Angewandte Chemie Int Ed.)*

distribution within an organism is differently controlled than the transport of ions or the corresponding bulk material, indicating that ENM may reach cells/tissue in an organism which were normally not affected. As indicated below, the body has a very efficient reticuloendothelial system to remove particulate matter from the circulation.

HEALTH IMPLICATIONS OF ENGINEERED NANOMATERIALS

Biodistribution, Biodegradation and Biopersistence

The biodistribution, biodegradation and biopersistence, in short, the kinetic properties of ENM can be described by four processes: absorption, distribution, metabolism and excretion (ADME), although, especially for solid ENM, it may be doubtful whether metabolism (M) really occurs. In general, four main routes of exposure can be distinguished, i.e. oral, inhalation, dermal and by systemic intravenous (IV) injection. Although the route of human exposure, with the exception of medical treatment, will not be by the IV route, many experimental studies for worst case (exposure) scenario purposes, have been performed using IV administration in order to avoid problems with (lack of) barrier passage in the gastrointestinal (GI) tract, lung or skin, and to evaluate which organs may be at risk for ENM toxicity based on the tissue distribution after systemic application. It should be realized that the route of entry itself also may affect the tissue distribution. Especially when comparing intravenous with inhalation administration, the translocation through the air/blood barrier may result in alterations of the ENM [58,59]. Systemic IV administration results in direct contact with blood, in which the particles' surfaces become coated with numerous proteins including immunoglobulins that can enhance recognition and uptake by cells of the reticuloendothelial system (RES) [60–62]. Such formation of a serum protein "corona" is suggested to initiate the mechanisms of recognition and internalization by cells of the RES [28,29], causing rapid clearance of the ENM from the blood into organs of the RES. To date, most particulate materials in the systemic circulation, including ENM, are taken up by the RES. This results in entrapment of the ENM mainly in organs with a high content of mononuclear phagocytes (macrophages) such as liver and spleen [63–65]. The sequestration of the particles into the liver and spleen is rapid, and can occur within minutes [66–68]. For more specific applications, i.e. drug targeting, modification of the surface area by coatings or antibodies may be used. Cellular binding may be increased by the use of carbohydrates [69], whereas specific antibodies can enable a more specific immunologically directed targeting of the ENM [70,71]. Cellular entrapment in macrophages can be avoided by surface modification of the ENM. Coupling of polyethylene glycol (PEG) onto the surface of ENM increases their circulation time in the blood by inhibiting recognition and phagocytosis by macrophages [66,68,72,73]. The use of PEG as a particle coating may also result in a change in local tissue exposure [66,68].

In addition, the size of ENM can be a significant determinant of particle tissue distribution, as shown for spherical AuNPs, gold composite nanodevices [63,74–76] and polystyrene ENM [77]. In contrast, for silver NPs (AgNPs) no such differences were observed, as three sizes of ENM showed a similar organ distribution, although 20 nm AgNPs appeared to be excreted (or dissolved) at a faster rate compared to larger (80 nm and 110 nm) AgNPs [67] (Figure 1.3). For IV administered quantum dots, a final hydrodynamic diameter <5.5 nm

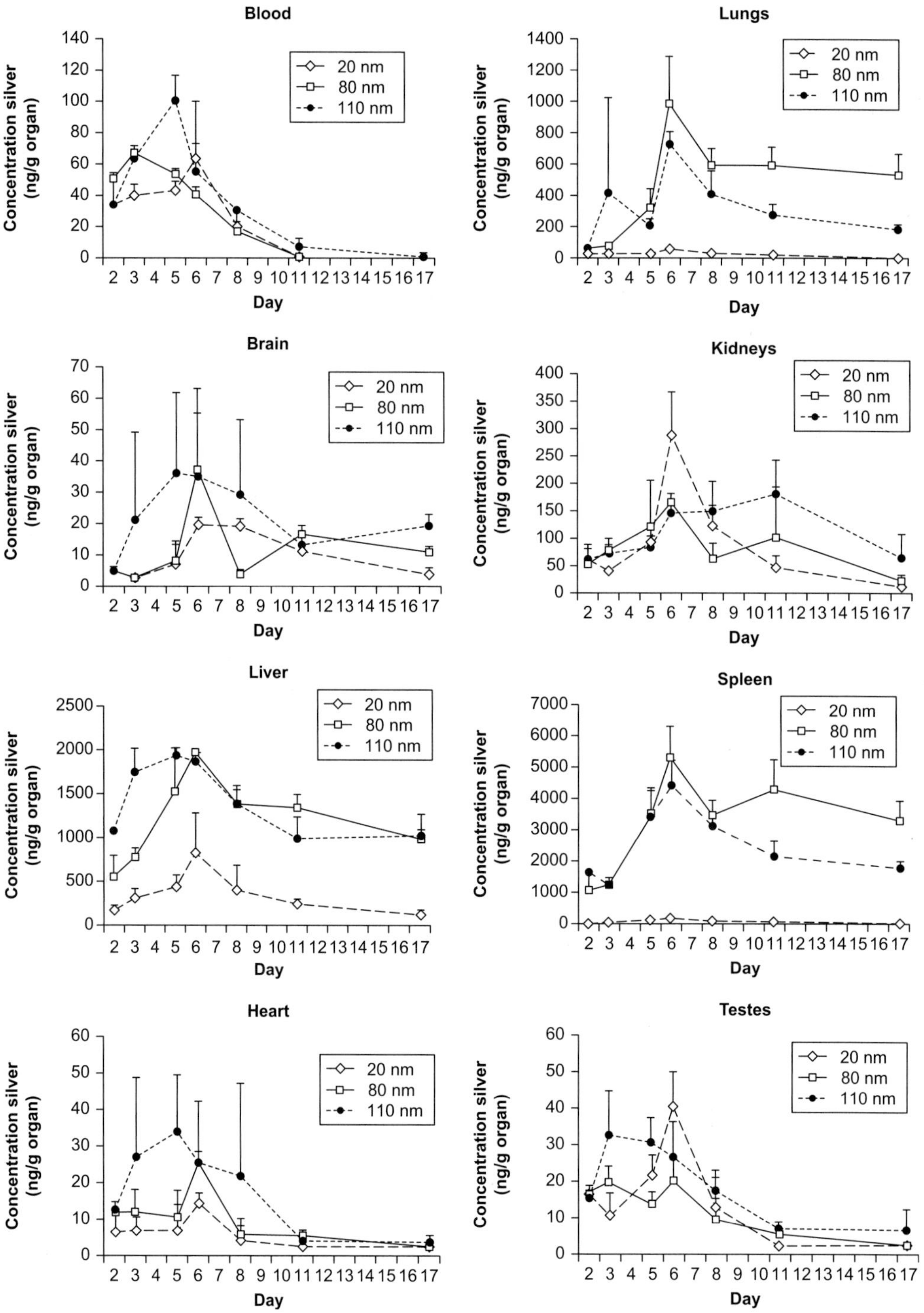

FIGURE 1.3 Tissue distribution and kinetics of silver nanoparticles of 20 nm, 80 nm and 110 nm. Animals received 5 consecutive IV administrations, $n=3$, error bar indicates standard deviation. *(From [67] with permission from Elsevier.)*

resulted in rapid and efficient urinary excretion and elimination of the quantum dots from the body, whereas for larger quantum dots, excretion was reduced and retention in the body was increased [78].

After exposure of the lung by either inhalation or instillation, a small but significant fraction of the dose of ENM was found to be absorbed into the body [59,66,79–81]. Such translocation may be rapid and size dependent [79]. With prolonged inhalation exposure, a notable increase, indicative of accumulation, in systemic organ distribution was observed between 5 days and 15 days of inhalation exposure to gold ENM [82]. However, the majority of the nanoparticles after inhalation (or instillation) exposure remains in the lung [59, 66]. The elimination half-time for both fine and ultrafine (nano)particles in rats was found to be around 65 days [83,84]. Another issue that needs to be considered with regard to inhalation exposure is the possible translocation from the nasal epithelium into the brain, although information on toxic effects resulting from exposure via this route is scarce.

There is a considerable effort to increase the systemic availability of food components and drugs after oral uptake by using nano-sized formulations both in the food area and for medical purposes [85–87]. For these applications, the degradation of the nanoformulations is inherent to their use. Although a risk assessment is needed for these biodegradable ENM as well, it is the potential adverse effects of the *non*-degrading ENM that may accumulate in the body which are of major concern. For several types of ENM, uptake from the GI tract was demonstrated after oral administration [88–91]. Although, in general, uptake of smaller ENM might be expected to be easier compared to larger ENM, as demonstrated for polystyrene latex and silver nanoparticles [88,92], also uptake of relatively large (500 nm) titanium particles was demonstrated [93].

Several *in vitro* studies have reported contradicting results for the penetration of intact skin by ENM [94–97] (see Chapter 11 for an extensive discussion on skin penetration by ENM). *In vivo*, the uptake via the skin of silica nanoparticless (70 nm), quantum dots (35 nm) and TiO_2 nanoparticles (4 nm and 60 nm) was reported [95,98], although oral uptake via grooming could not be excluded in these studies. After subcutaneous injection, silver ENM were found to distribute to main organs like liver, kidney, spleen, lung and brain [99]. In humans, a minimal uptake of Zn from a sunscreen preparation was observed but whether this was taken up as Zn ions or ZnO ENM could not be identified [100]. The indication of systemic availability of ENM implies that a careful consideration of skin or GI-tract uptake is needed when evaluating the safety of ENM.

Some ENM, such as polymeric constructs and phospholipd liposomal formulations, can be specifically designed to be degradable for release of incorporated drugs [101–103]. Potential solubility or insolubility is the key factor that determines whether a nanomaterial has the potential for bioaccumulation in cells and organs. However, for most metal and metal-oxide ENM, degradability is questionable or very slow indeed. Ag, CuO and ZnO nanoparticles are known to release ions that are considered to be responsible for part of their toxicity, indicating that degradation does occur to some extent [104–106].

The potential biopersistence of non-degradable ENM is considered one of the concerns in relation to possible long-term chronic effects of ENM. However, in the absence of chronic studies, there is a lack of information on the biopersistence of ENM. For silver, aluminum oxyhydroxides and titanium dioxide nanoparticles results indicated that these types of materials may indeed be biopersistent, warranting further long-term kinetic studies [67,83,107].

Nanotoxicology

Due to the relatively recent introduction of nanotechnology in industry, epidemiological studies, case studies and other human data on potential adverse effects of ENM are virtually absent. Any existing information on potential adverse effects is largely based on animal and *in vitro* studies with (human) cell lines. Especially the relevance of *in vitro* studies for *in vivo* effects may be limited, although they are useful for screening purposes and mechanistic studies [108]. Figure 1.4 shows the potential interactions of ENM with the cell and subcellular structures, and indicates mechanisms of action of ENM.

Since inhalation exposure is thought to be the most relevant exposure route to ENM, adverse effects are mainly expected to occur in the lung (see Chapter 7 for further information). A large body of literature already exists on inflammatory effects in the lung as a result of exposure to ambient particulate matter and, not coincidentally, this is also one of the most investigated and reported effects for engineered ENM in the published literature [16,109–111]. Numerous short-term studies report the induction of inflammatory markers

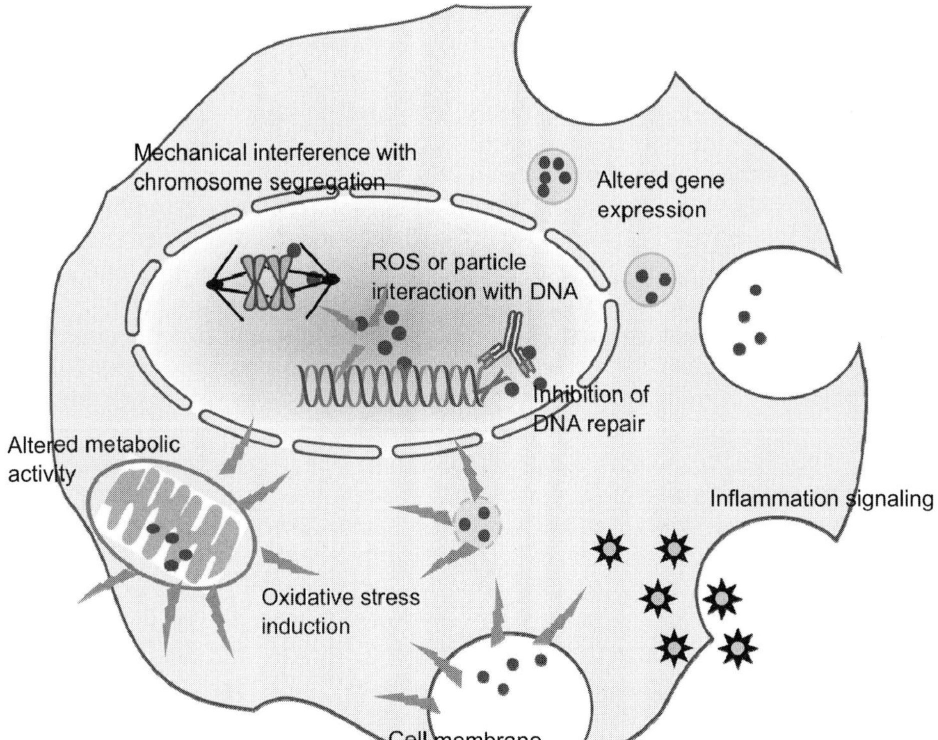

FIGURE 1.4 Possible interactions of ENM with the cell and subcellular structures. Suggested mechanisms underlying nanoparticle-induced responses at the cellular level which, at sufficiently high or persistent levels, potentially can lead to altered tissue function and damage.

in rats after instillation with and inhalation of metal (oxide) ENM and in many *in vitro* models representing the human lung [112–117]. The pulmonary inflammatory properties of low-solubility ENM, such as titanium dioxide (TiO_2) and carbon black, have been suggested to be related to particle surface area rather than mass [115], although other physicochemical properties, such as ENM surface chemistry and crystallinity, also play a role [118,119]. Local persistence in the lung of low toxicity particulates resulting in chronic inflammation may be a cause for non-genotoxic induction of lung tumors in the rat as suggested for both fine and ultrafine (nano) TiO_2 [120]. In addition to eliciting inflammatory responses in the lung by themselves, ENM have also been shown to aggravate endotoxin-elicited lung inflammation and airway hypersensitization in mice by acting as adjuvants [121,122]. Conversely, immunosuppressive effects have been reported in mice after inhaling carbon nanotubes [123].

As discussed previously, exposure routes other than inhalation are also relevant for ENM, for example when applied in cosmetics and in food, leading to exposure via the skin and the digestive tract, respectively. Especially the use of ENM for biomedical applications, such as intravenous diagnostic tools or drug carriers, will lead to relatively high levels of systemic bioavailability, which often is even targeted at a specific organ or tissue. The ideal design of an ENM that is used for these medical purposes is one that can be quickly removed from the body after it has completed its task and, as such, the potential risk of systemic bioavailability of these particular ENM may not outweigh the benefits [101–103,124,125]. Nevertheless, any application that may lead to systemic bioavailability of ENM will need to address potential systemic toxicity in their safety assessments.

An example of a toxic endpoint relevant for systemic exposure is the potential of ENM to cause DNA damage (see Chapter 14 for further information). ENM have been reported to induce genotoxic effects *in vitro* [126,127], but very few *in vivo* studies are available to evaluate these effects at relevant exposure levels [128–130]. Similarly, information on the effects of ENM on reproduction and embryonic development are very scarce and mostly limited to *in vitro* studies, or *in vivo* studies using very high doses, for which the relevance of the results to humans is difficult to interpret [131,132] (see Chapter 13 for further information). One study in pregnant mice inhaling surface coated TiO_2 at concentrations relevant to human occupational exposure limits resulted in lung inflammation in adult females and neurobehavioral changes in offspring [133].

In general, while acute effects resulting from short-term exposure to high levels of ENM, such as the inflammatory effects or DNA damage *in vitro*, have been investigated to a reasonable extent, little information exists on the effects of chronic exposure to low levels of persistent ENM [130,134]. Epidemiological studies have associated exposure to ambient ultrafine particular matter with systemic effects, such as cardiovascular and neurodegenerative diseases and adverse birth outcomes [134–136]. The generation of oxidative stress (discussed below) that is believed to play an important role in at least some of these effects has also been observed in studies with ENM and suggest that these effects may be relevant for ENM, which warrants further investigation [134,137].

Mechanism of Adverse Effects of Engineered Nanomaterials

The generation of reactive oxygen species (ROS) and oxidative injury is considered to play a significant role in many of the observed biological responses to ENM (see Figure 1.4) (oxidative stress injury by ENM is also discussed in Chapter 7). Due to the large surface area of ENM, relatively more reactive groups are located on the surface and, as such, may be available for electron transfer to oxygen, forming superoxide, a precursor of other ROS [14]. ROS may also be generated as a result of ENM–cell interactions, even in the absence of reactive groups on the surface [138]. During normal cellular processes, such as mitochondrial respiration, low levels of ROS are also generated, but these are immediately neutralized by the cell's antioxidant machinery. However, when the generation of ROS exceeds the cell's antioxidant capacity, a state of oxidative stress is induced in the cell. ENM size, surface area and surface chemistry (i.e. reactive groups) are thought to play a role in the generation of ROS. Apart from the damaging effects to cellular proteins, lipids and DNA, an increasing level of ROS triggers the cell to respond by inducing pro-inflammatory signaling cascades, and ultimately induces programmed cell death [14].

Additionally, ENM may induce or aggravate inflammatory and allergic responses by directly influencing immune-related cell populations in the lung [122]. After lung exposure, ENM recruit neutrophilic granulocytes to the lung and partially end up in lung macrophages [16,75,117]. In turn, persistent oxidative stress and inflammation in lungs and brain tissue after exposure to particulate matter are thought to be the underlying cause for lung fibrosis and neurodegenerative diseases, respectively [134,135].

As discussed previously, the very same properties that make ENM interesting to enhance target delivery of pharmaceutical drugs, antigens for vaccination or diagnostic contrast agents may also enhance transport of undesirable compounds to tissues and cell components. For example, cytotoxicity of Ag and Cu ENM was reported to be due in part by shuttling their respective ions into the cell [104–106,139]. In addition, trace amounts of TiO_2 ENM synergistically increased the genotoxic effects of dichlorodiphenyltrichloroethane (p,p′-DDT) in human hepatocytes, presumably by adsorption of p,p′-DDT to the surface of the ENM [140]. In a similar way, ENM may act as adjuvants for antigens, aggravating allergic responses, or they may act as antigens or haptens themselves [141]. Conversely, the immunosuppressive effects of inhaled carbon nanotubes on B-cell function in mice was reported to be mediated by the production of transforming growth factor (TGF)-β by alveolar macrophages, which activated cyclo-oxygenase enzymes in the spleen [123]. The changes in neurobehavioral patterns observed in offspring of mice exposed to TiO_2 are another example of potential indirect toxicity of ENM. Since the transfer of TiO_2 across the placenta after inhalation was expected to be low, the effects may have been caused by release of the coatings of the ENM or by cytokines released in the inflammatory response in the lungs of the pregnant mice [133].

Effects of ENM on the integrity of DNA can result from both direct interaction of ENM with the DNA, the generation of ROS or as a consequence of persistent inflammation. In addition, it has been shown, especially for carbon nanotubes, that the presence of production residues such as metal ions (used as catalysts), structural defects and the aspect ratio are important properties relevant to (geno)toxicity (see Chapter 14 for further information). The mechanisms underlying the toxicity of fiber-like ENM are discussed in more detail in the next paragraph.

Fiber Paradigm

The experience with certain HARN (fibers), such as asbestos, highlighted that types of ENM similar to asbestos may pose an additional hazard to the lung beyond that produced by conventional compact particles and gave rise to the discipline of fiber toxicology [142,143]. Several decades of fiber toxicology have led to an overarching fiber toxicology structure activity paradigm involving length (>20 μm), diameter (<3 μm) and biopersistence. Such fibers can be deposited in the lung beyond the ciliated airways but then they cannot be completely removed by the alveolar macrophages resulting in what is now designated as frustrated phagocytosis. If the fibers, due to a rigid structure, retain their fibrous shape during residence in the lung, the total fiber dose may accumulate [144]. Single- and multi-walled carbon nanotubes are used increasingly in different materials science applications. They have been attracting major attention owing to their alleged hazards to health. These hazards have been attributed to their morphological properties based on their likeness to asbestos. Recent studies demonstrated that aerosolized carbon nanotubes (CNTs) could reach the subpleura and induced fibrosis in mice up to 6 weeks following a single exposure of 6 hours [145]. Although such findings are of interest towards understanding the human health effects of CNTs, extreme caution should be taken in regard to the conclusions drawn since the doses applied are relatively high, probably causing an overload situation and, furthermore, are arguably unrealistic in terms of possible human exposure, especially considering the findings of Maynard and coworkers [146].

The potential underlying mechanisms of the behavior of long fibers in the lung and parietal pleura is extensively reviewed by [143,144] and is discussed in further detail in Chapter 7.

CONCLUSIONS

The possibility to design ENM with preferential physicochemical properties such as, for example, high electric or heat conductivity, increased catalytic activity, or mechanical resistance, makes nanomaterials the future investments for companies in order to implement their usage both for industrial, commercial and biomedical applications. The increasing use of ENM has also resulted in an increasing awareness of the potential adverse effects of ENM. There are clear indications that, after inhalation and oral exposure, ENM can become systemically available. The uptake as percentage of administered dose is not always determined in the various studies, but seems in general to be rather low with exception of the oral route, for which the uptake might be considerable depending on the actual composition of the ENM preparation. Some studies indicate that smaller ENM are more easily taken up and show a more widespread organ distribution. However, for the GI tract, the effect of size is less clear. When systemically available, the main target organs are those with a high RES content, being mainly liver, spleen and lung, but ENM were also demonstrated to distribute to other organs as well.

The potential toxicity of ENM has been demonstrated in numerous *in vitro* studies. Information on *in vivo* toxicity and human epidemiological studies is mainly limited to inhalation exposure. There is sufficient evidence that acute inhalation exposure to high levels of engineered ENM may lead to adverse inflammatory effects in the lung. At the moment, insufficient information exists on potential effects of chronic exposure to low levels of ENM, by exposure via routes other than inhalation, and toxicity endpoints such as carcinogenicity, reproductive toxicity and immunotoxicity.

TAKE-HOME MESSAGES

1. ENM have different biological activities due to their physicochemical properties compared to bulk material and should therefore be assessed separately.
2. There are clear indications that after inhalation and oral exposure ENM can become systemically available.
3. These particles have the capability to cross biological barriers and by that may reach cells and tissues which were not exposed by their macroscopic form.
4. Biochemical processes at the nano–bio interface are responsible for the ENM–cell interactions and the potential adverse effects of ENM which are not completely understood so far.

References

[1] Speiser P, Khanna SC. Perlpolymerisate, eine neue perorale Darreichungsform und ihre Beeinflussung durch Arzneistoffe. Praepar Pharm 1970;6:1–4.
[2] Zolle I, Hosein F, Rhodes BATK, Wagner Jr HN. Human serum albumin microspheres for studies of the reticuloendothelial system. J Nucl Med 1973;11:379.
[3] Scheffel U, Rhodes BA, Natarajan TK, Wagner Jr HN. Albumin microspheres for study of the reticuloendothelial system. J Nucl Med 1973;13:498–503.
[4] Kramer PA. Albumin microspheres as vehicles for achieving specificity in drug delivery. J Pharm Sci 1974;63:1646–7.
[5] Sugibayashi K, Morimoto Y, Nadai T, Kato Y, Hasegawa A, Arita T. Drug-carrier property of albumin microspheres in chemotherapy. II. Preparation and tissue distribution in mice of microsphere-entrapped 5- fluorouracil. Chem Pharm Bull 1979;27:204–9.
[6] ISO/TS 80004–1. Nanotechnology—Vocabulary—Part 1. Core terms. Geneva, Switzerland: International Organization for Standardization (ISO); 2010.
[7] SCENIHR (Scientific Committee on Emerging and Newly Identified Health Risks). Opinion on the scientific basis for the definition of the term "nanomaterial", 8 December 2010. Brussels, Belgium: European Commission, http://ec.europa.eu/health/scientific_committees/emerging/docs/scenihr_o_032.pdf; 2010.
[8] Health Canada. Interim policy statement on Health Canada's working definition for nanomaterials; 2010. Available from: URL, http://www.hc-sc.gc.ca/sr-sr/alt_formats/pdf/consult/_2010/nanomater/draft-ebauche-eng.pdf; 2010.
[9] OECD. OECD Working Party on Manufactured Nanomaterials. Guidance for the use of the OECD database on research into the safety of manufactured nanomaterials. Ver 1 October 2008. Available from: URL, http://www.oecd.org/dataoecd/21/19/44033847.pdf; 2008.
[10] NICNAS (National Industrial Chemicals Notification and Assessment Scheme). Guidance on new chemical requirements for notification of industrial nanomaterials. Available from URL, http://www.nicnas.gov.au/Current_Issues/Nanotechnology/Guidance%20on%20New%20Chemical%20Requirements%20for%20Notification%20of%20Industrial%20Nanomaterials.pdf; 2010.
[11] FDA (Food and Drug Administration). Draft Guidance: Considering Whether an FDA-Regulated Product Involves the Application of Nanotechnology. Silver Spring, MD, USA: US Food and Drug Administration, http://www.fda.gov/ScienceResearch/SpecialTopics/Nanotechnology/ucm257926.htm; 2011.
[12] European Commission. Official Journal of the European Commission L342/59. Regulation (EC) No 1223/2009 of the European Parliament and of the Council of 30 November 2009 on cosmetic products. Brussels, Belgium: European Commission, http://eur-lex.europa.eu/LexUriServ/LexUriServ.do?uri=OJ;2009. L:2009:342:0059:0209:EN: PDF.
[13] Auffan M, Rose J, Bottero J-Y, Lowry GV, Jolivet J-P, Wiesner MR. Towards a definition of inorganic nanoparticles from an environmental, health and safety perspective. Nat Nanotechnol 2009;4:634–41.
[14] Nel A, Xia T, Madler L, Li N. Toxic potential of materials at the nanolevel. Science 2006;311:622–7.

[15] Buzea C, Pachecob II, Robbie K. Nanomaterials and nanoparticles: Sources and toxicity. Biointerphases 2007;4:MR17–70.
[16] Oberdorster G, Oberdorster E, Oberdorster J. Nanotoxicology: An emerging discipline evolving from studies of ultrafine particles. Environ Health Perspect 2005;113:823–39.
[17] Tarantola M, Pietuch A, Schneider D, et al. Toxicity of gold-nanoparticles: Synergistic effects of shape and surface functionalization on micromotility of epithelial cells. Nanotoxicology 2010;5:254–68.
[18] Tsuji JS, Maynard AD, Howard PC, et al. Research strategies for safety evaluation of nanomaterials, part IV: risk assessment of nanoparticles. Toxicol Sci 2006;89:42–50.
[19] Borm PJ, Robbins D, Haubold S, et al. The potential risks of nanomaterials: A review carried out for ECETOC. Part Fibre Toxicol 2006;14:3–11.
[20] Wick P, Manser P, Limbach LK, et al. The degree and kind of agglomeration affect carbon nanotube cytotoxicity. Toxicol Lett 2007;168:121–31.
[21] Schaeublin NM, Braydich-Stolle LK, Schrand AM, et al. Surface charge of gold nanoparticles mediates mechanism of toxicity. Nanoscale 2011;3:410–20.
[22] Carlson C, Hussain SM, Schrand AM, et al. Unique cellular interaction of silver nanoparticles: Size-dependent generation of reactive oxygen species. J Phys Chem B 2008;112:13608–19.
[23] Pal S, Tak YK, Song JM. Does the antibacterial activity of silver nanoparticles depend on the shape of the nanoparticle? A study of the gram-negative bacterium *Escherichia coli*. Appl Environ Microbiol 2007;73:1712–20.
[24] El Badawy AM, Silva RG, Morris B, Scheckel KG, Suidan MT, Tolaymat TM. Surface charge-dependent toxicity of silver nanoparticles. Environ Sci Technol 2011;45:283–7.
[25] Kvitek L, Vanickova M, Panacek A, et al. Initial study on the toxicity of silver nanoparticles (NPs) against *Paramecium caudatum*. J Phys Chem C 2009;113:4296–300.
[26] Johnston HJ, Hutchison GR, Christensen FM, Aschberger K, Stone V. The biological mechanisms and physicochemical characteristics responsible for driving fullerene toxicity. Toxicol Sci 2010;114:162–82.
[27] Johnston HJ, Hutchison GR, Christensen FM, et al. A critical review of the biological mechanisms underlying the *in vivo* and *in vitro* toxicity of carbon nanotubes: The contribution of physico-chemical characteristics. Nanotoxicology 2010;4:207–46.
[28] Nel AE, Mädler L, Velegol D, et al. Understanding biophysicochemical interactions at the nano–bio interface. Nat Mater 2009;8:543–57.
[29] Lynch I, Dawson KA. Protein–nanoparticle interactions. Nano Today 2008;3:40–7.
[30] Cedervall T, Lynch I, Lindman S, et al. Understanding the nanoparticle–protein corona using methods to quantify exchange rates and affinities of proteins for nanoparticles. Proc Natl Acad Sci USA 2007;104:2050–5.
[31] Lynch I, Salvati A, Dawson KA. Protein–nanoparticle interactions: What does the cell see? Nat Nanotechnol 2009;4:546–7.
[32] Mu Q, Liu W, Xing Y, et al. Protein binding by functionalized multiwalled carbon nanotubes is governed by the surface chemistry of both parties and the nanotube diameter. J Phys Chem 2008;112:3300–7.
[33] Wasdo SC, Barber DS, Denslow ND, et al. Differential binding of serum proteins to nanoparticles. Int J Nanotechnol 2008;5:92–115.
[34] Dutta D, Sundaram SK, Teeguarden JG, et al. Adsorbed proteins influence the biological activity and molecular targeting of nanomaterials. Toxicol Sci 2007;100:303–15.
[35] Lundqvist M, Stigler J, Elia G, Lynch I, Cedervall T, Dawson KA. Nanoparticle size and surface properties determine the protein corona with possible implications for biological impacts. Proc Natl Acad Sci USA 2008;105:14265–70.
[36] Bertolazzi P, Guerra C, Liuzzi G. A global optimization algorithm for protein surface alignment. BMC Bioinformatics 2010;11:488.
[37] Dell'Orco D, Lundqvist M, Oslakovic C, Cedervall T, Linse S. Modelling the time evolution of the nanoparticle–protein corona in a body fluid. PloS one 2010;5:e10949.
[38] Cabaleiro-Lago C, Lynch I, Dawson KA, Linse S. Inhibition of IAPP and IAPP(20-29) fibrillation by polymeric nanoparticles. Langmuir 2010;26:3453–61.
[39] Linse S, Cabaleiro-Lago C, Xue WF, et al. Nucleation of protein fibrillation by nanoparticles. Proc Natl Acad Sci USA 2007;104:8691–6.
[40] Krug HF, Wick P. Nanotoxicology: An interdisciplinary challenge. Angew Chem Int Ed Engl 2011;50:1260–78.

[41] Gasser M, Rothen-Rutishauser B, Krug HF, et al. The adsorption of biomolecules to multi-walled carbon nanotubes is influenced by both pulmonary surfactant lipids and surface chemistry. J Nanobiotechnol 2010;8:31.
[42] Schleh C, Rothen-Rutishauser B, Kreyling WG. The influence of pulmonary surfactant on nanoparticulate drug delivery systems. Eur J Pharm Biopharm 2011;77:350–2.
[43] Hofmann-Amtenbrink M, Hofmann H, Montet X. Superparamagnetic nanoparticles—a tool for early diagnostics. Swiss Med Wkly 2010;140:w13081.
[44] Li J, Zhang C, Yang K, Liu P, Xu LX. SPIO-RGD nanoparticles as a molecular targeting probe for imaging tumor angiogenesis using synchrotron radiation. J Synchrotron Radiat 2011;18:612–6.
[45] Beyersmann D, Hartwig A. Carcinogenic metal compounds: Recent insight into molecular and cellular mechanisms. Arch Toxicol 2008;82:493–512.
[46] Conner SD, Schmid SL. Regulated portals of entry into the cell. Nature 2003;422:37–44.
[47] Limbach LK, Wick P, Manser P, Grass RN, Bruinink A, Stark WJ. Exposure of engineered nanoparticles to human lung epithelial cells: Influence of chemical composition and catalytic activity on oxidative stress. Environ Sci Technol 2007;41:4158–63.
[48] Geiser M, Rothen-Rutishauser B, Kapp N, et al. Ultrafine particles cross cellular membranes by nonphagocytic mechanisms in lungs and in cultured cells. Environ Health Perspect 2005;113:1555–60.
[49] Limbach LK, Li Y, Grass RN, et al. Oxide nanoparticle uptake in human lung fibroblasts: effects of particle size, agglomeration, and diffusion at low concentrations. Environ Sci Technol 2005;39:9370–6.
[50] Oberdorster G, Maynard A, Donaldson K, et al. Principles for characterizing the potential human health effects from exposure to nanomaterials: Elements of a screening strategy. Part Fibre Toxicol 2005;2:8.
[51] Dobrovolskaia MA, McNeil SE. Immunological properties of engineered nanomaterials. Nat Nanotechnol 2007;2:469–78.
[52] Kanno S, Furuyama A, Hirano S. A murine scavenger receptor MARCO recognizes polystyrene nanoparticles. Toxicol Sci 2007;97:398–406.
[53] Rothen-Rutishauser BM, Schurch S, Haenni B, Kapp N, Gehr P. Interaction of fine particles and nanoparticles with red blood cells visualized with advanced microscopic techniques. Environ Sci Technol 2006;40:4353–9.
[54] George S, Pokhrel S, Xia T, et al. Use of a rapid cytotoxicity screening approach to engineer a safer zinc oxide nanoparticle through iron doping. ACS Nano 2010;4:15–29.
[55] Xia T, Kovochich M, Liong M, et al. Comparison of the mechanism of toxicity of zinc oxide and cerium oxide nanoparticles based on dissolution and oxidative stress properties. ACS Nano 2008;2:2121–34.
[56] Xia T, Zhao Y, Sager T, et al. Decreased dissolution of ZnO by iron doping yields nanoparticles with reduced toxicity in the rodent lung and zebrafish embryos. ACS Nano 2011;5:1223–35.
[57] Kocbach A, Li Y, Yttri KE, Cassee FR, Schwarze PE, Namork E. Physicochemical characterisation of combustion particles from vehicle exhaust and residential wood smoke. Part Fibre Toxicol 2006;3:1.
[58] Oberdörster G. Safety assessment for nanotechnology and nanomedicine: concepts of nanotoxicology. J Intern Med 2010;267:89–105.
[59] Semmler-Behnke M, Kreyling WG, Lipka J, et al. Biodistribution of 1.4- and 18-nm gold particles in rats. Small 2008;4:2108–11.
[60] Aggarwal P, Hall JB, McLeland CB, Dobrovolskaia MA, McNeil SE. Nanoparticle interaction with plasma proteins as it relates to particle biodistribution, biocompatibility and therapeutic efficacy. Adv Drug Del Rev 2009;61:428–37.
[61] Dobrovolskaia MA, Patri AK, Zheng J, et al. Interaction of colloidal gold nanoparticles with human blood: Effects on particle size and analysis of plasma protein binding profiles. Nanomed Nanotechnol Biol Med 2009;5:106–17.
[62] Lacerda SH, Park JJ, Meuse C, et al. Interaction of gold nanoparticles with common human blood proteins. ACS Nano 2010;4:365–79.
[63] De Jong WH, Hagens WI, Krystek P, Burger MC, Sips AJAM, Geertsma RE. Particle size dependent organ distribution of gold nanoparticles after intravenous administration. Biomaterials 2008;29:1912–9.
[64] Demoy M, Gibaud S, Andreux JP, Weingarten C, Gouritin B, Couvreur P. Splenic trapping of nanoparticles: Complementary approaches for *in situ* studies. Pharm Res 1997;14:463–8.
[65] Sadauskas E, Wallin H, Stoltenberg M, et al. Kupffer cells are central in the removal of nanoparticles from the organism. Part Fibre Toxicol 2007;4:10.

[66] Lipka J, Semmler-Behnke M, Sperling RA, et al. Biodistribution of PEG-modified gold nanoparticles following intratracheal instillation and intravenous injection. Biomaterials 2010;31:6574–81.
[67] Lankveld DPK, Oomen AG, Krystek P, et al. The kinetics of the tissue distribution of silver nanoparticles of different sizes. Biomaterials 2010;31:8350–61.
[68] Lankveld DPK, Rayavarapu RG, Krystek P, et al. Blood clearance and tissue distribution of pegylated and non-pegylated gold nanorods after intravenous administration in rats. Nanomedicine 2011;6:339–49.
[69] Weissenbrock A, Wirth M, Gabor F. WGA grafted PLGA-nanospheres: Preparation and association with Caco-2 cells. J Contr Release 2004;99:383–92.
[70] Nobs L, Buchegger F, Gurny R, Allémann E. Poly(lactic acid) nanoparticles labeled with biologically active NeutravidinTM for active targeting. Eur J Pharm Biopharm 2004;58:483–90.
[71] Prinzen L, Miserus RJ, Dirksen A, et al. Optical and magnetic resonance imaging of cell death and platelet activation using annexin A5-functionalized quantum dots. Nano Lett 2007;7:93–100.
[72] Akiyama Y, Mori T, Katayama Y, Niidome T. The effects of PEG grafting level and injection dose on gold nanorod biodistribution in the tumor bearing mice. J Contr Release 2009;139:81–4.
[73] Von Maltzahn G, Park J-H, Agrawal A, et al. Computationally guided photothermal tumor therapy using long-circulating gold nanorod antennas. Cancer Res 2009;69:3892–900.
[74] Balogh L, Nigavekar SS, Nair BM, et al. Significant effect of size on the *in vivo* biodistribution of gold composite nanodevices in mouse tumor models. Nanomed Nanotechnol Biol Med 2007;3:281–96.
[75] Wang FH, Kim DK, Yoshitake T, et al. Diffusion and clearance of superparamagnetic iron oxide nanoparticles infused into the rat striatum studied by MRI and histochemical techniques. Nanotechnology 2011;22. 015103.
[76] Hirn S, Semmler-Behnke M, Schleh K, et al. Particle size dependent and surface charge dependent biodistribution of gold nanoparticles after intravenous administration. Eur J Pharm Biopharm 2011;77:407–16.
[77] Wick P, Malek A, Manser P, et al. Barrier capacity of human placenta for nanosized materials. Environ Heath Perspect 2010;118:432–6.
[78] Choi HS, Liu W, Misra1 P, et al. Renal clearance of quantum dots. Nat Biotech 2007;25:1165–70.
[79] Kreyling WG, Semmler M, Erbe F, et al. Translocation of ultrafine insoluble iridium particles from lung epithelium to extrapulmonary organs is size dependent but very low. J Toxicol Environ Health, Part A 2002;65:1513–30.
[80] Ji JH, Jung JH, Kim SS, et al. Twenty-eight-day inhalation toxicity study of silver nanoparticles in Sprague-Dawley rats. Inhal Toxicol 2007;19:857–71.
[81] Sung JH, Ji JH, Park JD, et al. Subchronic inhalation toxicity of gold nanoparticles. Part Fibre Toxicol 2011;8:16.
[82] Yu LE, Yung L-YL, Ong C-N, et al. Translocation and effects of gold nanoparticles after inhalation exposure in rats. Nanotoxicology 2007;1:235–42.
[83] Pauluhn J. Retrospective analysis of 4-week inhalation studies in rats with focus on fate and pulmonary toxicity of two nanosized aluminum oxyhydroxides (boehmite) and pigment-grade iron oxide (magnetite): The key metric of dose is particle mass and not particle surface area. Toxicology 2009;259:140–8.
[84] Pauluhn J. Poorly soluble particulates: Searching for a unifying denominator of nanoparticles and fine particles for DNEL estimation. Toxicology 2011;279:176–88.
[85] Bowman K, Leong KW. Chitosan nanoparticles for oral drug and gene delivery. Int J Nanomed 2006;1:117–28.
[86] Muchow M, Maincent P, Muller RH. Lipid nanoparticles with a solid matrix (SLN, NLC, LDC) for oral drug delivery. Drug Dev Ind Pharm 2008;34:1394–405.
[87] Powell JJ, Faria N, Thomas-McKay E, Pele LC. Origin and fate of dietary nanoparticles and microparticles in the gastrointestinal tract. J Autoimmun 2010;34:J226–33.
[88] Jani P, Halbert GW, Langridge J, Florence AT. The uptake and translocation of latex nanospheres and microspheres after oral administration to rats. J Pharm Pharmacol 1990;42:821–6.
[89] Wang J, Zhoua G, Chena C, et al. Acute toxicity and biodistribution of different sized titanium dioxide particles in mice after oral administration. Toxicol Lett 2007;168:176–85.
[90] Kim YS, Kim JS, Cho HS, et al. Twenty-eight-day oral toxicity, genotoxicity, and gender-related tissue distribution of silver nanoparticles in Sprague-Dawley Rats. Inhal Toxicol 2008;20:575–83.
[91] Park EJ, Yi J, Kim Y, Choi K, Park K. Silver nanoparticles induce cytotoxicity by a Trojan-horse type mechanism. Toxicol In Vitro 2010;24:872–8.
[92] Park EJ, Bae E, Yi J, et al. Repeated-dose toxicity and inflammatory responses in mice by oral administration of silver nanoparticles. Environ Toxicol Pharmacol 2010;30:162–8.

[93] Jani PU, McCarthy DE, Florence AT. Titanium dioxide (rutile) particle uptake from the rat GI tract and translocation to systemic organs after oral administration. Int J Pharmaceut 1994;105:157–68.
[94] Ryman-Rasmussen J, Riviere JE, Monteiro-Riviere NA. Penetration of intact skin by quantum dots with diverse physicochemical properties. Toxicol Sci 2006;91:159–65.
[95] Wu J, Liu W, Xue C, et al. Toxicity and penetration of TiO2 nanoparticles in hairless mice and porcine skin after subchronic dermal exposure. Toxicol Lett 2009;191:1–8.
[96] Larese Filon F, Crosera M, Adami G, Bovenzi M, Rossi F, Maina G. Human skin penetration of gold nanoparticles through intact and damaged skin. Nanotoxicology 2011 (in press).
[97] Smijs TG, Bouwstra JA. Focus on skin as a possible port of entry for solid nanoparticles and the toxicological impact. J Biomed Nanotechnol 2010;6:469–84.
[98] Nabeshi H, Yoshikawa T, Matsuyama K, et al. Systemic distribution, nuclear entry and cytotoxicity of amorphous nanosilica following topical application. Biomaterials 2011;32:2713–24.
[99] Tang J, Xiong L, Wang S, et al. Distribution, translocation and accumulation of silver nanoparticles in rats. J Nanosci Nanotechnol 2009;9:4924–32.
[100] Gulson B, McCall M, Korsch M, et al. Small amounts of zinc from zinc oxide particles in sunscreens applied outdoors are absorbed through human skin. Tox Sci 2010;118:140–9.
[101] Duncan R. Polymer conjugates as anticancer nanomedicines. Nat Rev Cancer 2006;6:688–701.
[102] De Jong WH, Borm PJA. Drug delivery and nanoparticles: Applications and hazards. Int J Nanomed 2008;3:133–49.
[103] Irache JM, Esparza I, Gamazo C, Agüeros M, Espuelas S. Nanomedicine: Novel approaches in human and veterinary therapeutics. Vet Parasitol 2011. doi:10.1016/j.vetpar.2011.05.028.
[104] Wijnhoven SWP, Peijnenburg WJGM, Herberts CA, et al. Nano-silver—a review of available data and knowledge gaps in human and environmental risk assessment. Nanotoxicology 2009;3:109–38.
[105] Dimkpa CO, Calder A, Britt DW, McLean JE, Anderson AJ. Responses of a soil bacterium, *Pseudomonas chlororaphis* O6 to commercial metal oxide nanoparticles compared with responses to metal ions. Environ Pollut 2011;159:1749–56.
[106] Park EJ, Yi J, Kim Y, Choi K, Park K. Silver nanoparticles induce cytotoxicity by a Trojan-horse type mechanism. Toxicol In Vitro 2010;24:872–8.
[107] Fabian E, Landsiedel R, Ma-Hock L, Wiench K, Wohlleben W, Van Ravenzwaay B. Tissue distribution and toxicity of intravenously administered titanium dioxide nanoparticles in rats. Arch Toxicol 2008:82,151–157.
[108] Park MVDZ, Lankveld DPK, van Loveren H, de Jong W. The status of *in vitro* toxicity studies in risk assessment of nanomaterials. Nanomedicine 2009;4:669–85.
[109] Ayres JG, Borm P, Cassee FR, et al. Evaluating the toxicity of airborne particulate matter and nanoparticles by measuring oxidative stress potential—a workshop report and consensus statement. Inhal Toxicol 2008;20:75–99.
[110] Madl AK, Pinkerton KE. Health effects of inhaled engineered and incidental nanoparticles. Crit Rev Toxicol 2009;39:629–58.
[111] Napierska D, Thomassen LC, Lison D, Martens JA, Hoet PH. The nanosilica hazard: Another variable entity. Part Fibre Toxicol 2010;7:39.
[112] Cho WS, Duffin R, Poland CA, et al. Metal oxide nanoparticles induce unique inflammatory footprints in the lung: Important implications for nanoparticle testing. Environ Health Perspect 2010;118:1699–706.
[113] Hsiao IL, Huang YJ. Effects of various physicochemical characteristics on the toxicities of ZnO and TiO nanoparticles toward human lung epithelial cells. Sci Total Environ 2011;409:1219–28.
[114] Kasper J, Hermanns MI, Bantz C, et al. Inflammatory and cytotoxic responses of an alveolar–capillary coculture model to silica nanoparticles: Comparison with conventional monocultures. Part Fibre Toxicol 2011;8:6.
[115] Sager TM, Castranova V. Surface area of particle administered versus mass in determining the pulmonary toxicity of ultrafine and fine carbon black: Comparison to ultrafine titanium dioxide. Part Fibre Toxicol 2009;6:15.
[116] Sung JH, Ji JH, Park JD, et al. Subchronic inhalation toxicity of silver nanoparticles. Toxicol Sci 2009;108:452–61.
[117] Gosens I, Post JA, De la Fonteyne LJJ, et al. Impact of agglomeration state of nano- and submicron sized gold particles on pulmonary inflammation. Part Fiber Toxicol 2011;7:37.

[118] Duffin R, Tran L, Brown D, Stone V, Donaldson K. Proinflammogenic effects of low-toxicity and metal nanoparticles *in vivo* and *in vitro*: Highlighting the role of particle surface area and surface reactivity. Inhal Toxicol 2007;19:849–56.
[119] Warheit DB. Debunking some misconceptions about nanotoxicology. Nano Lett 2010;10:4777–82.
[120] NIOSH. Occupational exposure to titanium dioxide. Current Intellegence Bulletin 63. Department of Health and Human Services, Public Health Service Centers for Disease Control and Prevention, National Institute for Occupational Safety and Health, (NIOSH), DHHS (NIOSH) Publication No 2011-160. Cincinnati, USA. http://www.cdc.gov/niosh/docs/2011-160/pdfs/2011-160.pdf; 2011
[121] de Haar C, Hassing I, Bol M, Bleumink R, Pieters R. Ultrafine but not fine particulate matter causes airway inflammation and allergic airway sensitization to co-administered antigen in mice. Clin Exp. Allergy 2006;36:1469–79.
[122] Inoue K, Takano H. Aggravating impact of nanoparticles on immune-mediated pulmonary inflammation. Sci World J 2011;11:382–90.
[123] Mitchell LA, Lauer FT, Burchiel SW, McDonald JD. Mechanisms for how inhaled multiwalled carbon nanotubes suppress systemic immune function in mice. Nat Nanotechnol 2009;4:451–6.
[124] McNeil SE. Nanoparticle therapeutics: A personal perspective. Rev Nanomed Nanobiotechnol 2009; 1:264–71.
[125] Hahn MA, Singh AK, Sharma P, Brown SC, Moudgil BM. Nanoparticles as contrast agents for in-vivo bioimaging: Current status and future perspectives. Anal Bioanal Chem 2011;399:3–27.
[126] Colognato R, Bonelli A, Ponti J, et al. Comparative genotoxicity of cobalt nanoparticles and ions on human peripheral leukocytes *in vitro*. Mutagenesis 2008;23:377–82.
[127] Migliore L, Saracino D, Bonelli A, et al. Carbon nanotubes induce oxidative DNA damage in RAW 264.7 cells. Environ Mol Mutagen 2010;51:294–303.
[128] Gonzalez L, Lison D, Kirsch-Volders M. Genotoxicity of engineered nanomaterials: A critical review. Nanotoxicology 2008;2:252–73.
[129] Landsiedel R, Kapp MD, Schulz M, Wiench K, Oesch F. Genotoxicity investigations on nanomaterials: methods, preparation and characterization of test material, potential artifacts and limitations—many questions, some answers. Mutat Res 2009;681:241–58.
[130] Singh N, Manshian B, Jenkins GJ, et al. NanoGenotoxicology: The DNA damaging potential of engineered nanomaterials. Biomaterials 2009;30:3891–914.
[131] Ema M, Kobayashi N, Naya M, Hanai S, Nakanishi J. Reproductive and developmental toxicity studies of manufactured nanomaterials. Reprod Toxicol 2010;30:343–52.
[132] Hougaard KS, Fadeel B, Gulumian M, Kagan VE, Savolainen K. Developmental toxicity of engineered nanoparticles. In: Gupta RC, editor. Reproductive and developmental toxicology. London: Elsevier; 2011.
[133] Hougaard KS, Jackson P, Jensen KA, et al. Effects of prenatal exposure to surface-coated nanosized titanium dioxide (UV-Titan). A study in mice. Part Fibre Toxicol 2010;7:16.
[134] Simko M, Mattsson MO. Risks from accidental exposures to engineered nanoparticles and neurological health effects: A critical review. Part Fibre Toxicol 2010;7:42.
[135] Brook RD, Rajagopalan S, Pope 3rd CA, et al. Particulate matter air pollution and cardiovascular disease: An update to the scientific statement from the American Heart Association. Circulation 2010;121:2331–78.
[136] Shah PS, Balkhair T. Air pollution and birth outcomes: A systematic review. Environ Int 2011;37:498–516.
[137] Xia T, Li N, Nel AE. Potential health impact of nanoparticles. Annu Rev Publ Health 2009;30:137–50.
[138] Xia T, Kovochich M, Brant J, et al. Comparison of the abilities of ambient and manufactured nanoparticles to induce cellular toxicity according to an oxidative stress paradigm. Nano Lett 2006;6:1794–807.
[139] Studer AM, Limbach LK, Van Duc L, et al. Nanoparticle cytotoxicity depends on intracellular solubility: Comparison of stabilized copper metal and degradable copper oxide nanoparticles. Toxicol Lett 2010;197:169–74.
[140] Shi Y, Zhang JH, Jiang M, Zhu LH, Tan HQ, Lu B. Synergistic genotoxicity caused by low concentration of titanium dioxide nanoparticles and p, p'-DDT in human hepatocytes. Environ Mol Mutagen 2010;51:192–204.
[141] Zolnik BS, Gonzalez-Fernandez A, Sadrieh N, Dobrovolskaia MA. Nanoparticles and the immune system. Endocrinology 2010;151:458–65.
[142] Donaldson K, Poland CA, Schins RP. Possible genotoxic mechanisms of nanoparticles: criteria for improved test strategies. Nanotoxicology 2010;4:414–20.

[143] Donaldson K, Murphy F, Duffin R, Poland C. Asbestos, carbon nanotubes and the pleural mesothelium: A review of the hypothesis regarding the role of long fibre retention in the parietal pleura, inflammation and mesothelioma. Part Fibre Toxicol 2010;7:5.
[144] Donaldson K. The inhalation toxicology of p-aramid fibrils. Crit Rev Toxicol 2009;39:487–500.
[145] Ryman-Rasmussen JP, Cesta MF, Brody AR, et al. Inhaled carbon nanotubes reach the subpleural tissue in mice. Nat Nanotechnol 2009;4:747–51.
[146] Maynard AD, Baron PA, Foley M, Shvedova AA, Kisin ER, Castranova V. Exposure to carbon nanotube material: Aerosol release during the handling of unrefined single-walled carbon nanotube material. J Toxicol Environ Health A 2004;67:87–107.

CHAPTER

2

Exposure Assessment

Antonio Bergamaschi[1], *Ivo Iavicoli*[1], *Kai Savolainen*[2]

[1]Institute of Occupational Medicine, Università Cattolica del Sacro Cuore, Rome, Italy
[2]Nanosafety Research Centre, Finnish Institute of Occupational Health, Helsinki, Finland

OUTLINE

Adverse Effects of Engineered Nanomaterials
DOI: 10.1016/B978-0-12-386940-1.00002-7

INTRODUCTION

Research of nanoscale technologies is growing rapidly worldwide, and the National Science Foundation [1] estimates that by 2015 nanotechnology will have a $1 trillion impact on the global economy and will employ at least 2 million workers, 1 million of which may be in the USA. Lux Research [2] projects that new emerging nanotechnology applications will affect nearly every type of manufactured product through the middle of the next decade, becoming incorporated into 15% of global manufacturing output, totalling $2.6 trillion in 2014. Consequently, it appears evident that exposure to engineered nanomaterials (ENM) may occur in several activities including construction, energy, industry, chemistry, energy, health care, electronics and communication, manufacturing of instruments and tools, and defense. Therefore, several workers categories are at risk of coming in contact with ENM both during their usual occupational task and in accidental events, and even more will be in the near future [3]. Moreover, human exposure to nanoparticles does not involve only nano researchers, nano manufacturers and nano product users exposed to ENM but also other workers categories exposed to ultrafine or incidental nanoparticles. The term "ultrafine particle" has traditionally been used by the aerosol research and occupational and environmental health communities to describe airborne particles smaller than 100 nm in diameter. However, in the context of nanometer-diameter particles, it is frequently used to indicate nanoparticles that have not been intentionally produced. In fact, nanoparticles are also incidentally produced by secondary sources as technical products and by-products of different industrial or combustion processes (engines, power plants, incinerators, cigarette smoke, diesel engine exhaust, cutting and welding fumes, open fire).

At the current status of knowledge there is no definition of "overexposure", since no occupational exposure limits specific for nanoparticles have yet been developed. An additional challenge is that, for example, the European Union has not yet an accepted definition for "nanoparticle" rendering legislative actions problematic [4] (for an extensive discussion on nanoparticle definitions, see Chapter 1). Thus, the main goal of exposure assessment is currently to locate any source from which nanoparticles may be released in the workplace, to implement measures aimed to eliminate the release, if necessary, and to control the effectiveness of these measures. In fact, it is not possible to apply to nanoparticle exposure the traditional model of workplace risk assessment, since reliable evaluation of personal exposure is not yet available.

Reducing particle size increases surface area and modifies unique physicochemical properties, such as high conductivity, strength, durability, and chemical reactivity [5]. Hence, consequently, the small size and unique physicochemical properties of ENM, the toxicological profiles of nanoparticles are likely to differ considerably from those of larger particles composed of the same materials [5,6]. For this reason, in the last decade, nanotoxicology has emerged to elucidate the relationship of the physical and chemical properties, such as size, shape, surface chemistry, composition, aggregation and surface area, of nanomaterials with induction of toxic biological responses [7]. Furthermore, there are a number of manufactured nanoparticles whose assessment is considered a high priority. The Organisation for Economic Cooperation and Development (OECD)'s Working Party on Manufactured Nanomaterials has agreed upon a prioritized list of nanomaterials which are to be addressed [8]. The list includes: single-walled carbon nanotubes (SWCNTs), multiwalled carbon nanotubes (MWCNTs), polystyrene, carbon nanotubes (CNTs), fullerene (C60), aluminum oxide, dendrimers, titanium dioxide, zinc oxide, cerium oxide, iron, silver, gold, layered silicates (nanoclays) and silicon dioxide. Only a fraction of this list has been checked in workplace surveys (see below).

THE STATE OF KNOWLEDGE ON ENM CHARACTERISTICS AND RELATED EFFECTS

Many of the most important health and safety concerns regarding ENM and nanotechnologies are due to lack of knowledge of potentially harmful effects and levels of occupational exposure to ENM during their production and use. The reason for the health concerns arises from the fact that many of the harmful effects associated with exposure to given ENM have been identified in relevant experimental models, often after relevant exposure levels to some of these materials, especially considering occupational environments. As indicated above, many ENM have been shown to be more reactive than their larger but chemically identical counterparts [9,10], emphasizing the importance of size when assessing potential health effects of nanoparticles. ENM have also been shown to have a potential to induce, for example, the formation of reactive oxygen species more effectively than otherwise identical but larger particles thereby rendering them *a priori* harmful to biological systems [11–14]. Long et al. [15] exposed cultured immortalized mouse microglia (BV2), rat dopaminergic (DA) and primary cultures of embryonic rat striatum to Degussa P25, a commercially available TiO_2 nanomaterial, at concentrations between 25 and 120 mg/L of cell culture medium, and found an immediate and prolonged release of reactive oxygen species in BV2 cells. This is an *in vitro* study without a direct relevance to risk assessment, but it provides insight into the possible mechanisms of effects of nano-sized TiO_2 on microglial cells. The agglomeration of particles though adds to the complexity of the assessment of the true effects of various ENM. However, the effective surface area of the agglomerates may not be that different from that of the primary particles forming the agglomerate. In an *in vivo* study, Rossi and co-workers [16] exposed BALB/c mice to commercial TiO_2 nanoparticles, SiO_2 nanoparticles, nano-sized TiO_2 generated in a gas-to-particle conversion process at 10 mg/m^3, and to SiO_2-coated TiO_2 nanoparticles. They found that only nano-sized titanium dioxide doped with amorphous silica induced a strong pulmonary inflammation in the exposed mice, whereas the other nano-sized particles were, in practice, without an effect.

These findings indicate that, in addition to the size of the particles, other features such as the surface chemistry, potentially adding to the reactivity, matter. Furthermore, recent observations on the translocation of ENM through the circulation due to their small size almost to any organ in the body have provided evidence that also systemic effects due to exposure to ENM are possible [17–19]. The issue of size of particles and the possible harm to human health by nano-scale particles is further illustrated by a recent study by Dankovic et al. [20] who assessed risks of cancer in rats subsequent to exposure to different types of titanium dioxide (fine TiO_2 <2.5 μm; ultrafine TiO_2 <0.1 μm in diameter). They used available rat dose–response data on these materials. The results of the modeling suggested that the maximum likelihood estimate was much lower for the nano-sized than for fine TiO_2 particles. At this stage, the reliability of any quantitative health risk estimates is limited due to the small amount of data on nano-sized TiO_2. Nevertheless, it is quite obvious that nano-sized particles seem to have more potential than chemically identical but larger particles to harm human health as the example of TiO_2 suggests. These observations also strongly emphasize the importance of assessment of exposure to ENM to enable the assessment of the significance of the results from experimental studies in occupational settings.

RELEVANT EXPOSURE ROUTE IN OCCUPATIONAL ENVIRONMENTS

There is little doubt that lungs are the prominent exposure route for ENM in the occupational environment [6,21]. There are a number of studies which clearly demonstrate that a number of ENM, including carbon nanotubes [6,21–23] and nano-sized titanium dioxide [16], readily reach the lungs. Both can induce pulmonary inflammation and carbon nanotubes also induce granuloma formation and interstitial fibrosis [21,23]. Inhaled carbon nanotubes can also induce subpleural fibrosis and collagen accumulation under pleura and in tissue adjacent to pleura [24].

On the other hand, skin is the largest organ in the human body, and hence the potential contribution of dermal exposure is hard to ignore even though studies show that dermal penetration could be important for systemic exposure. However, in a recent study, an Australian group [25] explored in human volunteers dermal penetration of ZnO-containing sunblock cream. They used a "traditional" course ZnO-containing sunblock cream and a cream that contained nano-sized ZnO nanoparticles (in both cases the concentration of ZnO was 18%). In this study, a standardized area on the back was treated with the sunblock cream and then the volunteers sunbathed for 30 min on six consecutive days. The results showed that at a group level the nano-sized Zn penetrated the skin better than the coarse Zn. However, when males and females were studied separately, there was no difference in the penetration of the coarse and nano-sized ZnO through the skin in males, whereas in females the nano-sized ZnO more readily penetrated the skin than the coarse material. (For additional information on the effect on skin of sunblock cream containing ZnO ENM see also Chapters 5 and 11). This study shows some evidence that the skin can also serve as a route for entry of nanomaterials into the circulation. The assessment of the role of the skin is problematic due to the small amount of data.

As to other exposure routes, Elder et al. [26] have demonstrated in rodents that when rats inhale manganese oxide nanoparticles via the nose there is entry of these particles to the olfactory bundle in the bottom of the forebrain through the axons of the olfactory nerve.

In assessing the significance of different exposure routes in occupational environments, it is evident that the lung is the most important route, followed by the skin when the exposure is extensive. In humans, the nose, as an exposure route in occupational environment is mainly of academic interest.

CHARACTERIZATION OF BEHAVIOR OF ENM

One major uncertainty in the safety and risk evaluation of ENM arises from a lack of systematic knowledge about the physicochemical characteristics of the material arriving at the receptor (in this context the receiving organ or tissue), be it the nose, the skin or the mouth of an exposed human. This is true for all forms of ENM, whether they exist in the form of macroscopic solid objects, as powders, emulsions or suspensions, or as aerosols, i.e. in the form of a mixture of air and the particles. All these ENM are essentially constituted of nanoparticles, or at least nano-structured building blocks such as agglomerates, with the potential of being released by some kind of mechanism into a transport chain from the "source", meaning an industrial process or handling of nanomaterials, to the receptor meaning, in this context, the exposed individual, whether a worker or a consumer [15].

An issue requiring special and thorough consideration is possible exposure to products containing ENM during the entire life cycle of the products. Many products are not likely to cause exposure of workers or consumers as long as the ENM are embedded in the polymers or other matrices in a given product [27]. The situation may change when a given material needs further processing, becomes waste, is worn during use, or is recycled and, in that context, for example, crushed or, when dealing with surfaces painted with nanoparticle-containing paints, sanded with a sanding machine that generates dust also containing ENM. A detailed discussion of this issue is beyond the scope of this chapter but will require a thorough life-cycle analysis as the number and volume of products containing ENM rapidly increases.

The general release and transport mechanisms are perhaps best researched for aerosols, highly dynamic systems, and this information is also relevant to exposure. Nano-aerosols often consist of strongly agglomerated primary particles, already close to the source. Depending on their origins and transport history, these nano-structured agglomerates have a potential for subsequent break up [28]. Kuhlbusch and Fissan [29] have characterized the behavior of carbon black particles in reactor and pelleting areas in an occupational environment.

The tendency of airborne ENM to agglomerate or become attached to the ubiquitous background aerosol is of special importance because this may very rapidly change their specific characteristics that depend on the size of the particles. Such transformations can be modeled for given scenarios provided enough information is available about the system [30]. The changes in airborne ENM characteristics during transport have several consequences. Indeed, a change in airborne size could alter the deposition mechanisms of ENM and hence the effective dose of exposure. For the time being, it is not understood whether the agglomeration process affects the toxicological mechanisms as well. Additionally, growth and/or attachment of airborne ENM to other particles challenge the existing methods for their adequate characterization. Separation and identification of ENM against the submicron background aerosol originating from different sources is therefore a difficult or impossible task facing ENM monitoring and characterization in workplaces and other environments [31]. At present, there are no rapid techniques available that would allow online distinction between background nanoparticles and ENM. Instead, one has to collect the aerosol and do an off-line image or compositional analysis, e.g. by transmission electron microscope (TEM). Therefore, TEM remains the gold standard for this type of analysis [31,32].

Novel materials introduce entirely new particle shapes, such as CNTs which, upon release into the air, can further form larger fiber-like structures resembling, e.g. crocidolite fibers with potentially serious health consequences [24,33]. The addition of fullerene structures to the backbone of SWCNT produces nano-buds [34] which may be associated with unexpected biological effects. The effects of functionalization of ENM add to the complexity of their risk assessment, and are beyond this discussion, and for the most part, have not been done so far.

The thorough characterization of airborne ENM is thus complicated by the dynamic behavior of ENM as an aerosol as well as the structural complexity of the individual particles.

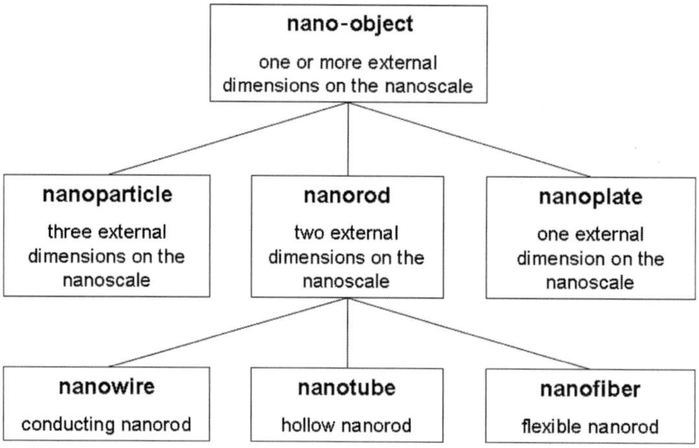

FIGURE 2.1 Hierarchical relationship of terms for engineered nanomaterials.

The large set of parameters required for their complete characterization, the range of ENM materials already in use, and the multitude of biological responses create a special challenge for this undertaking. At present, there is no robust set of devices which could be used for monitoring, measuring and characterizing ENP in workplace environments. A recent publication of the European Agency for Safety and Health at Work (EU-OSHA) [35] provides a hierarchical set of terms for engineered nanomaterials. Nano-object is the highest term divided into nanoparticles, nanorods and nanoplates from which nanorods can be further subdivided into nanowires, nanotubes and nanofibers (Figure 2.1).

CHALLENGES TO ASSESS EXPOSURE TO ENM

Measurement and monitoring of ENM present in the air of workers' breathing zones means capturing all relevant information about the amount, i.e. number, surface area or mass concentration and size distribution, as well as shape, composition and chemical reactivity of airborne ENM in a given size class or a broad size range [15,27]. Selection of the most relevant metric(s) for health-related sampling of ENM is an important component in the development of the concepts, methods and technology for ENM monitoring at workplaces [4,30,36–38]. For this purpose, understanding the relationship between ENM metrics and toxicological effects of ENM is necessary, but as yet unavailable. This is because a consensus on the correct metrics to be measured to assess exposure to ENM has not yet been reached internationally [8,39,40].

The multitude of relevant ENM metrics, in combination with the different possible release mechanisms for ENM into workplace air as well as the poorly defined transport pathways between source and receptor, i.e. the exposed individual, make it imperative to establish and define typical ENM exposure scenarios. The complexity of the ENM metrics is further challenged by the fact that all of the factors listed above have an impact on particle characteristics. Regarding different exposure scenarios for ENM, one may need to assess exposure to "fresh ENM" emitted from different production processes, as well as for settings in which prevalent exposure to "aged" and "attached" ENM takes place [30]. These work processes

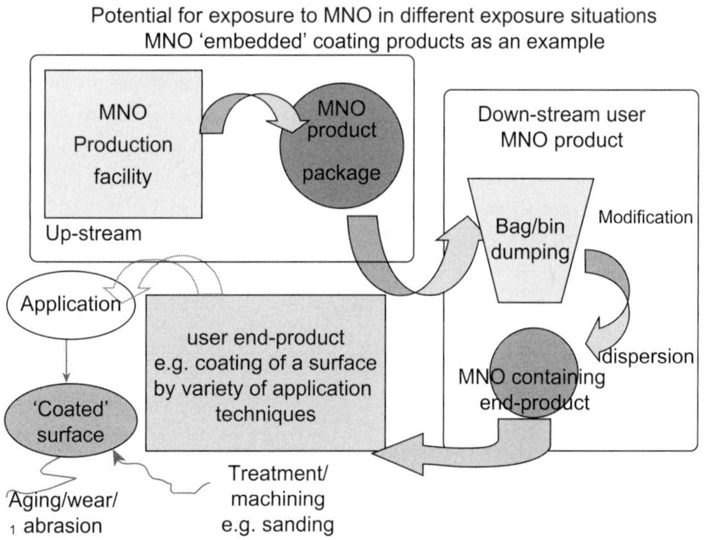

FIGURE 2.2 Schematic diagram of the potential for exposure to nanomaterials in different exposure situations. The figure depicts scenarios for a hypothetical coating containing nano-objects or MNOs. MNO, manufactured structured nano-objects. *(Reprinted from Brouwer D, Exposure to manufactured nanoparticles in different workplaces,* Toxicology *269, 120–127, 2010, with permission from Elsevier.)*

may include for example, handling and packing of partly agglomerated and aggregated ENM. It is important to obtain data from these types of exposure settings to be able to assess true exposure levels of ENM in workplaces [27,31,38,41]. Figure 2.2 depicts different exposure scenarios and the path of ENM (or manufactured nano-objects, see Figure 2.1) through the different stages of the life cycle of these materials whether as primary particles in aerosols or incorporated in products, or when recycled [42].

Currently, aerosol technology uses a range of methods for monitoring of ENM-containing aerosols. Broadly speaking, they include online as well as offline methods to capture a number of different metrics. The array of techniques and concepts for obtaining physical aerosol information related to concentration or size is relatively large, with a general trend away from mass toward number or surface area based techniques [29].

Some devices are stationary and capable of providing detailed information about size or composition, while others are "personal" at the expense of detailed information. The benefit of acquiring detailed size distribution data online (especially of number size distributions) is that such information can easily and accurately be converted to almost any other physical metric (except shape, which requires additional parameters). Keeping in mind the open ended discussion about appropriate metrics, number size distribution data are thus the most valuable record which leaves the door open to *a posteriori* tests of new hypotheses concerning health risks of ENM.

Over the past two decades, methods providing direct chemical, compositional, or biological information of airborne particles, usually on the basis of optical or mass spectroscopy, have received a major push in the field of atmospheric sciences and, most recently, also in response to global security threats [43]; for an excellent summary see also http://cires.colorado.edu/~jjose/ams.html. However, these devices are often quite expensive and complex and the technology has not yet filtered down to affordable levels for health hazard evaluation.

Generally, some of the real challenges ahead for ENM monitoring and health risk assessment are: (1) to redesign "ENM-capable" instruments already in laboratory use into portable or personal and affordable devices; (2) to expand the sensing technology available for ENM detection by adopting new options with realistic potential for real-time measurement and compact design; and (3) to extend the metrics into new areas such as CNT shape identification as well as surface chemical or catalytic properties. These latter concepts are especially interesting, since they have the potential of distinguishing ENM against background particles via special morphological features or function (not composition).

The current inability to separate ENM from the background nanoparticles by straightforward concentration and size distribution measurements makes it highly problematic to set occupational exposure limits (OEL) for ENM. A mandatory prerequisite for such background distinction of nano-sized particles from ENM would require the ability for online chemical characterization of particles, a feature that the currently available measuring technologies do not enable, at least in miniaturized and affordable form.

Harmful health effects, such as increased cardiac and pulmonary mortality, of ambient fine particles have been emphasized [44]. Wide utilization of nanotechnologies globally is a novel phenomenon and is likely to have an impact on human health on a global scale in future. Besides, exposure to particle aerosols outdoors or indoors is, in most cases, exposure to a mixture of particles with a wide range of diameters, and hence the importance of assessing the impact the particle size range has on human health is a special challenge. There are, though, data which suggest that exposure to ultrafine particles, the ubiquitous background nano-sized particles, may be especially harmful in inducing health effects such as pulmonary inflammation, effects on circulation and even increased cardiac mortality [6,16,45]. When assessing the health impact of ENM in occupational and other environments, the distinction between background ultrafine or nanoparticles from ENM becomes especially important, because the ability to dissect these effects is the prerequisite of setting of OEL.

Each of the above avenues addresses an important demand: (1) making current ENM monitoring technology more compact, more affordable and more versatile will provide imminent short-term solutions required by toxicologists and the inhalation exposure community; (2) new sensing technology will have a mid-term effect by providing sophisticated measurement options for very small particles which can be adapted to the needs of aerosol monitoring technology; (3) finally, the need for devices capable of capturing entirely new properties will provide new tools to characterize airborne ENM. It will be important for these new devices to provide real-time and online data. However, the foregoing discussion also makes it clear that an ideal, all-purpose monitoring method will only become available (if it ever becomes available), once a clear link between health effects and ENM characteristics is well established for a majority of exposure scenarios; and this will only happen after sufficient data have been collected and analyzed. After this introduction to exposure assessment of ENM, issues and measurement approaches and technologies will be discussed in depth below.

REASONS FOR THE ASSESSMENT OF EXPOSURE TO ENM IN WORKPLACES

The considerations presented above suggest the need to characterize carefully exposure to ENM at workplaces. In fact, they may be released in the workplace and consequently exposure to them may involve a number of activities and workers. Furthermore, their chemical, optical, magnetic, biological and structural properties (size, shape, surface chemistry, reactivity and area, number concentration, mass concentration, state of agglomeration) make them potentially dangerous, indeed, harmful effects to workers have been reported.

Regarding the assessment of ENM in the occupational setting, there are several still unresolved or partially resolved issues concerning the metrics and instruments that should be used to measure nanoparticles, the ENM modifications after their release and what ENM in the workplace may imply in terms of workers' protection.

The possible links between the ENM physicochemical characteristics and their detailed biological activity and toxicity are discussed in detail in other chapters of this book. However, issues related to ENM characteristics and associated health effects have also been briefly discussed above.

Evidence for Release in the Workplace

Several workplace air monitoring studies for manufactured nano-objects are now available, most of them published since 2008 [29,31,46–60]. They regard both commercial and research scale, and the investigated nanoparticles were carbon nanotubes and nanofibers, carbon black, fullerenes, metal and metal oxide nanoparticles. Many activities have been investigated such as workplace cleaning, packaging/bag filling, reactor cleanout, mixing, cutting composite, transfer, etc. Almost all studies found some release of ENM in the workplace in comparison to background levels. Therefore, extensive workplace investigations should be performed in order to identify the work activities that cause a significant exposure to ENM and, consequently, to eliminate or to reduce to the lowest possible level the potential workers' exposure to nanoparticles, until reference exposure levels are available.

Harmful Effects on Workers

Two recently published case reports suggest that serious adverse effects may be observed in workers exposed to ENM.

The first report regards seven workers who worked in the same room of a plant where they sprayed paste containing nano-sized polycyclic ester particles onto polystyrene boards. They were admitted to hospital with clinical findings of pleural effusion and pericardial effusion. The same nanoparticles to which they were exposed at work were found in the chest fluid and lung biopsies of the workers, two of whom subsequently died from respiratory failure [61]. The results of this paper have been challenged, and the contribution of poor occupational hygiene at that workplace in general has been claimed as the reason for the illnesses and deaths. The fact though remains that the exposure to ENM on this occasion was high. However, further studies are required to determine which levels of these or other ENM as aerosols may endanger the health of workers, and hence OELs and their reliable monitoring will be increasingly important in the future.

In the second report is described a worker who inhaled an estimated gram of nickel nanoparticles over an approximate 90-minute period, and who died from adult respiratory distress syndrome [62]. However, when compared to the OSHA permissible exposure limit for nickel of 1 mg/m^3 over an 8-hour period, the worker's exposure was large regardless of whether the nickel was composed of nanoparticles.

How to Measure Nanoparticles: Metrics

It should first of all be emphasized that exposure assessment is not just concentration measurement. Information on the following parameters is in fact needed: magnitude, physical form (e.g. dry powder or slurry formulation), route of absorption, frequency and duration of exposure, distribution (spatial, within worker-between workers) of control measures. In addition, an observational walkthrough to get familiar with the processes and work practice is also recommended. These factors can be used to identify lower or higher risk situations. Current methods of measurement of nanoparticles are far from being considered optimal. There are theoretical and practical limitations to a reliable workplace evaluation. So far as metrics are concerned, the three most widely used metrics (mass concentration, number concentration, and surface concentration) have weak points: the *mass* of nanoparticles is too small. To collect a sample that could be detected gravimetrically, the sample would most likely require a longer sampling time than an 8-hour work sample. If larger particles were also collected, the mass of the larger particles would mask any increase from the nanoparticles. *Number concentration* may have large spatial and temporal variations, making difficult discrimination with background nanoparticle number concentration. Finally, *surface concentration* alone cannot in general be used to demonstrate the presence or absence of nanoparticles in the breathing zone of workers, due to the coagulation/aggregation phenomena (see below).

On the other hand, the concomitant testing of mass, number and surface concentration is impractical, because no single instrument giving reliable concomitant information on the three parameters currently exists, and the concomitant routine use of several instruments is too cumbersome (among others, for economic reasons) to be applied in workplaces.

How to Measure Nanoparticles: Measurement Strategies

There is increasing consensus that, at the current status of knowledge, the best approach to the assessment of nanoparticles in the workplace is represented by the *nanoparticle emission assessment technique (NEAT)* which has been developed by the US National Institute of Occupational Health and Safety (NIOSH) [63]. NEAT is based on the detection of airborne nanomaterials by means of portable instruments measuring their number concentration, supplemented by filters which allow offline sample analysis of particle morphology, size, count, and elemental composition.

The first step of the technique is to quantify the average level of background particle concentration. Indeed, it is critical for exposure monitoring that measurements be taken before production or processing of a nanomaterial to obtain background nanoparticle exposure data to compare with the number concentration during the work activity. With the system/process off, nanoparticle number concentration is measured at suspected or potential emission points; the same procedure is performed again with the system/process on. Measurements made during production or processing can then be evaluated to determine if there has been an increase in particle number concentrations in relation to background measurements and whether that change represents worker exposure to the nanomaterial. A good example of this approach has been the study by Peters et al. [31] in which they measured nanoparticle concentrations before and after the workday and throughout the day, and overlaid the particle number concentration on daily work activities trying to find associations between those activities and alterations in nanoparticle number concentrations in the workplace air. The association was not particularly good, but some association was found.

If there is at least 25% difference between the two measurements, a release of particles is considered to be present; otherwise, no further evaluation is necessary. Particle number concentration is evaluated by using together a Condensation Particle Counter (CPC) and an Optical Particle Counter (OPC) (for a more detailed description of these instruments see the paragraph on instruments). The rationale for the concomitant use of these two instruments is that differences between results of CPC and OPC may give insights on the presence of nanoparticles in the workplace. In fact, CPC identifies particles in the range 10 to 1000 nm, whereas OPC has six specific cut points at 0.3, 0.5, 1, 3, 5 and 10 μm. If a high particle number concentration on the CPC, in combination with a high particle number concentration in the small size ranges (300–500 nm) on the OPC is obtained, probably nanoscale particles will be present. Conversely, a low CPC particle number concentration, in combination with a high OPC particle number concentration in the larger size ranges (>10 000) may indicate the presence of larger particles and/or nanoparticle agglomerates (see below).

During the second step of the procedure, particles are collected on filters: background sampling, source specific sampling, and personal sampling are performed. The collected material is thereafter transferred for scanning electron microscopy (SEM), or transmission electron microscopy (TEM), which may be coupled with energy dispersive x-ray analyzer (EDS) for a complete structural and chemical analysis. Elemental analysis may also be performed by means of inductively coupled plasma-mass spectrometry (ICP-MS) (third step).

In order to validate the procedure, the NIOSH recommends a fourth step in which background particle number concentration is measured again: if subtracting the average of before/after background from process-specific measurements makes still at least 25% difference, the findings are considered reliable. Successful application of the NEAT strategy has been reported in several facilities [64]. A recently suggested modification to the NEAT strategy is the addition of surface area analyzers (for research purposes only).

The appeal of the NEAT is that it is relatively simple and may be performed in the field with portable instruments of relatively low cost. It cannot be considered, however, the definitive response to the issue of workplace evaluation of nanoparticles. As an example, the upper detection limit of CPC (10^5 particles/mL) may be a bit restrictive for some workplace settings. Furthermore, at some time in the not too distant future, it is likely that workers' exposure standards will be promulgated, and therefore a quick, simple repeatable and inexpensive exposure assessment will have to be developed in order to compare it with a permissible exposure limit (PEL). In the meantime, however, assessments in occupational settings should be performed, adopting uniform, standardized strategies.

How to Measure Nanoparticles: Instruments

Instruments Measuring Number Concentration

1. CPCs provide real time number concentration measurements between their particle diameter detection limits. Without a nanoparticle pre-separator they are not specific to the

nanometer size range. Some models have a diffusion screen to limit top size to 1 μm. Consequently, they can provide real time measurements in the range 10 nm to 1 μm. To count the particles, the CPC increases the size of submicron particles by vapor (alcohol/water) supersaturation so the enlarged particles can then be detected by a laser. Depending on the model, particle counts can be logged every second. Many models are handheld, but due to the liquid needed to create the supersaturation environment, the CPC cannot be tipped. If tipped, the liquid will drip off the wick and flood the area in front of the laser interrupting measurements.

2. OPCs provide real time number concentration measurements between their particle diameter detection limits. Particle size diameters begin at 300 nm and may go up to 10 μm. The optical particle counters alone are not appropriate for nanoparticles, since they are insensitive to particles smaller than approximately 300 nm in diameter. However, they may give useful information if used in combination with CPC (see the paragraph on measurement strategies).
3. A scanning mobility particle sizer (SMPS) is composed of a differential mobility analyzer and a CPC. It works detecting real time size-selective (mobility diameter) number concentration, giving aerosol size distribution. It measures the size of the particles by using positive electrostatic charges on the particles to attract them to a negatively-charged rod. The possible detectable particle size range is from 3 nm to 800 nm (although during monitoring this must be narrowed down to a smaller range). The SMPS utilizes a radioactive source (Kr 85). Depending on the model, it may require up to two minutes to conduct a measurement which may be too slow to identify nanoparticle sources. Even if the recently introduced fast mobility particle sizer (FMPS) allows a much better time resolution, the size, expense and radioactivity emission of both SMPS and FMPS limit their application in the workplace.

Instruments Measuring Surface Concentration

1. Diffusion chargers (DCs) give real-time measurement of aerosol active surface area. Active surface area does not scale directly with geometric surface area above 100 nm. It is important to note that not all commercially available DCs have a response that scales with particle active surface area below 100 nm. Therefore, DCs are only specific to nanoparticles if used with an appropriate inlet pre-separator. For particles >100 nm, DCs underestimate the surface area. Depending on the model, measurements can be logged every 10 seconds. DCs vary in size with some small enough to be carried by a worker. Results are reported as $\mu m^2/cm^3$ or similar. Although they have the advantages of having no working fluids, no radioactive sources and the possibility of any orientation, they have limitations (e.g. signal interpretation is a bit confusing) that need to be evaluated with respect to measuring workplace aerosols. A recent improvement in DCs is represented by diffusion size classifiers (DiSCs) which allow obtaining both total number concentration and average particle size. The time resolution is about 2 seconds, and they are battery-powered and portable. Validation of DiSCs is still at the initial stage.
2. Electrical low pressure impactor (ELPI). This gives real-time size-selective (aerodynamic diameter) detection of active surface area concentration of particles between 7 nm and 10 μm, combining electrical charging, inertial collection and electrical detection. In the ELPI,

a real-time cascade impactor, particles first pass through a diffusion charger, are positively charged, and then they enter a Berner-type low pressure impactor made up of 13 stages where the particles are collected onto electrically isolated stages by aerodynamic diameter. The current on each stage, measured by electrometers, is further converted into a particle number concentration. Depending on the model, measurements can be logged every 5 seconds. The offline characterization (TEM, ICP-MS, etc.) of samples is also possible. Widespread application in the workplace is limited by size, expense and complexity of operation, and sensitivity to harsh field conditions. Technical improvements in precision and resolution power of ELPI have been recently suggested [65].

3. Nanoparticle surface area monitor (NSAM) is a recently launched instrument aimed at determining lung-deposited particle surface area concentrations, based on the widely accepted model developed by the International Commission on Radiological Protection (ICRP). The limitation and perspective of this kind of measurement have been extensively discussed by Asbach et al. [66].
4. Modified electrical aerosol detector (MEAD). The same principles as NSAM, but less expensive, less bulky, and easy to use. Validation is at the initial stage [67].

Instruments Measuring Mass Concentration

1. Size selective sampler (SSS). Cascade impactors are currently the only devices giving a direct estimation of mass concentration with a cut-off point around 100 nm, i.e. can measure a total mass above 100 nm and also below 100 nm (down to approximately 20 nm). They are easy to use and provide a mass concentration for a defined range of particle sizes. The only disadvantage is that they are not too portable but if a trolley is provided they can be used in an industrial setting.
2. Tapered element oscillating microbalance (TEOM) measures nano-aerosol mass concentration online with a suitable size selective inlet. The main disadvantage is that the TEOM does not provide size information. There is a version of the TEOM that can be used as a personal monitor.

Instruments Giving Indirect Estimates

Some of the above listed instruments may give, in addition to the measured metric, an indirect estimate of other parameters. The reliability of the assumptions and of the calculations needed to perform such estimations is questionable:

1. indirect estimation of mass concentration is given by ELPI and SMPS
2. indirect estimation of number concentration: ELPI and DiSC
3. indirect estimation of surface area: combined use of SMPS and ELPI may be performed to infer particle fractal dimensions.

Sample Collection and Characterization

Several relevant parameters can only be determined through offline analytical techniques (electron microscopy, ICP-MS etc.). Each technique requires specific methods for sample collection and preparation. For electron microscopy (EM), uniform deposit, minimum overlapping and good collection efficiency are required. Scanning electron microscopy (SEM) has a 10–50 nm resolution, whereas transmission electron microscopy (TEM) has a resolution <10 nm. TEM analysis provides direct information on the projected area of collected particles which may be related to geometric area for some particle shapes. TEM can provide an estimate of the particle size distribution and, if equipped with an EDS, a determination of elemental composition with the examination of one sample. Samples are collected onto filters or other substrates (e.g. TEM grid), for subsequent offline analysis. Recent developments in sampling techniques for EM have been performed: handled TEM sampler [68], and sample with automatic determination of sampling time [69] will probably be available in the near future.

In this context it is important to note that using particle counters in combination with sample collection for chemical analysis allows a good evaluation of worker exposure to ENM but, if this assessment strategy continues to rely only on static or area sampling, some uncertainty will always exist in estimating worker exposures.

A specific problem is posed by the high aspect ratio nanoparticles (HARN) (e.g. carbon nanotubes). Owing to their very small diameter, high magnifications are necessary for their detection, decreasing the likelihood of finding countable fibers (i.e. with both ends counted). The highly agglomerated nature of some types of HARN (e.g. SWCNT) adds further problems to the identification of single fibers.

ENM Modifications After their Release

In the workplace, workers may come in contact with ENM present in aerosols (generation of nanoparticles in non-enclosed systems, incidental leakage from gas-phase reactors), powders (milling, pouring or mixing, handling, e.g. weighing, spraying), liquids (pouring or mixing operations with suspensions of ENM, any agitation with some energy, cleaning operations), materials (machining, cutting, grinding, sanding, drilling, or any other mechanical operation on materials containing ENM or nano-structured material).

A potentially relevant property of ENM in the workplace is represented by their dustiness (nano-dustiness); this is especially true for workers handling nano-powders. Tsai et al. determined the dustiness of two nanoscale TiO_2 and ZnO in standard 1-minute and 30-minute tests and found that very few particles below 100 nm were generated. Thus, nano-dustiness does not seem to be of major concern in workplace exposure [70]. However, this work, along with other reports and conceptual models [71–73] confirms the dominance of nanoparticles agglomerates or nanoparticles attached to background particles in the air during production and handling of nano-powders. These larger particles should be taken into account when measuring the presence of nanoparticles in the workplace, because all particles in the respirable size range can enter and deposit deep in the lung, including aggregates and agglomerates. As an example, Cena and Peters found negligible release of nanoparticles during the processing of CNT-epoxy nanocomposites materials; nevertheless, there was a substantial release of micron-sized CNT containing particles, which were poorly controlled by the fume hood [74].

Indeed, most of "in the field" studies reveal the formation of larger agglomerates after the release of ENM. However, methodological issues may render difficult the interpretation of the data in the light of workers' exposure [72]. A consensus on the optimal methods of assessing the dynamics of nanoparticles/agglomerates equilibrium in the workplaces is urgently needed.

Workers' Protection

Until exposure limits are established, protective measures should be based on the precautionary principle (i.e. any exposure to nanoparticles in the workplace should be reduced to background levels).

In general, there are three main complementary approaches to risk and exposure control: engineering techniques, administrative means and personal protective equipment.

Although it is frequently assumed that the usual protective measures would not be effective, this assumption is not consistent with physical principles: particles larger than 300 nm are most effectively separated by impaction on the filter material, entrapment by the filter material, or sedimentation owing to gravity. For particles smaller than 300 nm, separation by diffusion (Brownian movement) and electrostatic forces becomes increasing relevant with decreasing size. It can therefore be asserted as a rule that technical measures which are effective against micrometric particles are also suitable for the elimination of nanoparticles. Of course, the engineering control systems, such as enclosure and ventilation, must be designed according to the gaseous and particulate properties of the nanoparticles and consider several parameters such as the quantity of the bulk ENM that is synthesized or handled in the manufacture of a product, the physical form of the ENM and task duration and frequency. Moreover, all ventilation systems should be evaluated, approved and maintained by workplace health and safety officers. It is therefore not surprising that available data seem to indicate that well-designed engineered protective measures are effective for nanoparticles [75]. In particular, airborne exposure to nanomaterials can most likely be controlled at most processes and job tasks using a wide variety of engineering control techniques similar to those used in reducing exposures to general aerosols [76,77]. The most widely used engineering control of nanoparticle exposure is represented by fume hoods and it has been recently reported that the air-curtain hood, a new design with significantly different airflow pattern from traditional hoods, may effectively control the release of nanoparticles under all operating conditions [78].

Administrative means of control constitute an additional approach when other methods have not achieved the expected control levels. In these situations, reduction of work periods, modification of work practices, personal hygiene measures, housekeeping and preventive maintenance constitute other ways of limiting the occupational exposure risks. As examples of good practices, hand washing, showering, changing and cleaning clothes facilities should be provided to prevent the inadvertent contamination of other areas (including take-home) caused by the transfer of nanoparticles on clothing and skin. The storage and consumption of food or beverages in workplaces should be prevented where nanomaterials are handled. Dry mopping, sweeping, dusting, cleaning using compressed air or portable blowers or fans should be prohibited. Work areas should be cleaned at the end of each work shift (at a minimum) using either a high-efficiency particulate air (HEPA)-filtered vacuum cleaner or wet wiping methods. Furthermore, it should be forbidden for workers to handle ENM in the open air and they should store dispersible ENM, whether suspended in liquids or in a dry particle form, in closed containers whenever possible. There are also some good practices for management, the most important of which are educating workers on the safe handling of ENM to minimize the likelihood of inhalation exposure and skin contact, providing information on the hazardous properties of ENM and providing additional control measures to ensure that ENM are not transported outside the work area.

Nanoparticle exposure may often be attributable to the wearing of inadequate personal protective equipment (PPE). Protective clothing that would typically be required for a wet-chemistry laboratory would be appropriate and could include but not be limited to:

- gauntlet-type gloves or nitrile gloves with extended sleeves
- chemical splash goggles and laboratory coats.

Standard tests showed that non-woven fabrics, like high density polyethylene textile (Tyvek type), seem to be much more efficient against nanoparticle penetrations. Indeed, recent data indicate that their efficiency in inhibiting the penetration of nanoparticles may be even higher than for larger material [79].

The use of respirators is often required when engineering and administrative controls do not adequately keep worker exposures to an airborne contaminant below a regulatory limit or an internal control target. Rengasamy et al. [80] used silver nanoparticles (4–30 nm) and sodium chloride particles (20–400 nm) in order to test NIOSH-tested (N95, P100) and CE-compliant (FFP2, FFP3) respiratory filter masks. The filter materials employed complied with the requirements stated in the standards.

TAKE-HOME MESSAGES

1. Nanoscale particles are ubiquitous in many workplaces; consequently, human exposure to nanoparticles does not involve only nano-researchers, nano-manufacturers and nano-product users exposed to engineered nanomaterials but also other worker categories exposed to ultrafine or incidental nanoparticles.
2. Currently, there is not one sampling method that can be used to characterize exposure to nanoparticles; however, there is increasing consensus that the best approach to the assessment of nanoparticles in the workplace is represented by the *nanoparticle emission assessment technique (NEAT).*
3. At the present time, no occupational exposure limits specific for nanoparticles have yet been developed. Until exposure limits are established, protective measures should be based on the precautionary principle.

Acknowledgments

The authors wish to thank Ms Leila Ahlström for excellent editorial assistance. This undertaking was partially supported by the European Union 7th Framework Programme Grant CP-IP-211464-2 (NANODEVICE Project).

References

[1] National Science Foundation (NSF). Societal implications of nanoscience and nanotechnology. Arlington, VA: Nanoscale Science, Engineering, and Technology (NSET) Workshop Report; 2001.
[2] Lux Research. 5th ed. New York: The nanotech report; 2007.
[3] Kaluza S, Balderhaar JK, Orthen B, et al. Workplace exposure to nanoparticles. Johanna Kosk-Bienko, European Agency for Safety and Health at Work (EU-OSHA) Ed; 2009.
[4] Scientific Committee on Emerging and Newly Identified Health Risks (SCENIHR). Scientific basis for the definition of the term "nanomaterial". Brussels, Belgium; 2010.
[5] Nel A, Xia T, Madler L, Li N. Toxic potential of materials at the nanolevel. Science 2006;311:622–7.

[6] Borm PJ, Robbins D, Haubold S, et al. The potential risks of nanomaterials: A review carried out for ECETOC. Part Fibre Toxicol 2006;3:11.
[7] Lynch I, Dawson KA, Linse S. Detecting cryptic epitopes created by nanoparticles. Sci STKE 2006;327:14.
[8] Environment Directorate Joint Meeting of The Chemicals Committee and Working Party on Chemicals, Pesticides and Biotechnology. Series on the Safety of Manufactured Nanomaterials No. 27. List of manufactured nanomatrials and list of endpoints for phase one of the sponsorship programme for the testing of manufactured nanomaterial: Revision; 2010
[9] Seipenbusch M, Binnig J, Heim M, Weber AP, Kasper G. Aerosol catalysis: The influence of particle structure on the catalytic activity of platinum-nanoparticle agglomerates. Helv Chim Acta 2001;84:3683−701.
[10] Weber AP, Seipenbusch M, Kasper G. Size effects in the catalytic activity of unsupported metallic nanoparticles. J Nanopart Res 2003;5:293−8.
[11] Long TC, Saleh N, Tilton RD, Lowry GV, Veronesi B. Titanium dioxide (P25) produces reactive oxygen species in immortalized brain microglia (BV2): Implications for nanoparticle neurotoxicity. Environ Sci Technol 2006;40:4346−52.
[12] Cedervall T, Lynch I, Lindman S, et al. Understanding the nanoparticle-protein corona using methods to quantify exchange rates and affinities of proteins for nanoparticles. Proc Natl Acad Sci USA 2007; 104:2050−5.
[13] Cedervall T, Lynch I, Foy M, et al. Detailed identification of plasma proteins adsorbed on copolymer nanoparticles. Angew Chem Int Ed Engl 2007;46:5754−6.
[14] Nel AE, Mädler L, Velegol D, et al. Understanding biophysicochemical interactions at the nano−bio interface. Nat Mater 2009;8:543−57.
[15] Long TC, Tajuba J, Sama P, et al. Nanosize titanium dioxide stimulates reactive oxygen species in brain microglia and damages neurons *in vitro*. Environ Health Perspect 2007;115:1631−7.
[16] Rossi EM, Pylkkänen L, Koivisto AJ, et al. Airway exposure to silica-coated TiO_2 nanoparticles induces pulmonary neutrophilia in mice. Toxicol Sci 2010;113:422−33.
[17] Kreyling WG, Semmler M, Erbe F, et al. Translocation of ultrafine insoluble iridium particles from lung epithelium to extrapulmonary organs is size dependent but very low. J Toxicol Environ Health A 2002;65:1513−30.
[18] Kuempel ED, Tran CL, Castranova V, Bailer AJ. Lung dosimetry and risk assessment of nanoparticles: Evaluating and extending current models in rats and humans. Inhal Toxicol 2006;18:717−24.
[19] Nurkiewicz TR, Porter DW, Hubbs AF, et al. Nanoparticle inhalation augments particle-dependent systemic microvascular dysfunction. Part Fibre Toxicol 2008;12:1.
[20] Dankovic D, Kuempel E, Wheeler M. An approach to risk assessment for TiO2. Inhal Toxicol 2007;19:205−12.
[21] Shvedova AA, Kisin ER, Mercer R, et al. Unusual inflammatory and fibrogenic pulmonary responses to single-walled carbon nanotubes in mice. Am J Physiol Lung Cell Mol Physiol 2005;289:L698−708.
[22] Lam CW, James JT, McCluskey R, Arepalli S, Hunter RL. A review of carbon nanotube toxicity and assessment of potential occupational and environmental health risks. Crit Rev Toxicol 2006;36:189−217.
[23] Donaldson K, Murphy FA, Duffin R, Poland CA. Asbestos, carbon nanotubes and the pleural mesothelium: A review of the hypothesis regarding the role of long fibre retention in the parietal pleura, inflammation and mesothelioma. Part Fibre Toxicol 2010;22:5.
[24] Ryman-Rasmussen JP, Cesta MF, Brody AR, et al. Inhaled carbon nanotubes reach the subpleural tissue in mice. Nat Nanotechnol 2009;4:747−51.
[25] Gulson B, McCall M, Korsch M, et al. Small amounts of zinc from zinc oxide particles in sunscreens applied outdoors are absorbed through human skin. Toxicol Sci 2010;118:140−9.
[26] Elder A, Gelein R, Silva V, et al. Translocation of inhaled ultrafine manganese oxide particles to the central nervous system. Environ Health Perspect 2006;114:1172−8.
[27] Savolainen K, Alenius H, Norppa H, Pylkkänen L, Tuomi T, Kasper G. Risk assessment of engineered nanomaterials and nanotechnologies—a review. Toxicology 2010;269:92−104.
[28] Rothenbacher S, Messerer A, Kasper G. Fragmentation and bond strength of airborne diesel soot agglomerates. Part Fibre Toxicol 2008;4:9.
[29] Kuhlbusch TA, Fissan H. Particle characteristics in the reactor and pelletizing areas of carbon black production. J Occup Environ Hyg 2006;3:558−67.
[30] Seipenbusch M, Binder A, Kasper G. Temporal evolution of nanoparticle aerosols in workplace exposure. Ann Occup Hyg 2008;52:707−16.

[31] Peters TM, Elzey S, Johnson R, et al. Airborne monitoring to distinguish engineered nanomaterials from incidental particles for environmental health and safety. J Occup Environ Hyg 2009;6:73–81.
[32] Kuhlbusch TA, John AC, Quass U. Sources and source contributions to fine particles. Biomarkers 2009;14:23–8.
[33] Sakamoto Y, Nakae D, Fukumori N, et al. Induction of mesothelioma by a single intrascrotal administration of multi-wall carbon nanotube in intact male Fischer 344 rats. J Toxicol Sci 2009;34:65–76.
[34] Nasibulin AG, Queipo P, Shandakov SD, et al. Studies on mechanism of single-walled carbon nanotube formation. J Nanosci Nanotechnol 2006;6:233–46.
[35] European Agency for Safety and Health at Work (EU-OSHA). Workplace exposure to nanoparticles. 2010.
[36] Maynard AD. Nanotechnology: A research strategy for addressing risk. Washington, DC: Woodrow Wilson International Center for Scholars; 2006.
[37] Maynard AD, Aitken RJ. Assessing exposure to airborne nanomaterials: Current abilities and future requirements. Nanotoxicology 2007;1:26–41.
[38] Schulte P, Geraci C, Zumwalde R, et al. Sharpening the focus on occupational safety and health in nanotechnology. Scand J Work Environ Health 2008;34:471–8.
[39] Scientific Committee on Emerging and Newly Identified Health Risks (SCENIHR). The appropriateness of the risk assessment methodology in accordance with the Technical Guidance Documents for new and existing substances for assessing the risks of nanomaterials. Brussels, Belgium; 2007.
[40] Scientific Committee on Emerging and Newly Identified Health Risks (SCENIHR). Risk assessment of products of nanotechnologies. Brussels, Belgium; 2009.
[41] Schulte PA, Schubauer-Berigan MK, Mayweather C, Geraci CL, Zumwalde R, McKernan JL. Issues in the development of epidemiologic studies of workers exposed to engineered nanoparticles. J Occup Environ Med 2009;51:323–35.
[42] Brouwer D. Exposure to manufactured nanoparticles in different workplaces. Toxicology 2010;269:120–7.
[43] Noble CA, Prather KA. Real-time single particle mass spectrometry: A historical review of a quarter century of the chemical analysis of aerosols. Mass Spectrom Rev 2000;19:248–74.
[44] Dockery DW, Pope 3rd CA, Xu X, et al. An association between air pollution and mortality in six U.S. cities. N Engl J Med 1993;329:1753–9.
[45] Duffin R, Mills NL, Donaldson K. Nanoparticles—a thoracic toxicology perspective. Yonsei Med J 2007;48: 561–72.
[46] Kuhlbusch TAJ, Neumann S, Fissan H. Number size distribution, mass concentration, and particle composition of PM1, PM2.5, and PM10 in bag filling areas of carbon black production. J Occup Environ Hyg 2004;1:660–71.
[47] Maynard AD, Baron PA, Foley M, Shvedova AA, Kisin ER, Castranova V. Exposure to carbon nanotube material: Aerosol release during the handling of unrefined single-walled carbon nanotube material. J Toxicol Environ Health A 2004;67:87–107.
[48] Methner MM, Birch ME, Evan DE, Ku B-K, Hoover MD. Identification and characterization of potential sources of worker exposure to carbon nanofibers during polymer composite laboratory operations. J Occup Environ Hyg 2007;4:D125–30.
[49] Bello D, Hart AJ, Ahn K, et al. Particle exposure levels during CVD growth and subsequent handling of vertically-aligned carbon nanotube films. Carbon 2008;266:974–81.
[50] Fujitani Y, Kobayashi T, Arashidani K, Kunugita N, Suemura K. Measurement of the physical properties of aerosols in a fullerene factory for inhalation exposure assessment. J Occup Environ Hyg 2008;5:380–9.
[51] Han JH, Lee EJ, Lee JH, et al. Monitoring multiwalled carbon nanotube exposure in carbon nanotube research facility. Inhal Toxicol 2008;20:741–9.
[52] Tsai S-J, Ashter A, Ada E, Mead JL, Barry CF, Ellenbecker MJ. Airborne nanoparticle release associated with the compounding of nanocomposites using nanoalumina as fillers. Aerosol Air Qual Res 2008;8:160–77.
[53] Yeganeh B, Kull CM, Hull MS, Marr LC. Characterization of airborne particles during production of carbonaceous nanomaterials. Environ Sci Technol 2008;42:4600–6.
[54] Bello D, Wardie BL, Yamamoto N, et al. Exposure to nanoscale particles and fibres during machining of hybrid advanced composites containing carbon nanotubes. J Nanopart Res 2009;11:231–49.
[55] Park J, Kwak BK, Bae E, et al. Characterization of exposure to silver nanoparticles in a manufacturing facility. J Nanopart Res 2009;11:1705–12.
[56] Tsai S-J, Ada E, Isaacs JA, Ellenbecker MJ. Airborne nanoparticle exposures associated with the manual handling of nanoalumina and nanosilver in fume hoods. J Nanopart Res 2009;11:147–61.

[57] Tsai S-J, Hofmann M, Hallock M, Ada E, Kong J, Ellenbecker M. Characterization and evaluation of nanoparticle release during the synthesis of single-walled and multiwalled carbon nanotubes by chemical vapour deposition. Environ Sci Technol 2009;43:6017–23.
[58] Vorbau M, Hillemann L, Stintz M. Method for the characterization of the abrasion induced nanoparticle release into air from surface coatings. Aerosol Sci 2009;40:209–17.
[59] Huang CH, Tai CY, Huang CY, et al. Measurement of respirable dust and nanoparticle concentrations in a titanium dioxide pigment production factory. J Environ Sci Health A Tox Hazard Subst Environ Eng 2010;45:1227–33.
[60] Vosburgh DJH, Boysen DA, Oleson JJ, Peters TM. Airborne nanoparticle concentration in the manufacturing of polytetrafluoroethylene (PTFE) apparel. J Occup Environ Health 2011;8:139–46.
[61] Song Y, Li X, Du X. Exposure to nanoparticles is related to pleural effusion, pulmonary fibrosis and granuloma. Eur Resp J 2009;34:559–67.
[62] Phillips JI, Green FY, Davies JC, Murray J. Pulmonary and systemic toxicity following exposure to nickel nanoparticles. Am J Ind Med 2010;53:763–7.
[63] Methner M, Hodson L, Geraci C. Nanoparticle emission assessment technique (NEAT) for the identification and measurement of potential inhalation exposure to engineered nanomaterials—Part A. J Occup Environ Health 2010;7:127–32.
[64] Methner M, Hodson L, Dames A, Geraci C. Nanoparticle emission assessment technique (NEAT) for the identification and measurement of potential inhalation exposure to engineered nanomaterials—Part B: results from 12 field studies. J Occup Environ Health 2010;7:163–76.
[65] Yli-Ojanperä J, Kannosto J, Marjamäki M, Keskinen J. Improving the nanoparticle resolution of ELPI. Aerosol Air Qual Res 2010;10:360–6.
[66] Asbach C, Fissan H, Stahlmecke B, Kulbush TAJ, Pui DYH. Conceptual limitations and extentions of lung-deposited nanoparticle surface area monitor (NSAM). J Nanopart Res 2009;11:101–9.
[67] Wang YF, Tsai PJ, Chen CW, Chen DR, Hsu DJ. Using a modified electrical aerosol detector to predict nanoparticle exposures to different regions of the respiratory tract for workers in a carbon black manufacturing industry. Environ Sci Technol 2010;44:6767–74.
[68] Miller A, Frey G, King G, Sunderman C. A handling electrostatic precipitator for sampling airborne particles and nanoparticles. Aerosol Sci Technol 2010;44:417–27.
[69] Azong-Wara N, Asbach C, Stahlmecke B, et al. Optimization of a thermophoretic personal sampler for nanoparticle exposure studies. J Nanopart Res 2010;12:1611–24.
[70] Tsai C-J, Wu Ch-H, Leu M-L, et al. Dustiness test of nanopowders using a standard rotating drum with a modified sampling train. J Nanopart Res 2008;10:121–31.
[71] Schneider T, Jensen KA. Relevance of aerosol dynamics and dustiness for personal exposure to manufactured nanoparticles. J Nanopart Res 2009;11:1637–50.
[72] Brouwer D. Exposure to manufactured nanoparticles in different workplaces. Toxicology 2010;269:120–7.
[73] Schneider T, Brouwer DH, Koponen IK, et al. Conceptual model for assessment of inhalation exposure to manufactured nanoparticles. J Expo Sci Environ Epidemiol 2011. Epub ahead of print.
[74] Cena G, Peters TM. Characterization and control of airborne particles emitted during production of epoxy/carbon nanotube nanocomposites. J Occup Environ Health 2011;8:86–92.
[75] US National Institute for Occupational Safety and Health, Centers for Disease Control and Prevention. Approaches to safe nanotechnology—Publication No. 2009–125. 2009.
[76] Ratherman S. Methods of control. In: Plog B, editor. Fundamentals of industrial hygiene. Itasca, IL: National Safety Council; 1996.
[77] Burton J. General methods for the control of airborne hazards. In: DiNardi SR, editor. The occupational environment—its evaluation and control. Fairfax, VA: American Industrial Hygiene Association; 1997.
[78] Tsai SJ, Huang RF, Ellenbecker MJ. Airborne nanoparticle exposures while using constant-flow, constant-velocity, and air-curtain-isolated fume hoods. Ann Occup Hyg 2010;54:78–87.
[79] Gao P, Jacques PA, Hsiao T-C, et al. Evaluation of nano and submicron particle penetration through ten non-woven fabrics using a wind-driven approach. J Occup Environ Health 2011;8:13–22.
[80] Rengasamy S, Eimer BC, Shaffer RE. Comparison of nanoparticle filtration performance of NIOSH-approved and CE-marked particulate filtering facepiece respirators. Ann Occup Hyg 2009;53:117–28.

CHAPTER

3

Biomonitoring

Enrico Bergamaschi[1], Andrea Magrini[2]

[1]Laboratory of Industrial Toxicology, University of Parma Medical School, Parma, Italy

[2]Department of Biopathology and Imaging, University of Rome "Tor Vergata", Rome, Italy

OUTLINE

INTRODUCTION

Biological monitoring (BM) has been defined as "a systematic, continuous or repetitive activity for collection of biological samples for analysis of concentrations of pollutants, metabolites or specific non-adverse biological effect parameters, with the objective to assess exposure and health risk to exposed subjects, comparing the data observed with the reference level and—if necessary—leading to corrective actions" [1]. BM is aimed at

Adverse Effects of Engineered Nanomaterials
DOI: 10.1016/B978-0-12-386940-1.00003-9

quantifying the amount of chemical absorbed and transformed into an active metabolite, or accumulated in deposit or target organs, tissues or cells as a consequence of occupational or environmental exposure to industrial chemicals. BM has been suggested as a valid tool in the practice of occupational safety and health with the purpose of: (1) understanding biologic changes of low doses of chemicals; (2) identifying potential hazards of new and emerging chemicals, including engineered nanomaterials (ENM) [2]; (3) understanding the mechanism of some occupational hazards; (4) identifying groups at higher risk of health outcomes. In routine applications, BM is complementary to ambient monitoring and is used to verify the compliance with hygienic standards, and to assure the effectiveness of risk management and prevention programs [3]. BM is becoming more popular as the number of biomarkers increases and the number of successful examples of field application proves the validity of such biomarkers. Biomarkers are increasingly used as surrogate indicators of designated events in a biological system due to the inaccessibility of target organs. Biomarkers used to model the interaction between a xenobiotic and the individual are indeed more directly related to the adverse effects which one attempts to prevent than any ambient measurement [4].

According to the National Research Council [5], biomarkers can be classified as: biomarkers of exposure, biomarkers of effect and biomarkers of susceptibility. A biomarker of exposure has been defined as "an exogenous substance or its metabolite or the product of an interaction between a xenobiotic agent and some target molecule or cell that is measured in a compartment within an organism", whereas a biomarker of effect is "any measurable biochemical, physiological or other alteration within an organism that, depending on magnitude, can be recognized as an established or potential health impairment or disease". Biomarkers of susceptibility, including both genetic (e.g. genetic polymorphisms of drug metabolizing and DNA repair enzymes) and acquired conditions [5], should help to distinguish groups at risk at the same exposure levels, but they can be regarded as effect modifiers rather than as risk factors.

From a pragmatic point of view, the use of a biomarker rather than its intrinsic properties may define its classification and biological meaning [6]. For instance, from a toxicological perspective, a biomarker of exposure should predict the likelihood of adverse effects or biological responses, whereas a biomarker of effect should be consistently correlated with the dose. From an epidemiological perspective, a biomarker of exposure should be consistently correlated with exposure (considering both background and interfering factors), whereas a biomarker of effect should be correlated with the risk of disease (considering spontaneous incidence) [6]. In order to become useful tools in risk assessment, biomarkers must be relevant, i.e. appropriate to provide information on important issues concerning health risks, and valid from both analytical and epidemiological aspects. In particular, external validity is required to obtain results which can be generalized to other populations [7].

CRITICAL ISSUES IN THE DEVELOPMENT OF BIOMONITORING FOR ENGINEERED NANOMATERIALS

There are a number of papers summarizing the state of the art of human BM of conventional chemicals (for a review, see [8,9]) but few foreseeing its applicability to ENM [2,10].

For a given chemical, BM requires the knowledge of pharmacokinetic and pharmacodynamic data, which are lacking for many classes of ENM, in spite of the great development of *in silico* models [11] (see Chapter 1 for an extensive discussion of ENM biokinetics). The unpredictable behavior of several ENM would require to assess the dose–response relationship of the most sensitive body organ rather than solely the organ through which ENP enter the body. The data generated by toxicological tests can be used to establish a quantitative relationship between the different physicochemical characteristics of the ENM and the response elicited. Those tests undertaken within the dose–response assessment of the ENM must further undergo both validation—assessing the test reproducibility—and verification, to ensure that the results of those *in vitro* tests carried out correspond to actual observations obtained from animal experiments (or human clinical situations) [12].

For nanoparticles, controversies exist about the best descriptor of dose related with biological effects. Particle number [13] and surface area [14–16] are the best dose metrics for ENM both *in vitro* and *in vivo*. Owing to their distribution in different target organs, it is also essential to understanding target organ dose, but current models and methods are insufficient [17,18]. However, new experimental methods for tracking and measuring ENM *in vivo*, currently applied to assess the effectiveness of delivered dose at target cells, could provide effective tools for obtaining quantitative dose data that are essential for dosimetry model validation [19] and, ultimately, for ascertaining the range of dose with no observed adverse effects (NOAEL).

Consistent data describing dose–response relationships between health effects and the accumulated mass following administration of ENM are available only for a limited number of particles (e.g. titanium dioxide, carbon black, amorphous silica and carbon nanotubes).

Research has identified the most significant biological responses and target organs affected by different ENM [12,20–22], but most of the published studies have focused on the adverse effects of nanoparticles on the respiratory system. Beyond the respiratory and cardiovascular systems, the impact of ENM is poorly understood. Experimental studies have characterized tissue response, e.g. of carbon nanotubes (CNT), and showed that deposition following instillation or inhalation leads to transient inflammatory changes, lung tissue remodeling and fibrosis [14–25]. Once inhaled, nanoparticles can deposit in lung cells and translocate through epithelial and endothelial cells into the blood and lymph circulation, potentially reaching sensitive target sites including bone marrow, lymph nodes, spleen, heart and central nervous system [26–31]. Translocation to many organs by ENM has been demonstrated, but the importance is not known, nor is there enough information to suggest what type of testing could be useful to assess dose and/or effects.

Short-term respiratory exposure to CNT can result not only in the increased lung expression of many gene encoding mediators of inflammation, oxidative stress, remodeling, and thrombosis, but also in increases in circulating biomarkers of inflammation and coagulation as well as activated blood cells, demonstrating a close cross-talk between the pulmonary and systemic circulation [32]. In spite of the great expectation about the possibility to implement organ-specific BM, it should be carefully considered that the relationship between lung inflammation and acute cardiovascular outcomes induced by particulate matter has not been clearly elucidated, and it is thus mandatory to investigate alternative mechanisms, such as the involvement of vagal bronchopulmonary receptors and neuronal pathways initiated in the lung, leading to an imbalance between the sympathetic and parasympathetic components of the heart [33].

OVERCOMING THE MOST CRITICAL ISSUES IN THE DEVELOPMENT OF BIOMARKERS

Nanomaterial distribution to different body tissues can potentially affect multiple cellular functions, interfering with many components of the cell machinery [34]. It is difficult to determine with conventional assays what changes and adverse effects may occur, and whether biochemical and/or functional changes can be material specific/exposure route specific. Conventional toxicology has used surrogate markers that are correlated with toxic responses to monitor adverse outcomes in inaccessible tissues. An alternative approach is suggested by system toxicology, which postulates that there are size-specific, shape-specific, and surface-specific effects and effectors for particles at the quantum range (1–100 nm). These effects are different from the effects observed for micro-sized particulates, and these quantum properties unique to ENM play an important role in determining toxicity, affecting different cellular functions, with altered genomic, proteomic and metabolomic profiles, altered mutagenesis and carcinogenesis potentials [35,36]. The process of determining the relationship between a particular expression profile and the toxicological phenotype of the organism for a particular exposure or dose and at a particular time is called "phenotypic anchoring". Therefore, anchoring the molecular expression profile in phenotype, dose and time helps to define the sequence of key molecular events in the mode-of-action of a toxicant. Phenotypic anchoring can also be used in conjunction with lower doses of the toxicant to classify agents and to explore the mechanisms of toxicity, e.g. by assessing transcriptional changes occurring after exposure to both low- and high-dose of ENM, and detecting the biological responses by using transcriptome measurements. The toxicogenomics approach has been used in assessing specific mechanisms at the molecular level for silicon dioxide, dendrimers, carbon nanoparticles (single- and multiwalled CNT, carbon nano-onions, fullerenes), metallic nanoparticles (e.g. copper and silver), identifying patterns of cellular perturbations in specific pathways, through identification and quantification of global shifts in gene expression resulting within cell models challenged with ENM [37,38].

Preliminary observations support the existence of both nanoscale effects and specific molecular effectors differentiating nanotoxicology from conventional toxicology. The perspective of identifing fingerprints of selected ENM has opened a field of research to assess whether particular biological pathways are activated or perturbed by nanoparticles and to evaluate if the quantitative changes in these biomarkers can be used as indicators or predictors for toxicity in humans. Single-walled CNT may induce cell apoptosis, which is characterized by morphological changes, chromatin condensation and internucleosomal DNA fragmentation, accompanied by upregulation expression of apoptosis-associated genes such as p16, bax, hrk, bak1, p57, FGFR2, TGF beta receptor 1 and TNFAIP2 genes and downregulation expression of cell cycle-associated genes such as cyclin D1, cdk2, cdk4, and cdk6 as compared to normal cells [37]. Seventy nanometer silica nanoparticles (nSP70) induced a higher level of acute phase proteins such as haptoglobin, C-reactive protein (CRP) and serum amyloid A (SAA) than larger silica particles (diameter >100 nm) after intravenous injection. In addition, the level of these acute phase proteins was elevated in the plasma of mice after intranasal treatment with nSP30 [38].

Using a macrophage cell model, Waters and co-authors [39] demonstrated that the ability of unopsonized amorphous silica particles to stimulate inflammatory protein secretion and induce macrophage cytotoxicity scales closely with the total administered particle surface area across a wide range of particle diameters (7–500 nm). Whole genome microarray analysis of the early gene expression changes induced by 10- and 500-nm particles showed that the magnitude of change for the majority of genes affected correlated more tightly with particle surface area than either particle mass or number, and gene expression changes were particle size specific. However, the overall biological processes represented by all gene expression changes were nearly identical, irrespective of particle diameter, in the 10–500 nm range of particles.

These findings suggest that on an equivalent nominal surface area basis, common biological modes of action are expected for nano- and supranano-sized silica particles. The most robust transcriptional change observed was an induction of the CXCL2 chemokine (MIP-2a), implicated in crystalline silica-induced lung neutrophil recruitment [39]. Pulmonary exposure to CNT resulted in an elevated series of measurable potential biomarkers in the blood, including genes expressed in the circulating blood cells and/or soluble proteins not unique to the type of particles but representing the unity of local and systemic pathophysiological responses induced by the exposure [32]. In particular, MWCNT induced gene upregulation of more than half of the tested genes in the lung related to inflammation, oxidative stress, coagulation, and tissue remodeling and a significant increase in the circulating blood gene expression of several biomarkers of neutrophil response, such as CXCL2 (MIP-2a), S100a8, IL-8rβ, and Mac-1. Furthermore, several genes were activated in the circulating blood cells but not in the lung at least at 4 h after pulmonary exposure to MWCNT, e.g. osteopontin, colony stimulating factor-1 (CSF-1), and insulin growth factor receptor 1 (IGF-1R). Pulmonary exposure to CNT also triggered the induction of primary cytokines such as IL-6 and IL-1β, which regulate multiple pathways of the inflammatory cascade as well as secondary inflammatory mediators, chemokines (CCL2, 4, 19, 22; CXCL1, 2), which directly regulate leukocyte recruitment to the inflammatory site [32]. Following SWCNT instillation, Teeguarden and co-authors [40] identified a pattern of 109 proteins representing cellular processes affected by crocidolite asbestos treatment as well, further evidencing the broad similarity in the tissue-level response to both fibrous materials. In particular, two high-sensitivity markers of inflammation, one (S100a9) also observed in humans exposed to asbestos, were found and may be promising biomarkers of human response to SWCNT exposure [40].

BIOMARKERS OF EXPOSURE

Exposure biomarkers should provide objective demonstration of the absorption of chemicals in the body and can be used in occupational toxicology for a more accurate risk assessment, reducing misclassification in epidemiological studies and modeling internal dose. An ideal biomarker of exposure should be not only specific for the exposure of interest, but also detectable in small quantities, measurable by non-invasive techniques, provided of a positive predictive value to a specific health status [41]. Various factors can affect the actual absorbed dose, including physical workload, additional skin absorption due to wrong working practice or, conversely, the use of personal protection equipment. Moreover, uptake of ultrafine particles can increase in ill people (e.g. asthmatic versus healthy people) [42] and this could happen also for ENM.

To assess the internal dose, both the amount of the chemical and/or its metabolites and the products of interaction with target biomolecules (DNA, proteins) are usually investigated. For some ENM (e.g. metal nanoparticles, fullerenes and single-walled CNT), biokinetic data are available [43,44]. It appears that translocation rates from the portal-of-entry—the respiratory tract—to secondary organs, are very low. Although some ENM have been demonstrated to easily cross biological barriers, the magnitude of ENM into the systemic circulation could be negligible to lead significant internal dose as to provide appreciable amount of nanoparticles detectable in peripheral matrices of humans.

In the practical implementation of BM, it should be considered that physicochemical characteristics influence the biological behavior of ENM (see Chapter 1 for an extensive discussion of physicochemical properties of ENM in relation to their biologic effects); for instance, low-soluble ENM (e.g. CNT or TiO_2) show biopersistence and their slow clearance leads to the tendency to accumulate over time with exposure. As compared to chemicals, ENM do not undergo biotransformation; on the other hand, interaction with biological media and crossing biological membranes can dramatically change their surface properties ("corona effect"), e.g. coming into contact with surfactant phospholipids and proteins, and this could affect organ distribution [45]. Evidence exists of leaching of metal ions from metal nanoparticle suspensions, like cobalt [46] and iron oxide [47], and the role of dissolution in biological fluids has been rightly emphasized [48]. It is witnessed that for many particles, size and agglomeration state are influential in dictating toxicity; thus, particle dissolution and the release of ions—which would be expected to be greater for smaller particles—may affect organ distribution and, if they accumulate with time, might influence the critical concentration in organs. For instance, following liver accumulation, gold and silver nanoparticles, show a protracted elimination and slow release of particles from target organ into systemic circulation [49], giving rise to detectable trace amounts in body fluids.

To set up biomarkers of exposure, we should be able to assess the uptake and deposited dose of inhaled particles. There is a general consensus about the likelihood that particles whose aerodynamic diameter is between 0.1 and 1 μm reach the lower respiratory tract, but their deposition is low compared with particles with lower and higher aerodynamic diameter [43,50] (see Chapter 7 for an extensive discussion on the relationship between aerodynamic diameter and lung deposition). To assess the deposited dose, Goldoni and coworkers [51] have developed a portable device designed to monitor the online exhalation kinetics of particles in a single breath; this device was validated in the field in a group of workers occupationally exposed to cristobalite in the size range of 0.3–2.5 μm, showing significant differences in the deposited dose depending on the nose versus mouth breathing and health status. The tissue dose of pneumotoxic metals—Co, Ti and Cr(VI)—has been assessed in exhaled breath condensate collected in workers occupationally exposed [52–54]. Although the advantages of such approaches need further validation with different aerosols and model nanoparticles, the above methods could provide a tool to assess the dose of particles deposited within the respiratory tract.

BIOMARKERS OF EFFECTS

Research on biomarkers of effect is rapidly generating a large amount of data measuring intermediate end-points occurring after exposure and before clinically apparent disease occurs. To be useful in preventive strategies, effect biomarkers would reflect early and reversible biochemical events preceding progressive structural or functional damage at the molecular, cellular and tissue level. The following paragraphs describe the rationale for assessing some potential biomarkers of effects.

Biomarkers to Detect Inflammation at the Respiratory Level

Quantification of lung inflammation is currently based on invasive methods including the analysis of bronchoalveolar lavage fluid (BALF), bronchoscopy and bronchial biopsies, semi-invasive methods such as sputum induction, and the measurement of inflammatory biomarkers in plasma and urine which are likely to reflect systemic rather than lung inflammation (for a review, see [55]). The exhaled breath condensate (EBC) analysis is a promising non-invasive approach which has been proposed for clinical applications [55] as well as for investigating occupational lung diseases [56,57]. EBC analysis allows the detection of several biomarkers of oxidative stress and inflammation, including thiobarbituric acid reactive substances (TBARS), such as malondialdehyde—produced by lipoperoxidation of membranes—and 8-isoprostane, a peroxidation product of prostaglandin metabolism [55]. Changes in pH, proteins and pro-inflammatory cytokines, such as leukotrienes B_4 (LTB_4) can also be determined. Inflammatory response in the airways is characterized by an influx of neutrophils, whose activation is associated with a respiratory burst resulting in overproduction of hydrogen peroxide (H_2O_2), measurable in EBC. Various classes of volatile organic compounds (VOCs) can also be measured in exhaled breath, including the saturated hydrocarbons and oxygen-containing substances formed during the fatty acid lipid peroxidation of cell membranes [58]. EBC should therefore be suitable to be applied in field studies and for longitudinal assessments of pulmonary biology, in spite of some problems in standardization [59].

Additionally, a wide variety of carbonyl compounds are generated as secondary oxidation products during respiratory burst. In particular, saturated aldehydes such as hexanal, heptanal and nonanal are formed by the peroxidation of omega-3 and -6 fatty acids (PUFAs), the basic components of cell membrane phospholipids. Volatile aldehydes are poorly soluble in blood and are therefore excreted into the breath within minutes of their formation in tissues, making possible the study of pulmonary diseases associated with inflammatory processes (e.g. chronic obstructive pulmonary disease [COPD], asthma, cystic fibrosis, interstitial lung disease) and the early diagnosis of lung cancer [60–62]. It is worth mentioning that aldehydes, particularly 4-hydroxy-2-nonenal (HNE), are second messengers influencing several post-transcriptional pathways [63].

Among the broad spectrum of gaseous compounds detectable in exhaled air, nitric oxide (NO) is produced by all cellular components of pulmonary inflammation (macrophages, epithelial cells, mast cells, lymphocytes and granulocytes). Studying NO exhaled from the lower airways offers a unique possibility to study features of pulmonary NO metabolism non-invasively in all states characterized by lung inflammation [64]. Under physiologic conditions, NO is unstable, reacting with oxygen to form oxides of nitrogen (NOx), such as nitrite (NO_2^-) and nitrate (NO_3^-), and with superoxide anion to form the potent oxidant peroxynitrite ($ONOO^-$), which reacts with a wide variety of compounds, including DNA, cellular lipids, and sulfydryl groups on proteins thus promoting nitrosative stress. Nitrosothiols (RS-NOs), which are formed by interaction of nitric oxide (NO) with glutathione (GSH), can be quantified in EBC of humans and reliably used as biomarkers in different inflammatory airway diseases [65]. Nitrite and nitrate are relatively stable products of NO that have been measured in EBC and have been used in evaluating the role of oxidative stress in acute responses to exposures that occur in vehicles or during near-roadway activities [66].

BIOMARKERS OF SYSTEMIC EFFECTS

The relationship between exposure to quasi-ultrafine particles (0.25 μm in aerodynamic diameter—$PM_{0.25}$), organic carbon and elemental carbon from outdoor sources and changes in circulating biomarkers of systemic effect has been assessed by Delfino and co-workers [67,68]. Biomarkers of inflammation, platelet activation, and antioxidant capacity by the copper/zinc-superoxide dismutase (Cu, Zn-SOD) and glutathione peroxidase-1 (GPx-1) activity, as well as the relative strength of biomarker associations between various particle size fractions were investigated. The studies showed the inactivation of antioxidant enzymes within erythrocytes by traffic-related pollutant components, including polycyclic aromatic hydrocarbons, among a subgroup of people. Plasma interleukin-6 (IL-6) and soluble tumor necrosis factor-receptor II (sTNF-RII) were investigated during a longitudinal study considering both indoor and outdoor exposure and showed a positive association with vehicle emissions tracer. On the basis of the above reported data, high-sensitivity C-reactive protein (CRP) in plasma, plasma fibrinogen, and interleukin-6 (IL-6) could represent the best candidates for human BM following exposure to inflammogenic ENM.

The endothelium plays an important role in maintaining the vascular homeostasis by producing vasoactive factors that regulate the tone of the vascular system in response to cell surface receptor stimulation or mechanical stress, but also in the regulation of vascular inflammation and thrombosis. Radomski and co-authors [69] showed that both urban dusts and engineered carbon particles, such as CNT and carbon black, except $C_{60}CS$, stimulated platelet aggregation and accelerated the rate of vascular thrombosis in rat carotid arteries with a similar rank order of efficacy. All particles resulted in upregulation of GPIIb/IIIa in platelets. In contrast, particles differentially affected the release of platelet granules, as well as the activity of thromboxane-, ADP-, matrix metalloproteinase- and protein kinase C-dependent pathways of aggregation. Consistent with the *in vitro* results, exposure to nano Ag (0.05–0.1 mg/kg i.v. or 5–10 mg/kg intratracheal instillation) *in vivo* enhanced platelet aggregation and promoted venous thrombus formation in rats [70]. Therefore, assessment of platelet aggregation and expression of GPIIb/IIIa in platelets may be useful to evaluate possible prothrombotic effects induced by ENM.

The progression of atherosclerosis in susceptible mice models [71] is consistent with the ability of ENM to affect endothelial cells metabolism via oxidative stress generation and endothelial repair capacity and points to the use of endothelial dysfunction markers to assess ENM systemic and long-term effects. Interestingly, exposure to ultrafine TiO_2 particles can affect vasomotor function with minimal pulmonary effects [72]. Vesterdal and co-workers [73] showed that exposure of young and aged apolipoprotein E knockout mice (apo$E^{-/-}$) to carbon black (Printex 90, 14 nm average size) by intratracheal instillation, resulted in modest vasomotor impairment, with a lack in association with nitrosative stress (3-nitrotyrosine) and without increases in the expression of vascular adhesion molecule (VCAM-1) and intercellular adhesion molecule (ICAM-1) on endothelial cells or in plaque progression. In order to assess the systemic prothrombotic effects of particle exposure, quantification of tissue factor (TF) activity in plasma, which is expressed in subendothelial cells upon injury, or on the surface of circulating monocytes, macrophages and neutrophils, or levels of plasminogen activator inhibitor-1 (PAI-1), may also be considered as markers of cell damage.

PERIPHERAL BIOMARKERS OF EPITHELIAL BARRIER PERMEABILITY

Clara cells are non-ciliated secretory epithelial cells lining the pulmonary airways, distinct from mucous and serous secretory cells in morphology and secretory products. Among these, the Clara cell protein (CC16) plays an immunomodulatory activity, it can inhibit phospholipase A2 activity, while inflammatory cytokines like TNF-α can modulate CC16 production [74]. CC16 and surfactant-associated protein B (SP-B) in the serum have been validated as markers of alveolo-capillary barrier integrity/permeability in a number of human studies on gaseous/particulate pollutants [75,76]. Recently, CC16 determination in serum has been applied to assess response to inhaled particulates [77] and, for the first time, in workers occupationally exposed to ENM [78]. The great advantage of pneumoproteins is their great sensitivity in detecting epithelial effects at levels of pollutants below current air standards.

BIOMARKERS OF OXIDATIVE STRESS AND DNA DAMAGE

Oxidative stress is a central mechanism of action for both pulmonary and extrapulmonary health effects of particulate matter [79,80] and a number of studies of exposure to particles in humans have applied biomarkers of oxidative damage in the blood compartment or in terms of products excreted in urine or EBC. Meta-analysis of panel studies and cross-sectional investigations considering the application of such biomarkers have shown a consistent association between exposure to combustion particles with oxidative damage of DNA and lipids, suggesting that it is possible to use these measurements as biomarkers of biologically effective dose [81,82].

Among the DNA oxidation products, such as 8-oxo-7,8-dihydroguanine (8-oxoGua) or the corresponding deoxynucleoside 8-oxo-7,8-dihydro-2′-deoxyguanosine (8-oxodG) measured in DNA and urine, the exocyclic M1 adduct to guanine (M1dG) were the most studied, as well as the lesions detected as sites in DNA sensitive to formamidopyrimidine DNA glycosylase (FPG) and endonuclease III (ENDOIII). The biomarkers of lipid peroxidation products include conjugated dienes (CDs), lipid hydroperoxides, malondialdehyde (MDA), TBARS and F2-isoprostanes measured in EBC, plasma, serum or urine. Base excision repair of oxidative damage to DNA measured in urine, such as 8-hydroxy-2′-deoxyguanosine (8OHdG), seems to originate mostly from the oxidation of the deoxynucleotide pool and does not represent solely repairing/excretion of the oxidized-DNA guanine. Once produced, 8OHdG is very stable and is not further metabolized in the systemic circulation. After exposure to oxidants, the repair and final 8OHdG excretion in urine is rapid, i.e. within at least 24 hours [83]. Increased 8OHdG urinary levels within each shift and between days have been shown in workers exposed to low levels of respirable PM_4 (range 25–71 $\mu g/m^3$) [84].

The DNA damaging potential of ENM is well established. For environmental particles, oxidative stress-mediated changes to biomolecules may arise from direct generation of reactive oxygen species from the surface, soluble compounds such as transition metals or organic compounds, altered function of mitochondria or NADPH-oxidase, and activation of inflammatory cells capable of generating ROS and reactive nitrogen species [82]. On the other hand, ENM of different composition (metal/metal-oxide nanoparticles, silica, quantum dots, fullerenes, nanofibers) can affect DNA also by direct mechanisms, including a mechanical interference with cellular and nuclear components, like microtubules of the mitotic spindle [85,86]. Genotoxic effects, such as chromosome breakage and chromosome loss, can be detected and quantified in target cells using the cytokinesis-block micronucleus method (CBMN assay)—originated from acentric chromosome fragments or whole chromosomes lagging behind during cell division—and fluorescence *in situ* hybridization (FISH) using a DNA probe specific for the centromere (further information on micronucleus assay is available in Chapter 14). These assays are effective in assessing cytogenetic alterations following *in vitro* exposure to metal nanoparticles [46,87,88] and CNT [24,89,90] and represent good candidate biomarkers for human BM purpose. The majority of the ENM tested both *in vitro* and *in vivo* are able to cause DNA strand breaks or oxidative DNA lesions. DNA damage in peripheral blood leukocytes can be quantified by single cell gel electrophoresis (SCGE; comet assay), a simple and sensitive method for detecting DNA strand breaks and alkali-labile sites, while oxidatively damaged DNA can be analyzed using the enzyme formamidopyrimidine glycosylase (FPG). Although interactions between the particles and the assay cannot be totally excluded, the use of the comet assay in human biomonitoring studies could provide valuable information for hazard identification of ENM [91] (further information on the comet assay is available in Chapter 14). Modified SCGE provides parallel information on oxidative DNA damage caused by ENM, such as CNT, detecting oxidized/damaged pyrimidines and purines [90]. Some studies showed dose–effect and dose–response relationships between the mass concentration of ENM and the frequency of micronucleated lymphocytes, with a threshold for the above effects [87,90]. For a meaningful interpretation of effects, common polymorphisms in genes involved in the repair of oxidative DNA damage (hOGG1, XRCC1, APEX1) could be analyzed in subgroups at risk.

BIOMARKERS OF NEUROTOXICITY

The mechanisms of neuron injury are diverse for a wide variety of materials used and ENM produce different effects from environmental particles with potential effects both on peripheral and central nervous systems [92]. Adsorption and inhibition of acetylcholinesterase (AchE)—a key enzyme present in the brain, blood and nervous system—by ENM (namely, SiO_2, TiO_2, Al_2O_3, Al, Cu, Cu–C [carbon-coated copper], MWCNT and

SWCNT) suggests an interference of ENM with neurochemical processes [93] and a use of AChE as a potential biomarker of neurotoxicity. Dopaminergic neurons can be specifically targeted *in vitro* by a broad spectrum of ENM, including Mn, Ag or Cu nanoparticles [94], but *in vivo* studies showed SiO_2-NPs entered into the brain and especially deposited in the striatum and induced oxidative damage and inflammation in dopaminergic brain areas with subsequent depletion of neuromediators [95]. To assess early effects from chemicals targeting the dopaminergic systems, a neurochemical and neuroendocrine approach based on surrogate biomarkers in peripheral media has been developed [96]. Neurotoxicity of Mn to dopaminergic systems, investigated in workers occupationally exposed to Mn oxides containing fumes and metallurgic dusts falling in the range from 0.001 to 10 μm, was revealed by increase in serum PRL levels [97] and decrease in serum dopamine beta-hydroxylase and type B platelet monoamine oxidase activities [98]. In spite of the lack of specificity, the above approach to the neurotoxicity of specific ENM should be reconsidered for its sensitivity in detecting neurotoxicity of ENM in exposed populations.

It is unlikely that acute high dose causing CNS effects occurs, whereas the long-term accumulation of low doses of ENP cannot be excluded. Recent research showed that nanoparticles (copolymer particles, cerium oxide particles, quantum dots and CNT) enhance the rate of protein fibrillation by decreasing the lag time for nucleation, a common process in many human neurodegenerative diseases [99]. Although it remains to be seen whether the enormous enhancements in nucleation rate for protein fibrillogenesis demonstrated in this work also occurs *in vivo* or in more complex biological fluids where competitive binding may screen the nucleation surface [99], the possibility that chronic exposures to ENM can influence processes within the brain that are triggering or aggravating pathological processes cannot be ruled out [100].

INTERMEDIATE BIOMARKERS IN HEALTH SURVEILLANCE AND EPIDEMIOLOGY

BM is usually used as part of health surveillance to provide the correct interpretation of doubtful clinical tests, especially when environmental monitoring data are unavailable or difficult to obtain. Any decision to include BM in medical surveillance should be taken after careful assessment of exposure to a health hazard and the availability of tests with a known sensitivity and specificity to detect health effects in an early, preferably reversible, stage. Biochemical tests or functional parameters should be supported by consistent pathophysiological mechanisms; it should be recognized that potentially relevant health end-points ascribed to ENM (cardiovascular, pulmonary, inflammatory) are often unspecific, show high prevalence in the general population, and share multiple non-occupational risk factors [101]. Thus, to monitor short-term health effects, it is necessary to identify potential biomarkers, assess their relationship with exposure and differentiate subclinical effects and the related biomarkers as well as relevant health end-points potentially ascribed to ENM. In addition to oxidative stress and inflammation, it is important to consider that some of the NM interactions may also result in other forms of injury, such as protein denaturation, membrane damage, DNA damage and immune reactivity.

The literature on short-term effects of air pollutants [67,68,77–79,81,82,102] and the available literature on ENM, suggest identifying multiple biomarkers—a biomarker profile—to assess both effects at the "portal of entry" (e.g. inflammatory changes, short-term respiratory changes, respiratory, eye or skin irritation) and systemic effects (e.g. heart-rate variability, platelet aggregation and prothrombotic changes, acute phase proteins). These studies mostly concern NP of environmental origin, heterogeneous for composition and size, and not ENM of defined composition and surface chemistry. Although similar in size, ambient and engineered NP may significantly differ for biological activity and toxicological properties.

In spite of the lack of validation of "nano-specific" biomarkers, it is proposed that the sensitivity instead of the specificity of biomarkers should be privileged to assess the causality of exposure conditions and association with hazards (e.g. at workplace) and also to identify predictive biomarkers [103]. Epidemiological data specifically derived from nanoparticle-exposed humans are currently not available, owing to the short latency since the introduction of ENM on the market; as a result, possible target people or groups not have had long and consistent exposure leading to appreciable effects [104]. The heterogeneity of ENM, making difficult identifying and recruiting enough workers with the same exposure, could be addressed by the BM. In epidemiological research, BM is aimed at overcoming the growing frustration with the limitations of such crude markers of exposure as job titles, exposure registries or point estimates of airborne pollution. It may be that exposure to the same type of ENM is not the critical feature, but rather whether exposures to different nanoparticles may lead to the same pathway for disease, or share common mechanisms (e.g. inflammation). Workers could be monitored and tested for reactive oxygen species or inflammation markers by using non-invasive methods (e.g. EBC) and exposure assessment could be based on distribution of the magnitude of these biomarkers [104]. To avoid "epidemiological frustrations", it is mandatory to identify at least intermediate biological changes of effects that could be indicative or predictive of disease and apply them in cross-sectional studies combining exposure characterization to a given ENM and sufficiently validated biomarkers. Researchers from the National Health Research Institutes of Taiwan carried out a cross-sectional investigation with the aim of developing and identifying sensitive and specific biomarkers of health effects among workers manufacturing ENM [78]. The CC16 and NF-κB levels (inflammation markers), the activity of myeoloperoxydase (MPO) and aryl esterase, vascular cell adhesion molecule (VCAM) and heart-rate variability (HRV) parameters were assessed. Among these biomarkers, CC16, hsCRP, and inflammatory cell activation (increased ICAM-1 in macrophages) showed statistically significant changes among nanomaterial workers.

CONCLUSIONS

One of the strategies for preventing possible adverse effects to occur is to identify early biological events in target organs or accessible biological fluid associated with exposure to harmful ENM. Although biomarker studies reflecting such changes for ENM toxicity are currently at their early stage, our experience in biomarker research for the incidental or ambient ultrafine particles along with improved exposure characterization, can be used to implement effective preventive strategies [10,105].

Figure 3.1, originally developed by Albertini [106] shows the *continuum* of events between exposure and effects and the possible use of BM in this deterministic multistage process linking exposure to chemicals to disease. Questions remain about the appropriateness of biomarkers in assessing such a continuum between ENM exposure and possible health outcomes. Whereas transforming exposure into dose may take seconds to hours, the transformation of dose into biologically effective dose may require years of exposure to cumulative toxic chemicals. Subsequent steps show an even greater variability and are often reversible, early effects being repaired or compensated for, making temporal relationships with preceding events uncertain [6]. However, questions about lowered resistance against ENM-induced effects in susceptible populations, such as those with underlying disease, genetic susceptibility, previous coronary events, and those at an early developmental stage should be considered.

There are still a number of obvious gaps that have been identified that need to be filled in order to get a more complete understanding of ENM-induced effects and a meaningful implementation of BM for ENM exposure. BM should fulfil the following criteria: (1) the use of suitable and accessible biological matrices; (2) suitable parameters, able to reflect internal exposure, biochemical or (early) biological effects; (3) suitable and reliable analytical methods which are kept under control by quality assurance; and (4) reference and limit values which enable the interpretation of results. Advances in proteomic technologies, using mass spectrometry, gas-phase separation or microarrays that use antibodies to quantify

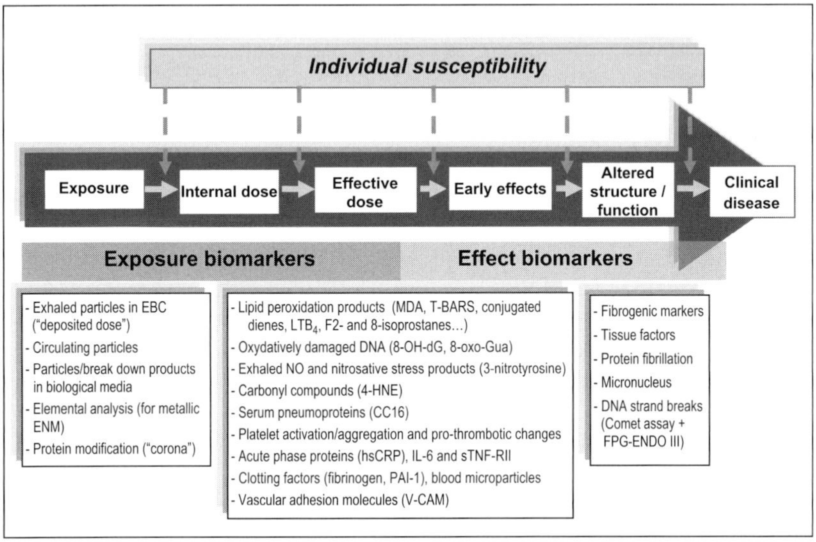

FIGURE 3.1 Use of biomarkers in a multifactorial model: possible application to ENM? Adapted from Albertini [106]. Abbreviations: ENM (engineered nanomaterial); EBC (exhaled breath condensate); MDA (malondialdehyde); T-BARS (thiobarbituric acid reactive substances); LTB_4 (leucotriene-B_4); NO (nitric oxide); 8-OH-dG (8-hydroxy-2′-deoxyguanosine); 8-oxo-Gua (8-oxo-7,8-dihydroguanine); 4-HNE (4-hydroxy-2-nonenal); CC16 (Clara cell protein); hsCRP (high sensitivity C-reactive protein); IL-6 (plasma interleukin 6); sTNF-RII (soluble tumor necrosis factor-receptor II); PAI-1 (plasminogen activator inhibitor-1); V-CAM (vascular adhesion molecule); FPG-ENDOIII (lesions detected as sites in DNA sensitive to formamidopyrimidine DNA glycosylase and endonuclease III).

protein levels, have greatly improved the ability to discover and validate protein profiles as biomarkers [107]. Analytical instrumentation now allows quantitization of "background levels" in people not knowingly exposed, but the health significance of these "background" doses needs to be assessed owing to the unusual dose–response relationships which characterize ENM [108]. In addition, practical issues, including cost, exposure levels, analytical requirements, and selection of an adequate sampling strategy, should be dealt with when planning a BM program. Biological materials should be easy to obtain in sufficient amounts under routine conditions and without unacceptable discomfort and health risk for the individual or worker; in that sense, EBC collection could be regarded as a suitable tool to assess respiratory health status in people exposed to effective doses of ENM.

It should, however, be borne in mind that dose is critical in determining toxicity and dose is the product of concentration and duration of exposure. The traditional paradigm for human health risk assessment of environmental chemicals involves comparing estimated daily doses to health-based criteria for acceptable, safe, or tolerable daily intakes (e.g. reference doses—RfDs—, tolerable daily intakes—TDIs—, or minimal risk levels—MRLs) to assess whether estimated doses exceed such health screening levels. Unfortunately, for many chemicals, including ENM, quantitative benchmarks of acceptable or safe concentrations in biological specimens needed to interpret these levels are still lacking. A fundamental method of avoiding harm is thus reduction or complete prevention of exposure, and this should be the aim of regulation of any ENM considered likely to be hazardous [109].

TAKE-HOME MESSAGES

Biological monitoring (BM) is a powerful tool, but its applicability to ENM requires the development of appropriate biomarkers of exposure and biological effect considering the unusual properties of ENM. Multiple biomarkers can be envisaged to assess effects at the "portal of entry" (e.g. by exhaled breath analysis) or systemic effects (e.g. platelet aggregation and prothrombotic changes, acute phase proteins, oxidative DNA damaged products). In order to get a meaningful implementation of BM for ENM, there are still a number of conceptual and practical issues, including: (1) track temporal trends in levels of exposure; (2) identify suitable parameters, able to reflect internal exposure or early/reversible biological changes; (3) reliable analytical methods which are kept under control by quality assurance; and (4) establishing reference ranges and limit values which enable the interpretation of results.

References

[1] Zielhuis RL, Henderson PT. Definitions of monitoring activities and their relevance for the practice of occupational health. Int Arch Occup Environ Health 1986;57:249–57.

[2] Schulte PA, Hauser JE. The use of biomarkers in occupational health research, practice, and policy. Toxicol Lett 2011 (doi:10.1016/j.toxlet.2011.03.027).

[3] Mutti A. Biomarkers of exposure and effect for non carcinogenic end-points. International programme on chemical safety. Environmental Health Criteria 222, Biomarkers in risk assessment: validity and validation, 104. Geneva: World Health Organization; 2001. pp. 130–136.

[4] Lauwerys RR, Hoet P. Industrial chemical exposure. Guidelines for biological monitoring. 2nd ed. vol. 1. Boca Raton: Lewis Publishers; 1993.

[5] NRC, National Research Council. Biological markers in environmental health research. Environ Health Perspect 1987;74:3–9.
[6] Watson WP, Mutti A. Role of biomarkers in monitoring exposure to chemicals: Present position, future prospects. Biomarkers 2004;9:211–42.
[7] Smolders R, Bartonova A, Boogaard PJ, et al. The use of biomarkers for risk assessment: Reporting from the INTARESE/ENVIRISK Workshop in Prague. Int J Hyg Environ Health 2010;213:395–400.
[8] Angerer J, Ewers U, Wilhelm M. Human biomonitoring: State of the art. Int J Hyg Environ Health 2007;210:201–28.
[9] Manno M, Viau C, Cocker J, et al. Biomonitoring for occupational health risk assessment (BOHRA). Toxicol Lett 2010;192:3–16.
[10] Li N, Nel A. Feasibility of biomarker studies for engineered nanoparticles. What can be learned from air pollution research. J Occup Environ Med 2011;53:S74–9.
[11] Riviere JE, Tran CL. Pharmacokinetics of nanomaterials. In: Nanotoxicology. Characterization, dosing and health effects. New York: Informa Healthcare; 2007. p. 127–40.
[12] Aschberger K, Johnston HJ, Stone V, et al. Review of carbon nanotubes toxicity and exposure. Appraisal of human health risk assessment based on open literature. Crit Rev Toxicol 2010;40:759–90.
[13] Wittmaack K. In search of the most relevant parameter for quantifying lung inflammatory response to nanoparticle exposure: Particle number, surface area, or what? Environ Health Perspect 2007;115:187–94.
[14] Oberdörster G, Sharp Z, Atudorei V, et al. Translocation of inhaled ultrafine particles to the brain. Inhal Toxicol 2004;16:437–45.
[15] Schmid O, Möller W, Semmler-Behnke M, et al. Dosimetry and toxicology of inhaled ultrafine particles. Biomarkers Suppl 2009;1:67–73.
[16] Tran CL, Buchanan D, Cullen RT, Searl A, Jones AD, Donaldson K. Inhalation of poorly soluble particles. II. Influence of particle surface area on inflammation and clearance. Inhal Toxicol 2000;12:1113–26.
[17] Kuempel ED, Tran CL, Castranova V, Bailer AJ. Lung dosimetry and risk assessment of nanoparticles: Evaluating and extending current models in rats and humans. Inhal Toxicol 2006;18:717–24.
[18] Kuempel E. Estimating nanoparticle dose in humans: Issues and challenges. In: Nanotoxicology. Characterization, dosing and health effects. New York: Informa Healthcare; 2007. p. 141–52.
[19] Chen HH, Chien CC, Petibois C, et al. Quantitative analysis of nanoparticle internalization in mammalian cells by high resolution X-ray microscopy. J Nanobiotechnol 2011;9:14–29.
[20] Helland A, Wick P, Koehler A, Schmid K, Som C. Reviewing the environmental and human health knowledge base of carbon nanotubes. Environ Health Perspect 2007;115:1125–31.
[21] Pappi T, Schiffmann D, Weiss D, Castranova V, Vallyathan V, Rahman Q. Human health implications of nanomaterial exposure. Nanotoxicology 2008;2:9–27.
[22] Brain JD, Curran MA, Donaghey T, Molina RA. Biologic responses to nanomaterials depend on exposure, clearance, and material characteristics. Nanotoxicology 2009;3:174–80.
[23] Muller J, Huaux F, Moreau N, et al. Respiratory toxicity of multi-wall carbon nanotubes. Toxicol Appl Pharmacol 2005;207:221–31.
[24] Muller J, Huaux F, Fonseca A, et al. Structural defects play a major role in the acute lung toxicity of multiwall carbon nanotubes: Toxicological aspects. Chem Res Toxicol 2008;21:1698–705.
[25] Shvedova AA, Kisin E, Murray AR, et al. Inhalation vs. aspiration of singlewalled carbon nanotubes in C57BL/6 mice: Inflammation, fibrosis, oxidative stress, and mutagenesis. Am J Physiol Lung Cell Mol Physiol 2008;295:L552–65.
[26] Choi HS, Ashitate Y, Lee JH, et al. Rapid translocation of nanoparticles from the lung airspaces to the body. Nat Biotechnol 2010;28:1300–3.
[27] Gibson RM. Understanding the potential neurotoxicology of nanoparticles. In: Nanotoxicology. Characterization, dosing and health effects. New York: Informa Healthcare; 2007. p. 299–316.
[28] Donaldson K, Newby DE, MacNee W, Duffin R, Lucking AJ, Mills NL. Pulmonary and cardiovascular effects of nanoparticles. In: Nanotoxicology. Characterization, dosing and health effects. New York: Informa Healthcare; 2007. pp. 267–98.
[29] Legramante JM, Valentini F, Magrini A, et al. Cardiac autonomic regulation after lung exposure to carbon nanotubes. Hum Exper Toxicol 2009;28:369–75.
[30] Oberdörster G, Elder A, Rinderknecht A. Nanoparticles and the brain: cause for concern? J Nanosci Nanotechnol 2009;9:4996–5007.

[31] Sharma HS, Sharma A. Conference scene: Nanoneuroprotection and nanoneurotoxicity: Recent progress and future perspectives. Nanomedicine (Lond) 2010;5:533–7.
[32] Erdely A, Hulderman T, Salmen R, et al. Cross-talk between lung and systemic circulation during carbon nanotube respiratory exposure. Potential biomarkers. Nano Lett 2009;9:36–43.
[33] Lotti M, Olivato I, Bergamo L. Inflammation and short-term cardiopulmonary effects of particulate matter. Nanotoxicology 2009:1–6. DOI: 10.1080/17435390802538763.
[34] Shvedova A, Kagan VE, Fadeel B. Close encounters of the small kind: Adverse effects of man-made materials interfacing with the nano-cosmos of biological systems. Annu Rev Pharmacol Toxicol 2010;50:63–88.
[35] Poma A, Di Giorgio ML. Toxicogenomics to improve comprehension of the mechanisms underlying responses of *in vitro* and *in vivo* systems to nanomaterials: A review. Curr Genomics 2008;9:571–85.
[36] Yang B, Wang Q, Lei R, et al. Systems toxicology used in nanotoxicology: Mechanistic insights into the hepatotoxicity of nano-copper particles from toxicogenomics. J Nanosci Nanotechnol 2010;10:8527–37.
[37] Chui D, Tian F, Ozkan CS, Wang MW, Gao H. Effect of single wall carbon nanotubes on human HEK293 cells. Toxicol Lett 2005;155:73–85.
[38] Higashisaka K, Yoshioka Y, Yamashita Y, et al. Acute phase proteins as biomarkers for predicting the exposure and toxicity of nanomaterials. Biomaterials 2011;32:3–9.
[39] Waters KM, Masiello LM, Zangar RC, et al. Macrophage responses to silica nanoparticles are highly conserved across particle sizes. Toxicol Sci 2008;104:155–62.
[40] Teeguarden JG, Webb-Robertson BJ, Waters KM, et al. Comparative proteomics and pulmonary toxicity of instilled single-walled carbon nanotubes, crocidolite asbestos, and ultrafine carbon black in mice. Toxicol Sci 2011;120:123–35.
[41] Henderson RF, Bechtold WE, Bond JA, Sun JD. The use of biological markers in toxicology. Crit Rev Toxicol 1989;20:65–82.
[42] Frampton MW, Utell MJ, Zareba W, et al. Effects of exposure to ultrafine carbon particles in healthy subjects and subjects with asthma. Res Rep Health Effect Inst 2004;126:1–47.
[43] Geiser M, Kreyling W. Deposition and biokinetics of inhaled nanoparticles. Part Fibre Toxicol 2010;7:2–17.
[44] Holgate ST. Exposure, uptake distribution and toxicity of nanomaterials in humans. J Biomed Nanotechnol 2010;6:1–19.
[45] Schleh C, Rothen-Rutishauser B, Kreyling WG. The influence of pulmonary surfactant on nanoparticulate drug delivery systems. Eur J Pharm Biopharm 2011;77:350–2.
[46] Borm P, Klaessig FC, Landry TD, et al. Research strategies for safety evaluation of nanomaterials, Part V: Role of dissolution in biological fate and effects of nanoscale particles. Toxicol Sci 2006;90:23–32.
[47] Colognato R, Bonelli A, Ponti J, et al. Comparative genotoxicity of cobalt nanoparticles and ions on human peripheral leukocytes *in vitro*. Mutagenesis 2008;23:377–82.
[48] Pettibone JM, Adamcakova-Dodd A, Thorne PS, O'Shaughnessy PT, Weydert JA, Grassian VH. Inflammatory response of mice following inhalation exposure to iron and copper nanoparticles. Nanotoxicology 2008;2:189–204.
[49] Johnston HJ, Hutchison G, Christensen FM, Peters S, Hankin S, Stone V. A review of the *in vivo* and *in vitro* toxicity of silver and gold particulates: Particle attributes and biological mechanisms responsible for the observed toxicity. Crit Rev Toxicol 2010;40:328–46.
[50] Broday DM, Agnon Y. Asymmetric human lung morphology induce particle deposition variation. J Aerosol Sci 2007;38:701–18.
[51] Goldoni M, Caglieri A, De Palma G, et al. Development and set-up of a portable device to monitor airway exhalation and deposition of particulate matter. Biomarkers 2009;14:326–39.
[52] Goldoni M, Catalani S, De Palma G, et al. Exhaled breath condensate as a suitable matrix to assess lung dose and effects in workers exposed to cobalt and tungsten. Environ Health Perspect 2004;112:1293–8.
[53] Goldoni M, Caglieri A, Poli D, et al. Determination of hexavalent chromium in exhaled breath condensate and environmental air among chrome plating workers. Anal Chim Acta 2006;562:229–35.
[54] Corradi M, Acampa O, Goldoni M, et al. Metallic elements in exhaled breath condensate of patients with interstitial lung diseases. J Breath Res 2009;3. 046003.
[55] Montuschi P. Analysis of exhaled breath condensate in respiratory medicine: Methodological aspects and potential clinical applications. Ther Adv Resp Dis 2007;1:5–23.
[56] Corradi M, Gergelova P, Mutti A. Use of exhaled breath condensate to investigate occupational lung diseases. Curr Opin Allergy Clin Immunol 2010;10:93–8.

[57] Gube M, Ebel J, Brand P, et al. Biological effect markers in exhaled breath condensate and biomonitoring in welders: Impact of smoking and protection equipment. Int Arch Occup Environ Health 2010;83:803–11.
[58] Horvath I, Lazar Z, Gyulai N, Kollai M, Losonczy G. Exhaled biomarkers in lung cancer. Eur Respir J 2009;34:261–75.
[59] Horvath I, Hunt J, Barnes PJ, et al. ATS/ERS Task Force on Exhaled Breath Condensate. Exhaled breath condensate: Methodological recommendations and unresolved questions. Eur Respir J 2005;26:523–48.
[60] Poli D, Carbognani P, Corradi M, et al. Exhaled volatile organic compounds in patients with non-small cell lung cancer: Cross sectional and nested short-term follow-up study. Respir Res 2005;6:71.
[61] Poli D, Goldoni M, Corradi M, et al. Determination of aldehydes in exhaled breath of patients with lung cancer by means of on fiber derivatisation SPME–GC/MS. J Chromatogr B Analyt Technol Biomed Life Sci 2010;878:2643–51.
[62] Corradi M, Rubinstein I, Andreoli R, et al. Aldehydes in exhaled breath condensate of patients with chronic obstructive pulmonary disease. Am J Respir Crit Care Med 2003;167:1380–6.
[63] Forman HJ. Reactive oxygen species and α,β-unsaturated aldehydes as second messengers in signal transduction. Ann NY Acad Sci 2010;1203:35–44.
[64] Pacher P, Beckman JS, Liaudet L. Nitric oxide and peroxynitrite in health and disease. Physiol Rev 2007;87:315–424.
[65] Corradi M, Montuschi P, Donnelly LE, Pesci A, Kharitonov SA, Barnes PJ. Increased nitrosothiols in exhaled breath condensate in inflammatory airway diseases. Am J Respir Crit Care Med 2001;163:854–8.
[66] Laumbach RJ, Kipen HM. Acute effects of motor vehicle traffic-related air pollution exposures on measures of oxidative stress in human airways. Ann NY Acad Sci 2010;1203:107–12.
[67] Delfino RJ, Staimer N, Tjoa T, et al. Air pollution exposures and circulating biomarkers of effect in a susceptible population: Clues to potential causal component mixtures and mechanisms. Environ Health Perspect 2009;117:1232–8.
[68] Delfino RJ, Staimer N, Tjoa T, et al. Association of biomarkers of systemic inflammation with organic components and source tracers in quasi-ultrafine particles. Environ Health Perspect 2010;118:756–62.
[69] Radomski A, Jurasz P, Alonso-Escolano D, et al. Nanoparticle-induced platelet aggregation and vascular thrombosis. Br J Pharmacol 2005;146:882–93.
[70] Jun E-A, Lim KM, Kim K, et al. Silver nanoparticles enhance thrombus formation through increased platelet aggregation and procoagulant activity. Nanotoxicology 2011;5:157–67.
[71] Li Z, Hulderman T, Salmen R, et al. Cardiovascular effects of pulmonary exposure to single-wall carbon nanotubes. Environ Health Perspect 2007;115:377–82.
[72] Nurkiewicz TR, Porter DW, Hubbs AF, et al. Nanoparticle inhalation augments particle-dependent systemic microvascular dysfunction. Part Fibre Toxicol 2008;5:1–16.
[73] Vesterdal LK, Folkmann JK, Jacobsen NR, et al. Pulmonary exposure to carbon black nanoparticles and vascular effects. Part Fibre Toxicol 2010;7:33–48.
[74] Hermans C, Bernard A. Lung epithelium-specific proteins: Characteristics and potential applications as markers. Am J Respir Crit Care Med 1999;159:646–78.
[75] Broeckaert F, Arsalane K, Hermans C, et al. Lung epithelial damage at low concentrations of ambient ozone. Lancet 2000;353:900–1.
[76] Bergamaschi E, De Palma G, Mozzoni P, et al. Polymorphism of quinone-metabolizing enzymes and susceptibility to ozone-induced acute effects. Am J Respir Crit Care Med 2001;163:1426–31.
[77] Alessandrini F, Weichenmeier I, van Miert E, et al. Effects of ultrafine particles-induced oxidative stress on Clara cells in allergic lung inflammation. Part Fibre Toxicol 2010;7:11.
[78] Liou SH, Lin MH, Hsu CH, et al. Pilot study of health hazards among engineered nanoparticles manufacturing workers. Occup Environ Med 2010;60(Suppl. 1) (A100).
[79] Mills NL, Donaldson K, Hadoke PW, et al. Adverse cardiovascular effects of air pollution. Nat Clin Pract Cardiovasc Med 2009;6:36–44.
[80] Møller P, Jacobsen NR, Folkmann JK, et al. Role of oxidative damage in toxicity of particulates. Free Radic Res 2010;44:1–46.
[81] Loft S, Danielsen PO, Mikkelsen L, Risom L, Forchhammer L, Møller P. Biomarkers of oxidative damage to DNA and repair. Biochem Soc Trans 2008;36:1071–6.
[82] Møller P, Loft S. Oxidative damage to DNA and lipids as biomarkers of exposure to air pollution. Environ Health Perspect 2010;118:1126–36.

[83] Cooke MS, Henderson PT, Evans MD. Sources of extracellular, oxidatively modified DNA lesions: Implications for their measurement in urine. J Clin Biochem Nutr 2009;45:255–70.

[84] Sauvain JJ, Setyan A, Wild P, et al. Biomarkers of oxidative stress and its association with the urinary reducing capacity in bus maintenance workers. J Occup Med Toxicol 2011;6:18.

[85] Gonzalez L, Lison D, Kirsch-Volders M. Genotoxicity of engineered nanomaterials: A critical review. Nanotoxicology 2008;2:252–73.

[86] Singh N, Manshian B, Jenkins GJS, et al. NanoGenotoxicology: The DNA damaging potential of engineered nanomaterials. Biomaterials 2009;30:3891–914.

[87] Ponti J, Sabbioni E, Munaro B, et al. Genotoxicity and morphological transformation induced by cobalt nanoparticles and cobalt chloride: An *in vitro* study in Balb/3T3 mouse fibroblasts. Mutagenesis 2009;24:439–45.

[88] Falck GC, Lindberg HK, Suhonen S, et al. Genotoxic effects of nanosized and fine TiO_2. Hum Exp Toxicol 2009;28:339–52.

[89] Muller J, Decordier I, Hoet PH, et al. Clastogenic and aneugenic effects of multi-wall carbon nanotubes in epithelial cells. Carcinogenesis 2008;29:427–33.

[90] Migliore L, Saracino D, Bonelli A, et al. Carbon nanotubes induce oxidative DNA damage in RAW 264.7 cells. Environ Mol Mutagen 2010;51:294–302.

[91] Karlsson HL. The comet assay in nanotoxicology research. Anal Bioanal Chem 2010;398:651–66.

[92] Hu YL, Gao JQ. Potential neurotoxicity of nanoparticles. Int J Pharm 2010;394:115–21.

[93] Wang Z, Zhao J, Li F, Gao D, Xing B. Adsorption and inhibition of acetylcholinesterase by different nanoparticles. Chemosphere 2009;77:67–73.

[94] Wang J, Rahman MF, Duhart HM, et al. Expression changes of dopaminergic system related genes in PC12 cells induced by manganese, silver, or copper nanoparticles. Neurotoxicology 2009;30:926–33.

[95] Wu J, Wang C, Sun J, Xue Y. Neurotoxicity of silica nanoparticles: Brain localization and dopaminergic neurons damage pathways. ACS Nano 2011;5:4476–89.

[96] Manzo L, Artigas F, Martinez E, et al. Biochemical markers of neurotoxicity. A review of mechanistic studies and applications. Hum Exp Toxicol 1996;S1:20–35.

[97] Smargiassi A, Mergler D, Bergamaschi E, Vettori MV, Lucchini R, Apostoli P. Peripheral markers of catecholamine metabolism among workers occupationally exposed to manganese (Mn). Toxicol Lett 1995;77:329–33.

[98] Mutti A, Bergamaschi E, Alinovi R, Lucchini R, Vettori MV, Franchini I. Serum prolactin in subjects occupationally exposed to manganese. Ann Clin Lab Sci 1996;26:10–7.

[99] Linse S, Cabaleiro-Lago C, Xue WF, et al. Nucleation of protein fibrillation by nanoparticles. Proc Natl Acad Sci USA 2007;104:8691–6.

[100] Simká M, Mattsson MO. Risks from accidental exposures to engineered nanoparticles and neurological health effects: A critical review. Part Fibre Toxicol 2010;7:42–57.

[101] Nasterlack M, Zober A, Oberlinner C. Considerations on occupational medical surveillance in employees handling nanoparticles. Int Arch Occup Environ Health 2008;81:721–6.

[102] Hildebrandt K, Rücker R, Koenig W, et al. Short-term effects of air pollution: A panel study of blood markers in patients with chronic pulmonary disease. Part Fibre Toxicol 2009;6:25–40.

[103] Schulte PA, Trout D, Zumwalde RD, et al. Options for occupational health surveillance of workers potentially exposed to engineered nanoparticles: State of the science. J Occup Environ Med 2008;50:517–26.

[104] Schulte PA, Schubauer-Berigan MK, Mayweather C, Geraci CL, Zumwalde R, McKernan JL. Issues in the development of epidemiologic studies of workers exposed to engineered nanoparticles. J Occup Environ Med 2009;51:323–35.

[105] Bergamaschi E. Occupational exposure to nanomaterials: Present knowledge and future development. Nanotoxicology 2009;3:194–201.

[106] Albertini RJ. Developing sustainable studies on environmental health. Mutat Res 2001;480-1:317–31.

[107] Roux A, Lison D, Junot C, Heilier JF. Applications of liquid chromatography coupled to mass spectrometry-based metabolomics in clinical chemistry and toxicology: A review. Clin Biochem 2011;44:119–35.

[108] Iavicoli I, Calabrese EJ, Nascarella MA. Exposure to nanoparticles and hormesis. Dose-Response 2010;8:501–17.

[109] Seaton A, Tran L, Aitken R, Donaldson K. Nanoparticles, human health hazard and regulation. J R Soc Interface 2010;7:S119–29.

CHAPTER

4

Critical Evaluation of Toxicity Tests

Maria Dusinska[1], *Elise Rundén-Pran*[1], *Sara Correia Carreira*[2], *Margaret Saunders*[2]

[1]NILU-Norwegian Institute for Air Research, Health Effects Laboratory, Departments of CEE and Environmental Chemistry, Kjeller, Norway

[2]BIRCH, Biophysics Research Unit, Department of Medical Physics & Bioengineering, University Hospitals Bristol NHS Foundation Trust, St Michael's Hospital, Bristol, UK

OUTLINE

INTRODUCTION

Nanotoxicology is an emerging field, dealing with human and environmental hazard and risk assessment caused by structures smaller than 100 nm [1–3]. Progress within nanotechnology is important, but the benefits of nanotechnology must be weighed against its

Adverse Effects of Engineered Nanomaterials
DOI: 10.1016/B978-0-12-386940-1.00004-0

risks. People have been exposed to ambient nanoparticles (NPs) such as volcano ashes or diesel exhaust for a long time. Although this field provides some knowledge, engineered nanomaterials (ENM) are designed with structural benefits and specific modifications that introduce totally different features, and thus should be handled accordingly [4]. Data obtained for the bulk compound cannot be directly applied to the scaled-down nano-size material [5]. Toxicity testing of ENM is not straightforward and needs special attention and consideration [6]. Generally, chemicals are regulated under REACH (registration, evaluation, authorization and restriction of chemicals) and nanopharmaceuticals and nanomedical products and devices under regulatory directives of the European Medicines Agency (EMA) in Europe, or the Food and Drug Administration (FDA) in the USA. However, ENM probably require adjustments to the legislation [7] (and see Chapter 6).

In vitro and *in vivo* tests used to assess toxicity of chemicals and underlying mechanisms (e.g. oxidative stress, immunotoxicity, genotoxicity) can also be applied to ENM [8]. Some of these methods are already developed into OECD guidelines and can be used for regulatory purposes (Tables 4.1 and 4.2). However, there are methodological concerns about ENM regarding dosimetry, dispersion, lack of washout, uptake and interaction of ENM with cells

TABLE 4.1 OECD Guidelines for *In vitro* Tests for Cytotoxicity, Genotoxicity and Carcinogenicity

Type of Toxicity	*In vitro* Test	Test Guideline
Cytotoxicity (starting doses for acute oral systemic toxicity tests)	Neutral Red Uptake (NRU) test with human cells (normal human epidermal keratinocytes [NHK]) (NHK NRU assay) and mouse BHK 3T3fibroblasts 3T3 NRU tests	OECD Draft Guidance 129 (2009)
Phototoxicity	3T3 Neutral Red Uptake for identifying substances with acute oral LD50 >2000 mg/kg b.w	ECVAM validated—(2009)
	3T3 NRU Phototoxicity Test (photo-irritation); 3T3 NRU Phototoxicity Test: Application to UV Filter Chemicals	OECD TG 432 (2004) EU Test Method B.41
	Tiered testing strategy to predict phototoxicity (3T3 NRU PT and reconstructed human epidermis models)	ECVAM feasibility study completed
	Produced reactive oxygen species (ROS) and photostability study	JaCVAM
	Test method battery to predict phototoxicity (yeast growth inhibition; Phototoxicity assay and red blood cell photohemolysis assay)	JaCVAM
Genotoxicity	Bacterial Reverse Mutation Test	OECD 471
	In vitro Mammalian Chromosome Aberration Test	OECD 473
	In vitro Mammalian Cell Gene Mutation Test	OECD 476
	Genetic Toxicology: *In vitro* Sister Chromatid Exchange Assay in Mammalian Cells	OECD 479

TABLE 4.1 OECD Guidelines for *In vitro* Tests for Cytotoxicity, Genotoxicity and Carcinogenicity—cont'd

Type of Toxicity	*In vitro* Test	Test Guideline
	Genetic Toxicology: *Saccharomyces cerevisiae*, Gene Mutation Assay	OECD 480
	Genetic Toxicology: *Saccharomyces cerevisiae*, Mitotic Recombination Assay	OECD 481
	Genetic Toxicology: DNA Damage and Repair, Unscheduled DNA Synthesis in Mammalian Cells *in vitro*	OECD 482
	In vitro Mammalian Cell Micronucleus Test	OECD 487
	In vitro Comet assay (Single-Cell Gel Electrophoresis [SCGE])	JaCVAM ECVAM/ICVAM
Carcinogenicity	Cell Transformation Assays; Combination of different assays: 1. Syrian hamster embryo (SHE) assay; 2. Balb/c assay; 3. C3H/10T1/2 assay (Alternative to 451) http://www.oecd.org/dataoecd/59/29/36069919.pdf	OECD draft review paper (No 31) EU B.21

TABLE 4.2 *In vivo* OECD Guideline Tests for Chemical Compounds

Type of Toxicity	*In vivo* Test	OECD Test Guideline No
Toxicokinetics	ADME	417
	Skin absorption	427
Acute toxicity	Inhalation	403, 436
	Oral	401, 420, 423, 425
	Dermal	402
Subacute toxicity	Oral, 28 days	407
	Inhalation, 28 days	412
	Dermal, 21/28 days	410
Subchronic toxicity	Oral, 90-days rodents	408
	Inhalation, 90-days rodents	413
	Oral, 90-days non-rodents	409
	Combined repeated dose toxicity (54 days) with reproduction/developmental toxicity screening	422
	Dermal, 90-days	411

(Continued)

TABLE 4.2 *In vivo* OECD Guideline Tests for Chemical Compounds—cont'd

Type of Toxicity	*In vivo* Test	OECD Test Guideline No
Reproduction and teratogenicity	Reproduction and developmental toxicity	414, 415, 416
	1-generation reproduction	415
	2-generation reproduction	416
	Prenatal	414
	Combined repeated dose toxicity with reproduction/developmental toxicity screening	422
	Reproduction/developmental toxicity screening test	421
	Developmental neurotoxicity	426
Neurotoxicity	Rodents	424
Chronic	Carcinogenicity, 24 months	451
	Chronic toxicity, 12 months	452
	Chronic toxicity and carcinogenicity, 12/24 months	453
Genotoxicity	*In vivo* micronucleus	474
	In vivo comet assay	No approved OECD guideline yet, however international agreed protocols available (hptt://cometassay.com)
	Transgenic Rodent Gene Mutation Assay	Draft
	Mammalian Erythrocyte Micronucleus Test	474
	Mammalian Bone Marrow Chromosome Aberration Test	475
	Mammalial Spermatogonial Chromosome Aberration Test	483
	Unscheduled DNA synthesis test with mammalian liver cells in vivo	486
	Mouse Spot Test	484
	Rodent Dominant Lethal Test	478
Skin sensitization	Dermal	406, 429, 442A, 442B
Irritation/corrosion	Eye	405
	Dermal	404
Endocrine effects	Hershberger Bioassay: Short-term screening test for androgenic properties in rat	441
	Uterothropic Bioassay: Short-term screening test for estrogenic properties in rats	440

and tissues. It is becoming clear that for development of standardized and validated tests for ENM, OECD-recommended or scientifically validated tests need to be adapted to fit the unique features of nano-size material, and thus implement characterization of ENM, uptake by cells and detection in biological systems. Some of these issues will be addressed and discussed below.

NANOPARTICLE UPTAKE

Since uptake through barriers (e.g. skin, blood–brain barrier [BBB], pulmonary mucosa, placenta) and toxicokinetics for a compound may change considerably with a decrease in size, toxicological data must be obtained specifically for nano-sized particles [9–11]. Reactivity increases concomitantly with a decrease in size and corresponding increased surface area. Surface area and composition strongly determine reactivity, dispersion, interaction with biological environments including cellular macromolecules and thus toxicity of ENM [12–14]. ENM may be designed deliberately to overcome biological barriers (e.g. for drug delivery) or they may cross them unintentionally as a result of environmental exposure. Once ENM have been taken up, they can potentially be deposited in any region of the body [15,16]. Due to their small size or specific functionalization, ENM may cross important biological barriers like the placenta or the BBB and cause toxicity to very sensitive systems such as the brain or developing fetus, respectively (see Chapters 9 and 13, respectively).

METHODS FOR VISUALIZING CELLULAR UPTAKE AND BIODISTRIBUTION OF ENM

Methods of NP detection need to be established in order to study both targeted delivery and distribution of ENM in exposed cells and tissues, as well as in the body. Detection and accurate determination of NP concentrations are particularly important for *in vivo* toxicity studies, as the concentration of systemically applied NPs will vary between tissues. A summary of the most commonly used NP detection methods is presented in Table 4.3 and discussed below.

ENM uptake is commonly studied using microscopy. Confocal and electron microscopy (EM) deliver very high resolution images, making them suitable tools for visualizing nano-sized particles in cells or tissues. For confocal imaging, ENM are often labeled with a fluorescent tag which allows real time tracking of particles [29] or determining their localization in tissue samples [21,22]. However, bleaching or loss of fluorescence is a challenge for long-term experiments. Also, the tag can possibly alter the behavior and the toxic potential of the particle itself. Electron-dense ENM can be visualized label-free using EM techniques [17]. EM is useful to visualize the location of particles but also to characterize their shape in tissue samples [18–20]. Both these microscopic methods require *ex situ* sample preparation and are therefore mainly of use for *in vitro* and *ex vivo* imaging. Artifacts, such as particle agglomerates, can be induced through transition of particles from dispersed to dried state in sample preparation for EM. Also, visualization of cellular particle uptake using confocal microscopy struggles to distinguish between internalized particles and those merely adhering to the cell surface.

TABLE 4.3 Selection of Most Commonly Used Techniques for Nanoparticle Detection *In vitro* and *Ex vivo*

Method	Principle	Sample	Benefits	Limitations	Example
Electron microscopy	Electron beam illuminates sample to produce a magnified, high resolution image	Sample needs to be dried and placed on support grid; if NP exposed cells/tissues are imaged specific fixing/staining is required	Very high resolution; accurate determination of nanoparticle shape and intracellular localization possible	Drying of sample may alter behavior of the particle; quantification of NP difficult	Localization of silver NP in histological samples [17]; intracellular localization of carbon nanotubes (CNT) [181], TiO_2 [19] and CeO NP [20]
Confocal microscopy	Point illumination and a spatial pinhole used to eliminate out-of-focus light in specimen, increasing resolution of the image	Sample in suspension; if NP exposed cells/tissues are imaged specific fixing/staining is required	Very high resolution; 3-dimensional reconstruction of specimen possible	Particles may need labeling with fluorophore; quantification of NP difficult	Intracellular localization of fluorescently labeled TiO_2 [21] and cationic solid lipid NP [22]
Raman spectroscopy	Identification of molecular vibrations characteristic of distinct functional groups that comprise the molecules	Sample in suspension or associated with cells/tissues	Label-free NP tracking; minimal sample preparation; high sensitivity, spatial resolution	Poor penetration into body, therefore limited to *ex vivo* or *in vivo* surface analyses	Cellular uptake of gold [23] and PLGA NP [24], CNT distribution in tissues [25]
Spectrophotometry	Particle concentration estimated as a measure of amount of light transmission, reflection or fluorescence intensity of a sample	Sample in suspension; standard curve to be included for concentration determination	Simple, non-invasive technique; further analysis and quantification of sample possible	Relatively low sensitivity; not for qualitative use	Accumulation of silver NP in cells [26]; transport of fluorescently-labeled silica NP across BBB [27]
Mass spectrometry	Particles are ionized, ions separated on basis of their mass-to-charge ratio; signal proportional to concentration	Sample in suspension	High sensitivity; qualitative and quantitative use	Most suitable for metal-based NP; protocols may need adaption for nano-size material	Silver [17] and TiO_2 [28] NP in tissues; intracellular concentration of CeO NP [20]

A promising tool to overcome these issues is Raman spectroscopy which is more often associated with determination of molecular structure and qualitative chemical analysis. When combined with optical microscopy (often referred to as Raman microscopy), this is also a very powerful tool for fast, non-invasive and label-free study of cellular binding and uptake of NPs. The technique is based on the identification of molecular vibrations that are characteristic of distinct functional groups. Therefore, ENM can be detected on the basis of their intrinsic physical properties rather than relying on extrinsic labels such as fluorescent tags [23]. This direct detection method may lead to a more accurate assessment of NP behavior over both the short and long term. Xu et al. [24] compared cellular uptake of fluorescently labeled and unlabeled poly(lactic-*co*-glycolic acid) (PLGA) NP by confocal and Raman microscopy. Analysis with confocal microscopy revealed intracellular fluorescence which indicates particle uptake. However, coherent anti-Stokes Raman scattering (CARS) microscopy revealed that unlabeled PLGA NPs were not taken up by cells. The authors concluded that the observed intracellular fluorescence was a result of dye transfer rather than NP uptake. Similarly to EM and confocal microscopy, Raman microscopy is more adapted to *ex vivo* investigations. Confocal and Raman microscopy can also be applied *in vivo* for surface investigations, but have poor tissue penetration. For *in vivo* imaging of particle distribution, medical imaging techniques such as magnetic resonance imaging (MRI) may be used [25].

EVALUATION OF TRANSPORT AND UPTAKE OF ENM *IN VITRO*

It is important for both risk assessment and therapeutic advances to determine the mechanisms by which ENM may reach areas such as the brain and fetus. Preliminary detailed investigation of transport and uptake can be undertaken *in vitro*. Cells can be grown to confluence on semipermeable membranes in specialized inserts, e.g. Transwells®, placed in tissue culture wells to provide access to both apical and basal compartments. Samples can be collected at time intervals and ENM concentration determined using techniques as above. Integrity of the cell barrier can be determined by measuring marker transfer, e.g. Lucifer Yellow dye or radioactive mannitol, combined with measurement of transepithelial/endothelial electrical resistance (TEER) [30,31]. Careful preparation of the ENM to avoid agglomeration is important as this will impact on uptake and transport. In principle, any adherent cell line can be used as a barrier model, but selection requires careful thought. Some cell lines do not form an intact monolayer [32] and others have been shown not to be representative of the organ of origin [33].

Elucidation of the transport mechanisms and cellular uptake can be determined by energy inhibition studies under varying metabolic conditions and use of appropriate pathway inhibitors, e.g. sucrose or chlorpromazine (clathrin-dependent), filipin or nystatin (caveolin-dependent), 5-ethylisopropyl amiloride (EIPA) or cytochalasin D (macropinocytosis) in combination with imaging by confocal or electron microscopy as appropriate [34]. However, these approaches can only provide an insight into the potential mechanisms of biological interaction with ENM and should be considered as a screening approach prior to further investigation *in vivo*.

IN VIVO UPTAKE, TRANSPORT AND DETECTION

In vivo studies of NP biodistribution add further complexity to NP detection due to biological interactions, e.g. opsonization. Sampling from the sites of interest will provide data on uptake and transport, but the time-points selected need to be appropriate to the material and organism under study. The pharmacokinetics and biodistribution of NPs within tissues will be dependent upon the NP characteristics such as size, surface charge and physical chemistry, and the tissue concentration should be monitored for long enough to obtain the full profile (usually 3 × half-life) [35]. Longer retention times may be necessary, and should be considered in experimental design.

Inductively coupled plasma mass spectrometry (ICP-MS) can be used to determine tissue levels of a range of metal (and some non-metal) NPs down to ng/g [17,28]. The use of radiolabeled NPs also provides a quantitative approach to *in vivo* biodistribution studies. It is possible to determine the NP uptake not only in the key organs but also in the remainder of the tissue, thus enabling total mass balance calculations to be done. Dependent upon the isotope selected, further tissue processing may be required after sample removal, e.g. tissue solubilization or ashing. It is also possible to undertake real-time imaging and follow the biodistribution *in vivo* using techniques such as positron-emission tomography (PET), computed tomography (CT) and single photon emission computed tomography (SPECT) with appropriate labeling of NPs [36]. This is an area of great clinical interest for delivery of imaging and therapeutic agents.

Other imaging approaches can be used for biodistribution studies [37] but tend to be less sensitive compared with radiolabeling due to less specific target tissue uptake. Contrast agents such as iron oxide or gadolinium NPs, and gold NPs can be imaged using magnetic resonance imaging (MRI) or x-ray, respectively, and are of particular interest for tumor detection and therapy. Fluorescent contrast agents, such as quantum dots (QD), can be utilized for optical imaging with *ex-vivo* pharmacokinetic data obtained through cryosectioning of tissues followed by fluorescence microscopy [38]; however, the risk of systemic toxicity from QD administration is a concern for longer-term studies. Development of longer wavelength new generation fluorescent probes enables whole body imaging, e.g. near infrared imaging [39,40]. Advances in ENM chemistry mean that combined labeling techniques can be used with the same material allowing increased sensitivity of detection and reduction in toxicity. Examples include neutron activation combined with electron microscopy, fluorescence and MRI [41]; MRI, PET and fluorescence imaging [42]; multiplex Raman imaging [43]. Such advances will play a vital role in determining the precise uptake, transport and biodistribution of NPs *in vivo*. For further discussion of imaging techniques, see Chapter 15.

DETERMINING CONCENTRATIONS OF ENM

Concentrations of detected ENM can be determined in tissue samples, cells or cell culture supernatants. ENM are then often labeled with a fluorescent or radioactive tag, as fluorescence intensity and radioactivity are proportional to particle concentration. Fluorescence intensity of labeled particles [27] or light absorption through a non-fluorescent NP suspension can be detected by spectrophotometric methods [26]. Spectrophotometry is a simple and non-invasive method to determine particle concentration, but limited in terms of sensitivity. A more sensitive method is the use of mass spectrometry (MS). With MS, particles are ionized and the ions are separated on the basis of their mass-to-charge ratio. The produced ions generate a signal which is proportional to the concentration of the particles. MS may also be used to identify the chemical composition of an unknown sample. Although this is a widely used analytical technique, it should be borne in mind that existing protocols for MS analysis of bulk materials may need to be validated according to the specific features for NPs. Federici et al. [44] performed calibrations with TiO_2 NP and Ti metal standards for ICP-MS. They found that TiO_2 NP solutions had a shallower slope than calibrations performed with Ti metal standards, indicating that ICP-MS responds differently to NPs compared with pure metal solutions.

The techniques discussed above and summarized in Table 4.3 are not an exhaustive list but represent the most commonly used ENM detection methods in *in vitro* and *in vivo* nanotoxicology. Choosing the most appropriate detection method depends greatly upon what question is addressed. For short-term tracking of NPs *in vitro*, it may be most suitable to attach a fluorescent label and visualize particles with confocal microscopy or quantify samples spectrophotometrically. For long-term follow up, however, the use of Raman spectroscopy may be more appropriate. If an unknown sample is to be analyzed, e.g. for human exposure assessments, a method with both qualitative and quantitative use, like MS, would be the most suitable.

IN VITRO CYTOTOXICITY TESTS

Uptake, biodistribution and concentration are important parameters besides toxicity testing for risk assessment of ENM. Human toxicity can be assessed in mammalian *in vitro* assays with primary cell culture or transformed cell lines of different origin (e.g. human, monkey, mice, hamster or rat). The widespread use of cell lines for toxicity testing reflects the fact that they provide a more standardized system compared to primary cell cultures, as well as higher throughput. Primary cultures often diverge between different laboratories, and are much more susceptible to environmental changes. However, primary cultures have a normal karyotype and offer the advantage of being closer to the *in vivo* situation.

Novel three-dimensional (3D) cell culture systems and *ex vivo* organotypical slice cultures are very interesting in regard to toxicity testing *in vitro*, as cells are in an environment where they are better able to mimic their *in vivo* counterparts [45–49].

In vivo, cells are organized three dimensionally and very often the function of individual cells depends on interactions of proteins and neighboring cells within this 3D organization [50]. In common cell cultures used for toxicity testing, this important level of complexity is reduced to a 2D culture on a flat substrate that does not exactly recapitulate the structure, function, or physiology of living tissues. Therefore, toxicity tests carried out in 2D culture may not accurately reflect *in vivo* cellular response. There is evidence that toxic effects of gold and cadmium telluride (CdTe) NP are reduced in 3D spheroid cultures of liver cells compared with conventional 2D liver cell culture [51]. However, for existing 3D systems, especially appropriate complex scaffolds for 3D culture need to be further developed to overcome certain limitations with the spatial distributions of oxygen, metabolites, and signaling molecules found in tissues [52] and for broader use in *in vitro* toxicity testing.

So-called organ-on-a-chip microdevices represent even more sophisticated *in vitro* models, implementing not only structural but also mechanical features of the whole living organ. Huh et al. [53] developed a biomimetic microsystem that represents key structural, functional and mechanical properties of the alveolar–capillary interface of the human lung. Within this system, the authors were able to mimic the stretching of the alveolar epithelium during breathing, which proved to be a crucial factor influencing the toxic and inflammatory response of the lung to silica NPs.

To diminish species-specific responses to NPs with different thresholds, a battery of *in vitro* tests covering a wide range of species should be used. Existing OECD guidelines for *in vitro* tests (see Table 4.1) have limitations with respect to assessment of biological hazards of ENM and do not include characterization. The extent to which nanoparticles are dispersed will influence the outcome of cytotoxicity testing of ENM, and is affected by physicochemical properties of NPs, dispersion medium, temperature, ultrasound energy used for dispersion, dispersing agents and sequence of dispersion preparation steps [54]. NP dispersion/agglomeration is a concentration-dependent phenomenon [55], and hence agglomeration state, uptake mechanisms and toxic effects may vary non-linearly with dose. A currently unanswered question is whether tests should be performed on dispersed ENM or under the most natural conditions with minimal processing, allowing natural formation of aggregates and agglomerates. Additionally, ENM often interact with the cell growth medium, eliciting in itself a toxic response that may have little or no physiological relevance *in vivo*. Many assays need to be adapted for testing of ENM and there is also a need to develop NPs that can serve as standards and as positive and negative controls for a range of relevant biological end-points. Here, we briefly discuss cytotoxicity and genotoxicity tests.

Cytotoxicity Tests

Cell lines of different origins, such as intestine or lung cells, can be used to predict acute toxicity by estimation of LC_{50} values (lethal concentration for 50% of the cells) and usually both 2D and 3D systems can be used to address specific end-points. Most cytotoxicity assays are based on colorimetric assays involving optical activity of organic dyes. In the Trypan blue exclusion assay, only damaged cells take up the dye and the proportion of damaged cells can be counted. However, it should be noted that the assay detects cell membrane damage and not necessarily cell death. The lactate dehydrogenase (LDH) assay measures cellular enzyme activity as damaged cells release LDH into the supernatant. This assay also measures cell membrane damage and not necessarily total cell death. The result will be affected by cell number and so, if there is any inhibition of proliferation during exposure, the decrease in cell concentration must be taken into account to avoid a false estimate of cytotoxicity.

Tetrazolium assays such as MTT, XTT and WST are based on enzymatic formation of formazan salts. Cellular dehydrogenases catalyze conversion of a tetrazolium salt into formazan with a resultant color change which can be measured by spectrophotometry. As these assays occur inside cells, they are a measure of metabolic activity related to cell viability.

In the Alamar Blue assay a resazurin dye (dark blue, non-fluorescent) is reduced to resorufin (pink and highly fluorescent) through a redox reaction. Fluorescence intensity corresponds with the cellular metabolic activity and is proportional to the number of living cells. Damaged and dead cells have lower innate metabolic activity and thus generate a proportionally lower signal than viable cells.

Several OECD guidelines have been developed for the neutral red assay (see Table 4.1) based on neutral red uptake. Neutral red dye is weakly cationic and can diffuse across the membrane and accumulate in cellular lysosomes. Increased cell permeability and lysosome fragility are associated with the latter stages of cell death; thus the level of cellular dye is indicative of the level of viability. This assay is more sensitive than other cytotoxicity tests and dye uptake is readily measured by a plate reader. However, as internalized NPs can be engulfed in lysosomes, increased lysosomal activity does not always reflect decreased cell viability.

It is important to note that different endpoints represent distinct metabolic (dys-) functions. Thus the assays can differ significantly in terms of their sensitivity.

Unfortunately, all colorimetric assays are prone to interference due to interaction between the dye and NPs, and wash-out can be difficult due to interactions. Probably the most reliable assays for cytotoxicity testing of ENM to be used in both 2D and 3D systems appear to be the proliferation assay and clonogenic or plating efficiency assay (adapted from OECD test guideline 476). In the clonogenic assay, single cells are seeded at low density, and colonies (with minimum 50 cells) are counted at fixed time points after exposure. Thus both viability (colony number) and proliferative capacity (colony size) are measured in this assay [56].

Genotoxicity Tests

Genotoxicity assays have different endpoints, such as single- and double-strand breaks, point mutations, deletions, chromosomal aberrations, micronuclei formation, DNA repair and cell-cycle interactions [57]. These endpoints are related to genotoxicity *in vitro* (human and mammalian cells) as well as *in vivo* (population biomonitoring and animal studies). Use of the same end-points in various biological systems in different species enables comparison of interspecies effects. *In vitro* genotoxicity assays represent simple, robust and time- and cost-effective testing of targeted toxicity and underlying mechanisms (see Chapter 14). Better selection of appropriate cells can be based upon identified target organs in *in vivo* studies (see below).

The most commonly applied methods for detecting genotoxicity include the bacterial Ames test, DNA strand break measurements in cells (e.g. comet assay, alkaline unwinding and hydroxyapatite chromatography, alkaline elution), and cytogenetic assays (micronucleus and chromosomal aberration assays, including the use of fluorescence *in situ* hybridization and chromosome painting).

The Ames test (bacterial reverse mutation assay) [58] is the most commonly used genotoxicity assay in regulatory toxicology. It is based on induction of reverse mutation in the histidine gene, which enables the bacteria to synthesize histidine and form visible colonies in minimal histidine medium. There are a number of concerns regarding the suitability of the Ames test for NP testing due to the bacterial cell wall composition and the fact that larger NPs cannot cross the cell membrane. It is also unknown how ENM could possibly interfere with histidine synthesis and induce false-negative (downregulation) or positive (upregulation) results. The Ames test has largely been used to assess genotoxicity of NPs such as

C60 NPs [59,60], multiwalled carbon nanotubes (MWCNT) [61,62], aluminum oxide nanomaterials [63], iron–platinum FePt NPs [64,65], and ZnO NPs [66] and has so far given largely negative results for ENM [67].

The comet assay or single cell gel electrophoresis is one of the most common tests for exploring the genotoxic potential of NPs. It detects DNA damage at the level of a single cell. DNA breaks cause relaxation of supercoiled DNA which then extends to form comet-like tails under electrophoresis. The relative intensity of DNA in the tail can be quantified and indicates the DNA break frequency. This method represents a rapid, sensitive, reliable, robust and relatively inexpensive way to study DNA damage (including DNA oxidation), and repair in different cell types both *in vitro* as well as *in vivo*. The assay is currently under validation by ECVAM/JaCVAM, and a draft guideline protocol for regulatory purposes is being developed. Specific DNA lesions can be analyzed by incubation of agarose-embedded nuclei with lesion-specific DNA repair enzymes, such as bacterial endonuclease III, (Endo III), or formamidopyrimidine DNA glycosylase (Fpg). Fpg recognizes oxidized purine bases, and thus detects oxidative DNA damage [68,69]. Additionally, photogenotoxic effects of NPs can be measured in combination with ultraviolet radiation [70]. Any cell type with a nucleus can be used, and thus the comet assay can assess both cell- and tissue-specific DNA damage induced by NPs.

The micronucleus (MN) assay can be applied both *in vitro* and *in vivo* and has been recommended for assessing mutagenic and clastogenic effects of pollutants [71]. The MN assay can easily be applied in different target cells and tissues [72,73] and has been used also in nanotoxicology [67,74]. Micronuclei are formed during anaphase from chromosomal fragments or whole chromosomes that are left behind when the nucleus divides. After telophase, these fragments may not be included in the nuclei of daughter cells but instead form single or multiple micronuclei in the cytoplasm [71]. The MN test detects both clastogenic and aneugenic effects [75] and is suitable for detection of genotoxicity of a wide range of compounds, including NPs. Formation of nucleoplasmic bridges and binucleated cells provides a complementary detection of chromosome rearrangement. Increased assay sensitivity can be obtained with cytochalasin-B, a cytokinesis-blocking agent that inhibits cell division, and differential staining methods (e.g. use of fluorescent acridine orange). This method is now routinely used for measuring chromosome breakage, impaired DNA repair, chromosome loss, non-disjunction, necrosis, apoptosis and cytostasis [76]. The MN assay is easy to score and has advantages over other cytogenetic methods, such as sister chromatid exchanges, or chromosome aberrations, being less time consuming. In order to reduce the risk of falsely identifying ENM aggregates as MN fragments, histological staining of positively identified nucleic acids can be performed.

The chromosomal aberration test is often applied in mammalian systems, both *in vitro* and *in vivo*, and has been used also for testing of genotoxicity of ENM [65,77]. The *in vitro* chromosomal aberration test identifies agents that cause structural chromosomal or chromatid breaks in cultured mammalian cell lines or primary cultures [78,79]. Chromosomal aberrations are the cause of many human genetic diseases, and there is substantial evidence that chromosomal damage and related events causing alterations in oncogenes and tumor suppressor genes of somatic cells are involved in cancer induction in humans and experimental animals. Cell cultures are exposed to the test substance both with and without metabolic activation and are treated with a metaphase-arresting substance (e.g. colcemid or colchicine) before harvest. Metaphase cells are analyzed microscopically for the presence of chromosomal aberrations.

The most recent review on nanogenotoxicity [80] shows that the most frequently used genotoxicity test is the comet assay (42 studies), followed by the MN test (27 studies), chromosome aberrations and the Ames test (both 10 studies). Some of these assays have been applied in both 2D and 3D culture systems. However, the results from the Ames test were almost always negative, indicating that the method may not be suitable for ENM testing [67]. Overall, the most promising assays for nanogenotoxicology testing *in vitro* seem to be the MN and comet assays, but identifying the most sensitive cell types for these assays remains to be established for most ENM.

For hazard assessment of nanogenotoxicity in human/mammalian systems, cross-species comparisons are required for each assay. A range of cell types from different organs should be used unless target organs for the specific ENM have been identified. *In vitro* assays should incorporate appropriate standards and controls (dispersion controls, positive ENM relevant controls, nearest equivalent traditional chemical for bench marking). Current experiments have often used high μg or mg concentrations of ENM perhaps of little relevance in risk assessment given the lower μg or ng circulating levels of ENM more likely to occur *in vivo*.

For regulatory genotoxicity/mutagenicity testing of chemicals, three different *in vitro* tests with two to three different end-points are initially chosen. If two out of three tests are positive, the compound is considered genotoxic, and is also likely to be carcinogenic. If two tests are negative, then the substance is regarded as "non-genotoxic". Demonstrated genotoxic potential *in vitro* should be confirmed *in vivo* (often applying the MN assay and DNA repair assays to liver samples).

The tiered approach and the associated technical difficulties with ENM have been recently reviewed [67,81]. It is becoming clear that testing of ENM will have to be revised and an additional tier will be needed to incorporate information on particle characterization, dispersion and uptake, and to determine whether or not the nano form can be demonstrated to have a different hazard from the equivalent bulk chemical. The use of dispersion controls and an appropriate bulk powder or mineral salt control should also be incorporated in the test methods additionally to positive and negative controls for ENM.

EX VIVO MODELS

Ex vivo tissue cultures represent a model between *in vitro* and *in vivo* [82], where whole tissue slices are cultured, such as organotypical slice cultures [45]. In these *ex vivo* models based on tissue slices, the cytoarchitecture is retained as well as many of the intercellular connections and interplays. Thus, metabolic processes more closely represent the *in vivo* situation. Studies can also be performed *ex vivo* with blood samples and cultivation of lymphocytes. *Ex vivo* tests can be applied, e.g. for testing of ocular corrosives on bovine corneas from slaughtered animals (OECD test guideline 437). Use of co-culture systems enduring different cell types is also a possibility for enhanced resemblance of the *in vivo* approach. These culture systems can be constructed to resemble barriers and used to investigate mechanisms of uptake (see above).

IN VIVO TOXICITY TESTING

In addition to validated *in vitro* test methods, *ex vivo* and *in silico* approaches, validated *in vivo* tests are highly warranted in the field of ENM. Interactions of ENM with macromolecules and other biological components can only be explored properly in *in vivo* models, making *in vivo* testing a prerequisite for toxicological risk assessment for ENM [82–84].

Toxicokinetics

Toxicokinetic studies are important not least in regard to the differences in toxicokinetics between ENM and the bulk compound. Pharmacokinetic studies exploring absorption, distribution, metabolism and excretion (ADME) (OECD guideline 471) of NPs are needed as a basis to determine toxicity and perform risk assessments as well as for designing ENM for diagnostics and therapy. Only kinetics studies *in vivo* can provide data about concentrations in target organs and cells, uptake into blood and distribution in the body.

The concentration at the target cells is dependent on metabolism and elimination. ENM that are not easily excreted will accumulate in the body, and this needs to be tested *in vivo* with repeated administration of ENM, e.g. in subchronic 90 day study in rats (see Table 4.2). Toxicokinetic studies should preferably be carried out in the same species as used for other toxicological experiments with the ENM, most commonly rat.

Toxicity will generally be dose dependent, although some substances induce non-threshold low dose effects (e.g. genotoxic/carcinogenic compounds). Doses should be sufficient for detection of ENM in blood, plasma or excreta, and it is important to investigate retention in gut or lung epithelium in addition to distribution to secondary target organs or tissues. A minimum of two doses with four animals per group should be tested in an ADME study. Administration route and doses should be selected based upon relevant human exposure to the EMN. For investigations of accumulation potential in the body, both *in vivo* repeated dose 28 days or 90 days toxicity studies in rodents can be performed (see Table 4.2). As a minimum, the percentage of applied dose in tissues should be measured at the end of the study. Rate and extent of excretion, and thus potential for accumulation, should be determined by measuring the percent of administered dose in urine, feces and expired air (OECD guideline 417). Demonstrated toxicity *in vitro* is only relevant if it is verified that the cells represent a target organ for the NP, and at realistic concentrations [84]. Toxicokinetics can also to some extent be modeled and provide a basis for extrapolation across species, exposure routes, dosimetry and human risk assessment (OECD guideline 417).

Also, metabolism is not fully accounted for in cell lines or cell cultures, and information should be attained from toxicokinetics studies *in vivo* by use of radiolabeled test substance for mass balance and metabolite identification (OECD guideline 417). Some information can be obtained from *in vitro* studies in, for example, microsomal fractions of the target tissue or S9 mitochondrial fraction. ENM can be subjected to metabolic changes and thus encounter new features and mechanisms of actions, as well as changed distribution pattern. Though some metabolism will also be present in *in vitro* systems, it can be fully explored only in animal models. However, animals differ in metabolism from humans and thus extrapolation from animal models to humans could be also questioned.

Immunological Responses

Another aspect that should be addressed *in vivo* is immunological responses (see Chapter 10). Many of the serum proteins that interact with ENM, such as complement and immunoglobulins, are immunoactive [13]. In addition, many target organs, such as spleen, lymph node and bone marrow, contain large amounts of macrophages, which are part of the mononuclear phagocyte system (or reticuloendothelial system) together with monocytes. Many ENM are shown to activate this system, and it is important to investigate a possible activation of the immune system in its entirety in *in vivo* studies [84]. However, some aspects of immunotoxicity can also be addressed in *in vitro* models.

Chronic Toxicity and Carcinogenicity

The need to perform *in vivo* chronic toxicity and carcinogenicity studies should be considered on the basis of results from *in vivo* ADME and *in vitro* genotoxicity studies. If there is evidence of toxic effects and/or accumulation in organs or tissues, chronic toxicity testing may be appropriate in order to reveal progressive or delayed toxicity with repeated exposure. Data from these types of study are commonly used for determination of No-Observable-Adverse-Effect-Level (NOAEL) or No-Observable-Effect-Level (NOEL) and lowest observable adverse effect level/lowest observable effect level (LOAEL/LOEL) used in risk assessment [83]. If the EMN are shown to cross the placenta in ADME studies, testing for reproductive and developmental toxicity should be considered in light of existing toxicity data [85]. Positive results from *in vitro* genotoxicity tests normally require further testing *in vivo*, unless it can be demonstrated that the positive *in vitro* findings are not relevant for the *in vivo* situation. However, the results should be assessed and evaluated together with other relevant information on the substance, like chemical reactivity, bioavailability, metabolism, toxicokinetics and target organ specificity. If *in vivo* genotoxicity and carcinogenicity testing is considered necessary, the tests should be designed according to appropriate target organs or tissues as well as *in vitro* genotoxicity end-points. There are several possible *in vivo* tests for genotoxicity (see Table 4.2), and the type of test should be carefully selected [83].

CONSIDERATIONS FOR SELECTION OF TOXICITY TESTS FOR ENM—*IN VITRO* AND *IN VIVO* COMPARISON

OECD guidelines for testing of chemicals are generally used in regulatory safety testing and chemical registration. When applying established test methods (e.g. OECD test guidelines) for ENM, one should keep in mind that these methods may need methodological modifications to be applied to ENM [86]. ENM could induce toxic effects which are not caught by the end-points under the current protocols [6]. There might also be a need for additional endpoints, such as cardiovascular, immunological and neurological [4,9]. The fact that NPs can interact with the end-points of test assays, especially those based upon detection of light, highlights the need for careful analysis of these interactions as well as the benefit of applying different test systems to validate the findings [6,13]. NPs could also interact with components in the test system, such as absorption to the wall of cell culture flasks or drinking vessels in oral *in vivo* studies and, consequently, the dose will be considerably lower than expected. Thus, measurements of actual concentration of NPs in media as well as inside the cells should be performed. Similarly, in *in vivo* experiments, it is important to measure the amount taken up by the gastrointestinal tract as well as blood concentration and concentration in target organs. Attention should be given towards comprehensive characterization, both for *in vivo* experiments and *in vitro* studies [6]. Stability and physicochemical characteristics of the ENM should be determined after synthesis but also in the applied vehicle [7,87]. This is a prerequisite for appropriate testing *in vivo* as well as *in vitro* [87]. Currently, there are no validated methods for testing of ENM, and it can be difficult to draw conclusions across different *in vitro* and *in vivo* studies due to discrepancies in experimental conditions and lack of test standardization [6,88,89]. An example of a change in physicochemical features when shrinking a compound to nano-size, is given by TiO_2 which appears in sunscreen formulations for UV protection. Nano-sized TiO_2, on the other hand, makes the formulation transparent and less thick, and thus more convenient for the consumer [12].

Data from *in vivo* studies with bulk compound should always be taken into consideration in view of the need for long-term *in vivo* studies. The scientific testing strategy should always be in accordance with the three3 Rs—refine, replace and reduce—for animal experiments. We should always use as few animals as possible, and refine the methods to minimize the suffering of the animals involved in the study. The replacement of *in vivo* studies with *in vitro* studies in cell cultures should always be considered. Use of human blood is also an alternative. An intermediate strategy would be to apply co-culture systems, 3D cell cultures systems or *ex vivo* tissue slice cultures [82].

There are advantages and disadvantages of both *in vitro* and *in vivo* studies. Mechanistically, investigations can better be explored in cell culture systems, which can be more easily controlled and yield more reproducible factors than *in vivo* systems. Although high-throughput analysis for nanotoxicology can best be performed in monolayer or suspension cultures, information obtained on interactions between NPs and tissues and organs are limited. Such high-throughput toxicity screening models for ENM will not be able to replace fully *in vivo* studies, but could be useful for selection of ENM for further testing, including animal testing.

For design of both *in vivo* and *in vitro* experiments, exposure assessment is of utmost importance. Relevant exposure routes should always be taken into consideration for selection of administration route in studies and choice of cell lines or cultures for *in vitro* testing. Dosimetry is still a challenge for *in vivo* studies, as the dose should reflect the actual dose to which humans are exposed. It is technically very difficult to determine the concentration or number of NPs in air, water, soil or any consumer products due to their small size and quantity. When no information is available, a worst case scenario should be applied, assuming presence, absorption and systemic distribution of ENM [83]. However, if the dose is too high, agglomeration will occur, and this can introduce a different toxicological profile [90].

In vivo testing is expensive and time consuming and also presents an ethical issue. Nonetheless, *in vivo* studies have to be performed at least for ADME, interactions with cells, tissues and proteins, hormonal effects and long-term chronic effects. It is also important to ensure that the vehicle is suitable for good dispersion of the NPs to avoid intrinsic toxicity. NPs are prone to agglomerate at physiological conditions due to their large surface area. Selection of an appropriate vehicle will also depend upon the route of exposure and source of NPs and standardization of tests should be performed [14].

For proper hazard and risk assessment, all available toxicity data from *in vitro* and *in vivo* experiments should be evaluated together with exposure data and other relevant information from read-across of other ENM or from bulk chemicals. The output should be the overall assessment of safety of ENM in their intended use. A multitechnique approach should be utilized to analyze the materials from as many aspects as possible to ensure that the results reflect intrinsic toxicity of the ENM and do not interact with the selected test assay. When it comes to exposure, the whole life cycle of the ENM should be taken into consideration [5]. There will always be uncertainties when extrapolating from animal models to humans, as well as from laboratory experiments to complex systems [91]. The risk of ENM will be determined by interaction with cells and tissues, ADME and potential exposure levels, and all these factors are likely influenced by physicochemical properties (such as shape, reactivity, size, surface coating, surface charge) of the ENM [83].

We should also investigate whether *in silico* modeling and quantitative structure–activity relationships (QSAR) can be applied successfully to ENM, and thus determine which features influence toxicity the most. This approach could be used as a basis for grouping of NPs [92]. Gene expression profiles could also aid in hazard assessment of ENM [93]. Until this can be done, ENM should be treated case by case when it comes to toxicity testing and risk assessment. The size-specific properties of ENM introduce a need for validated screening tests for toxicity, as interference with end-points in general cytotoxicity tests must be precluded. Use of new endpoints with marker-free detection (such as the colony formation assay) is a possibility. Another challenge is dosimetry as this is not yet standardized when it comes to mass, number, surface area or other metrics, and this makes comparisons and objective analysis difficult [3]. Reference NPs and standardized test protocols are highly desirable.

CONCLUSIONS

Challenges arise from the fact that the unique features of ENM may be coupled with toxicity. Existing validated and standardized tests for toxicity testing of NPs both *in vitro* and *in vivo* need to be adopted for ENM testing while new methods also need to be developed. Physicochemical characterization of the test substance is an important step, as well as agreement on validated reference material and appropriate controls. Also, detection techniques for studying uptake and biodistribution of the NPs need to be an integral part of the testing strategy and have to be improved and validated.

In vitro and *in vivo* studies should always be closely linked. Cytotoxicity tests should be chosen to preclude interaction of NPs with the end-points of the assay, as could occur with colorimetric assays. A battery of tests with cells from different species will reduce the risk of interspecies differences. Genotoxicity can be tested with different *in vitro* assays addressing several end-points, and positive results should be obtained in several *in vitro* systems for the ENM to be classified as genotoxic.

Toxicity testing of ENM should always be in accordance with the three Rs (reduce, refine, replace). Minimal numbers of animals should be used, and the method should be refined to reduce any suffering. Also, the use of cell culture systems that better reflect *in vivo* toxicity parameters, like co-cultures, 3D cell culture systems and *ex vivo* models and primary human cells such as blood would be in accordance with the strategy of limiting the number of animal studies. Use of high-throughput screening with validated tests, combined with *in silico* data handling could be a possible strategy to narrow the gap between nanotechnology and nanotoxicology and simultaneously reduce the number of animal experiments.

TAKE-HOME MESSAGES

The rapid development of nanotechnology outstrips our current understanding of the potential risks to human and environmental health posed by nanoscale materials. Altered behavior at the nanoscale compared with bulk material requires the development of improved tiered testing strategies to determine the impact of biological interaction with nanomaterials both *in vitro* and *in vivo*. The use of more *in vivo*-like cell culture systems combined with validated high throughput screening and *in silico* data handling will go some way to improving our knowledge.

Acknowledgments

Supported by EC FP7 [Health-2007-1.3-4], Contract no: 201335, www.nanotest-fp7.eu.

References

[1] Donaldson K, Stone V, Tran CL, Kreyling W, Borm PJ. Nanotoxicology. Occup Environ Med 2004;61:727–8.
[2] Haynes CL. The emerging field of nanotoxicology. Anal Bioanal Chem 2010;398:587–8.
[3] Maynard AD, Warheit DB, Philbert MA. The new toxicology of sophisticated materials: Nanotoxicology and beyond. Toxicol Sci 2011;120(Suppl. 1):S109–29.
[4] Seaton A, Tran L, Aitken R, Donaldson K. Nanoparticles, human health hazard and regulation. J R Soc Interface 2010;7(Suppl. 1):S119–29.
[5] Feliu N, Fadeel B. Nanotoxicology: No small matter. Nanoscale 2010;2:2514–20.
[6] Kroll A, Pillukat MH, Hahn D, Schnekenburger J. Current *in vitro* methods in nanoparticle risk assessment: Limitations and challenges. Eur J Pharm Biopharm 2009;72:370–7.
[7] Gwinn MR, Tran L. Risk management of nanomaterials. Wiley Interdiscip Rev Nanomed Nanobiotechnol 2010;2:130–7.
[8] Dusinska M, Fjellsbo L, Magdolenova Z, et al. Testing strategies for the safety of nanoparticles used in medical applications. Nanomedicine (Lond) 2009;4:605–7.
[9] Simko M, Mattsson MO. Risks from accidental exposures to engineered nanoparticles and neurological health effects: A critical review. Part Fibre Toxicol 2010;7:42.
[10] Schleh C, Semmler-Behnke M, Lipka J, et al. Size and surface charge of gold nanoparticles determine absorption across intestinal barriers and accumulation in secondary target organs after oral administration. Nanotoxicology 2011 [Epub ahead of print].
[11] Lehr CM, Daum N, Schneider M, Schafer UF. Biological barriers—a need for novel tools in nanotoxicology and nanomedicine. Preface Eur J Pharm Biopharm 2011;77:337.
[12] Warheit DB, Sayes CM, Reed KL, Swain KA. Health effects related to nanoparticle exposures: Environmental, health and safety considerations for assessing hazards and risks. Pharmacol Ther 2008;120:35–42.
[13] Kunzmann A, Andersson B, Thurnherr T, Krug H, Scheynius A, Fadeel B. Toxicology of engineered nanomaterials: Focus on biocompatibility, biodistribution and biodegradation. Biochim Biophys Acta 2011;1810:361–73.
[14] Dhawan A, Sharma V. Toxicity assessment of nanomaterials: Methods and challenges. Anal Bioanal Chem 2010;398:589–605.

[15] Borm PJ, Muller-Schulte D. Nanoparticles in drug delivery and environmental exposure: Same size, same risks? Nanomedicine (Lond) 2006;1:235–49.
[16] Oberdorster G, Oberdorster E, Oberdorster J. Nanotoxicology: An emerging discipline evolving from studies of ultrafine particles. Environ Health Perspect 2005;113:823–39.
[17] Garza-Ocanas L, Ferrer DA, Burt J, et al. Biodistribution and long-term fate of silver nanoparticles functionalized with bovine serum albumin in rats. Metallomics 2010;2:204–10.
[18] Monteiro-Riviere NA, Nemanich RJ, Inman AO, Wang YY, Riviere JE. Multi-walled carbon nanotube interactions with human epidermal keratinocytes. Toxicol Lett 2005;155:377–84.
[19] Wang Y, Wang J, Wu M, et al. Internalization, translocation and biotransformation of silica-coated titanium dioxide nanoparticles in neural stem cells. J Nanosci Nanotechnol 2010;10:7121–5.
[20] Raemy DO, Limbach LK, Rothen-Rutishauser B, et al. Cerium oxide nanoparticle uptake kinetics from the gas-phase into lung cells *in vitro* is transport limited. Eur J Pharm Biopharm 2011;77:368–75.
[21] Thurn KT, Arora H, Paunesku T, et al. Endocytosis of titanium dioxide nanoparticles in prostate cancer PC-3M cells. Nanomedicine 2011;7:123–30.
[22] Kuo YC, Chen HH. Effect of electromagnetic field on endocytosis of cationic solid lipid nanoparticles by human brain-microvascular endothelial cells. J Drug Target 2010;18:447–56.
[23] Shah NB, Dong J, Bischof JC. Cellular uptake and nanoscale localization of gold nanoparticles in cancer using label-free confocal Raman microscopy. Mol Pharm 2011;8:176–84.
[24] Xu P, Gullotti E, Tong L, et al. Intracellular drug delivery by poly(lactic-co-glycolic acid) nanoparticles, revisited. Mol Pharm 2009;6:190–201.
[25] Al Faraj A, Fauvelle F, Luciani N, et al. *In vivo* biodistribution and biological impact of injected carbon nanotubes using magnetic resonance techniques. Int J Nanomed 2011;6:351–61.
[26] Trickler WJ, Lantz SM, Murdock RC, et al. Silver nanoparticle induced blood–brain barrier inflammation and increased permeability in primary rat brain microvessel endothelial cells. Toxicol Sci 2010;118:160–70.
[27] Ragnaill MN, Brown M, Ye D, et al. Internal benchmarking of a human blood–brain barrier cell model for screening of nanoparticle uptake and transcytosis. Eur J Pharm Biopharm 2011;77:360–7.
[28] Chen J, Dong X, Xin Y, Zhao M. Effects of titanium dioxide nano-particles on growth and some histological parameters of zebrafish (*Danio rerio*) after a long-term exposure. Aquat Toxicol 2011;101:493–9.
[29] Seo JH, Cho K, Lee SY, Joo SW. Concentration-dependent fluorescence live-cell imaging and tracking of intracellular nanoparticles. Nanotechnology 2011;22:235101.
[30] Liu F, Soares MJ, Audus KL. Permeability properties of monolayers of the human trophoblast cell line BeWo. Am J Physiol 1997;273:C1596–604.
[31] Ampasavate C, Chandorkar GA, Vande Velde DG, Stobaugh JF, Audus KL. Transport and metabolism of opioid peptides across BeWo cells, an *in vitro* model of the placental barrier. Int J Pharm 2002;233:85–98.
[32] Saunders M. Transplacental transport of nanomaterials. Wiley Interdiscip Rev Nanomed Nanobiotechnol 2009;1:671–84.
[33] Brown J, Reading SJ, Jones S, et al. Critical evaluation of ECV304 as a human endothelial cell model defined by genetic analysis and functional responses: A comparison with the human bladder cancer derived epithelial cell line T24/83. Lab Invest 2000;80:37–45.
[34] Saovapakhiran A, D'Emanuele A, Attwood D, Penny J. Surface modification of PAMAM dendrimers modulates the mechanism of cellular internalization. Bioconjug Chem 2009;20:693–701.
[35] Li SD, Huang L. Pharmacokinetics and biodistribution of nanoparticles. Mol Pharm 2008;5:496–504.
[36] Zheng J, Jaffray D, Allen C. Quantitative CT imaging of the spatial and temporal distribution of liposomes in a rabbit tumor model. Mol Pharm 2009;6:571–80.
[37] Skotland T, Iversen TG, Sandvig K. New metal-based nanoparticles for intravenous use: Requirements for clinical success with focus on medical imaging. Nanomedicine 2010;6:730–7.
[38] Robe A, Pic E, Lassalle HP, Bezdetnaya L, Guillemin F, Marchal F. Quantum dots in axillary lymph node mapping: Biodistribution study in healthy mice. BMC Cancer 2008;8:111.
[39] Tasciotti E, Godin B, Martinez JO, et al. Near-infrared imaging method for the *in vivo* assessment of the biodistribution of nanoporous silicon particles. Mol Imaging 2011;10:56–68.
[40] Burns AA, Vider J, Ow H, et al. Fluorescent silica nanoparticles with efficient urinary excretion for nanomedicine. Nano Lett 2009;9:442–8.

[41] Vaccaro DE, Yang M, Weinberg JS, Reinhardt CP, Groman EV. Cell tracking using nanoparticles. J Cardiovasc Transl Res 2008;1:217–20.
[42] Stelter L, Pinkernelle JG, Michel R, et al. Modification of aminosilanized superparamagnetic nanoparticles: Feasibility of multimodal detection using 3T MRI, small animal PET, and fluorescence imaging. Mol Imaging Biol 2010;12:25–34.
[43] Wang Y, Seebald JL, Szeto DP, Irudayaraj J. Biocompatibility and biodistribution of surface-enhanced Raman scattering nanoprobes in zebrafish embryos: *In vivo* and multiplex imaging. ACS Nano 2010;4:4039–53.
[44] Federici G, Shaw BJ, Handy RD. Toxicity of titanium dioxide nanoparticles to rainbow trout (*Oncorhynchus mykiss*): Gill injury, oxidative stress, and other physiological effects. Aquat Toxicol 2007;84:415–30.
[45] Runden E, Seglen PO, Haug FM, et al. Regional selective neuronal degeneration after protein phosphatase inhibition in hippocampal slice cultures: Evidence for a MAP kinase-dependent mechanism. J Neurosci 1998;18:7296–305.
[46] Noraberg J. Organotypic brain slice cultures: An efficient and reliable method for neurotoxicological screening and mechanistic studies. Altern Lab Anim 2004;32:329–37.
[47] Ring A, Tanso R, Noraberg J. The use of organotypic hippocampal slice cultures to evaluate protection by non-competitive NMDA receptor antagonists against excitotoxicity. Altern Lab Anim 2010;38:71–82.
[48] Prestwich GD. Evaluating drug efficacy and toxicology in three dimensions: Using synthetic extracellular matrices in drug discovery. Acc Chem Res 2008;41:139–48.
[49] Legendre A, Froment P, Desmots S, Lecomte A, Habert R, Lemazurier E. An engineered 3D blood–testis barrier model for the assessment of reproductive toxicity potential. Biomaterials 2010;31:4492–505.
[50] Abbott A. Cell culture: Biology's new dimension. Nature 2003;424:870–2.
[51] Lee J, Lilly GD, Doty RC, Podsiadlo P, Kotov NA. *In vitro* toxicity testing of nanoparticles in 3D cell culture. Small 2009;5:1213–21.
[52] Derda R, Laromaine A, Mammoto A, et al. Paper-supported 3D cell culture for tissue-based bioassays. Proc Natl Acad Sci USA 2009;106:18457–62.
[53] Huh D, Matthews BD, Mammoto A, Montoya-Zavala M, Hsin HY, Ingber DE. Reconstituting organ-level lung functions on a chip. Science 2010;328:1662–8.
[54] Byrne HJ, Lynch I, de Jong WH, et al. Protocols for assessment of biological hazards of engineered nanoparticles. Eur Network Health Environ Impact Nanomaterials 2010; report on www.nanoimpact.eu.
[55] Priya BR, Byrne HJ. Investigation of sodium dodecyl benzene sulphonate assisted dispersion and debundling of single wall carbon nanotubes. J Phys Chem C 2008;12:332–7.
[56] Herzog E, Casey A, Lyng FM, Chambers G, Byrne HJ, Davoren M. A new approach to the toxicity testing of carbon-based nanomaterials—the clonogenic assay. Toxicol Lett 2007;174:49–60.
[57] Ng CT, Li JJ, Bay BH, Yung LY. Current studies into the genotoxic effects of nanomaterials. J Nucl Acids 2010; pii: 947859.
[58] Mortelmans K, Zeiger E. The Ames Salmonella/microsome mutagenicity assay. Mutat Res 2000;455:29–60.
[59] Shinohara N, Matsumoto K, Endoh S, Maru J, Nakanishi J. *In vitro* and *in vivo* genotoxicity tests on fullerene C60 nanoparticles. Toxicol Lett 2009;191:289–96.
[60] Mori T, Takada H, Ito S, Matsubayashi K, Miwa N, Sawaguchi T. Preclinical studies on safety of fullerene upon acute oral administration and evaluation for no mutagenesis. Toxicology 2006;225:48–54.
[61] Wirnitzer U, Herbold B, Voetz M, Ragot J. Studies on the *in vitro* genotoxicity of baytubes, agglomerates of engineered multi-walled carbon-nanotubes (MWCNT). Toxicol Lett 2009;186:160–5.
[62] Di Sotto A, Chiaretti M, Carru GA, Bellucci S, Mazzanti G. Multi-walled carbon nanotubes: Lack of mutagenic activity in the bacterial reverse mutation assay. Toxicol Lett 2009;184:192–7.
[63] Balasubramanyam A, Sailaja N, Mahboob M, Rahman MF, Hussain SM, Grover P. *In vitro* mutagenicity assessment of aluminium oxide nanomaterials using the Salmonella/microsome assay. Toxicol In Vitro 2010;24:1871–6.
[64] Maenosono S, Suzuki T, Saita S. Mutagenicity of water-soluble FePt nanoparticles in Ames test. J Toxicol Sci 2007;32:575–9.
[65] Maenosono S, Yoshida R, Saita S. Evaluation of genotoxicity of amine-terminated water-dispersible FePt nanoparticles in the Ames test and *in vitro* chromosomal aberration test. J Toxicol Sci 2009;34:349–54.

[66] Yoshida R, Kitamura D, Maenosono S. Mutagenicity of water-soluble ZnO nanoparticles in Ames test. J Toxicol Sci 2009;34:119–22.
[67] Landsiedel R, Kapp MD, Schulz M, Wiench K, Oesch F. Genotoxicity investigations on nanomaterials: methods, preparation and characterization of test material, potential artifacts and limitations—many questions, some answers. Mutat Res 2009;681:241–58.
[68] Collins AR, Dusinska M, Gedik CM, Stetina R. Oxidative damage to DNA: Do we have a reliable biomarker? Environ Health Perspect 1996;104(Suppl. 3):465–9.
[69] Dusinska M, Collins AR. Detection of oxidised purines and UV-induced photoproducts in DNA, by inclusion of lesion-specific enzymes in the comet assay (single cell gel electrophoresis). Altern Lab Anim 1996;24:405–11.
[70] Jha AN. Ecotoxicological applications and significance of the comet assay. Mutagenesis 2008;23:207–21.
[71] OECD. Guidelines for testing chemicals. Mammalian erythrocyte micronucleus test, Vol. Guideline; 1997;474.
[72] Laingam S, Froscio SM, Humpage AR. Flow-cytometric analysis of *in vitro* micronucleus formation: Comparative studies with WIL2-NS human lymphoblastoid and L5178Y mouse lymphoma cell lines. Mutat Res 2008;656:19–26.
[73] Fenech M, Kirsch-Volders M, Natarajan AT, et al. Molecular mechanisms of micronucleus, nucleoplasmic bridge and nuclear bud formation in mammalian and human cells. Mutagenesis 2011;26:125–32.
[74] Doak SH, Griffiths SM, Manshian B, et al. Confounding experimental considerations in nanogenotoxicology. Mutagenesis 2009;24:285–93.
[75] Gonzalez L, Sanderson BJ, Kirsch-Volders M. Adaptations of the *in vitro* MN assay for the genotoxicity assessment of nanomaterials. Mutagenesis 2011;26:185–91.
[76] Fenech M. Cytokinesis-block micronucleus cytome assay. Nat Protoc 2007;2:1084–104.
[77] Hackenberg S, Scherzed A, Kessler M, et al. Silver nanoparticles: Evaluation of DNA damage, toxicity and functional impairment in human mesenchymal stem cells. Toxicol Lett 2011;201:27–33.
[78] Galloway SM, Aardema MJ, Ishidate Jr M, et al. Report from working group on *in vitro* tests for chromosomal aberrations. Mutat Res 1994;312:241–61.
[79] Galloway SM, Armstrong MJ, Reuben C, et al. Chromosome aberrations and sister chromatid exchanges in Chinese hamster ovary cells: Evaluations of 108 chemicals. Environ Mol Mutagen 1987;10(Suppl. 10):1–175.
[80] Magdolenova Z, Dhawan A, Collins A, Dusinska M. (submitted). Review on nanoparticle genotoxicity. J Biomedical Nanotechnol.
[81] Singh N, Manshian B, Jenkins GJ, et al. NanoGenotoxicology: The DNA damaging potential of engineered nanomaterials. Biomaterials 2009;30:3891–914.
[82] Clift MJ, Gehr P, Rothen-Rutishauser B. Nanotoxicology: A perspective and discussion of whether or not *in vitro* testing is a valid alternative. Arch Toxicol 2011;85:723–31.
[83] Commitee ES. Guidance on the risk assessment of the application of nanoscience and nanotechnologies in the food and feed chain. EFSA J 2011;9:2140–76.
[84] Fischer HC, Chan WC. Nanotoxicity: The growing need for *in vivo* study. Curr Opin Biotechnol 2007;18:565–71.
[85] Ema M, Kobayashi N, Naya M, Hanai S, Nakanishi J. Reproductive and developmental toxicity studies of manufactured nanomaterials. Reprod Toxicol 2010;30:343–52.
[86] Savolainen K, Alenius H, Norppa H, Pylkkanen L, Tuomi T, Kasper G. Risk assessment of engineered nanomaterials and nanotechnologies—a review. Toxicology 2010;269:92–104.
[87] Sayes CM, Warheit DB. Characterization of nanomaterials for toxicity assessment. Wiley Interdiscip Rev Nanomed Nanobiotechnol 2009;1:660–70.
[88] Khlebtsov N, Dykman L. Biodistribution and toxicity of engineered gold nanoparticles: A review of *in vitro* and *in vivo* studies. Chem Soc Rev 2011;40:1647–71.
[89] Sayes CM, Reed KL, Warheit DB. Assessing toxicity of fine and nanoparticles: comparing *in vitro* measurements to *in vivo* pulmonary toxicity profiles. Toxicol Sci 2007;97:163–80.
[90] Tang M, Zhang T, Xue Y, et al. Dose dependent *in vivo* metabolic characteristics of titanium dioxide nanoparticles. J Nanosci Nanotechnol 2010;10:8575–83.
[91] Canady RA. The uncertainty of nanotoxicology: Report of a Society for Risk Analysis Workshop. Risk Anal 2010;30:1663–70.
[92] George S, Xia T, Rallo R, et al. Use of a high-throughput screening approach coupled with *in vivo* zebrafish embryo screening to develop hazard ranking for engineered nanomaterials. ACS Nano 2011;5:1805–17.
[93] Poma A, Di Giorgio ML. Toxicogenomics to improve comprehension of the mechanisms underlying responses of *in vitro* and *in vivo* systems to nanomaterials: A review. Curr Genomics 2008;9:571–85.

CHAPTER

5

Computational Approaches

Vidana Epa[1], Dave Winkler[2,3], Lang Tran[4]

[1]CSIRO Materials Science & Engineering, Parkville, Australia,
[2]CSIRO Materials Science & Engineering, Clayton, Australia,
[3]Monash Institute of Pharmaceutical Science, Parkville, Australia
[4]Institute of Occupational Medicine, Research Avenue North, Edinburgh, UK

OUTLINE

INTRODUCTION

Our understanding of the novel properties of materials and particles with dimensions in the nanometer range is growing rapidly, and commercial nanomaterials are reaching the market place at an increasing rate (>1000 "nano" products available in the USA alone). However, we still understand very little about how nanomaterials interact with mammalian systems or the environment. The precautionary principle demands care in handling and distributing nanomaterials in the workplace, marketplace, or environment until we better understand their interactions with biological systems [1–3]. For example, titanium dioxide

Adverse Effects of Engineered Nanomaterials
DOI: 10.1016/B978-0-12-386940-1.00005-2

nanoparticles generate free radicals in light, a property exploited for water photolysis reactions and for developing self-cleaning surfaces. However, titanium dioxide and zinc oxide nanoparticles also are used in sunscreens as they absorb UV radiation and effectively prevent sunburn [4,5]. The potential for generating free radicals that would adversely affect skin cells is a cause for concern, particularly as sunscreen is mandated for childcare centers in some countries, such as Australia. A further example is high aspect ratio carbon nanotubes that have been shown to have detrimental health effects similar to asbestos [6,7]. There is also increasing general concern about the occupational health and safety and public health impacts of nanoparticles. Their size allows them to be inhaled, penetrate the skin, mucosa, or lung tissue, be transported around the body, and enter cells where they could interact with biomolecules and interfere with cellular biochemistry. We have grossly inadequate knowledge of how cells, organs, and tissues interact with nanoparticles, particularly for chronic exposure. There is, therefore, an urgent need to determine whether nanoparticles impact negatively on human and environmental health, and to develop robust and reliable methods for predicting nanotoxicity proactively, rather than reactively once a serious public health or environmental problem appears.

As experiments on biological effects of nanomaterials are absolutely necessary but time consuming and expensive, there is a strong need to develop cheaper, predictive methods for rapidly assessing potential toxicities of new and modified nanomaterials. Computational modeling techniques have a relatively long history of performance, and have obvious synergy with experimental toxicology, in assessing the risk of industrial chemicals, agrochemicals and drugs. Chemical regulatory agencies around the world are increasingly using computational tools, particularly statistical modeling and machine learning methods embodied in QSAR (quantitative structure–activity relationships) modeling [8,9]. Recent reviews and commentaries in high impact journals [1–3] have called for computational methods to be applied to understanding nanotoxicology, and regulators are calling for these tools to be developed as a matter of urgency. As with industrial chemicals, high throughput methods can provide large volumes of data on the biological effects of nanomaterials, although experimental programs are just starting to embrace these technologies. Increased used of gene arrays and related technologies to assess more rapidly the impact of chemical and nanomaterials on biological systems should also be of major significance. Computational tools, and the necessary informatics and ontological tools that must accompany them, will be necessary to manage nanotoxicological data in the future. Hopefully large-scale scientific consortia will be formed (similar to ToxCAST [10] and the HPV challenge [11] with industrial chemicals) that will provide data for risk assessment and modeling. Other initiatives for data consolidation similar to ChEMBL, ACToR [12], data fusion, MIAME [13,14] for small molecules and gene arrays, and embracing new information technologies like FaceBook, RSS feeds will also be important for efficient data sharing between research groups and regulatory agencies.

SPECIFIC DIFFICULTIES WITH MODELING NANOPARTICLE BIOLOGICAL EFFECTS

Unlike small organic molecules, which have been studied computationally for a long period, a number of important issues arise when attempting to model the biological effects of complicated materials like nanoparticles. Nanoparticles are complex materials with diverse and important surface features that play a dominant role in their biological properties. Microscopic surface properties, such as surface charge densities (and related zeta potentials), surface coatings and modifications, crystal faces and edges, electrochemical properties etc., are all likely to affect not only how a nanoparticle interacts with cells or the environment, but also its photochemical properties, aggregation potential, and type of "corona" formed in biological fluids. In addition, the possibility of soluble metal ion leaching and the difficulties in accurately characterizing manufactured nanoparticle sizes, shapes, morphologies, crystal phases etc. adds an extra dimension of complexity in understanding the biological effects and identifying what is in fact the biologically relevant entity. Finally, lack of knowledge about how nanoparticles are taken up, distributed and transformed in the body, how they enter cells and affect cellular processes, and how they are eliminated, make traditional methods of pharmacokinetic and pharmacodynamic models, that work well for small discrete molecules, very problematic. Clearly, *in vitro* assays also need to be chosen that generate results that are indicative of *in vivo* effects of nanomaterials (see Chapter 4 for a critical discussion).

Mechanism-based toxicity models are the ideal way of understanding and predicting likely biological effects. However, given the significant increase in complexity of nanoparticles and their interaction with their environment and the paucity of knowledge described above, it is clear that very robust modeling tools are required. The modeling methods summarized in the following sections fall into two main types. First are those that provide precise information on likely modes of toxicity (quantum chemical and molecular dynamics) but which cannot deal with large data sets or make predictions about most *in vitro* or *in vivo* effects. The second are the pattern recognition or machine learning methods (QSAR, classifiers and fingerprints) that provide limited mechanistic information but can extract useful information from large data sets and make predictions of properties of new nanomaterials not yet assessed experimentally.

EXAMPLES OF NANOTOXICOLOGY MODELING APPROACHES

There have been a very small number of publications reporting models for the adverse biological effects of nanoparticles, in spite of the importance of such models to regulators. Clearly, the issues discussed in the previous section, coupled with the relatively small number of expert molecular-level modeling groups in the materials area, are the main reasons for this dearth of published work. Four main approaches have been taken in the literature:

- the use of quantum chemical methods to understand the fundamental principles of nanoparticle structure, surface properties, photocatalytic effects, and likely toxicity
- statistical modeling or machine learning (QSAR) methods applied to existing nanotoxicological data
- development of nanotoxicological fingerprints or indices to classify groups of nanomaterials with similar biological properties
- the application of molecular modeling and molecular dynamics methods to understand the kinetics and mechanisms of the interaction of nanomaterials with biological systems at the molecular level.

Recent reviews have captured some of this work, but the field is moving ahead rapidly so a summary of recent work in these four areas is provided here.

Quantum Chemical Studies of Toxicity from Engineered Nanoparticles

Barnard's group has been very active in the application of quantum chemical methods to understanding and predicting the adverse affects of nanomaterials. Recent reviews by the group provide accessible summaries of the methods and recent results [15,16]. We provide an example of a recent study of the biological effects of nanoparticles used in sunscreens. ZnO and TiO_2 nanoparticles in sunscreens are successful examples of engineered nanoparticles in increasing use in everyday life. Here the nanometer-sized particles contribute to the high sun protection factors and transparency afforded by these sunscreens when applied on the skin. This makes them much more popular with consumers compared to more conventional sunscreens which appear as "unsightly" opaque white creams when applied to the skin and explains their commercial success. However, the high surface to volume ratios of these nanoparticles means also that large numbers of potentially reactive moieties on the particles may be exposed. As Nel et al. point out [17], the decreased size of the nanoparticles could expose reactive functional groups by increasing structural defects and altering the electronic configuration through the creation of discontinuous crystal planes.

The TiO_2 nanoparticles in sunscreen can, on exposure to UV light, produce electrons and positive "holes" by excitation of valence band electrons. These electrons and holes are capable of reacting with adsorbed oxygen and water on the particle surface to produce hydroxyl, OH•, and superoxide, HO_2• free radicals [18]. These radicals can then go on to produce more reactive oxygen species (ROS) in a cascade of Fenton chemistry-like reactions.

The seminal paper in using computational simulations to estimate the toxicity due to the production of ROS from titanium dioxide (TiO_2) nanoparticles was published by Barnard [4]. In this paper, Barnard calculated the quantity of ROS production by these nanoparticles as a function of the particle diameter, and obtained a very striking "saw-toothed" graph with a maximum in the ROS production at 33 nm (Figure 5.1).

Barnard estimates the amount of ROS produced by these nanoparticles by simply counting the number of reactive sites on the most reactive faces of the two TiO_2 polymorphs, anastase [001], and rutile [110]. The fractional surface areas of these polymorphs present on a nanoparticle at a given size, temperature, and surface adsorbate is computed using a thermodynamic nanomorphology model [19,20]:

$$G_x(T) = \Delta G_f^0(T) + \frac{M}{\rho x}\langle 1 - \frac{2\sum_i f_i \sigma_{xi}(T,\mu_y)}{B_0 < R >} + \frac{P_{ex}}{B_0}\rangle \left[q \sum_i f_i \gamma_{xi}(T,\mu_y) \right]$$

where the free energy of TiO_2 in phase x at temperature T is dependent on isotropic surface stresses $\sigma_{xi}(T,\mu_y)$ external pressure P_{ex}, bulk modulus B_0, average particle radius $< R >$, and total surface-to-volume ratio q. This equation contains the information necessary for what is in effect an "environmentally sensitive" phase map of TiO_2 [20]. Density functional theory is used to calculate the energies from this equation for all the relevant facets of anatase and rutile in the presence of the adsorbates OH and H_2O.

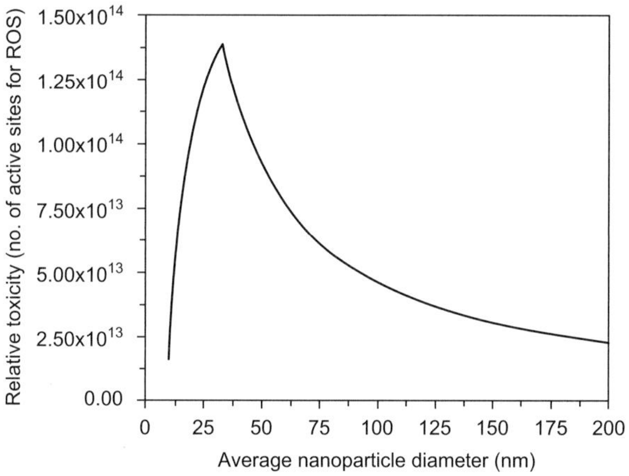

FIGURE 5.1 Relative phototoxicity of TiO_2 nanoparticles as a function of particle size. *(Reprinted from [4] with permission from Nature Publishing Group.)*

Acknowledging the fact that the reason TiO_2 nanoparticles are used in sunscreens that are commercially popular is their transparency as well as the sun protection factor, Barnard also computes the latter two quantities as a function of particle size. Combining these two quantities with the toxicity (assumed proportional to ROS levels), Barnard concludes that the TiO_2 in the sunscreen will need to be of high concentration of nanoparticles smaller than 13 nm in diameter for maximal efficacy and transparency and minimal toxicity. However, the body has significant capacity to tolerate ROS and it is only when this is overwhelmed that toxicity is observed.

The major caveat of this work is, of course, the assumption that the only source of toxicity from these nanoparticles comes from their photo-induced production of ROS; other mechanisms of toxicity or adverse effects may become operational particularly at these very small sizes. We note that the majority of published literature indicates that nanoparticles do not penetrate normal healthy skin [18,21] (and see Chapter 11), although other reports from human trials show that zinc from ZnO particles in sunscreens does enter the body but could not determine whether it was still in a nanoparticulate form [5]. However, damaged skin, hair follicles, and even flexing of normal skin are potential routes for such penetration [4,17,18]. We also note in passing that nanoparticles have been shown to penetrate the blood–brain barrier and affect the aggregation of proteins, as well as potentially causing ROS-associated damage [22]. Both of these mechanisms have been implicated in the neurotoxicity occurring in Alzheimer's disease [23]; however, this is normally caused by Cu^{2+} bound to amyloid-beta peptide reacting with endogenous reducing agents.

Statistical Modeling and Machine Learning Methods

Recently, Fourches et al. reported one of the first investigations of the potential of cheminformatics methods to establish statistically significant relationships between the biological profiles of nanoparticles and their physical, chemical, and geometric properties [24]. They studied the two data sets reported by Harvard groups. Weissleder et al. [25] used fluorescent magnetic nanoparticles to screen a library of nanoparticles decorated with various small molecules against a number of different cell lines, i.e. human umbilical vein endothelial cells (HUVEC), human pancreatic adenocarcinoma cells (PaCa2), a human macrophage-like cell line (U937), primary resting human macrophages (RestMph), and human macrophages activated by granulocyte macrophage colony stimulating factor (GMCSF_Mph). Shaw et al. [26] published a proof-of-concept study of the effects of 50 different nanomaterials at four different doses in four cell types, using four physiologic assays, a total of 3200 data points. The 50 nanoparticles possessed varying core compositions, coatings, and surface attachments. For the first data set consisting of biochemical assay data for 50 nanoparticles, they conducted a binary classification study by transforming the 64 assay results for each nanoparticle into a single value by calculating their arithmetic mean. They defined two binary classes using an arbitrary threshold to split the set into two groups each containing the same number of nanoparticles. Support vector machines were used to develop a classification model that had an external prediction accuracy of 56–88%. For the second set consisting of cellular uptake data for 109 nanoparticles with different small molecules covalently attached to their surfaces, they modeled uptake of the particles using molecular descriptors and a nearest neighbors modeling approach. These models had prediction accuracies expressed as the r^2 value of 0.72 on average. This seminal paper showed that nanoparticle properties could be modeled successfully using classification and quantitative methods.

In a similar vein, Puzyn et al. [27] reported a very simple nano-QSAR model relating the toxicity of 17 types of metal oxide nanoparticles towards *Escherichia coli*. They found an excellent relationship between a single nanoparticle property, ΔH_{Me+}, and the concentration of nanoparticles causing 50% loss in viability of the bacteria, EC_{50}.

$$\log EC_{50} = 2.59 - 0.50\,\Delta H_{Me+}$$

ΔH_{Me+} is the enthalpy of formation of a gaseous cation having the same oxidation state as the metal ion structure. This descriptor combines information about two processes involved in toxicity, dissolution, and the sum of the ionization potential of the metals. The model performed very well on the training data (predicting 85% of the variance in the data), and internal and external validation sets (77 and 83% of the variance in the data).

Two of us (Epa and Winkler) have employed sparse feature selection methods, and parsimonious non-linear methods for quantitative modeling of structure–property relationships [28–30]. We conducted a more intensive study of the nanoparticle data sets from the Harvard groups using sparse non-linear methods employing Bayesian neural networks and sparse Bayesian feature selection methods reported recently in the literature [31,32]. We studied all of the data quantitatively, rather than using a single measure of biological activity (for the biochemical assay data) or a linear cluster analysis method for uptake in a single cell type as the previous modeling study did [24]. These techniques generated high quality, non-linear, predictive quantitative structure–activity models for both data sets. Apart from improving model predictivities, we also elucidated the mechanisms by which nanoparticles are taken up by cells, or influence cellular physiology [33].

The models based on data from Shaw et al. [26] describing the effect of 50 diverse hard nanoparticles on apoptosis of four cells types yielded a simple QSAR model. This explained smooth muscle apoptosis quantitatively in terms of the relaxivities and zeta potentials of the

50 nanoparticles. The models of the second nanoparticle uptake data set [25] yielded excellent quantitative nano-QSAR models for uptake in pancreatic cancer and human umbilical endothial cell (HUVEC) cell lines. Uptake in macrophages and macrophage-like cells was smaller than in the cancer and HUVEC lines, and showed little variation with the identity of the small molecules functionalizing the nanoparticle surfaces. The models were able to predict successfully the cellular uptake of nanoparticles in an independent test set of nanoparticles. Figure 5.2 shows the performance of the pancreatic cancer nanoparticle uptake model for 109 nanoparticles.

These three studies have shown that statistical modeling and machine learning methods embodied in QSAR methods are capable of modeling data on the biological effects of nanoparticles, sometimes with relative simplicity. One reason for this may be that methods such as neural networks are capable of modeling very complex biological phenomena using pattern recognition, albeit with loss of mechanistic clarity.

Nanotoxicological Fingerprints or Indices to Classify Nanomaterials

Regulatory agencies assessing the risk of chemicals in the workplace, marketplace or environment use classification schemes to group chemicals with similar biological profiles. Two groups have reported studies aimed at generating similar classification thresholds or indices for nanomaterials. If suitable indices could be found, the hazard (intrinsic toxicity) and risk (intrinsic toxicity and likely exposure) assessment of nanomaterials for regulatory purposes could be simplified. While the two recent reports represent the first attempts to achieve these aims, and they relate to relatively narrow biological effects, they show that such a scheme may be achievable.

Xia et al. described an index for characterizing adsorption of small molecules onto the surfaces of nanoparticles [34]. They defined five indices or descriptors that described the adsorption properties of 28 small organic molecules onto multiwalled carbon nanotubes (MWCNT).

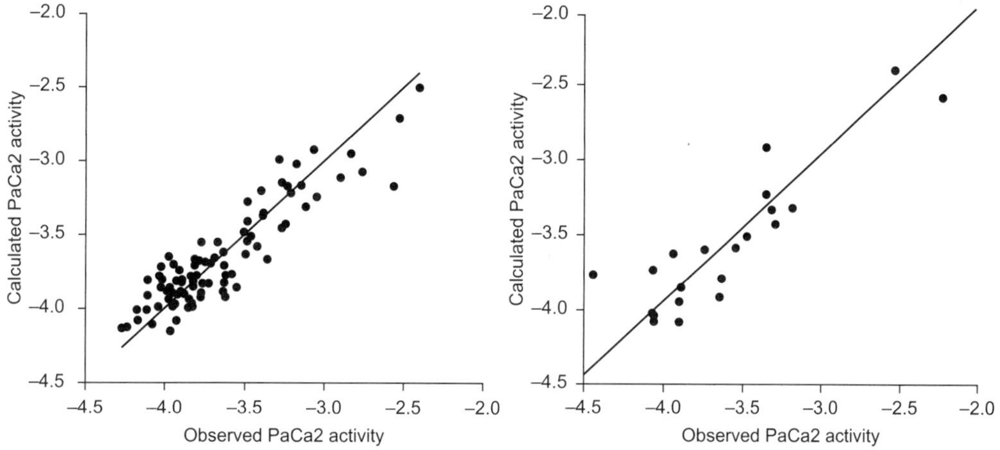

FIGURE 5.2 Cellular uptake of nanoparticles. Performance of the PaCa2 (human pancreatic adenocarcinoma) cell uptake model for the training set (left) and test set (right) of nanoparticles.

The indices described a wide range of physical interactions between the molecules and the nanotube surface: molar refractivity R, the lipophilicity π, effective hydrogen bond acidity α, effective hydrogen bond basicity β, and the molecular volume V. The aim of the study was to define a set of "probe" molecules whose interactions with nanoparticles could be used as indices or classification descriptors for predicting various specific biological effects of nanoparticles. The study effectively reported a quantitative structure–property relationship that could predict 93% of the variance in the data used to build the model, and 88% of the variance in an independent test set of molecules not used in training. The adsorption of the small molecules onto the MWCNTs correlated strongly with the lipophilicity (logP octanol/water partition coefficient).

More recently, Liu at al. described classification nanoSAR models for the effects of nine metal oxide nanoparticles against bronchial epithelial cells. They used a high throughput screening approach to generate the data at nanoparticle concentrations between 0.375 and 200 mg/L, and a set of 14 intuitive physicochemical descriptors to model the toxicity [35]. Logistic regression was used to determine the probability that a given nanoparticle at a specified concentration will be toxic or non-toxic. The best performing model, containing only three descriptors, had 100% classification accuracy for both internal and external validation data sets. These key descriptors were the nanoparticle size and the nanoparticle metal period and oxide atomization energy. The authors propose that, although this simple model performed well, a much larger data set would be required to increase the domain of applicability (range of nanomaterials types) and confidence of data-driven nanoSAR models.

Molecular Modeling and Molecular Dynamics Methods

Molecular dynamics is a physics-based modeling method that provides detailed information on the fluctuations and conformational changes of atoms and molecules in materials. Liu and Hopfinger used molecular dynamics simulations to explore four possible sources of cellular toxicity due to the insertion of a carbon nanotube into a dimyristoylphosphatidylcholine (DMPC) membrane bilayer [36]. Comparisons of (1) the structural organization of the membrane bilayer, (2) dynamical features of the membrane bilayer, and (3) transport of small polar molecules across the membrane bilayer were carried out with, and without, a carbon nanotube inserted into the bilayer (Figure 5.3). A fourth study was performed to determine how the transport of solvated ions through the inserted nanotube might alter the structure of the membrane bilayer.

Other Modeling Approaches for Nanotoxicology

Other quantitative approaches exist for modeling the biodistribution of internalized nanomaterials and the consequential adverse responses. They are the physiologically-based-pharmacokinetics (PBPK) and pharmacodynamics (PBPD) models [37]. The PBPK models described the pathways from exposure (via inhalation, ingestion or dermal exposure) to the internal dose and the distribution of this internal dose beyond the portal of entry—the lung, gut, and skin—to secondary organs. The PBPD models describe the pathways from the deposited dose at the target tissues to the responses at the cellular, molecular level. These models are based mainly on data from animal experiments. However, recent advances in *in vitro* techniques have allowed the estimation of some key model parameters (e.g. partition coefficients).

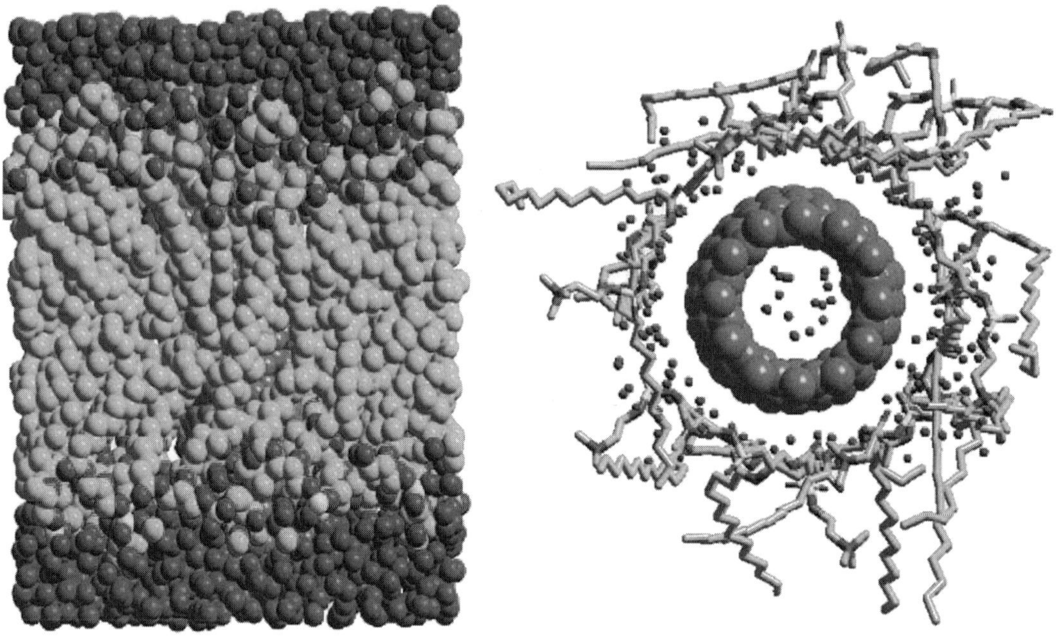

FIGURE 5.3 Effect of a carbon nanotube inserted into a phospholipid bilayer (left), showing exclusion of lipids from the zone around the nanotube (right). *(Used with permission from ACS [36].)*

These models have captured the most important aspect of toxicology, namely—the dose that causes the effect. The subsequent dosimetric question, "what metric(s) can best describe the dose?", will help in integrating QSAR and PBPK models. Indeed, since the kinetics of nanoparticles in the body is also dependent on their physicochemical characteristics, it is also possible to use QSAR techniques to determine the kinetics parameters of PBPK models.

PBPK/PD models can be extrapolated from their basis in animal data to humans by means of allometric scaling [38]. The resulting "human-based" model can be used for risk assessment. The model can be further extended to capture the uncertainty and variations observed in the data. This is done by assigning different probabilistic distributions to the model parameters resulting in a probabilistic distribution for each model output instead of a point prediction as in the case of a deterministic model [39].

PBPK models are also used to extrapolate to lower exposure levels and derive a no effect level of exposure. This derived no effect level (DNEL) is compared to the measured exposure level (in the workplace or ambient air) in order to assess risk. The integration of QSAR and PBPK/PD will help to extend PBPK/PD models to a wider class of engineered nanomaterials.

FUTURE OF NANOTOXICOLOGY

The few publications summarized above provide substantial encouragement that computational modeling methods can provide a valuable and synergistic element to the understanding and prediction of biological effects, particularly adverse effects, of engineered nanomaterials. Even at this early stage, it has been shown that several of the computational tools that have been invaluable for design of biologically active molecules, or predicting their properties or toxicities, are useful for modeling properties of nanomaterials. A recent review has summarized the still relatively small amount of published work on the application of machine learning and statistical modeling tools to complex non-biological materials more generally [40]. This reinforces the likelihood that these and the quantum chemical and molecular dynamics methods exemplified above will find a useful role in assessing the risk of nanomaterials in the workplace and environment.

There are a number of potential roadblocks and challenges that computational nanotoxicology needs to address. Statistical modeling and machine learning methods require nanomaterials to be represented mathematically by a set of *descriptors*, mathematical objects that encapsulate the molecular or physicochemical properties of nanomaterials. Development of new *nano-specific* descriptors is a major research need. Unlike chemicals, the structures of nanomaterials change in an environment-specific manner. Nanoparticles may be coated with ions, proteins, or other molecules like humic substances. The physical form nanoparticles adopt in a given environment is poorly understood. Modeling biological effects of nanomaterials requires measurement or prediction of the actual biologically active entity. QSAR approaches can play a valuable role in modeling and predicting the *in situ* form of nanoparticles. Similar computational methods can help predict the *in vivo* effects of nanomaterials based on the results of relevant *in vitro* assays and knowledge of the transport of nanomaterials through various biological or environmental compartments, thus extending the limited experimental *in vivo* data collected. The key requirement for using these computational methods is having sufficient relevant experimental data to allow understanding of mechanisms and transport to allow robust predictive computational models to be developed that extend these data and provide a useful framework for regulation of engineered nanomaterials.

TAKE-HOME MESSAGES

1. Nanomaterials are reaching the market place in increasing numbers but we have insufficient knowledge of their biological effects.
2. Nanoparticles are changed by the environment in which they exist, taking on a corona of ions and/or complex molecules that affects their physicochemical and biological properties.
3. Machine learning methods show great promise in modeling the changes that occur in nanoparticle structure in different environments, and in relating these to *in vitro* and *in vivo* biological responses. Lack of sufficient experimental data and need for new descriptors for some types of nanomaterials are holding back the field.
4. Other computational approaches using quantum chemistry or molecular dynamics can yield valuable insight into mechanisms of biological interactions of nanoparticles.
5. Questions of how nanoparticles enter the body, how they are distributed, metabolized and excreted are still largely unanswered.
6. If the existing resources on nanoparticle manufacture, characterization, biological screening, modeling and analysis can be harnessed by international collaboration, there

are good prospects for making useful predictions of risk profiles of nanomaterials within a decade. High throughput technologies will be pivotal in driving this progress. Predictive models will guide regulators in how to minimize adverse effects of nanomaterials on humans and the natural environment, while maximizing useful application of these novel materials.

References

[1] Hansen SF, Maynard A, Baun A, Tickner JA. Late lessons from early warnings for nanotechnology. Nat Nanotechnol 2008;3:444–7.
[2] Maynard A, Rejeski D. Too small to overlook. Nature 2009;460:174.
[3] Morris J, Willis J, De Martinis D, et al. Science policy considerations for responsible nanotechnology decisions. Nat Nanotechnol 2011;6:73–7.
[4] Barnard AS. One-to-one comparison of sunscreen efficacy, aesthetics and potential nanotoxicity. Nat Nanotechnol 2010;5:271–4.
[5] Gulson B, Wong H, McCall M, et al. Dermal absorption of ZnO nanoparticles in sunscreen using the stable isotope approach. Toxicol Lett 2008;180:S222.
[6] Sakamoto Y, Nakae D, Fukumori N, et al. Induction of mesothelioma by a single intrascrotal administration of multi-wall carbon nanotube in intact male Fischer 344 rats. J Toxicol Sci 2009;34:65–76.
[7] Takagi A, Hirose A, Nishimura T, et al. Induction of mesothelioma in p53+/− mouse by intraperitoneal application of multi-wall carbon nanotube. J Toxicol Sci 2008;33:105–16.
[8] Ahlers J, Stock F, Werschkun B. Integrated testing and intelligent assessment-new challenges under REACH. Environ Sci Pollut Res 2008;15:565–72.
[9] Nabholz JV. Environmental-hazard and risk assessment under the United-States Toxic-Substances Control Act. Sci Total Environ 1991;109:649–65.
[10] Dix DJ, Houck KA, Martin MT, Richard AM, Setzer RW, Kavlock RJ. The ToxCast program for prioritizing toxicity testing of environmental chemicals. Toxicol Sci 2007;95:5–12.
[11] Nicholson A, Sandler J, Seidle T. An evaluation of the US High Production Volume (HPV) chemical-testing programme: A study in (Ir)relevance, redundancy and retro thinking. Altern Lab Anim 2004;32:335–41.
[12] Judson R, Richard A, Dix D, et al. ACTOR—Aggregated computational toxicology resource. Toxicol Appl Pharmacol 2008;233:7–13.
[13] Brazma A. Minimum information about a microarray experiment (MIAME)—successes, failures, challenges. Sci World J 2009;9:420–3.
[14] Brazma A, Hingamp P, Quackenbush J, et al. Minimum information about a microarray experiment (MIAME)—toward standards for microarray data. Nat Genet 2001;29:365–71.
[15] Barnard AS. Computational strategies for predicting the potential risks associated with nanotechnology. Nanoscale 2009;1:89–95.
[16] Barnard AS. How can *ab initio* simulations address risks in nanotech? Nat Nanotechnol 2009;4:332–5.
[17] Nel A, Xia T, Madler L, Li N. Toxic potential of materials at the nanolevel. Science 2006;311:622–7.
[18] Brayner R. The toxicological impact of nanoparticles. Nano Today 2008;3:48–55.
[19] Barnard AS. A thermodynamic model for the shape and stability of twinned nanostructures. J Phys Chem B 2006;110:24498–504.
[20] Barnard AS, Xu HF. An environmentally sensitive phase map of titania nanocrystals. ACS Nano 2008;2:2237–42.
[21] Osmond MJ, McCall MJ. Zinc oxide nanoparticles in modern sunscreens: An analysis of potential exposure and hazard. Nanotoxicology 2010;4:15–41.
[22] Oberdorster G, Elder A, Rinderknecht A. Nanoparticles and the brain: cause for concern? J Nanosci Nanotechnol 2009;9:4996–5007.
[23] Cappai R, Barnham KJ. Delineating the mechanism of Alzheimer's disease A beta peptide neurotoxicity. Neurochem Res 2008;33:526–32.
[24] Fourches D, Pu DQY, Tassa C, et al. Quantitative nanostructure-activity relationship modeling. ACS Nano 2010;4:5703–12.

[25] Weissleder R, Kelly K, Sun EY, Shtatland T, Josephson L. Cell-specific targeting of nanoparticles by multivalent attachment of small molecules. Nat Biotechnol 2005;23:1418–23.
[26] Shaw SY, Westly EC, Pittet MJ, Subramanian A, Schreiber SL, Weissleder R. Perturbational profiling of nanomaterial biologic activity. Proc Natl Acad Sci USA 2008;105:7387–92.
[27] Puzyn T, Rasulev B, Gajewicz A, et al. Using nano-QSAR to predict the cytotoxicity of metal oxide nanoparticles. Nat Nanotechnol 2011;6:175–8.
[28] Burden FR, Winkler DA. Robust QSAR models using Bayesian regularized neural networks. J Med Chem 1999;42:3183–7.
[29] Burden FR, Winkler DA. An optimal self-pruning neural network and nonlinear descriptor selection in QSAR. QSAR Combin Sci 2009;28:1092–7.
[30] Burden FR, Winkler DA. Optimal sparse descriptor selection for QSAR using Bayesian methods. QSAR Combin Sci 2009;28:645–53.
[31] Figueiredo MAT. Adaptive sparseness for supervised learning. IEEE Transact Pattern Anal Machine Intell 2003;25:1150–9.
[32] Mackay DJC. A practical Bayesian framework for backpropagation networks. Neur Comput 1992;4:448–72.
[33] Epa VC, Burden FR, Shaw S, Tassa C, Weissleder R, Winkler DA. Modelling cellular uptake and apoptosis by nanoparticles. Nat Nanotechnol submitted 2011.
[34] Xia XR, Monteiro-Riviere NA, Riviere JE. An index for characterization of nanomaterials in biological systems. Nat Nanotechnol 2010;5:671–5.
[35] Liu R, Rallo R, George S, et al. Classification NanoSAR development for cytotoxicity of metal oxide nanoparticles. Small 2011;7:1118–26.
[36] Liu JZ, Hopfinger AJ. Identification of possible sources of nanotoxicity from carbon nanotubes inserted into membrane bilayers using membrane interaction quantitative structure–activity relationship analysis. Chem Res Toxicol 2008;21:459–66.
[37] MacCalman L, Tran CL, Kuempel E. Development of a bio-mathematical model in rats to describe clearance, retention and translocation of inhaled nano particles throughout the body. Inhaled Part X 2009;151:677.
[38] Feng MR, Lou XC, Brown RR, Hutchaleelaha A. Allometric pharmacokinetic scaling: Towards the prediction of human oral pharmacokinetics. Pharm Res 2000;17:410–8.
[39] Marino DJ, Starr TB. Probabilistic dose-response modeling: Case study using dichloromethane PBPK model results. Reg Toxicol Pharmacol 2007;49:285–300.
[40] Le TC, Epa VC, Burden FR, Winkler DA. Towards predictive modelling of diverse materials properties. Accounts Chem Res Submitted 2011.

CHAPTER

6

Regulation and Legislation

Maureen R. Gwinn[1], *Birgit Sokull-Klüttgen*[2]

[1]US Environmental Protection Agency, Washington, DC, USA

[2]European Commission, Joint Research Centre, Institute for Health and Consumer Protection, Unit IO5 Nanobiosciences, Ispra, Italy

OUTLINE

Disclaimer: The views expressed in this review are thoset of the authors and do not represent the views and/or policies of the US Environmental Protection Agency, the European Commission or a European Agency. The authors are not responsible for any use which might be made of this text.

DOI: 10.1016/B978-0-12-386940-1.00006-4

ISSUES WITH CURRENT STUDIES FOR RISK ASSESSMENT OF NANOMATERIALS

The Organization for Economic Cooperation and Development (OECD) established a Working Party on Manufactured Nanomaterials (WPMN) [1] in 2006 in order to collaborate internationally on the safety research of nanomaterials. This workgroup is a key example of the interdisciplinary nature of the work needed in nanomaterial research, including scientists (e.g. chemists, physicists, biologists and toxicologists) and regulators working on environmental health and safety, exposure, and testing of nanomaterials; in addition, the WPMN covers a vast geographical area that enables it to harvest many different approaches. Subcommittees (Steering Groups, SG) of the WPMN are looking at different issues such as databases; test guidelines and risk assessment (Table 6.1). Particularly important is SG 3 on "Testing a Representative Set of Manufactured Nanomaterials", the so-called sponsorship program, organized with a sponsor group for each of the nanomaterials under study, with various countries including industry serving as sponsors (Table 6.2). The objective of SG 3 is to develop and test a representative set of manufactured nanomaterials in accordance with the Guidance Manual for sponsors [2]. The outcome will be an understanding of the kind of information on intrinsic properties that may be relevant for hazard assessment of nanomaterials, and practical experience and evidence on whether current OECD Test Guidelines and other methods are applicable. There is also an ongoing National Toxicology Program (NTP) bioassay on a variety of nanomaterials that will inform some of the issues related to the effects of nanomaterials [3]. These efforts will help to design standardized tests, which require well-controlled and defined conditions and the availability of well-characterized manufactured nanomaterials will help to design standardized tests, which require well-controlled and defined conditions and the availability of well-characterized manufactured nanomaterials.

Various attempts to analyze nanomaterials have focused on combining the risk assessment paradigm with a life-cycle framework. The use of a comprehensive environmental assessment (CEA) entails all life-cycle stages in the context of hazard identification and risk assessment. CEA involves both qualitative and quantitative approaches to risk characterization of human or ecological risk [4]. Recent case studies and workshops have addressed the issues of nanomaterial risk assessment using this methodology[5,6].

Risk assessment is generally performed based on the chemical composition of the material in question. In the case of nanomaterials, however, this may not be sufficient as chemicals

TABLE 6.1 OECD Working Party on Manufactured Nanomaterials

SG 1/2 merged: OECD Database on Manufactured Nanomaterials to Inform and Analyse EHS Research Activities
SG 3: Safety Testing of a Representative Set of Manufactured Nanomaterials
SG 4: Manufactured Nanomaterials and Test Guidelines
SG 5: Cooperation on Voluntary Schemes and Regulatory Programmes
SG 6: Cooperation on Risk Assessment
SG 7: The Role of Alternative Methods in Nano Toxicology
SG 8: Exposure Measurement and Exposure Mitigation
SG 9: Environmentally Sustainable Use of Manufactured Nanomaterials

SG = steering group

TABLE 6.2 Nanomaterials Investigated in the "Sponsorship Programme" (SG 3) of the OECD Working Party for Manufactured Nanomaterials

Nanomaterial	Lead Sponsors	Co-Sponsors	Contributors
Fullerenes	USA/Japan	DK/China	—
SWCNT	USA/Japan	—	CDN/FR/DE/EC/China/BIAC
MWCNT	USA/Japan	Korea/BIAC	CDN/FR/DE/EC/China/ BIAC
Silver NP	USA/Korea	AUS/CDN/DE/Nordic	FR/EC/China
Iron NP	BIAC/China	—	CDN/USA/EC/Nordic
TiO_2	FR/DE	AT/CDN/ES/Korea/USA/ EC/ BIAC	China/DK/Japan/UK
Gold NP	South Africa	USA	EC/NL
Al_2O_3	USA	—	—
SiO_2	EC/FR	BE/Korea/BIAC	DK/Japan
Nanoclays	USA/BIAC	—	EC
Dendrimers	—	ES/USA	—
CeO_2	USA/UK/BIAC	AUS/NL/ES	DK/DE/CH/EC/Japan
ZnO	UK/BIAC	AUS / ES / USA / BIAC	CDN/DK / Japan

SWCNT ≡ single-walled carbon nanotubes; MWCNT ≡ multi-walled carbon nanotubes; NP ≡ nanoparticle; TiO_2 ≡ titanium dioxide; Al_2O_3 ≡ aluminium oxide; SiO_2 ≡ silicon dioxide; CeO_2 ≡ cerium dioxide; ZnO ≡ zinc oxide; AT ≡ Austria; AUS ≡ Australia; BE ≡ Belgium; BIAC ≡ Business and Industry Advisory Committee; CH ≡ Switzerland; CND ≡ Canada; DE ≡ Germany; DK ≡ Denmark; EC ≡ European Commission; ES ≡ Spain; FR ≡ France; NL ≡ The Netherlands; Nordic ≡ Nordic Council of Ministers; UK ≡ United Kingdom; USA ≡ United States of America.

generally considered non-toxic in macro-scale ranges may become toxic in the nano-scale range. The most studied example is titanium dioxide which, in the larger size range, does not appear to exhibit toxicity to humans or the environment. Exposures to nano-scale titanium dioxide have been associated with a variety of pulmonary effects in rats, including inflammation, pulmonary damage, fibrosis, and lung tumors [7,8]. Also, nano-scale titanium dioxide particles can impair macrophage function and increase pulmonary retention, enter the epithelium faster and translocate to the subepithelium space more readily than fine particles [9]. These have also been shown to lead to mitotic disturbances, DNA damage and apoptosis.

A key area of nanomaterial safety research is characterization of physicochemical properties. While most researchers agree that characterization of nanomaterials is a necessary step in toxicological analysis, needs regarding the level, type and timing of characterization are not as clear. The increasing usage of nanomaterials takes advantage of the differences between these materials and their larger scale counterparts. Research has focused on many of these characteristics and their potential role in toxicity generally compared to their larger scale counterparts. Studies have examined the impact of their smaller size, increased surface area and increased surface reactivity, and some examples in relation to environmental and toxicological behavior are given (see Chapters 1 and 2).

REGULATION AREAS FOR NANOMATERIALS

As nanomaterials are used in an increasing range of high-tech applications and consumer products there is growing concern about their safe use and possible impact on human health and the environment. Current and future legislation will have to take this into account and address nanomaterials more specifically. To date, there is no internationally accepted definition for nanomaterials, and the question remains whether or not a definition for regulatory purposes is needed and useful [10].

In the recent update of the US National Nanotechnology Initiative Environmental Health and Safety Research Strategy [11], nanomaterials are defined to include more specifically all materials *designed* to have at least one dimension less than 100 nm and exhibit unique qualities determined by this size. This definition may not apply to regulatory statutes or policies relevant to nanotechnology.

In the European Union, the European Parliament considered it important to address nanomaterials explicitly within the scope of certain legislation and called for the introduction of a comprehensive science-based definition of nanomaterials in community legislation [12]. Based on a Reference Report on "Considerations on a Definition of Nanomaterial for Regulatory Purposes" [13] of the European Commission's Joint Research Centre (JRC) and the opinion of the Scientific Committee on Emerging and Newly Identified Health Risks (SCENIHR) [14], entitled "Scientific Basis for the Definition of the Term 'Nanomaterial'" [15], the European Commission published a recommendation on the definition of nanomaterial [16]: "'Nanomaterial' means a natural , incidental or manufactured material containing particles, in an unbound state or as an aggregate or as an agglomerate and where, for 50% or more of the particles in the number size distribution, one or more external dimensions is in the size range 1 nm−100 nm." In specific cases and where warranted by concerns for the environment, health, safety or competitiveness the number size distribution threshold of 50% may be replaced by a lower threshold. Fullerenes, graphene flakes and single wall carbon nanotubes with one or more external dimensions below 1 nm should also be considered as nanomaterials. The terms 'particle', 'agglomerate' and 'aggregate' are defined. For a further discussion of the definition of nanomaterials, see Chapter 1.

Some of the key questions surrounding the regulation of nanomaterials relate to whether or not these should be considered a new material, or if they can be regulated based on their macroscale counterparts (if these exist). The following section will describe the current legislation that may play a role in the regulation of nanomaterials in the USA (Table 6.3) and the European Union.

THE USA

Nanotechnology policy in the USA is primarily coordinated through the National Science and Technology Council [17] through the Nano-scale Science, Engineering, and Technology (NSET) Subcommittee [18]. The NSET Subcommittee and its working groups, with support from the National Nanotechnology Coordination Office (NNCO), prepare the multiagency planning, budget and assessment documents related to the National Nanotechnology Initiative (NNI). NNI is a United States government-wide effort to coordinate federally-funded

TABLE 6.3 Legislative Authorities that may Encompass Nanomaterial Usage in the USA

Regulation	Description	Potential Use	Governing Body	Website
Toxic Substances Control Act (TSCA)	TSCA provides EPA with the authority to require reporting, record-keeping and testing requirements, and restrictions relating to chemical substances and/or mixtures	Proposed case-by-case determination of nanomaterials as new materials	United States Environmental Protection Agency (US EPA)	http://www.epa.gov/lawsregs/laws/tsca.html
Federal Insecticide, Fungicide and Rodenticide Act (FIFRA)	FIFRA provides the basis for regulation, sale, distribution and use of pesticides in the USA	Determination of unreasonable adverse human health risks with the addition of nanomaterials to existing pesticides	United States Environmental Protection Agency (US EPA)	http://www.epa.gov/opp00001/regulating/laws.htm
Safe Drinking Water Act (SDWA)	The main federal law that ensures the quality of America's drinking water	Currently no maximum contaminant level goals set for nanoparticles	United States Environmental Protection Agency (US EPA)	http://www.epa.gov/OGWDW/sdwa/
Comprehensive Environmental Response, Compensation, and Liability Act and Resource Conservation and Recovery Act (CERCLA/RCRA)	Commonly known as Superfund, CERCLA was enacted in 1980 and created a tax on the chemical and petroleum industries and provided a broad Federal Authority to respond directly to releases or threatened releases of hazardous substances that may endanger public health or the environment	This would cover the evaluation and control of nanomaterials in waste sites	United States Environmental Protection Agency (US EPA)	http://www.epa.gov/superfund/policy/cercla.htm
Clean Air Act (CAA)/Clean Water Act (CWA)	The Clean Air Act and Clean Water Act both allow the EPA to set limits on certain air and water pollutants in the USA. Under these Acts, the EPA also review and approve permit applications for industries or chemical processes	Release of nanomaterials into the air or water would fall under these Acts	United States Environmental Protection Agency (US EPA)	http://www.epa.gov/air/caa/peg/understand.html http://www.epa.gov/oecaagct/lcwa.html

(*Continued*)

TABLE 6.3 Legislative Authorities that may Encompass Nanomaterial Usage in the USA—cont'd

Regulation	Description	Potential Use	Governing Body	Website
Food, Drug and Cosmetic Act (FDCA)	The FDCA is a set of laws giving FDA the authority to oversee the safety or food, drugs and cosmetics in the USA	Already in use for premarket testing of food additives, this may cover regulation of the use of nanomaterials in food additives and cosmetics	United States Food and Drug Administration (US FDA)	http://www.fda.gov/nanotechnology/regulation.html
Occupational Safety and Health Act of 1970 (29 US 654)	Section 5(a)(1) of the OSH Act requires employers to "furnish each of his employees employment and a place of employment which are free from recognized hazards that are causing or likely to cause death or serious physical harm to his employees"	OSHA has approved plans for 24 states to adopt federal safety standards for workers exposed to nanomaterials in private industry	United States Occupational Safety and Health Administration (OSHA)	http://www.osha.gov/dsg/nanotechnology/nanotech_standards.html

nanomaterial research focusing on multiple topics, including metrology, environmental health and safety and exposure to nanomaterials. Coordination of efforts is an important step in making sure the most useful research is done in an efficient manner.

The NNI releases an Environmental, Health and Safety Strategy document approximately every 2 years to provide guidance to federal agencies involved in the development of nanotechnology. Written by members of the various federal agencies involved in NNI, these strategies are based on the state of the science, ongoing research activities and input from the public through workshops and public review [11]. The NNI is not a regulatory body; however, one goal of the NNI is to give recommendations on the research needed for risk assessment and risk management of nanomaterials to ensure responsible development of nanotechnology.

Environmental/Human Health

Toxic Substances Control Act (TSCA)

The Toxic Substances Control Act (TSCA) [19] is the main legislative authority under which the US EPA requires manufacturers of new chemicals to submit specific information on the effects of these materials [20]. Under TSCA, the US EPA has taken steps to limit the exposure to nanomaterials, including carbon nanotubes (CNTs). The EPA has required the use of personal protective equipment and limitations on the use and environmental exposures of nanomaterials. The US EPA's Office of Pollution Prevention and Toxics recently released a fact sheet describing the plans to study nanomaterials on a case-by-case basis as new chemicals, rather than existing chemicals.

Under TSCA, the US EPA has reviewed new nano-scale chemical premanufacture notices for carbon nanotubes, nanosilver and other materials required prior to the manufacture or introduction of new chemicals. This process allows for the US EPA to control and/or limit exposures to these chemicals through a variety of mechanisms (e.g. limiting use and release of materials, requiring use of personal protective equipment). The US EPA has permitted the manufacture of new nano-scale chemical materials through the use of the Significant New Use Rules (SNUR) under TSCA. Through this process, the US EPA is provided with a basic set of information on nano-scale materials, including the chemical properties, toxicity data and production volume, which would allow the US EPA to determine an unreasonable risk of these new materials to human health or the environment.

Federal Insecticide, Fungicide, and Rodenticide Act (FIFRA)

Nanomaterials added to pesticides would fall under the mandates of the Federal Insecticide, Fungicide, and Rodenticide Act (FIFRA) to determine whether their addition leads to adverse health risks [21]. Under FIFRA, the US EPA held its first Science Advisory Panel (SAP) review of nanomaterials in November 2009, specifically examining nanosilver and other nanometal pesticide products. The panel was designed to address issues related to whether nanosilver and other nanometals in the 1–100 nm have unique behavior under conditions relevant to human health and ecological risk. The SAP review focused mainly on use of nanosilver in pesticide products, with recommendations on how to conduct risk assessments of pesticide products containing nanosilver, particularly related to what types of data would be needed to evaluate the risks. The panel determinations related to the role of the silver ion release in the toxicity of silver (macro-scale and nano-scale), and the lack of available data demonstrating any variability in the release of silver ions from nanomaterials as opposed to macro-scale silver, and the specific biological impacts of particle size (both individual or agglomerated particles). In general, for risk assessment, the SAP review recommended the need for a case-by-case evaluation of pesticide products containing nano-silver [22].

Other Regulations

There are other environmental regulations that may play a key role in regulating the use and disposal of nanomaterials. The Comprehensive Environmental Response, Compensation, and Liability Act (CERCLA) and Resource Conservation and Recovery Act (RCRA) would be used to evaluate the risks of nanomaterials at waste sites, although this has not yet been performed [23]. The 1990 Clean Air Act Amendments (CAAA) [24] and the 1996 Clean Water Act Amendments (CWAA) [25] would be used for regulation of nanomaterials in ambient air or ambient water and wastewater. For CAAA, the levels transported in air would be regulated based on their danger to public health. The CWAA would regulate the effluent limits of nanomaterials in the wastewater.

Nano-Scale Materials Stewardship Program (NMSP)

The Nano-scale Materials Stewardship Program (NMSP) [26] of the US EPA was launched in 2008 in order voluntarily to acquire information on the use and development of nanomaterials. However, to date, limited information has been acquired through this program. Following the first year of the program, information on only approximately 100 nanomaterials has been submitted, mostly from smaller companies with limited health and safety data.

Food Safety and Medical Applications

The Food, Drug and Cosmetic Act (FDCA) has been used by the FDA to address the impact and use of nanomaterials in food, including additives and contaminants, medicines and cosmetics [27]. The FDCA requires premarket testing for food and color additives, as well as drugs, but not for cosmetics. The FDA can require information on the identity and properties of materials, regardless of the particle size of a material. The FDA also regulates cosmetics, including sunscreens, e.g. the use of nano-sized titanium dioxide and other nanomaterials in sunscreens. The FDA generally does not require premarket safety data on cosmetics.

There are currently no regulations specific for nanomaterials related to food or drug safety. Currently, all medical products that contain nanomaterials go through the traditional regulatory process and include requirements for clinical, toxicological and other studies for determining the safety and efficacy of these materials.

Consumer Products

The use of nanomaterials in consumer products in the USA is under the oversight of the Consumer Product Safety Commission (CPSC) [28]. The CPSC mission is protecting the public from unreasonable risks of serious injury or death from the products under their jurisdiction. Although there is no specific ruling for nanomaterials, if any of these products contain nanomaterials, they would fall under the regulation of CPSC.

Occupational

The purpose of the Occupational Safety and Health Act of 1970 [29] is to assure safe and healthy working conditions in all occupational settings, generally based on recommendations from the National Institute for Occupational Safety and Health (NIOSH) and enforced by the Occupational Safety and Health Administration (OSHA). Generally, occupational exposure to nanomaterials falls under OSHA's General Industry standards. OSHA has approved plans for 24 states to adopt federal safety standards for workers exposed to nanomaterials in private industry [30].

In 2011, NIOSH released a Current Intelligence Bulletin on the scientific hazards of occupational exposure to titanium dioxide (TiO_2), including engineered nano-sized particles [31]. This bulletin performed a risk assessment of the currently available scientific information on TiO_2, including experimental animal toxicity and human health data, dose–response information and exposure monitoring techniques and control strategies. This NIOSH bulletin recommends exposure limits for both fine TiO_2 ($2.4\,mg/m^3$) and ultrafine TiO_2 (including engineered nano-scale TiO_2; $0.3\,mg/m^3$) as time-weighted average (TWA) concentrations for up to 10 hours per day during a 40-hour work week. Further, NIOSH has categorized ultrafine TiO_2 as a potential occupational carcinogen, but there was insufficient data to conclude on carcinogenicity of fine TiO_2.

Recent examples of occupational exposures and human health effects reported in the literature include two pulmonary disease occurrences in conjunction with exposure to nanomaterials: workplace exposures to multiple chemicals with limited ventilation and personal protective equipment [32] and exposures of volunteers to dust following the World Trade Center collapse on September 11, 2001 [33]. Although limited, these are examples of potential adverse human health effects following exposure to nanomaterials in the workplace. However, these two examples demonstrate the need for clear investigation of the causal factors for these events which can provide insight into whether production and use of nano-scale engineered materials are creating undue risk that needs to be controlled.

EUROPEAN UNION

The main legal instruments in the European Union (EU) are Regulations and Directives. A regulation is directly applicable in all the Member States, while a Directive is addressed to the Member States and leaves them the choice how to adopt the Community objectives within their national legal order. EU legislation is accessible and searchable on-line at http://EUR-LEX.europa.eu/.

With respect to regulatory aspects of nanomaterials within the EU, the Commission of the European Communities (the European Commission) undertook in 2008 a first regulatory review of EU legislation in relevant sectors. The outcome is available in a Communication from 2008 [34], which comes to the result that: "Overall, it can be concluded that current legislation [in the EU] covers to a large extent risks in relation to nanomaterials and that risks can be dealt with under the current legislative framework. However, current legislation may have to be modified in the light of new information becoming available, for example as regards thresholds used in some legislation". It was recognized that knowledge on essential questions such as characterization of nanomaterials, their hazards, exposure, risk assessment and risk management needs to be improved.

Regulatory aspects of nanomaterials were also discussed in the European Parliament. In its resolution of 2009 [12], the parliament considered it important to address nanomaterials explicitly within the scope of certain legislation for areas such as chemicals, biocides and food. Besides the introduction of a definition of nanomaterials in Community legislation, the parliament called to compile an inventory of the different types and uses of nanomaterials on the European market, and for better information to consumers on the use of nanomaterials in consumer products: all ingredients present in the form of nanomaterials should be clearly indicated on the label of the product.

The European Commission (July 2009) informed the Parliament that it intends to review relevant legislation and instruments of implementation and will propose regulatory changes where necessary, agrees on the need to develop a definition, and intends to present information on types and uses of nanomaterials [35]. These reports are planned for the end of 2011.

The following section summarizes the main outcome of the above mentioned first review by the European Commission concerning legislation relevant for health, safety and environment of nanomaterials [34], using also information from an accompanying Commission staff working document [36], which provides additional details, and information from other relevant documents, updated where necessary. Further reviews and other activities are expected by 2011/2012, and thus the amount of information is increasing in a very dynamic way.

Chemicals

Industrial Chemicals

The REACH (Registration, Evaluation, Authorisation and Restriction of Chemicals) Regulation [37] applies to chemical substances on their own, in preparations or in articles. It is based on the principle that manufacturers, importers and downstream users have to ensure that they manufacture, place on the market or use substances that do not adversely affect human health or the environment. Its provisions are underpinned by the precautionary principle. The "European Chemicals Agency" (ECHA) [38] based in Helsinki (Finland) is responsible for the implementation of REACH. Guidance documents [39] for industry and authorities support the implementation.

There are no provisions in REACH referring explicitly to nanomaterials. However, nanomaterials are covered by the "substance" definition in REACH: "Substance means a chemical element and its compounds in the natural state or obtained by any manufacturing process, including any additive necessary to preserve its stability and any impurity deriving from the process used, but excluding any solvent which may be separated without affecting the stability of the substance or changing its composition" (Article 3 of the REACH Regulation). Thus, REACH addresses chemical substances, in whatever size, shape or physical state [40]. Substances at the nano-scale are therefore covered by REACH and its provisions apply.

Under REACH, manufacturers and importers have to submit a registration dossier to ECHA for substances that they manufacture or import at or above 1 tonne per year. REACH has staggered information requirements, increasing at the thresholds at or above 10, 100 and 1000 tonnes/year. At or above 10 tonnes/year, the registrant is also required to submit a chemical safety report.

If a nanomaterial is considered to be a nanoform of a non-nano (bulk) substance, it can be registered within the registration dossier of this substance; the dossier must then contain all relevant information for all forms of the substance. The *total* volume (i.e. the sum of the nano and the non-nano volume) determines the need and the timing for the registration and the information requirements. The registration dossier has to include all relevant information on the bulk and the nanomaterial covering the properties, uses, effects and exposure related information. REACH does currently not include specific information requirements for nanomaterials; however, as the listed information is stated to be minimum information requirements (REACH Art. 12(1)) additional testing or information to determine the specific properties and hazards of the nanomaterial may therefore be required. Furthermore, standard test guidelines may not be fully suitable for the testing of nanomaterials. The information on the nanoform needs to be adequate for its Chemical Safety Assessment, if required (>10 tonnes/ga), as well as for the classification and labeling (according to CLP Regulation) [41]. Any different classification and labeling of the nanoform and additional risk management measures, if needed, have to be included in the registration dossier and have to be communicated to the supply chain via the safety data sheet [42]. If a nanomaterial is considered to be a substance on its own, a registration dossier addressing the nanomaterial needs to be submitted.

While the content of the registrations dossier is the responsibility of the registrant (Industry), a minimum of 5% of the dossiers will be evaluated by ECHA to ensure compliance with REACH (dossier evaluation). Furthermore, EU Member State Authorities may examine any information on a substance (substance evaluation). If deemed necessary, ECHA can require any information on the substance, i.e. nanomaterial, in addition to the standard information requirements of REACH. Authorizations for substances of very high concern or restrictions for substances which pose an unacceptable risk are the regulatory instruments for risk management under REACH; they are independent from tonnage triggers or registration duty and could therefore also apply to nanomaterials of low tonnage levels as for any other substance.

In conclusion, nanomaterials are substances and thus covered by the REACH Regulation, though there are no provisions in REACH referring explicitly to nanomaterials. Furthermore, there is currently no guidance for industry (or authorities) on how to implement REACH for nanomaterials. Therefore, the European Commission initiated REACH Implementation Projects for Nanomaterials (RIP-oNs) to provide scientific and technical advice on key aspects of the implementation of REACH with regard to nanomaterials, in particular on substance identification, information requirements, and chemical safety assessment. The final reports of these projects can be downloaded from the following website: http://ec.europa.eu/environment/chemicals/nanotech/index.htm#ripon. They will provide advice on how the current REACH guidance documents could be updated.

As for all substances, REACH is fundamental to obtain information on properties, including hazardous properties, of nanomaterials, and the CLP Regulation is essential to transfer this knowledge along the supply chain. The resulting hazard identification has consequences for many other legal instruments such as occupational and environmental legislation.

Plant Protection Products and Biocidal Products

The placing on the EU market of Plant Protection Products (PPP) and Biocidal Products is regulated by the Plant Protection Product Regulation [43] and the Biocidal Products Directive [44]. The PPP Regulation covers classical "'pesticides" such as insecticides and fungicides used in agriculture by professional users. The Biocidal Products Directive covers 23 different product types, varying from disinfectants to antifouling agents and rodenticides, and the biocidal products may be used by both professional users and by consumers. As PPPs and Biocidal Products contain active substances, which due to their nature have the potential to be toxic and/or ecotoxic, they are evaluated and approved at EU level (i.e. inclusion in Annex I). Only PPPs and Biocidal Products containing these approved active substances can be authorized.

The Biocidal Products Directive uses the same and the PPP Regulation a similar substance definition as REACH. Thus, both legal instruments address chemical substances, in whatever size, shape or physical state, and substances at the nano-scale are therefore covered.

Nanomaterials could also be used as co-formulants in PPPs and Biocidal Products. Full composition of the products needs to be presented when applying for product authorization, and additional data can be requested for substances of concern, and certain co-formulants can be prohibited in case of concern for human and animal health and/or the environment.

The implementation of the PPP Regulation on EU level is guided and monitored under the responsibility of the "European Food Safety Authority" (EFSA) [45] located in Parma (Italy). The scientific aspects and the evaluation of the active substances for the Biocidal Products Directive is currently managed by the "Institute for Health and Consumer Protection" of the European Commission's Joint Research Centre (JRC-IHCP) [46], but will be handed over in the coming years to ECHA [38].

In conclusion, nanomaterials are covered by the PPP Regulation and the Biocidal Products Directive, although there are no provisions referring explicitly to nanomaterials. Therefore, current guidance documents (on data requirements, risk assessment and decision making) may need to be reviewed and amended where relevant in order to highlight that the legislative acts cover nanomaterials and to ensure that all aspects of risks of nanomaterials are properly addressed. It should be mentioned that the Biocidal Products Directive is currently being reviewed and the revised version is expected specifically to address nanomaterials.

Worker Protection

Framework Directive 89/391/EEC on the introduction of measures to encourage improvements in the safety and health of workers [47] places a number of obligations on employers to take measures necessary for the safety and health protection of workers. It applies to all substances and work activities including manufacturing and use of chemicals at all levels of the production process. Although this Directive does not specifically address nanomaterials, it fully applies to nanomaterials. Employers, therefore, must carry out a risk assessment and, where a risk is identified, take measures to eliminate this risk. The Directive foresees the possibility to adopt individual "daughter Directives" laying down more specific provisions with respect to particular aspects of safety and health. Relevant directives thus adopted relate, for example, to risks related to chemical agents at work [48], the use of work equipment by workers at work [49], and the use of personal protective equipment at the workplace [50]. As these directives introduce minimum requirements, national authorities have the possibility to introduce more stringent rules.

Directive 98/24/EC on the protection of the health and safety of workers from the risks related to chemical agents [48] is aimed at controlling and reducing occupational exposure risks. This is achieved, among others, by establishing indicative and binding occupational exposure limit (OEL) values, as well as biological limit values for specific substances. The term "occupational exposure limit value" is defined in the Directive, article 2(d), as "... the limit of the time-weighted average of the concentration of a chemical agent in the air within the breathing zone of a worker in relation to a specified reference period". In this directive the term "biological limit value" is defined as "... the limit of the concentration in the appropriate biological medium of the relevant agent, its metabolite, or an indicator effect". While OEL values address the concentration of the chemical in the air inhaled by workers, the biological limit is an internal dose of a chemical or its metabolite for which the route of entry is not defined (so it may be oral, dermal and/or inhalation). OEL values are given as 8-hours time weighted average for airborne concentrations and/or as a short-term period of 15 minutes. Directive 98/24/EC has been amended several times including lists with specific substances for which limit values have been agreed. Specific OEL values for nanoparticles do not yet exist. Currently, the limit values for amorphous silica ($2-6\ mg/m^3$) [51] and silver [52] are also applicable to their nanoforms, although it would be relevant to confirm whether these mass-based OELs established for "bulk" substances would be appropriate for nanomaterials.

The European Agency for Safety and Health at Work (EU-OSHA) [53] located in Bilbao in Spain, which deals with nanomaterials and worker protection on Community level, has published a series of expert forecasts on emerging risks for the working environment including a report on chemical risks related to occupational safety and health [54]. The experts agreed that nanoparticles and ultrafine particles pose the most significant emerging risk.

Furthermore, EU-OSHA has published a literature review on workplace exposure to nanoparticles [55]. This report focuses on exposure measurements and assessments, health effects of nanomaterials, and possible subsequent activities to manage the risk.

Consumer Products

Products legislation lays down requirements regarding products, such as plant protection products, biocidal products, cosmetic products and medical products. Consumer products not covered by specific regulations are covered by the Directive on General Safety of Consumer Products [56]. In many cases, more than one piece of legislation applies to the same product. Where regulation contains requirements of a general nature, they will cover also risks related to nanotechnology, even if they have been adopted without specifically intending to address risks associated with nanomaterials and nanotechnologies.

Cosmetic Products Regulation

The Cosmetic Products Regulation from December 2009 is the first legal act in the EU which includes specific provisions for nanomaterials [57]. Most of these provisions will come into force in 2013.

A "cosmetic product" means any substance or mixture intended to be placed in contact with the external parts of the human body (such as skin, hair, nails and lips) or within the oral cavity with a view to cleaning, perfuming, changing the appearance, protecting or keeping in good condition.

The introduction of provisions specific to nanomaterials requires the adoption of a definition. Thus, the new Cosmetic Products Regulation introduced the definition: "'nanomaterial' means an insoluble or biopersistent and intentionally manufactured material with one or more external dimensions, or an internal structure, on the scale from 1 to 100 nm". This definition may be revised in line with the Commission recommendation on the definition of nanomaterial [16]. Article 16 of the Regulation concerns nanomaterials. Cosmetic products containing nanomaterials shall be notified 6 months prior to being placed on the market. This notification shall contain additional information on the nanomaterial such as the specification of the nanomaterial including size of particles; physical and chemical properties; an estimate of the quantity of nanomaterial contained in cosmetic products; the toxicological profile of the nanomaterial; safety data and the reasonably foreseeable exposure conditions. Furthermore, the Cosmetic Products Regulation includes a labeling obligation for ingredients present in the form of nanomaterials, i.e. in the list of ingredients the names of such substances shall be followed by the word *nano* in brackets. This is not meant to be a hazard labeling, but will allow consumers to make an informed choice.

The Scientific Committee on Consumer Safety (SCCS) [58] provides guidance for the testing of cosmetic ingredients and their safety evaluation. The latest revision of its Notes of Guidance [59] addresses besides many other issues also basic requirements for nanomaterials.

Food Legislation

Regulation 178/2002 lays down the general principles and requirements of food law, establishes the European Food Safety Authority (EFSA) and applies horizontally to all foods and feed [60]. More particularly relevant in relation to risks associated with nanomaterials are the following Regulations and Directives:

- Novel Food Regulation No 258/97 concerning novel foods and novel food ingredients [61]
  "Novel food" is food not consumed to any significant degree in the EU prior to May 1997 (when the first Novel Foods legislation entered into force). "Novel food" can be newly developed, innovative food, food produced using new technologies and production processes or food which has been traditionally consumed outside of the EU. A revision of the Novel Food Regulation from 1997, including specific requirements for food containing nanomaterials (such as a labeling requirement) was discussed, but has not been endorsed.
- Food additives
  Directive 89/107/EEC on food additives [62] covers substances added to food for a "technological purpose in the manufacture, processing, preparation, treatment, packaging,

transport or storage of such food". Some of the currently used particulate food additives, such as anticaking powders (SiO_2, $CaCO_3$) may contain a fraction in the nano-size.

- Food supplements
  Framework Directive 2002/46/EC (and its daughter Directives) on food supplements provides a definition of food supplements, specific rules on labeling, presentation and advertising of these products and regulates their compositions in vitamins and minerals [63]. Positive lists of vitamins and minerals and of their sources that can be used in food supplements are given in the Annexes of the Directive. Prior to the authorization and inclusion in the positive lists, every vitamin and mineral and/or every source needs to undergo a safety assessment carried out by EFSA. It can be envisaged that substances produced in "nano" form or packed into nanoglobules (such as vitamins) to improve their solubility or bioavailability could be used as food supplements.
- Food contact materials
  Regulation (EC) 1935/2004 relates to materials and articles intended to come into contact with food [64]. It covers also articles for contact with water intended for human consumption. Prior to authorization for their use in food contact materials substances need to undergo a safety assessment carried out by EFSA.
  Products deriving from nanotechnology seem to be promising for food contact materials such as materials with barriers or coatings against humidity, UV light or oxygen, or antibacterial materials. The new Regulation (EC) 10/2011 on plastic materials and articles intended to come into contact with food [65] clarifies that authorizations which are based on the risk assessment of the conventional particle size of a substance do not cover engineered nanoparticles. Nanoparticles' risk should be assessed on a case-by-case basis. Nanoparticles are not covered by the functional barrier concept (i.e. the use of non-authorized substances behind a functional barrier, which separates the food from plastic by a multilayer material or article). An example for an authorized nanomaterial in a food contact material on the EU market is nano titanium nitrite for use in PET bottles up to 20 ng/kg, with a defined primary particle size (20 nm).

EFSA [45] has published a "Guidance on the risk assessment of the application of nano-science and nanotechnologies in the food and feed chain" [66]. The guidance covers risk assessments for food and feed applications including food additives, enzymes, flavorings, food contact materials, novel foods, feed additives and pesticides. It concludes that the risk assessment paradigm (hazard identification and hazard characterization followed by exposure assessment and risk characterization) is appropriate also for nanomaterials. The report provides guidance on nano-specific considerations that need to be assessed in addition to conventional aspects. It stipulates the additional data needed for the physical and chemical characterization of nanomaterials in comparison with conventional applications and outlines different toxicity testing approaches.

In its resolution of 6 July 2011, the European Parliament adopted new rules to provide more and better information to consumers so they can make informed choices when buying food [67]. Besides information such as nutrition values, allergens or country of origin, all ingredients present in the form of engineered nanomaterials shall be clearly indicated in the list of ingredients. The names of such ingredients shall be followed by the word "nano" in brackets. Once the legislation is published in the EU Official Journal, food companies will have 3 years to adapt to most of the new rules.

Medicinal Products and Medical Devices

Nanotechnology and nanomaterials are promising for new medicinal products, medical devices, and diagnostic agents. There are currently no specific provisions for "nano" products used in the above areas. However, the usual comprehensive legal frame applies and details are published on the homepage of the "European Medicines Agency" (EMA) [68] in London (UK). EMA has created an Innovation Task Force to ensure coordination of scientific and regulatory competence in the field of emerging therapies and technologies, including nanotechnologies, and to provide a forum for an early dialogue with applicants on regulatory, scientific or other issues that may arise from the development. In the absence of specific guidance, applicants are encouraged to contact the EMA from the early stages of the development of their products.

Environmental Protection

Environmental regulation relevant in this context relates in particular to integrated pollution prevention and control (IPPC), the control of major accident hazards involving dangerous substances (Seveso II), the water framework directive and a number of waste directives.

The IPPC Directive [69] covers approximately 52 000 industrial installations across the EU and requires installations falling under its scope to operate in accordance with permits including emission limit values based on the application of best available techniques (BAT). In principle, the IPPC Directive could be used to control environmental impacts of nanomaterials.

The Seveso II Directive [70] applies to companies where named dangerous substances (or substances falling within certain CLP classification categories) are present above specific quantities. It imposes a general obligation on operators to take all measures necessary to prevent major accidents and to limit their consequences for humans and the environment. If certain nanomaterials are found to demonstrate a major accident hazard, they may be categorized, together with appropriate thresholds, in the context of the Directive.

The Water Framework Directive 2000/60/EC [71] sets common principles and an overall framework for action to improve the aquatic environment and progressively to reduce the pollution from priority substances and phasing out emissions, discharges and losses of priority hazardous substances to water. Nanomaterials could be included among the priority substances depending on their hazardous properties. Environment Quality Standards would in these cases be proposed by the Commission.

Directive 2006/12/EC on waste [72] sets the general framework and imposes an obligation on EU Member States to ensure that waste treatment does not adversely affect health and the environment. The Hazardous Waste Directive [73] defines which wastes are hazardous and lays down stricter provisions regarding such waste. Wastes containing nanomaterials could be classified as hazardous if the nanomaterial displays relevant properties which render the waste hazardous.

TAKE-HOME MESSAGES

Risk assessment and regulation of nanomaterials would benefit from continued efforts to address the many research needs remaining to inform the differential responses to nano-scale materials in the environment. This includes comparative analysis of nanomaterials and their macroscale counterparts in both studies for human health and the environment. Further comparative analysis of various nanomaterials to each other may be necessary to understand which characteristics play a role in the toxic response to nanomaterials, and therefore should be monitored in the production of new nanomaterials.

As the use of nanomaterials increases rapidly, it is important that the ongoing research is designed to address the risk assessment needs involving manufacture, use and disposal of nanomaterials. Risk assessment of nanomaterials is complicated by limited exposure and hazard information, no clear dose metric and the lack of characterization of nanomaterials in many early studies. There are clear data gaps in exposure assessment of nanomaterials, particularly moving beyond occupational exposures [11]. Very limited information is available on the exposure to consumers through the use, transport, and disposal of multiple products. The lack of a dose metric for the analysis of toxic effects of nanomaterials is another limitation. Once a risk assessment framework for nanomaterials can be designed, it will be possible to conclude on the adequacy of the legislative framework for nanomaterials.

Acknowledgments

The authors would like to thank Bengt Fadeel, Antonio Pietroiusti and Anna Shvedova for the invitation to contribute to this book. Furthermore, they are grateful to their colleagues in the USA and in Europe for their input and useful suggestions, with special thanks to Kristen Rasmussen, Karin Aschberger, Hermann Stamm, Babasaheb Sonawane and Charles Ris.

References

[1] Organization for Economic Cooperation and Development (OECD) [homepage on the internet]. Paris: Safety of Manufactured Nanomaterials. Available at, http://www.oecd.org/about/0, 3347, en_2649_37015404_1_1_1_1_37465,00.html; [accessed 27.07.11].

[2] Organization for Economic Cooperation and Development (OECD) [publication on the Internet]. Paris: OECD 2010. Guidance manual for the testing of manufactured nanomaterials: OECD's Sponsorship programme; first revision. ENV/JM/MONO(2009)20/REV. Available at, http://www.oecd.org/officialdocuments/displaydocumentpdf/?cote=env/jm/mono(2009)20/rev&doclanguage=en; [cited 02.06.10].

[3] NTP. Nanotechnology Safety Initiatives. National Toxicology Program, National Institute of Environmental Health Sciences, Research Triangle Park, NC. Available at, http://ntp.niehs.nih.gov/index.cfm?objectid=30577CE2-F1F6-975E-7E3CC99ACBAB0E67; 2009 [cited 28.07.11].

[4] Davis JM. How to assess risks of nanotechnology: Learning from past experience. J Nanosci Nanotechnol 2007;37:186–91.

[5] US EPA. Nanomaterial case study: Nanoscale titanium dioxide in water treatment and topical sunscreen (final). Available at, http://cfpub.epa.gov/ncea/cfm/recordisplay.cfm?deid=230972; 2011 [cited 23.06.11].

[6] US EPA. Nanomaterial case study: nanoscale silver in disinfectant spray (external review draft). Available at, http://cfpub.epa.gov/ncea/cfm/recordisplay.cfm?deid=226723; 2011 [cited 23.06.11].

[7] Bermudez E, Mangum JB, Asgharian B, et al. Long-term pulmonary responses of three laboratory rodent species to subchronic inhalation of pigmentary titanium dioxide particles. Toxicol Sci 2002;70:86–97.

[8] Bermudez E, Mangum JB, Wong BA, et al. Pulmonary responses of mice, rats, and hamsters to subchronic inhalation of ultrafine titanium dioxide particles. Toxicol Sci 2004;77:347–57.

[9] Churg A, Stevens B, Wright JL. Comparison of the uptake of fine and ultrafine TiO_2 in a tracheal explant system. Am J Physiol 1998;274:L81–6.

[10] Maynard AD. Don't define nanomaterials. Nature 2011;475:31.

[11] NNI. National Nanotechnology Initiative 2011 Environmental, Health, and Safety Strategy. Available at, http://www.steptoe.com/assets/attachments/Draft-2011-NNI-EHS-Research-Strategy.pdf; 2011 [cited 24.05.11].

[12] European Parliament. [document on the Internet], Strasbourg. European Parliament resolution of 24 April 2009 on regulatory aspects of nanomaterials (2008/2208(INI)). P6_TA(2009)0328. Available at, http://www.europarl.europa.eu/sides/getDoc.do?type=TA&reference=P6-TA-2009-0328&language=EN; 2009 [cited 24.04.09].

[13] Lövestam G, Rauscher H, Roebben G, et al. Considerations on a Definition of Nanomaterial for Regulatory Purposes. Luxembourg: Publications Office of the European Union. EUR 24403 EN; 2010. Available at, http://publications.jrc.ec.europa.eu/repository/handle/111111111/14270; 2010.

[14] European Commission [homepage on the Internet]. Brussels: Public Health. Scientific Committees. Scientific Committee on Emerging and Newly Identified Health Risks (SCENIHR). Available at, http://ec.europa.eu/health/scientific_committees/emerging/index_en.htm; [accessed 03.08.11].

[15] European Commission [publication on the Internet]. Brussels: Directorate-General for Health & Consumers. Scientific Committee on Emerging and Newly Identified Health Risks (SCENIHR). Scientific Basis for the Definition of the Team nanomaterial. European Union. ISBN 978-92-79-12757. Available at, http://ec.europa.eu/health/scientific_committees/emerging/docs/scenihr_o_032.pdf; 2010.

[16] European Union. Luxembourg 2011: Commission Recommendation of 18 October 2011 on the definition of nanomaterial. Publications Office of the European Union, Offical Journal of the European Union (20.10.2011): L 275/38.

[17] OSTP. Available at, http://whitehouse.gov/administration/eop/ostp/nstc; 2011.

[18] NNI. National Nanotechnology Initiative. Available at, www.nano.gov; 2011.

[19] US EPA. The Toxic Substances Control Act (TSCA). Available at, 15 USC (C. 53) 1976:2601–92.

[20] US EPA. Available at, http://www.epa.gov/oppt/nano/; 2010.

[21] US EPA. Federal Insecticide, Fungicide, and Rodenticide Act (FIFRA). Available at, (P.L. 75–717) 7 USC § 136;1972.

[22] http://www.epa.gov/scipoly/sap/meetings/2009/november/110309ameetingminutes.pdf

[23] US EPA. Comprehensive Environmental Response, Compensation, and Liability Act and Resource Conservation and Recovery Act (CERCLA/RCRA) 2008.

[24] US EPA. Clean Air Act Amendments (CAAA) 1990.

[25] US EPA. Clean Water Act Amendments (CWAA) 1996.

[26] US EPA. Nanoscale Materials Stewardship Program. Available at, http://www.epa.gov/oppt/nano/stewardship.htm; 2011.

[27] FDA. The Food, Drug and Cosmetic Act (FDCA) Amendments Available at, PL 87–781, 76 Stat 780;1962.

[28] CPSC. Consumer Product Safety Commission. Available at, www.cpsc.gov/; 2011.

[29] OSHA. Occupational Safety and Health Act. Available at, 29 USC 651(b). 1970.

[30] OSHA. Nanotechnology OSHA Standards. Available at, www.osha.gov/dsg/nanotechnology/nanotech_standards.html; 2010.

[31] NIOSH. Occupational Exposure to Titanium Dioxide. Current Intelligent Bulletin. Available at, http://www.cdc.gov/niosh/docs/2011-160/pdfs/2011-160.pdf, 2011;63.

[32] Song Y, Li X, Du X. Exposure to nanoparticles is related to pleural effusion, pulmonary fibrosis and granuloma. Eur Respir J 2009;34:559–67.

[33] Wu M, Gordon RE, Herbert R, et al. Environ. Health Perspect 2010;118:499–504.

[34] European Commission [document on the Internet]. Brussels 2008: Communication from the Commission to the European Parliament, the Council and the European Economic and Social Committee on regulatory aspects of nanomaterials. COM (2008) 366 final, SEC(2008) 2036 of 17.6. 2008 Available at, http://ec.europa.eu/nanotechnology/pdf/comm_2008_0366_en.pdf; 2008.

[35] European Commission [document on the Internet]. Brussels 2009: Follow-up to the European Parliament resolution on regulatory aspects of nanomaterials (adopted by the Commission on 14 July 2009). Available at, http://www.google.it/search?q=Follow-up+to+the+European+Parliament+resolution+on+regulatory+aspects+

of+nanomaterials%2C+adopted+by+the+Commission+on+14+July+2009&rls=com.microsoft:*&ie=UTF-8&oe=UTF-8&startIndex=&startPage=1&redir_esc=&ei=0hvyTZ7NFZHOswbNzoSIBw

[36] European Commission [document on the Internet]. Brussels 2008: Commission staff working document, Accompanying document to Communication from the Commission to the European Parliament, the Council and the European Economic and Social Committee on regulatory aspects of nanomaterials, COM (2008) 366 final, SEC(2008) 2036 of 17.6.2008. Available at, http://ec.europa.eu/nanotechnology/pdf/com_regulatory_aspect_nanomaterials_2008_en.pdf

[37] European Union. Luxembourg 2007: Regulation (EC) No 1907/2006 of the European Parliament and the Council of 18 December 2006 concerning the Registration, Evaluation, Authorisation and Restriction of Chemicals (REACH), establishing a European Chemicals Agency, amending Directive 1999/45/EC and repealing Council Regulation (EEC) No 793/93 and Commission Regulation (EC) No 148894 as well as Council directive 76/769/EEC and Commission Directives 91/155/EEC, 93/67/EEC, 93/105/EC and 2000/21/EC. Publications Office of the European Union, Official Journal of the European Union (29.5.2007): L 136/3.

[38] European Chemicals Agency (ECHA) [homepage on the Internet]. Helsinki: Available at, http://ec.europa.eu/echa/home_en.html; [accessed 03.08.11].

[39] European Chemicals Agency (ECHA) [publication on the Internet]. Helsinki 2011: Guidance documents. Available at, http://guidance.echa.europa.eu/guidance_en.htm; [accessed 03.08.11].

[40] European Commission [publication on the Internet]. Brussels 2008: Nanomaterials in REACH. Document CA/59/2008 rev.1 dated 2008 December 16. Available at, http://ec.europa.eu/environment/chemicals/reach/pdf/nanomaterials.pdf or http://ec.europa.eu/enterprise/sectors/chemicals/files/reach/nanomaterials_en.pdf

[41] European Union. Luxembourg 2008: Regulation (EC) No 1272/2008 of the European Parliament and the Council of 16 December 2008 on classification, labelling and packaging of substances and mixtures, amending and repealing Directives 67/548/EEC and 1999/45/EC, and amending Regulation (EC) No 1907/2006. Publications Office of the European Union, Official Journal of the European Union (31.12.2008): L 353/1

[42] European Commission [publication on the Internet]. Brussels 2009: Classification, labelling & packaging of nanomaterials in REACH and CLP. Document CA/90/2009 Rev.2 dated 2009 December 3. Available at, http://ec.europa.eu/environment/chemicals/reach/pdf/classif_nano.pdf or http://ec.europa.eu/enterprise/sectors/chemicals/files/reach/nanos_in_reach_and_clp_en.pdf

[43] European Union. Luxembourg 2009: Regulation (EC) No 1107/2009 of the European Parliament and the Council of 21 October 2009 concerning the placing of plant production products on the market and repealing Council Directives 79/117/EEC and 91/414/EEC. Publications Office of the European Union, Official Journal of the European Union (24.11.2009): L 309/1

[44] European Communities. Luxembourg 1998: Directive 98/8/EC of the European Parliament and the Council of 16 February 1998 concerning the placing of biocidal products on the market. Publications Office of the European Union, Official Journal of the European Communities (24.4.1998): L 123/41

[45] European Food Safety Authority (EFSA) [homepage on the Internet]. Parma. Available at, http://www.efsa.europa.eu/; [accessed 03.07.11].

[46] European Commission [homepage on the Internet]. Ispra: Joint Research Centre, Institute for Health and Consumer Protection. Risk Assessment of Biocides. Available at, http://ihcp.jrc.ec.europa.eu/our_activities/health-env/risk_assessment_of_Biocides; [accessed 03.08.11].

[47] European Communities. Luxembourg 1989: Council Directive 89/391/EEC of 12 June 1989 on the introduction of measures to encourage improvements in the safety and health of workers at work. Publications Office of the European Union, Official Journal of the European Communities (29.6.1989): L 183/1

[48] European Communities. Luxembourg 1998: Council Directive 98/24/EEC of 7 April 1998 on the protection of the health and safety of workers from the risks related to chemical agents at work (fourteenth individual Directive within the meaning of Article 16(1) of Directive 89/391/EEC). Publications Office of the European Union, Official Journal of the European Communities (5.5.1998): L 131/11

[49] European Communities. Luxembourg 1989: Council Directive 89/655/EEC of 30 November 1989 concerning the minimum safety and health requirements for the use of work equipment by workers at work (second individual Directive). Publications Office of the European Union, Official Journal of the European Communities (30.12.1989): L 393/13

[50] European Communities. Luxembourg 1989: Council Directive 89/656/EEC of 30 November 1989 on the minimum health and safety requirements for the use of personal protective equipment at the workplace (third individual Directive). Publications Office of the European Union, Official Journal of the European Communities (30.12.1989): L 393/18

[51] European Communities. Luxembourg 2000: Commission Directive 2000/39/EC of 8 June 2000 establishing a first list of indicative occupational exposure limit values in implementation of Council Directive 98/24/EC on the protection of the health and safety of workers from the risks related to chemical agents at work. Publications Office of the European Union, Official Journal of the European Communities (16.6.2000): L 142, p. 47−50

[52] European Union. Luxembourg 2006: Commission Directive 2006/15/EC of 7 February 2006 establishing a second list of indicative occupational exposure limit in implementation of Council Directive 98/24/EC and amending Directives 91/322/EEC and Directive 2000/39/EC. Publications Office of the European Union, Official Journal of the European Union (9.2.2006): L 38, p. 36−39

[53] European Agency for Safety and Health at Work (EU-OSHA) [homepage on the Internet]. Bilbao. Available at, http://osha.europa.eu/en; [accessed 03.08.11].

[54] European Agency for Safety and Health at Work (EU-OSHA). Bilbao: Report—Expert forecast on emerging chemical risks related to occupational safety and health. ISBN: 978-92-9191-171-4. Available at, http://osha.europa.eu/en/publications/reports/TE3008390ENC_chemical_risks/view; 2009.

[55] European Agency for Safety and Health at Work (EU-OSHA) [publication on the Internet]. Bilbao. Workplace exposure to nanoparticles. Available at, http://osha.europa.eu/en/publications/literature_reviews/workplace_exposure_to_nanoparticles/view; 2009.

[56] European Communities. Luxembourg 2002: Directive 2001/95/EC of the European Parliament and of the Council of 3 December 2001 on general product safety. Publications Office of the European Union, Official Journal of the European Communities (15.1.2002): L 11, p. 36−39

[57] European Union. Luxembourg 2009: European Commission (2009). Regulation (EC) No 1223/2009 of the European Parliament and the Council of 30 November 2009 on cosmetic products. Publications Office of the European Union, Official Journal of the European Union (22.12.2009): L 342, p. 59−209.

[58] European Commission [homepage on the Internet]. Brussels: Public Health. Scientific Committees. Scientific Committee on Consumer Safety (SCCS). Available at, http://ec.europa.eu/health/scientific_committees/consumer_safety/index_en.htm; [accessed 03.08.11].

[59] European Commission [publication on the Internet]. Brussels: Directorate-General for Health & Consumers. Scientific Committee on Consumer Safety (SCCS): The SCCS's notes of guidance for the testing of cosmetic ingredients and their safety evaluation, 7th revision. SCCS/1416/11 (2010), p. 51−52. Available at, http://ec.europa.eu/health/scientific_committees/consumer_safety/docs/sccs_s_004.pdf

[60] European Communities. Luxembourg 2002: Regulation (EC) No 178/2002 of the European Parliament and of the Council of 28 January 2002 laying down the general principles and requirements of food law, establishing the European Food Safety Authority and laying down procedures in matters of food safety. Publications Office of the European Union, Official Journal of the European Communities (1.2.2002): L 31/1

[61] European Communities. Luxembourg 1997: Regulation (EC) No 258/97 of the European Parliament and of the Council of 27 January 1997 concerning novel foods and novel food ingredients. Publications Office of the European Union, Official Journal of the European Communities (14.2.1997): L 43/1

[62] European Communities. Luxembourg 1989: Council Directive 89/107/EEC of 21 December 1988 on the approximation of the laws of the Member States concerning food additives authorized for use in foodstuffs intended for human consumption. Publications Office of the European Union, Official Journal of the European Communities (11.2.1989): L 40, p. 27−33

[63] European Communities. Luxembourg 2002: Directive 2002/46/EC of the European Parliament and of the Council of 10 June 2002 on the approximation of the laws of the Member States relating to food supplements. Publications Office of the European Union, Official Journal of the European Communities (12.7.2002): L 183/5

[64] European Union. Luxembourg 2004: Regulation (EC) No 1935/2004 of the European Parliament and of the Council of 27 October 2004 on materials and articles intended to come into contact with food and repealing Directives 80/590/EEC and 89/109/EEC. Publications Office of the European Union, Official Journal of the European Union (13.11.2004): L 338/4

[65] European Union. Luxembourg 2011: Commission Regulation (EU) No 10/2011 of 14 January 2011 on plastic materials and articles intended to come into contact with food. Publications Office of the European Union, Official Journal of the European Union (15.1.2011): L 12, p. 1–89.
[66] European Food Safety Authority (EFSA) [publication on the Internet]. Parma 2011: Guidance on the risk assessment of the application of nanoscience and nanotechnologies in the food and feed chain. EFSA Journal 2011;9(5):2140. Available at, http://www.efsa.europa.eu/en/efsajournal/pub/2140.htm
[67] European Parliament [document on the Internet]. Strasbourg 2011. European Parliament legislative resolution of 6 July 2011 on the Council position at first reading with a view to the adoption of a regulation of the European Parliament and of the Council on the provision of food information to consumers, amending Regulations (EC) No 1924/2006 and (EC) No 1925/2006 and repealing Directives 87/250/EEC, 90/496/EEC, 1999/10/EC, 2000/13/EC, 2002/67/EC, 2008/5/EC and Regulation (EC) No 608/2004 (17602/1/2010 – C7-0060/2011 – 2008/0028(COD)). Available at, http://www.europarl.europa.eu/sides/getDoc.do?type=TA&reference=P7-TA-2011-0324&language=EN&ring=A7-2011-0177#BKMD-5; [accessed 06.07.11].
[68] European Medicines Agency (EMA) [homepage on the Internet]. London. Available at, www.ema.europa.eu; [accessed 03.08.11].
[69] European Union. Luxembourg 2008: Directive 2008/1/EC of the European Parliament and the Council of 15 January 2008 concerning integrated pollution prevention and control. Publications Office of the European Union, Official Journal of the European Union (29.1.2008): L 24/8
[70] European Communities. Luxembourg 1997: Council Directive 96/82/EC of 9 December 1996 on the control of major-accident hazards involving dangerous substances. Publications Office of the European Union, Official Journal of the European Communities (14.1.1997): L 10/13
[71] European Communities. Luxembourg 2000: Directive 2000/60/EC of the European Parliament and the Council of 23 October 2000 establishing a framework for Community action in the field of water policy. Publications Office of the European Union, Official Journal of the European Communities (22.12.2000): L 327/1
[72] European Union. Luxembourg 2006: Directive 2006/12/EC of the European Parliament and the Council of 5 April 2006 on waste. Publications Office of the European Union, Official Journal of the European Union (27.4.2006): L 114/9
[73] European Communities. Luxembourg 1991: Council Directive 91/689/EEC of 12 December 1991 on hazardous waste. Publications Office of the European Union, Official Journal of the European Communities (31.12.1991): L 377, p. 20–27.

SECTION B

ENGINEERED NANOMATERIALS: IMPACT ON HUMAN HEALTH

CHAPTER

7

Respiratory System

Ken Donaldson[1], *Craig A. Poland*[2]

[1]Queen's Medical Research Institute, University of Edinburgh, Edinburgh, UK

[2]Institute of Occupational Medicine, Edinburgh, UK

OUTLINE

Adverse Effects of Engineered Nanomaterials
DOI: 10.1016/B978-0-12-386940-1.00007-6

INTRODUCTION

The lungs and the pleura are major targets for the effects of airborne nanoparticles that enter the lungs by inhalation. Following aerodynamic size-dependent deposition in the bronchial tree there are a number of routes whereby particles can translocate or be cleared from their original site of deposition. All three anatomical compartments of the lung—the airways, parenchyma and pleura—are likely to receive nanoparticle dose, albeit to different degrees, following deposition. However, clearance and particle handling in these different compartments differs greatly and the subsequent pathological outcome depends on the nature of this handling and how the particles subvert it. The lung is defended in these different compartments by different processes—in airways mucociliary defense is the principle defense system, while in the alveolar region, alveolar macrophage-mediated clearance dominates and, in the pleura, rapid transport of the pleural fluid through the stomata in the parietal pleural wall is the main mechanism whereby particles which reach the pleural space are removed. There are also pleural macrophages which may phagocytize particles in the pleural space. There are a number of ways whereby nanoparticles may overcome or modulate these defenses which may lead to their translocation, retention and/or activation of the immuno-inflammatory response in these regions leading to further injury and pathological change. Large-scale inhalation studies that took place in the 1980s and 1990s with nanoparticles of carbon black, photocopier toner and titanium dioxide among other nanoparticles (then called ultrafine particles), provide a substantial body of evidence as to how low toxicity nanoparticles affect the lungs. In this regard, the unique response of the rat lung to high particulate load and the induction of "rat lung overload" is an important issue that should be considered. Recent studies with nanoparticles following inhalation have shown that certain types, such as forms of carbon nanotubes, are highly inflammogenic and pathogenic while other forms of carbon nanotubes are less so. It is useful to consider a number of exemplar nanoparticles and consider how these are known to affect the lungs and the pleural space to set the scene for a greater and broader understanding of how nanoparticles, in general, affect the pleuro-pulmonary target tissues.

DEPOSITION AND CLEARANCE FOLLOWING PULMONARY DEPOSITION OF NANOPARTICLES

The concept of dose and response are the fundamental processes in toxicology and, in the case of inhalation exposure of the lung to particles, the dose is delivered by deposition to various parts of the lung dependent on their aerodynamic diameter [1]. Larger particles deposit higher up in the nose and upper respiratory tract, while there is an ever greater elutriation of particles as the branches of the bronchial tree are penetrated and flow of air slows down, resulting in only the smaller size fractions being deposited in the more peripheral bronchioles and proximal alveolar region. In the air, nanoparticles undergo dynamic interactions with each other, resulting in aggregation so that the size distribution of the nanoparticles is variable over time [2]. Particle deposition is affected by three different mechanisms: (1) sedimentation; (2) impaction; and (3) diffusion (see fiber section below for interception). Sedimentation affects particles $>0.5\ \mu m$ in aerodynamic diameter, while impaction acts on particles >1.5 μm aerodynamic diameter, and diffusion regulates diffusion of particles <0.5 μm in thermodynamic diameter [3]. These mechanisms operate in different parts of the bronchial tree resulting in hotspot deposition at airway bifurcations because of turbulent flow leading to increased impaction; there is also large-scale deposition by Brownian motion around the proximal alveolar/terminal bronchiolar region where the net movement of air becomes zero and there is time for particles to encounter the airspace walls by diffusion. Following deposition in these different regions of the lung, the clearance mechanisms that are normally operative can act on the particles to remove them:

1. *Airways* In the airways, ciliated and mucus-producing cells combine together to form the mucociliary escalator which propels particles trapped in the mucus upwards towards the throat for swallowing. This system becomes evermore depleted peripherally until, by the time terminal bronchioles are encountered, mucociliary clearance is much slower [4].
2. *The proximal alveolar region* Beyond the ciliated airways, macrophage action is the principal mechanism of clearance of deposited particles. This site of interest has been defined as the proximal alveolar region (PAR) and comprises the terminal bronchiole and the first 400–600 μm beyond the bronchiolar–alveolar duct junction [5]. This same region is also referred to as centri-acinar, i.e. the opening into the pulmonary acinus, the unit of gas exchange, defined as the composite set of alveoli arising from a single terminal bronchiole, also known as the ventilatory unit of the lungs [6]. The PAR is the transitional zone between the terminal airways and the alveoli where gas exchange occurs and is the prime site for particle deposition and accumulation [7]. Macrophages are important cells in integrating the immunological and inflammatory responses and so any modulating effect of phagocytosed particles on macrophage function can have a profound effect on initiation of immuno-inflammatory responses. The epithelial cells in the airways, terminal bronchioles and in the alveolar region are also involved in immuno-inflammatory response by their ability to produce chemokines and other inflammatory mediators. Therefore the interaction between deposited particles in the fixed epithelial cells and the mobile alveolar macrophages is important in dictating the local responses to particle deposition during clearance and retention. Particles have been observed to clear through the epithelium in and around the terminal bronchioles and enter the interstitium where they can lodge over a protracted period of time, e.g. coal macules in coal miners [8]. Nanoparticles are also considered to cross cellular and tissue barriers [9] (cited also in Chapter 1) where they can accumulate in the interstitium [10] and reach the blood where they can translocate to other organs (not dealt with here). Once across the airspace epithelial barrier they may also enter the sumps of the lymphoid system which terminate around the terminal bronchioles. The particles are then swept cranially in the lymph to the mediastinal lymph nodes which are well known to be loaded with dust following any kind of dust exposure.
3. *The pleura* In addition, it has become apparent that particles deposit peripherally and in alveoli that are subpleural can be cleared to the pleural space (reviewed in [11] cited also in Chapters 1 and 2). Normally, only very small particles reach this region due to the elutriating effect of negotiating the airways in a bifurcating airstream that becomes progressively slower as it moves into the lungs. However, because of the unusual aerodynamics of fiber-shaped particles, long fibers can also reach this site (see below).

Particles that deposit subpleurally appear to be able to migrate through the subpleural tissues and enter the pleural space [12]. Particles entering the pleural space are cleared via the outflow through the stomata in the parietal pleura to the mediastinal lymph nodes. However, long fibers unable to negotiate the stomata, which are size-restricted to $<\approx 10\ \mu m$, will be trapped at parietal pleura where protracted interaction with the mesothelial cells will occur along with elicitation of an inflammatory response. The pleura is thus only a target for fiber-shaped particles.

4. *The Lymph nodes* The consequences of accumulation of particles in the lymph nodes is not known, but particle-loaded lymph nodes are common in all industrial-related lung diseases, with little evidence of lymph adenopathy or immunopathological sequelae.

Therefore, the sites of particle retention where the dose is concentrated over a period of time include alveolar macrophages, epithelial cells, the interstitium, the lymph nodes and, in the case of high aspect ratio nanoparticles (HARN), around the stomata on the parietal pleura. In these sites, where the dose of nanoparticles (NP) is delivered over a protracted time, there is potential for adverse effects via oxidative stress and inflammation if the particles are sufficiently reactive or are long fibers. There is also clearance of a small fraction of deposited nanoparticles to the blood and their translocation around the body for deposition in other target organs, but this is not dealt with here.

ANTIOXIDANT DEFENSE

Oxidative stress is considered to be a major process in the pathobiological effects of particles and, indeed, is the dominant hypothesis for the adverse effects of NP at the cellular level [13]. The ability of NP to generate oxidative stress has formed the basis of a hypothetical structure/toxicity relationship for NP [14]. Oxidative stress is defined as an imbalance between oxidants and antioxidants inside cells, in lung lining fluid or tissue fluid, such that there is more oxidation. Reactive oxygen species (ROS) are produced by particles themselves from intrinsic chemical reactions, such as the Fenton reaction [15], and by cells as part of normal respiration. Inflammatory cells are particularly active at producing free radicals as part of their microbicidal arsenal and so during inflammatory responses these cells can contribute considerably to oxidative stress. The ROS released from phagocytes are thought to be a major source of cell and tissue damage associated with chronic inflammation and the development of disease. Therefore, the effects of deposited particles and the oxidative leukocytic response to them can be important in causing oxidative stress that leads to further inflammation and pathogenic change. The redox sensitive transcription factors, such as NF-κB, are important in the inflammatory effects following particle deposition [16]. When these transcription factors are activated they lead to transcription of pro-inflammatory genes leading to the production of cytokines, chemokines and adhesion molecules. Antioxidants are found in two broad categories—enzymatic and non-enzymatic. The non-enzymatic antioxidants are important in lung lining fluid and include uric acid, ascorbic acid and alpha-tocopherol (vitamin E). Together these antioxidants can break the oxidation chain in the aqueous and lipid phases of the cell. Antioxidant enzymes include superoxide dismutase, which catalyzes the dismutation of superoxide to hydrogen peroxide and oxygen, and catalase which scavenges H_2O_2 to water. Glutathione (GSH) is a tripeptide which incorporates two thiol or sulfhydryl groups which, on oxidation, are converted to a sulfhydryl bridge to make glutathione disulfide. GSH is a major antioxidant in the lungs and is found at 100 times more concentration in the epithelial lining fluid of the lungs than in the plasma and intercellular levels of GSH are also very high. There are differences between rodents and humans in their antioxidant status and their ability to synthesize and regulate the compounds involved and so comparisons across species bounds should be executed with care.

PROCESSES AND MECHANISMS UNDERLYING PARTICLE-INDUCED EFFECTS IN THE PLEURO-PULMONARY SYSTEM

Cellular Stress

The ability of NP to induce cellular stress in epithelial cells and macrophages following uptake can be seen as a key switching event in determining whether a nanoparticle is likely to cause inflammation. This section outlines the potential pathways and mechanisms for NP to cause cellular stress. These pathways are summarized in Figure 7.1.

1. *Oxidative Stress*

When the antioxidant defenses are overwhelmed oxidative stress occurs. Oxidative stress is considered to be a major process in the pathobiological effects of particles and, indeed, is the dominant hypothesis for the adverse effects of NP at the cellular level. Nel et al. [17] have described a tiered response to differing degrees of nanoparticle-induced oxidative stress which may partly explain the pleiotropic effects of oxidative stress. The adverse effects can be seen only to occur at moderate level of oxidative stress when inflammation occurs and a mild oxidative stress is dealt with homeostatically and is not harmful. The challenge of

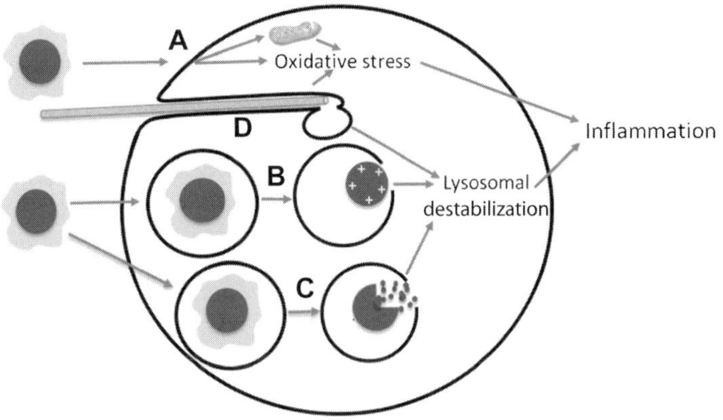

FIGURE 7.1 Pathways for nanoparticle-induced cellular stress. The compact nanoparticle is shown surrounded by its corona. (A) Nanoparticles induce oxidative stress directly or indirectly via disruption of normal redox function in the cell (e.g. mitochondrial effects). (B) Positively-charged nanoparticle with a high positive zeta potential is stripped of its corona in the lysosome and the positively-charged surface that is revealed interacts with the lysosomal membrane from within, destabilizing it. (C) A nanoparticle that is soluble at acid pH delivers a large intra-lysosomal dose of ions which destabilizes the lysosome. (D) A long thin high aspect ratio nanoparticle (HARN) induces frustrated phagocytosis with subsequent destabilization of the phagolysosomal process and chronic stimulation of NADPH oxidase leading to oxidative stress.

in vitro studies where oxidative stress is suggested to be predictive of pathogenic potential is to contextualize the extent of the stress to the kind of graded response seen in the diagram. Merely inducing oxidative stress does not indicate that there is necessarily a pathogenic outcome if the homeostatic antioxidant defenses can deal with the stress [13].

As stated above, ROS are produced by particles themselves from intrinsic chemical reactivity, such as the Fenton reaction [15]. Additionally, cells themselves produce oxidative stress from a number of sources but especially as part of normal respiration, as mitochondria leak electrons from the respiratory chain. Particles could increase this by directly affecting mitochondria or by depleting the normal antioxidant defense of the cell that protects against this endogenous source of oxidative stress [18]. Additionally, phagocytic cells which ingest particles undergo an oxidative burst which produces ROS and also reactive nitrogen species (RNS). This is part of the evolved microbicidal function of phagocytes but, when multiple sterile phagocytic events occur with particle exposure over time, there can be pathogenic consequences of continued oxidative burst activity and its subsequent oxidative stress.

Direct induction of cellular oxidative stress by particles has formed the basis of a hypothetical structure/toxicity relationship [14] although other studies have failed to show a clear relationship between intrinsic free radical generating activity and ability to cause inflammation in the lungs [19]. Studies with particles and some with nanoparticles have shown that there is upregulation of antioxidant enzymes [20] and depletion of non-enzymatic antioxidants following exposure to nanoparticles [15]. The induction of oxidative stress is directly linked to other key pathogenic processes namely inflammation, fibrosis, genotoxicity and carcinogenesis. This pathway for production of cellular stress is outlined in Figure 7.1 pathway A.

2. *Lysosomal Destabilization and Inflammasome Activation Via NP with a High Positive Zeta Potential*

The surface charge of NP, measured by zeta potential (ZP), has been suggested to be an important factor in the ability to cause cellular injury. Nel et al. [21] described a high positive ZP of particles as a factor in destabilizing the phagolysosome after uptake. The positive ZP on NP is immediately neutralized on formation of a "corona" (for further information about the corona see the Chapter 1) after deposition in lung lining fluid but inside the lysosomes the acid and protease/lipase activity may be able to remove this revealing the charged surface. This can the interact with the lysosomsal membrane leading to rupture and release of lysosomal contents with concomitant activation of the NALP3 inflammasome, as has been shown for quartz particles [22] (for a more extensive discussion on inflammasome see Chapter 10). Another hypothetical mechanism whereby positively-charged NP might destabilize membranes implicates accumulation of chloride ions and water as a consequence of chronic stimulation of the lysosomal membrane proton pump [21]. Figure 7.1 pathway B outlines this mechanism.

3. *Lysosomal Destabilization and Inflammasome Activation Via NP that are Acid Soluble*

For NP that are sparingly soluble at pH 7.0 but acid soluble, there is potential for dissolution in the acid milieu of the lysosomes after they have been taken up into this compartment. ZnO NP are exemplars of this mechanism as suggested by Nel et al. [21]. In this case, the key structural attributes are that the NP must be acid soluble and when such NP are in the acid milieu of the lysosomes they will undergo dissolution; the local high ion concentrations in the lysosomes then cause destabilization. This is illustrated in Figure 7.1 pathway C.

4. *Frustrated Phagocytosis of Long HARN*

The fiber pathogenicity paradigm (see below) is a model that predicts fiber pathogenicity and this has been found to be applicable to nanofibers, at least as far as carbon nanotubes [23,24] and nickel oxide nanowires (Poland, unpublished observations) are concerned. During frustrated phagocytosis, which occurs with long fibers, the phagosomes fail to close and phagolysosomal formation is affected leading to lysosomal destabilization and NALP3 inflammasome activation [25]. This pathway for stress is shown in Figure 7.1 pathway D.

Inflammation

Normal lung defense involves the uptake of particles into macrophages which are also part of the innate immune system, which are capable of mediating inflammation and immune responses. For completely phagocytosable particles and for particles that do not exert stress on the macrophage, uptake has little effect on macrophages secretion or motility and so is followed by translocation of the macrophage to the foot of the mucociliary escalator and eventually its clearance up the mucociliary escalator. However, if there is damage or stress to the macrophage, it can undergo differentiation and produce cytokines, chemokines and expression of other NFκB-dependent genes and production of lipid mediators leading to inflammation. A number of nanoparticle types have been shown to induce inflammation in the rat lung [14,19] and for the same material there is greater inflammation produced by an equal mass of nanoparticles compared to micron-sized particles of the same materials [26]. This is because on the whole for low-toxicity, low-solubility materials such as compact carbon, titanium dioxide and polystyrene beads, the surface area is the metric that dictates inflammation [27]. Surface area as a metric driving inflammation received considerable support in the form of studies which took place late in the 20th century using bulk manufactured nanoparticles such as titanium dioxide, photocopier toner and diesel soot [28]. These studies showed that surface area of low-toxicity materials was the driver for onset of inflammation in rat lung overload (see below). In the case of fiber-shaped particles, inflammation is induced by long particles via their ability to cause frustrated phagocytosis in macrophages which cannot fully enclose the fiber, resulting in an unstable, stressed macrophage. This paradigm [11] has been shown to operate for long carbon nanotubes [24]. Such "frustrated" macrophages are pro-inflammatory in terms of their cytokine profile and via activation of transcriptional factors such as NF-κB [25,29].

Fibrosis

Up until relatively recently, a clear link has been drawn between persistent inflammation in the lungs and fibrosis. In many models, including conventional particles, there is a distinct relationship between continuing inflammation following exposure to persistent pathogenic particles in the lungs and the development of fibrosis. Fibrosis is characterized by the accumulation of collagen and fibronectin matrix, fibroblast proliferation and myofibroblast transformation. This occurs along with an increased level of growth factors and cytokines expressed at the site of the developing lesion. More recently, however, the link between inflammation and fibrosis has been questioned and some kinds of pulmonary fibrosis, such as idiopathic pulmonary fibrosis, do not seem to be correlated with inflammation [30]. In the case of carbon nanotubes, there is evidence that they can cause fibrosis in the absence of inflammation [31], but the dominant paradigms still suggest that inflammation, and inflamed tissue in general, is a fertile ground for fibrosis. However, the type of inflammation might be sufficiently subtle not to be easily detectable, e.g. interstitial is not detected by bronchoalveolar lavage (BAL) or it may be macrophagic rather than neutrophilic. The extensive studies of Brody et al. using high levels of asbestos in a single exposure show fibrosis developing in and around their site of deposition around the first alveolar duct bifurcations without neutrophilic response [32–34]. However, the ability of nanoparticles to interstitialize close to the sites where the fibroblasts proliferate during fibrosis suggests that some kinds of nanoparticles may have direct effects on fibroblasts. This has been supported by studies with carbon nanotubes [35]. In these studies, single wall carbon nanotubes were added to cultures of lung fibroblasts and this was found to stimulate fibroblast proliferation and collagen production. MMP9, which is involved in lung fibrosis, was also increased in the cultures treated with the single wall carbon nanotubes. These data show that, in the absence of leukocytes or any kind of inflammatory response, there can be direct nanoparticle-induced stimulation of fibroblasts indicative of effects which, *in vivo*, would lead to fibrosis. It is not yet known whether other kinds of nanoparticles can directly activate fibroblasts.

Genotoxicity and Carcinogenesis

For particles, there are two models for genotoxic action—primary and secondary genotoxicity (see Chapter 14 for a detailed discussion). Primary genotoxicity involves direct interaction between the nanoparticle and the target cell, whereas secondary genotoxicity acts through inflammation to cause ROS to damage DNA [36]. Primary particle-induced genotoxicity may be direct where the nanoparticle enters the nucleus and directly interacts with the DNA or indirect by release of free radicals that directly damage the DNA or obstruct chromosome segregation during mitosis leading to aneuploidy. It might also be indirect where the particles deplete antioxidants so increasing the steady state of oxidative damage within the cell or increase oxidative damage that arises from mitochondrial activity or inhibition of DNA repair. Many *in vitro* studies have shown the ability of nanoparticles to cause direct genotoxicity [37]. There are much fewer data to demonstrate the mechanism of action or to demonstrate whether *in vivo* primary or secondary genotoxicity is the effective mechanism. There is reason to believe that the mechanism might differ between different nanoparticle types. Although none of the new engineered nanoparticles have been tested for carcinogenicity in full-term animal studies, in the 1980s, bulk manufactured nanoparticles, such as carbon black, titanium dioxide and diesel oil soot, were tested for their pathogenicity in rats over lifetime studies. These studies demonstrated that, for all ultrafine materials which were low toxicity/low solubility, at very high lung burden following a protracted period to high exposure, there was a carcinogenic and fibrogenic response. This phenomenon, which appears to be confined only to rats, was named "rat lung overload" and has questionable relevance for human risk assessment. It seems possible that high enough pulmonary dose of any nanoparticle that is reactive and causes inflammation and oxidative stress in rats could lead to genotoxicity and, ultimately, cancer.

Unusual Pulmonary Responses

It is possible that various engineered NP with unusual composition could, given sufficient exposure, produce a more unusual type of immunopathological response. In a study in rats, a single moderate exposure to metal oxide NP caused a range of effects including eosinophilia in the BAL (ZnO NP) [38] (cited also in Chapter1), delayed type hypersensitivity-like immune responses with lymphocytosis in the BAL (NiONP) [38] and fibrosis with alveolar lipo-proteinosis (ZnO NP) (manuscript in preparation). This highlights the potential for more unusual pulmonary responses, although these are not unique effects, since other particle types or metal ions elicit these responses, which could be explained by soluble components as well as solid NP effects.

Biopersistence

The ability of nanoparticles to persist in the lung is an important factor in sustaining the dose. This paradigm originates from fiber studies where the biopersistence of fiber is a major factor in determining their long-term pathogenic effects. Non-biopersistent fibers break down and are removed from the lung, whereas biopersistent fibers retain their shape and deliver long fiber dose to the lung over an extended period of time. In the case of nanoparticles, those which are able to retain their morphology continue to deliver dose in the compartment in which they are retained. There is a more complex aspect to biopersistence where the soluble products of the dissolution may be toxic, as in the case of soluble transition metal ions or zinc ions which are important in the toxicity of zinc [38].

EXEMPLAR NANOPARTICLES

In this section (and see Figure 7.2) examples of nanoparticles will be used to demonstrate how the factors described above come into play for different nanoparticle types in describing their hazard.

Titanium Dioxide, Low-Toxicity Nanoparticles

Titanium dioxide is one of the most studied particles in particle toxicology. In its micron-sized form it is used as a negative control because of its low solubility and low toxicity.

Given the reliance of inflammation in rat lungs on surface area, it has been quoted in several studies that fine TiO_2 is much less inflammogenic than nanoparticulate TiO_2 [39]. However, these effects are produced by instilled high lung mass burdens where the dose is delivered instantaneously at high dose rate. In benchmark studies, titanium dioxide always causes much less inflammation than, for example, quartz which has a highly reactive surface [27]. Compared with other types of nanoparticles, titanium dioxide has a very low intrinsic free radical-generating activity [20] and, while at high dose *in vitro*, macrophages and other cells can become stressed by nanoparticulates, at plausible TiO_2 *in vivo*, unless

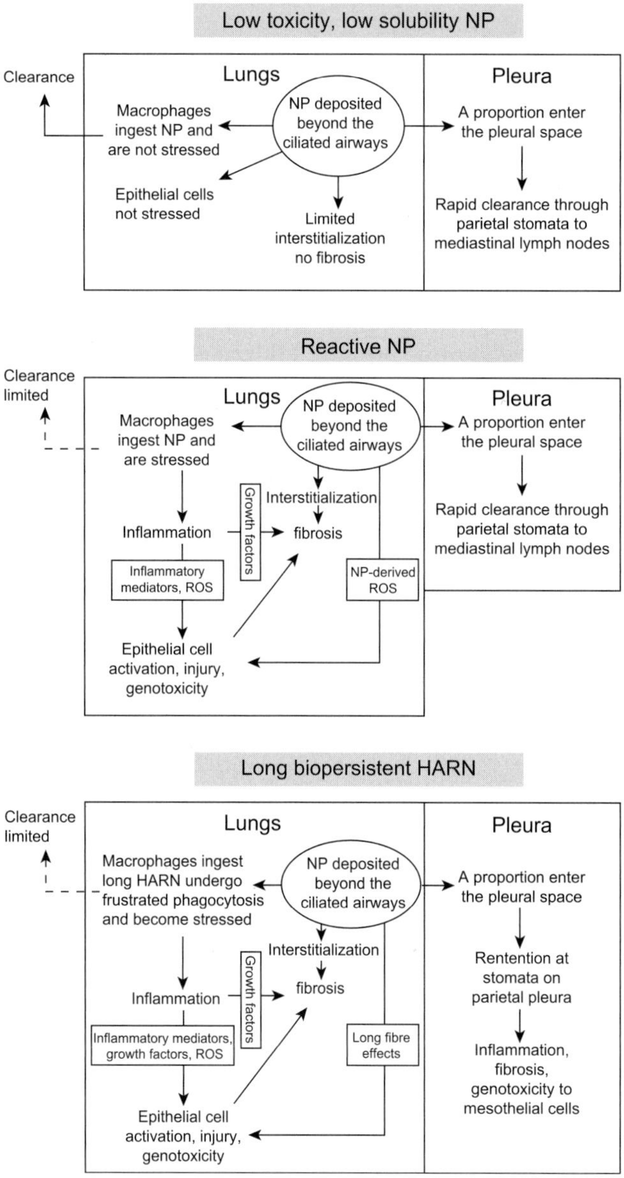

FIGURE 7.2 Toxicity mechanisms of nanoparticles according to their classification as low-toxicity/low-solubility NP, reactive NP, or long biopersistent HARN.

the doses are very high, the inflammatory effects are very small and resolve over the 16 days of recovery [40]. The dependence of inflammation in the rat lung for TiO_2 on surface area rather than number or mass has been confirmed in several different studies [41]. In inhalation studies, titanium dioxide has not caused lung tumors under non-overload conditions and there is no reason to believe that humans show the features of overload, even at the very high lung burdens seen in coal miners [42]. There is little doubt that titanium dioxide is a low-toxicity material but, in many *in vitro* tests at high dose, it has effects such as activating the inflammasome [43] or causing genotoxicity [44] (cited also in Chapters 3 and 14). However, the received dose for cells in culture can be orders of magnitude greater than that encountered by cells following inhalation in the lung. Therefore, finding of effects *in vitro* with TiO_2 NP such as genotoxicity need to be extrapolated with care to human exposures, since it is not necessarily the case that doses that are present *in vitro* would be replicated *in vivo* at any plausible exposure.

Nickel Oxide NP—Reactive Nanoparticles

Nickel nanoparticles are a good example of a reactive nanoparticle which is highly redox active [19] and which is highly inflammogenic in the rat lung, much more so than micron-sized NiO [45]. Following installation in the lung, the nickel oxide nanoparticles cause a progressive lung injury characterized by fibrosis and alveolar lipoproteinosis [38] when the same instilled surface area dose of titanium dioxide had no inflammatory effect, highlighting the role of the reactive surface. It should be noted that Ni ions were not pro-inflammatory on instillation into rat lungs [46]. A similar differential toxic effect was seen with nickel hydroxide and nickel sulfate nanoparticles alongside a similar exposure to titanium dioxide nanoparticles [47]. Oxidative stress was a feature of nickel oxide nanoparticle-induced lung injury as evidenced by high levels of lipid peroxide in the BAL fluid 24 hours after installation which was accompanied by increased LDH levels signifying cytotoxicity [48]. The pro-inflammatory and oxidative stressing effects of nickel oxide nanoparticles were seen up to 6 months in the lungs following instillation of a single dose of nickel oxide nanoparticles [49]. In chronic exposure studies to nickel hydroxide nanoparticles in C57 black 6 mice, exposures were 5 hours per day, 5 days per week for up to 5 months with an overall concentration of 79 $\mu g/m^3$. This exposure produced significant lung injury as shown by markers of inflammation in the bronchoalveolar lavage fluid, increased cytokines at 1 week and 5 months while, at 5 months, lungs were markedly damaged as viewed by thickening of the alveolar septae and inflammatory foci in the airspaces. Taken together, these data suggest that nickel oxide nanoparticles have a profound pro-inflammatory effect in the lungs which is chronic and sustained and therefore, in the long run, could lead to fibrosis and possibly tumor production. This nanoparticle serves as a model for more reactive and toxic forms of nanoparticles which are likely to produce similar kinds of changes.

Carbon Nanotubes in Compact (Non-Fibrous) Form

Carbon nanotubes (CNT) are long thin tubes of graphene that can present in essentially two different forms. Whether as single tubes or a number of tubes intertwined, the essential shape is that of a fiber which can vary in length. However, if the nanotubes grow with defects, as they more commonly do, they curl up and form assemblages that are compact, like a ball of string or wool. This can be looked upon as a compact particle, albeit a loose one and with a lot of space inside them, but they are not fibrous. If they grow with few defects they can form straight fibres. We refer below to these as compact CNT and fibrous CNT.

Carbon nanotubes in compact form are one of the most studied forms of nanoparticle. There are numerous studies, many of which are conflicting due to the differences in the structure and size, contaminating metal, defect levels, etc. in the different samples of carbon nanotube that are being used. In terms of lung effects, the effects range from severe inflammation and fibrosis produced following both inhalation and instillation of single wall carbon nanotubes [50] to very mild effects with virtually no inflammation [51]. In addition, some studies have shown extremely long-lived and severe toxic effects after similar 6-hour inhalation exposure at admittedly very high airborne mass concentrations [52]. In another subchronic study exposing for up to 3 months for 6 hours a day, exposures down to $0.5\,mg/m^3$ produced detectable pro-inflammatory effects and there was alveolar lipoproteinosis in the highest exposure groups. Some instillation studies have had problems due to the likely aggregation of nanotubes into bundles which are highly fibrogenic following lodging as boluses in and around the terminal bronchioles [31]. This has to be taken into account in assessing the overall toxicity. *In vitro*, CNT show a range of oxidative stressing, genotoxic and pro-inflammatory effects that differ in their extent in different studies, again no doubt reflecting the variable nature of the sample used [53]. It is unlikely that there are any general lessons to be learned from these studies apart from the fact that there are big differences between the different carbon nanotube samples used. At the moment, it is not possible to tell what the key structural features are that dictate toxicity and, until this is determined, it is impossible to gain any large picture of the toxicity of compact carbon nanotubes.

High Aspect Ratio Nanoparticles (HARN) in the Lungs and Pleura

High aspect ratio nanoparticles (HARN) represent a growth area in nanotechnology because of their very useful properties similar to the uses of other kinds of fibers in industry. Carbon nanotubes that are present in a fibrous shape are the best known and studied but there are a number of other nanorods, nanofibers and nanotubes other than carbon which are fibrous in habit. The resemblance of these materials to asbestos has precipitated fears that they might show some similar pathological effects. Several decades of fiber toxicology studies in the late 20th century into asbestos and synthetic vitreous fibers resulted in a "fiber pathogenicity paradigm" which highlights the fiber parameters that dictate whether or not any fiber will be pathogenic. Width or diameter is the factor which determines the aerodynamic diameter of a fiber and so has obvious importance for site of particle deposition [54]. Deposition of fibers is complex but aerodynamic diameter determines where in the respiratory tree any fiber will deposit. Fibers deposit by the usual factors of sedimentation, impaction and diffusion but, in addition, they have a function of interception which concerns the tip of the fiber making contact with the surface of the lung immediately causing deposition while the center of the fiber follows the airstream in the center of the airflow. Length is a very important feature in determining whether the fiber can be effectively cleared from the lung by macrophages which attempt to phagocytose deposited fibers. Very long fibers cannot be completely phagocytosed by the macrophages, while short fibers can be and are effectively cleared. The partial uptake of a long fiber is termed frustrated phagocytosis [29] and leads to phagosomal destabilization and activation of the inflammasome [25].

The third property is biopersistence which is the property of retaining structural integrity during residence in lung tissue. Biopersistence as a key concept arose when it was shown that different long fibers of about the same length had very different lung retention times and very different pathogenicities [55]. Thus, for a non-biopersistent fiber, components are leached or dissolved from the fiber such that the fiber is weakened, may partially or wholly dissolve, split longitudinally or break cross-wise to make short fibers which are easily cleared [11].

Additional to the effects of fibers in the lung, there is concern over the effect of fibers in the pleura. The pleural space is present between the chest wall and the lungs and the surface of the two pleurae, the parietal and visceral pleurae, are tightly opposed and each is lined by a single layer of mesothelial cells. Uniquely, asbestos and other fibers are a hazard to the pleural space, producing a range of pathologies ranging from fibrosis to pleural effusion to mesothelioma. The latter, mesothelioma, is a neoplasm which is virtually unique to fiber exposure and whose etiology has not been clear. The fact that long fibers are associated with mesothelioma and that short fibers are not is a clue which has led to research into the fate of fibers that reach the pleura [11]. The development of pleural pathologies obviously implicates delivery of fibers into the pleural space to these mesothelial cells. Although the exact mechanism is not understood, it has become clear that all particles that deposit in the periphery of the lung pass to some extent into the pleural space. They are rapidly cleared through stomata in the parietal pleural wall into the lymph which drains to the mediastinal lymph nodes. Previous studies have shown that long fibers do accumulate on the parietal pleura [56] and it was hypothesized that the longer fibers could not negotiate the stomata and so were retained at the pleura where they could deliver dose to the mesothelial cells in and around the stomata, leading to inflammatory, fibrotic and eventually neoplastic change [11]. This has been studied in relation to carbon nanotubes using first the peritoneal space, which has a very similar mesothelial surface and stomatal clearance [24] and then the pleural space [23]. It appears very clear that long nanofibers behave like long asbestos in that they are unable to negotiate the stomata and so are retained at the parietal pleura, while short and compact nanotube bundles can clear through the stomata [23]. Although no studies so far have been carried out demonstrating parietal pleura retention of fibers deposited in the airspaces, inhalation studies with chrysotile asbestos and amphibole asbestos have shown a clear effect of long biopersistent (amphibole) fiber retention and response at the parietal pleura. Preliminary studies in the laboratory of the authors suggest that other nanofibers, such as NiO and silver, comply with the fiber pathogenicity paradigm suggesting that any HARN that are long, thin and biopersistent will be pathogenic to the pleura. Further research on this is warranted.

POTENTIAL HUMAN DISEASES THAT MIGHT FOLLOW FROM EXPOSURE TO NANOPARTICLES AND FIBRES

From the above, it follows that there is potential for human lung disease from NP. Disease is unlikely to occur with plausible occupational exposure to low-toxicity insoluble NP like TiO_2. However, if there is high-level exposure, beyond some threshold, to more reactive NP, then there could be ongoing inflammation that worsens bronchitis or asthma in those who already have lung disease or even may cause fibrosis in the lung parenchyma as in the case of occupational exposures to other pathogenic dusts (pneumoconiosis). Likewise, ongoing inflammation or genotoxic effects of reactive NP could lead to lung cancer if the exposures are high enough and for a protracted period. Long, thin nanofibers that look like asbestos might cause diseases that asbestos causes based on the same structure–toxicity paradigm that controls asbestos pathogenicity—namely pathogenicity residing in fibers that are long, thin and biopersistent. The diseases that asbestos causes are pleural fibrosis, pleural plaques, mesothelioma in the pleura and peritoneal cavity, as well as fibrosis and lung cancer in the lung itself.

CONCLUSIONS

It is evident that the pleuro-pulmonary system is a target for airborne NP. The basic pathological processes that nanoparticles elicit that can be linked to long-term pathology have been shown for some nanoparticles but not for others. In other words, NP, like other particles, show differences and heterogeneity in their pathogenicity. Nanoparticles that are composed of low-toxicity materials, such as titanium dioxide, will be unlikely to show any kind of pathogenic effects at plausible exposures. However, reactive NP are likely to cause oxidative stress, inflammation, fibrosis and possibly cancer in the long term if thresholds are exceeded such that the various pathological processes are initiated. High aspect ratio NP pose a hazard to the lungs from the issue of frustrated phagocytosis and inflammation and its consequences in the lung. They will also pose a problem to the pleura because of their ability to be inspired deeply into the lung, even if they are long, then pass into the pleura and be retained at the stomata which they will be too long to negotiate. This will set up inflammatory effects and protracted interaction between longer fibers and the mesothelial cells with genotoxic and inflammatory outcomes. We may anticipate that a more profound understanding of the pleuro-pulmonary system in regard to the nanoparticle hazard will develop with the substantial body of research that is currently in train.

TAKE-HOME MESSAGES

The lungs and pleura are the major target for airborne NP. Inflammation is a central process in the effects of NP being the outcome of a range of different toxicities and so is an important process that toxicologists study. Not all NP will cause the same extent of lung and pleural inflammation and disease. Like other particles, NP will show a range of toxic potencies. The extent of the exposure is another important consideration in the likelihood that a nanoparticle exposure will be injurious to the lungs and pleura. For insoluble NP, surface and their reactivity will dictate the extent of the toxic potency. Low-toxicity soluble NP will be less toxic than soluble NP that dissociate to toxic ions (e.g. Zn^{2+}). Long, thin NP could behave like pathogenic asbestos if they are long enough and biopersistent. Further work on NP hazard and exposure is needed to establish fully good risk management towards the prevention of lung and pleural disease from NP exposures.

References

[1] Kreyling WG, Moller W, Semmler-Behnke M, Oberdorster G. Particle dosimetry: Deposition and clearance from the respiratory tract and translocaton to extra-pulmonary sites. In: Donaldson K, Borm P, editors. Particle Toxicology. Boca Raton: CRC Press; 2007. p. 47–74.

[2] Dennekamp M, Howarth S, Dick CA, Cherrie JW, Donaldson K, Seaton A. Ultrafine particles and nitrogen oxide generated by gas and electric cooking. Occup Environ Med 2001;58:511–6.
[3] Moller W, Kreyling WG, Schmid O, Semmler-Behnke M, Schulz H. Deposition, retention and clearance and translocation of inhaled fine and nano-sized particles in the respiratory tract. In: Gehr P, Muhlfeld C, Rothen-Rutishauser B, Blank F, editors. Particle–lung cell interactions. 2nd ed. New York: Informa Healthcare; 2010. p. 79–107.
[4] Schlesinger RB, Ben-Jebria A, Dahl AR, Snipes MB, Ultman J. Disposition of inhaled toxicants. In: Massaro EJ, editor. Handbook of Human Toxicology. Boca Raton: Florida CRC Press; 1997. p. 493–550.
[5] Pinkerton KE, Zhou YM, Teague SV, et al. Reduced lung cell proliferation following short-term exposure to ultrafine soot and iron particles in neonatal rats: Key to impaired lung growth? Inhal Toxicol 2004;16(Suppl. 1):73–81.
[6] Mercer RR, Crapo JD. Three-dimensional reconstruction of the rat acinus. J Appl Physiol 1987;63:785–94.
[7] Donaldson K, Borm PJ, Oberdorster G, Pinkerton KE, Stone V, Tran CL. Concordance between *in vitro* and *in vivo* dosimetry in the proinflammatory effects of low-toxicity, low-solubility particles: The key role of proximal alveolar region. Inhal Toxicol 2008;20:53–62.
[8] Gibbs AR, Wagner JC. Diseases due to silica. In: Churg A, Green FHY, editors. Pathology of Occupational Lung Disease. 2nd ed. Baltimore: Williams & Wilkins; 1998. p. 209–33.
[9] Geiser M, Rothen-Rutishauser B, Kapp N, et al. Ultrafine particles cross cellular membranes by nonphagocytic mechanisms in lungs and in cultured cells. Environ Health Perspect 2005;113:1555–60.
[10] Ferin J, Oberdorster G, Penney DP. Pulmonary retention of ultrafine and fine particles in rats. Am J Respir Cell Mol Biol 1992;6:535–42.
[11] Donaldson K, Murphy FA, Duffin R, Poland CA. Asbestos, carbon nanotubes and the pleural mesothelium: A review of the hypothesis regarding the role of long fibre retention in the parietal pleura, inflammation and mesothelioma. Part Fibre Toxicol 2010;22:5.
[12] Mercer RR, Hubbs AF, Scabilloni JF, et al. Distribution and persistence of pleura penetration by multi-walled carbon nanotubes. Part Fibre Toxicol 2010;7:28.
[13] Donaldson K, Borm PJ, Castranova V, Gulumian M. The limits of testing particle-mediated oxidative stress *in vivo* in predicting diverse pathologies; relevance for testing nanoparticles. Part Fibre Toxicol 2009;6:13.
[14] Rushton EK, Jiang J, Leonard SS, et al. Concept of assessing nanoparticle hazards considering nanoparticle dosemetric and chemical/biological response metrics. J Toxicol Environ Health A 2010;73: 445–61.
[15] Ahamed M, Siddiqui MA, Akhtar MJ, Ahmad I, Pant AB, Alhadlaq HA. Genotoxic potential of copper oxide nanoparticles in human lung epithelial cells. Biochem Biophys Res Commun 2010;396:578–83.
[16] Schins R, Donaldson K. Nuclear factor kappa-B activation by particles and fibres. Inhal Toxicol 2000;12(Suppl. 3):317–26.
[17] Nel A, Xia T, Madler L, Li N. Toxic potential of materials at the nanolevel. Science 2006;311:622–7.
[18] Li H, Haberzettl P, Albrecht C, et al. Inhibition of the mitochondrial respiratory chain function abrogates quartz induced DANN damage in lung epithelial cells. Mutat Res 2007;617:46–57.
[19] Lu S, Duffin R, Poland C, et al. Efficacy of simple short-term *in vitro* assays for predicting the potential of metal oxide nanoparticles to cause pulmonary inflammation. Environ Health Perspect 2009;117:241–7.
[20] Park EJ, Yi J, Chung KH, Ryu DY, Choi J, Park K. Oxidative stress and apoptosis induced by titanium dioxide nanoparticles in cultured BEAS-2B cells. Toxicol Lett 2008;180:222–9.
[21] Nel AE, Mädler L, Velegol D, et al. Understanding biophysicochemical interactions at the nano–bio interface. Nat Mater 2009;8:543–57.
[22] Hornung V, Bauernfeind F, Halle A, et al. Silica crystals and aluminum salts activate the NALP3 inflammasome through phagosomal destabilization. Nat Immunol 2008;9:847–56.
[23] Murphy F, Duffin R, Poland CA, et al. Length-dependent retention of carbon nanotubes in the pleural space of mice initiates sustained inflammation and progressive fibrosis on the parietal pleura. Am J Pathol 2011;178:2587–600.
[24] Poland CA, Duffin R, Kinlock I, et al. Carbon nanotubes introduced into the abdominal cavity of mice show asbestos-like pathogenicity in a pilot study. Nat Nanotech 2008;3:423–8.
[25] Hamilton RF, Wu N, Porter D, Buford M, Wolfarth M, Holian A. Particle length-dependent titanium dioxide nanomaterials toxicity and bioactivity. Part Fibre Toxicol 2009;6:35.

[26] Brown DM, Wilson MR, MacNee W, Stone V, Donaldson K. Size-dependent proinflammatory effects of ultrafine polystyrene particles: A role for surface area and oxidative stress in the enhanced activity of ultrafines. Toxicol Appl Pharmacol 2001;175:191–9.
[27] Duffin R, Tran L, Brown D, Stone V, Donaldson K. Proinflammogenic effects of low-toxicity and metal nanoparticles *in vivo* and *in vitro*: highlighting the role of particle surface area and surface reactivity. Inhal Toxicol 2007;19:849–56.
[28] Oberdorster G, Ferin J, Lehnert BE. Correlation between particle size, *in vivo* particle persistence and lung injury. Environ Health Perspect 1994;102(Suppl. 5):173–9.
[29] Ye J, Shi X, Jones W, et al. Critical role of glass fiber length in TNF-alpha production and transcription factor activation in macrophages. Am J Physiol 1999;276:L426–34.
[30] Stramer BM, Mori R, Martin P. The inflammation-fibrosis link? A Jekyll and Hyde role for blood cells during wound repair. J Invest Dermatol 2007;127:1009–17.
[31] Warheit DB, Laurence BR, Reed KL, Roach DH, Reynolds GA, Webb TR. Comparative pulmonary toxicity assessment of single walled carbon nanotubes in rats. Toxicol Sci 2004;77:117–25.
[32] Brody AR, Bonner JC, Overby LH, et al. Interstitial pulmonary macrophages produce platelet-derived growth factor that stimulates rat lung fibroblast proliferation *in vitro*. J Leukoc Biol 1992;51:640–8.
[33] Brody AR, McGavran PD, Overby LH. Brief inhalation of chrysotile asbestos induces rapid proliferation of bronchiolar–alveolar epithelial and interstitial cells. IARC Sci Publ 1989;90:93–9.
[34] Yin Q, Brody AR, Sullivan DE. Laser capture microdissection reveals dose-response of gene expression *in situ* consequent to asbestos exposure. Int J Exp Pathol 2007;88:415–25.
[35] Wang L, Mercer RR, Rojanasakul Y, et al. Direct fibrogenic effects of dispersed single-walled carbon nanotubes on human lung fibroblasts. J Toxicol Environ Health A 2010;73:410–22.
[36] Donaldson K, Poland CA, Schins RP. Possible genotoxic mechanisms of nanoparticles: Criteria for improved test strategies. Nanotoxicology 2010;4:414–20.
[37] Landsiedel R, Kapp MD, Schulz M, Wiench K, Oesch F. Genotoxicity investigations on nanomaterials: Methods, preparation and characterization of test material, potential artifacts and limitations—many questions, some answers. Mutat Res 2009;681:241–58.
[38] Cho WS, Duffin R, Poland CA, et al. Metal oxide nanoparticles induce unique inflammatory footprints in the lung: Important implications for nanoparticle testing. Environ Health Perspect 2010; 118:1699–706.
[39] Sager TM, Kommineni C, Castranova V. Pulmonary response to intratracheal instillation of ultrafine versus fine titanium dioxide: Role of particle surface area. Part Fibre Toxicol 2008;5:17.
[40] Ma-Hock L, Burkhardt S, Strauss V, et al. Development of a short-term inhalation test in the rat using nano-titanium dioxide as a model substance. Inhal Toxicol 2009;21:102–18.
[41] Hohr D, Steinfartz Y, Schins RP, et al. The surface area rather than the surface coating determines the acute inflammatory response after instillation of fine and ultrafine TiO2 in the rat. Int J Hyg Environ Health 2002;205:239–44.
[42] ILSI Risk Science Institute Workshop Participants. The relevance of the rat lung response to particle overload for human risk assessment: A workshop consensus report. Inhal Toxicol 2000;12:1–17.
[43] Yazdi AS, Guarda G, Riteau N, et al. Nanoparticles activate the NLR pyrin domain containing 3 (Nlrp3) inflammasome and cause pulmonary inflammation through release of IL-1{alpha} and IL-1{beta}. Proc Natl Acad Sci USA 2010;107:19449–54.
[44] Falck GC, Lindberg HK, Suhonen S, et al. Genotoxic effects of nanosized and fine TiO2. Hum Exp Toxicol 2009;28:339–52.
[45] Zhang Q, Kusaka Y, Zhu X, et al. Comparative toxicity of standard nickel and ultrafine nickel in lung after intratracheal instillation. J Occup Health 2003;45:23–30.
[46] Janoff A, Schaefer S. Mediators of acute inflammation in leukocyte lysosomes. Nature 1967;213:144–7.
[47] Kang GS, Gillespie PA, Gunnison A, Rengifo H, Koberstein J, Chen LC. Comparative pulmonary toxicity of inhaled nickel nanoparticles; role of deposited dose and solubility. Inhal Toxicol 2011;23:95–103.
[48] Horie M, Fukui H, Nishio K, et al. Evaluation of acute oxidative stress induced by NiO nanoparticles *in vivo* and *in vitro*. J Occup Health 2011;53:64–74.
[49] Morimoto Y, Ogami A, Todoroki M, et al. Expression of inflammation-related cytokines following intratracheal instillation of nickel oxide nanoparticles. Nanotoxicology 2010;4:161–76.

[50] Shvedova AA, Kisin E, Murray AR, et al. Inhalation versus aspiration of single-walled carbon nanotubes in C57BL/6 mice: Inflammation, fibrosis, oxidative stress and mutagenesis. Am J Physiol Lung Cell Mol Physiol 2008;295:L552–65.
[51] Elgrabli D, Abella-Gallart S, Robidel F, Rogerieux F, Boczkowski J, Lacroix G. Induction of apoptosis and absence of inflammation in rat lung after intratracheal instillation of multiwalled carbon nanotubes. Toxicology 2008;253:131–6.
[52] Ellinger-Ziegelbauer H, Pauluhn J. Pulmonary toxicity of multiwalled carbon nanotubes (Baytubes) relative to alpha-quartz following a single 6h inhalation of rats and a 3 month exposure period. Toxicology 2009;266:16–29.
[53] Aschberger K, Johnston HJ, Stone V, et al. Review of carbon nanotubes toxicity and exposure—assessment of the feasibility and challenges for human health risk assessment based on open literature. Crit Rev Toxicol 2010;40:759–90.
[54] Donaldson K. The inhalation toxicology of p-aramid fibrils. Crit Rev Toxicol 2009;39:487–500.
[55] Bernstein DM, Riego Sintes JM, Ersboell BK, Kunert J. Biopersistence of synthetic mineral fibers as a predictor of chronic intraperitoneal injection tumour response in rats. Inhal Toxicol 2001;13:823–49.
[56] Boutin C, Dumortier P, Rey F, Viallat JR, De Vuyst P. Black spots concentrate oncogenic asbestos fibers in the parietal pleura. Thoracoscopic and mineralogic study. Am J Respir Crit Care Med 1996;153:444–9.

CHAPTER

8

Cardiovascular System

Jennifer B. Raftis[1], *Nicholas L. Mills*[2], *Rodger Duffin*[1]

[1]Centre for Inflammation Research, Edinburgh University, Edinburgh, UK

[2]Centre for Cardiovascular Science, Edinburgh University, Edinburgh, UK

OUTLINE

INTRODUCTION

Nanotechnology aims for the development of nanoparticles (NP; particles with at least one dimension less than 100 nm) and other nanomaterials with novel properties, which cannot be obtained through use of the same material in bulk or molecular form. It is currently being applied in all manner of commercial industries to confer unique properties on products or improve existing ones [1]. Some of the most significant applications are likely to occur in the field of medicine. Many engineered nanoparticles and nanosurfaces have already been evaluated in the clinic [2] and it is likely nanomaterials will have a major impact in areas such as diagnostics, medical imaging and therapeutics.

Adverse Effects of Engineered Nanomaterials
DOI: 10.1016/B978-0-12-386940-1.00008-8

While the nanotechnologies develop novel engineered NPs, we are already exposed to numerous NPs in the environment through inhalation of ambient particles (particles less than 10 μm in diameter; PM_{10}) from combustion, industrial and occupational sources and through the ingestion of contaminated food and water. Epidemiological and experimental studies have identified a range of adverse health effects related to PM_{10}, and the key components and underlying mechanisms responsible are beginning to emerge [3–7]. Interestingly the "ultrafine" or nanoparticulate component of PM_{10} is considered to have the greatest potential for toxicity and harm. Findings from environmental research and particle toxicology studies have been extrapolated and used to guide our approach to the risk assessment of engineered nanoparticles. Combustion derived nanoparticles (CDNPs) share some characteristics with engineered NPs, such as size, charge and surface area that influence toxicity. Exposure to CDNPs has been shown to impair vascular function [8], promote myocardial ischemia [9], and induce prothrombotic changes in the circulation in humans [10] and may directly contribute to the formation of atherosclerotic plaque [11]. These observations highlight an important paradox between the harmful effects of CDNP and the potential for engineered NPs to be used in clinical applications; especially for those NPs designed for intravenous administration in medical imaging.

In this chapter, we review the clinical applications of engineered NPs focusing on currently available NPs with diagnostic and therapeutic roles in cardiovascular medicine (for a discussion of other diagnostic and therapeutic applications of NPs, see Chapters 15 and 16). The chemical and physical properties that are functionally important for each particle and application are outlined, and strategies to optimize performance through modification of these properties identified. We also highlight the potential for engineered NPs to adversely affect cellular function and have relevant adverse biological effects on the cardiovascular system.

DIAGNOSTIC APPLICATIONS IN CARDIOVASCULAR DISEASE

Atherothrombosis remains the leading cause of death in developed countries. Successful therapies and strategies for prevention are dependent on identification of at risk individuals using clinical risk scores based on age, sex, blood pressure, serum cholesterol and cigarette smoking [12,13]. While these risk scores are used widely and identification of traditional risk factors is the basis of most primary prevention screening policies, these calculators are imperfect; the majority of persons presenting with an adverse cardiac event have no prior history or any established risk factors [12]. The converse is also true, many patients with risk factors for cardiovascular disease do not go onto develop atherothrombosis or have an acute cardiovascular event.

An alternative and perhaps preferable strategy would be to identify individuals with early atherosclerotic disease in an attempt to modify the course of the disease and prevent the clinical consequences of plaque rupture, thrombosis and infarction. Clinical and experimental research has focused on the identification of novel biomarkers that could indicate the early development of atherothrombotic disease. Before a biomarker is introduced to the clinic as a diagnostic tool, it must satisfy the following criteria: it must be highly specific, easy to measure, provide novel information about the disease, and reflect disease burden accurately helping the clinician decide on an appropriate course of treatment (see also Chapter 3 for an extensive discussion on biomarkers). These basic criteria are not easily met for a heterogeneous condition like atherothrombosis.

Atherothrombosis is a complex inflammatory disease and is classified according to plaque composition [14]. Plaque rupture and subsequent thrombosis involve a number of inflammatory processes including endothelial dysfunction, leukocyte recruitment and migration, platelet activation, matrix metalloproteinase activity, and degradation of the fibrous cap resulting in plaque rupture and thrombosis [15]. Imaging strategies that target these pathological processes in atherosclerotic plaque development may provide more sensitive methods to monitor disease progression, permit early detection of the disease, and have a role in the assessment of treatment efficacy. The ability to identify plaque structure and function/activity would allow the clinician to identify those individuals at highest risk of plaque rupture.

ENGINEERED NANOMATERIALS

Nanoparticles are unique in terms of their size and structure. Their small size allows them to interact with the body on a molecular level and confers upon them a high ratio of surface area to volume. They can be engineered to have different shapes, sizes, structures, surface functionalization and coatings, allowing them to be specifically targeted to a biomarker of disease [16]. The ability to understand and manipulate the properties of nanomaterials will ultimately determine their usefulness in nanomedicine [17].

Table 8.1 adapted from [18], gives examples of nanomaterials intended for clinical use (see also Chapter 15 for an extensive discussion of nanomaterials used for imaging).

Nanomaterial synthesis has progressed rapidly in recent years and has generated complex state-of-the-art nano-scale structures with diverse and complex functionalities [17]. The ability to construct such nanomaterials has led to the development of multimodal and multifunctional nanoparticles to facilitate magnetic resonance imaging (MRI), computed tomography (CT) and positron emission tomography (PET) imaging (see also Chapters 4 and 15 for a discussion of these diagnostic imaging techniques) or for targeted drug delivery [19]. The development of such nanoparticles has led to the terms "theragnostic" or "theranostic" to describe a nanoparticle that can be used both for diagnostic and therapeutic applications [17,20].

Superparamagnetic Iron Oxide Nanoparticles (SPIONs)

In 1985, Stark et al. used MRI to determine the quantitative relationship between relaxation rates and iron levels in patients with iron overload [21] and it was quickly deduced that the increase in T2 relaxation rates in these patients could be artificially replicated and used to provide contrast in magnetic resonance imaging [22]. MRI imaging is used routinely to assess the extent of damage after a myocardial infarct or to monitor the extent of peripheral vascular disease [23]. SPIONs were originally used to provide contrast when imaging liver [24], spleen [25] and lymph nodes [26] as this is where these particles naturally accumulated. With improvements in nanoparticle production, iron oxide nanoparticles were designed to

TABLE 8.1 Nanomaterials in Clinical Use and Development for Imaging, Diagnosis and Therapy

Nanomaterial	Commercial Name	Application	Target	Adverse Effects	Manufacturer	Current Status
Metallic						
Iron oxide	Feridex	MRI	Liver	Back pain, vasodilation	Bayer Schering	FDA approved
	Resovist	MRI	Liver	None	Bayer Schering	FDA approved
	Combidex	MRI	Lymph nodes	None	Advanced Magnetics	In phase 3 clinical trials
	Nano Therm	Cancer therapy	Various	Acute urinary retention	MagForce	In phase 3 clinical trials
	Feraheme	Iron replacement therapy	RES macrophages of the liver, spleen and bone marrow	Diarrhea, nausea, dizziness, hypotension, constipation, peripheral edema anaphylactic-type reactions, cardiac arrest, tachycardia, angioedema, ischemic myocardial events.	AMAG Pharmaceuticals	FDA approved
Gold	Verigene	*In vitro* diagnostics	Genetic	N/A	Nanosphere	FDA approved
	Aurimmune	Cancer therapy	Various	Fever	CytImmune Sciences	In phase 2 clinical trials
Nanoshells	Auroshell	Cancer therapy	Head and neck	Under investigation	Nanospectra Biosciences	In phase 1 clinical trials
Semiconductor						
Quantum dot	Qdots, EviTags, semiconductor nanocrystals	Fluorescent contrast *In vitro* diagnostics	Tumors, cells, tissues and molecular sensing structures	N/A	Life Technologies, eBioscience, Nanoco, CrystalPlex, Cytodiagnostics	Research use only
Organic						
Protein	Abraxane	Cancer therapy	Breast	Cytopenia	Abraxis Bioscience	FDA approved
Liposome	Doxil/Caelyx	Cancer therapy	Various	Hand–foot syndrome, stomatitis	Ortho Biotech	FDA approved
Polymer	Oncaspar	Cancer therapy	Acute lymphoblastic leukemia	Urticaria, rash	Rhône-Poulenc Rorer	FDA approved
	CALAA-01	Cancer therapy	Various	Mild renal toxicity	Calando	In phase 2 clinical trials
Dendrimer	VivaGel	Microbicide	Cervicovaginal	Abdominal pain, dysuria	Starpharma	In phase 2 clinical trials
Micelle	Genexol-PM	Cancer therapy	Various	Peripheral sensory neuropathy, neutropenia	Samyang	For phase 4 clinical trials

be more biocompatible with increased circulation times to assist with MRI imaging [27]. Particle size and surface coating can be modified to target SPIONs to different sites. Larger particles, 50−150 nm in diameter, predominantly produce a signal decrease and are used as contrast media for imaging the liver and spleen. Smaller particles, less than 20 nm, are distributed more widely *in vivo* and can improve the identification of lymph nodes [28,29] or assist in the characterization of vulnerable atherothrombotic plaque. SPIONs have been used to enhance plaque imaging in a number of experimental models of atherosclerosis [30−33], however, the clinical studies published to date are less consistent [34−37].

Macrophages are the most common target for molecular imaging agents as there is a large body of evidence implicating them in the disease and, as a naturally phagocytic cell, they tend to scavenge foreign material without much manipulation. Several experimental studies have demonstrated that SPIONs or ultra-small paramagnetic iron oxide nanoparticles (USPIONs) can locate to atherosclerotic plaques and can be used to evaluate its metabolic state [31,35,38−40]. Targeted molecular imaging has been shown to result in localization of engineered nanoparticles within macrophages in atherosclerotic plaque [39,41,42].

Other applications of SPIONs include the assessment of myocardial perfusion [27,43] and characterization of aortic aneurysm formation as demonstrated in Figure 8.1 [35].

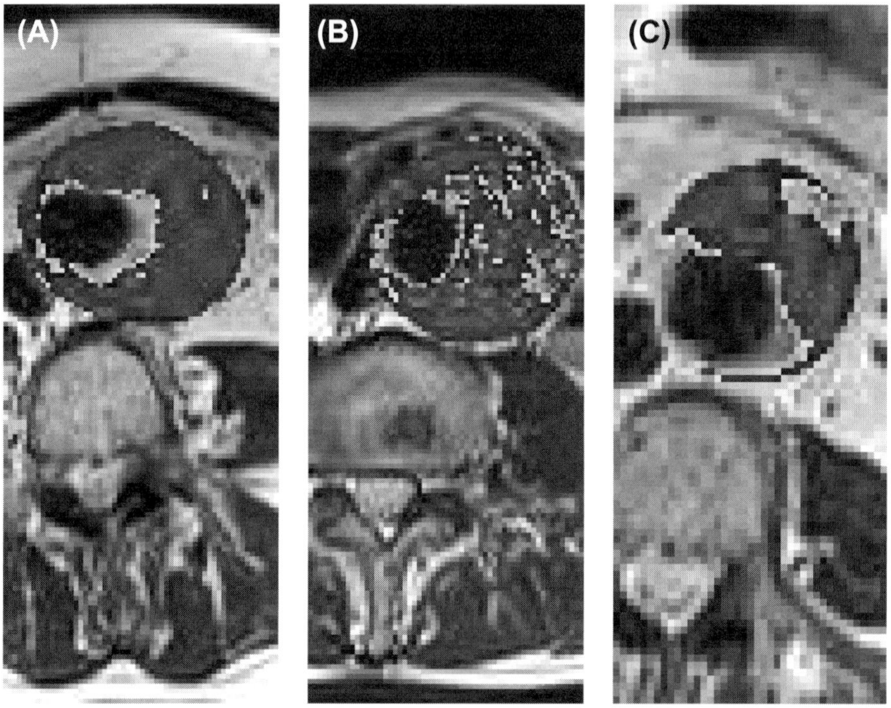

FIGURE 8.1 Representative abdominal aortic aneurysm (AAA) slices from patients showing differential patterns of uptake of 20 nm dextran-coated iron oxide nanoparticles corresponding to sites of cellular inflammation in sufficent amounts to cause a drop out in signal on an MRI scan. *(Reproduced from [35] with permission from the American Heart Association.)*

SPIONs with a modified coat have also been used in molecular imaging, such as receptor-directed imaging [44], cell labeling for *in vivo* monitoring of stem cell migration [45], and in labeling gene constructs for localization in gene therapy studies [46].

Examples of SPIONs in development or clinical use are given in Table 8.2 (adapted from [47]).

Lipid-Based Nanoparticles

Although more commonly associated with drug delivery, a number of studies have assessed the diagnostic potential of lipid-based nanoparticles [27,48,49]. Liposomes contain amphiphilic molecules, which have both hydrophobic and hydrophilic groups that self-assemble in water resulting in the formation of a hydrophilic core, whereas a micelle is a single layer of amphiphilic molecules with a hydrophobic core (Figure 8.2).

These constructs can be "loaded" with drugs, contrast agents, such as gadolinium, or gases for *in vivo* imaging [48,50,51]. Amirbekian et al. demonstrated that micelles containing gadolinium (Gd-micelles:100 nm) and targeted to the macrophage scavenging receptor CD204 enhanced MR imaging of atheroma compared to untargeted Gd-micelles, with increases in signal intensity directly related to the macrophage content of atheromatous plaque [52].

TABLE 8.2 Iron Oxide Based Nanomaterials in Clinical, Pre-Clinical and Development for Diagnosis and Therapy

Name	Status	Application	Administration	Relaxivities $mmol^{-1} sec^{-1}$	Coating	Core size (nm)
Lumirem and Gastromark	USA and Europe	GI	Oral	T2-72 T1-3.2	Silica	300
Abdoscan	Europe	GI	Oral	N/A	Polystyrene	300
Endorem/ Feriedex	USA	Liver/Spleen	IV	T2-98.3 T1-23.9	Dextran	5.6
Resovist	Withdrawn from some markets	Liver/Spleen	IV	T2-151.0 T1-25.4	Carbo-Dextran	4.2
Sinerem/ Combidex	Clinical trial	Lymph node Bone Marrow	IV	T2-44.1 T1-21.6	Dextran	4–6
Clariscan	Discontinued	Perfusion/ angiography	IV	T2-35 T1-20	Carbohydrate-PEG	5–7
Supravist	Preclinical	Perfusion Lymph node Bone Marrow	IV	T2-57 T1-7.3	Carbo-Dextran	3–5
MION 46	Preclinical	Angiography Lymph node Tumour Infarction	IV	N/A	Dextran	4–6
Feraheme	USA	Iron replacement therapy	IV	N/A	Polyglucose sorbitol carboxymethylether	2–10

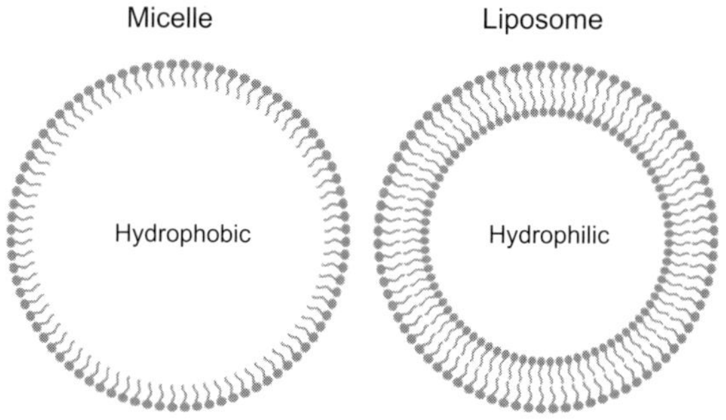

FIGURE 8.2 Lipid-based nano-constructs.

Briley-Saebo et al. used lipid-coated USPIONs and SPIONs functionalized to target oxidized phospholipids or malondialdehyde-lysine for plaque imaging and demonstrated signal loss on T2* weighted MR images of atherosclerotic lesions with preferential accumulation of the epitope-targeted PEG-nanoparticles within lipid-rich macrophage foam cells [49]. The authors suggest this was due to binding of particles to extracellular oxidized low density lipoproteins (LDL) at these sites, or binding to oxidized LDL on macrophage scavenger receptors and enhanced uptake. They conclude that these particles may provide a platform to target vulnerable atherosclerotic plaques [49]. Combining nanoparticle technologies may yield more versatile particles with both diagnostic and therapeutic applications.

Gold

Certain metal nanoparticles, such as gold, with the application of light, generate a strong surface plasmon field by oscillating electrons around the metal surface. The field strength may depend upon the metal type, particle size and the wavelength of light. Gold is a metal of interest as it is considered chemically inert and can generate an intense surface plasmon field. Therefore, several protocols have been generated in order to obtain optimal dispersion, size and shape of gold nanoparticles for diagnostic purposes. At present, most colloidal synthesis protocols are based around the reduction of chloroauric acid, $HauCl_4$, in the presence of citrate to produce monodisperse particles with controlled equivolume diameters of 10 to 60 nm [53]. Different reduction agents are used if other shaped nanoparticles are desired. Rod-shaped gold nanoparticles are useful for near-infrared imaging and in combination with photoacoustic imaging show promise in detection and diagnosis of cardiovascular disease [54]. van Schooneveld et al. developed a tri-modal gold nanoparticle for enhanced imaging using MRI, CT and FI of macrophage cells *in vitro* and in the liver *in vivo* [55] with potential applications for imaging chronic inflammatory diseases.

PARTICOKINETICS OF NANOMATERIALS FOR DIAGNOSTIC USE

Understanding the particokinetics of engineered nanomaterials is a critical consideration when designing them for biomedical uses (see also Chapter 1 for a discussion on this subject).

Engineered nanoparticles accumulate differentially throughout the body depending on their physiochemical properties; characteristics of the NP such as size, surface area, charge and surface chemistry as well as the method by which they are introduced (intravenous infusion, subdermal injection, inhalation or ingestion) determine the ultimate fate of the NP once inside the body.

Surface characteristics are important in determining whether or not a particle will be cleared by the recticuloendothelial system after intravenous administration [56]. When hydrophobic nanoparticles are injected directly into the circulation, they rapidly become opsonized by blood serum proteins and are cleared by the reticuloendothelial system. However, this opsonization and first pass clearance can be limited by alteration of the surface coatings of the nanoparticles. The most common method of prolonging circulation is conjugation of the nanoparticle with polyethylene glycol (PEG) polymer, providing the particle with steric hindrance to opsonization or phagocytosis [57] and allowing the PEGylated form to remain in the circulation for much longer, increasing interaction with blood components. A number of studies have demonstrated altered distribution patterns based on modification of the surface coating. Beduneau [25] covalently conjugated IgG molecules to the surface of SPIONs showing accelerated uptake by monocytes and enhanced retention *in vivo* without affecting monocyte viability. This technique could be very useful in the monitoring and tracking of disease progression. It is clear that surface modification has the ability to alter the distribution and clearance patterns of nanoparticles in the body, with small changes in surface moiety resulting in major differences in behavior of these particles in the body [58,59].

Hydrodynamic size also determines distribution [60]. Normally, in a healthy individual, the endothelium is very effective at preventing extravasation of nanoparticles due to the presence of tight gap junctions (<2 nm in diameter). However, when the endothelium is compromised, as is the case in atherosclerosis, these normally tight junctions become leaky and could potentially allow passage of nanoparticles to the underlying basement membrane. The final destination of particles in the body has been reported to be directly related to their size; larger particles, greater than 100 nm in diameter, deposit in the spleen due to its discontinuous endothelium with large fenestrations and smaller particles, less than 100 nm in diameter, deposit in the liver passing through the fenestrated endothelium and finally residing in the underlying parenchymal cells. Kinetic studies with gold nanoparticles have shown their pattern of distribution to be size dependent [61]. De Jong et al. [60] intravenously administered gold nanoparticles in rats and found that the biodistribution was dependent on particle size with the smallest particles (10 nm) having the most widespread organ distribution. Table 8.3 illustrates the size-dependent biodistribution of systemic nanoparticles.

ADVERSE EFFECTS OF COMBUSTION DERIVED NANOPARTICLES IN CARDIOVASCULAR DISEASE

While nanotechnology aims to develop new engineered NPs with novel properties, there are numerous sources of NPs in the environment from natural and anthropogenic sources. There have been many epidemiological, *in vitro*, *in vivo* and *ex vivo* studies demonstrating

TABLE 8.3 Size Based Biodistribution of Nanomaterials

Hydrodynamic Diameter	Tissue Uptake	Renal Excretion
>30nm	Rapidly cleared by reticuloendothelial system (RES)	Not excreted by kidneys, excretion involves degradation of NP in RES tissue
10–30nm	Longer circulation time leads to increased uptake by non RES tissues	Little, <5%, excretion by renal system excretion involves degradation of NP in RES tissue
4–10nm	Longer circulation time leads to increased uptake by non RES tissues	Renal clearance increases as particle size decreases
<4nm	Uptake in most cell types for particles not excreted by kidneys	Rapid clearance of nanoparticles by kidneys with retention of between 5-10% of injected dose

the adverse health effects of PM_{10} with the majority of investigators identifying the "ultrafine"' or nanoparticulate component as the most toxic of airborne pollutants [5–7,62–65]. These studies have raised concerns about the potential for engineered NPs to exert similar adverse health effects. Exposure to ambient particles is associated with more hospitalizations with acute cardiovascular events and an increase in cardiovascular deaths [66–69]. These effects are thought to be mediated by particle-induced oxidative stress and inflammation leading to vascular dysfunction, impaired fibrinolysis, and changes in the autonomic regulation of heart rate and rhythm [69,70]. Combustion derived nanoparticles (CDNPs) have been shown to be prothrombotic [9] and may contribute to the formation of atherosclerotic plaques. While CDNPs differ greatly in chemical composition compared to engineered nanoparticles, they share similarities in other characteristics considered toxicologically significant, such as size and surface area. It is therefore plausible that engineered nanoparticles will exert similar pro-inflammatory and prothrombotic effects; especially those manufactured for direct intravenous administration as a contrast agent in medical imaging [71,72].

Given the huge amount of published data implicating CDNPs in the pathogenesis of cardiovascular diseases [73], an understanding of the toxicology of nanoparticles intended for medical use is central to their future use. Mills et al. [3] provide an overview of the most widely accepted mechanisms for the adverse cardiovascular effects of CDNP.

TOXICITY OF ENGINEERED NANOPARTICLES IN CARDIOVASCULAR DISEASE

It is not surprising that a number of recent reviews focusing on the effects of occupational exposure to engineered nanomaterials have raised concerns that they could have similar health implications as CDNP [74–76]. Resolution of these issues will require a collaborative approach between industry, government, academic researchers and clinicians. In addition to occupational exposure to engineered nanomaterials, the direct exposures to engineered nanoparticles through medical use in susceptible populations is of immediate concern. Engineered nanomaterials unlike novel pharmacological therapies are not routinely tested to determine their effect on the cardiovascular system prior to clinical use. Tables 8.1 and 8.2 highlight the many different nanoparticles being developed for clinical use. All of these nanoparticles, not just those intended for the diagnosis and treatment of cardiovascular disease, have the potential to interact with cells associated with the progression of cardiovascular disease. To date, there are no observational studies linking exposure to engineered nanomaterials and cardiovascular events, however, a number of preclinical studies have reported that exposure to engineered nanomaterials, directly or indirectly, can cause endothelial dysfunction and platelet activation, and may plausibly contribute to ischemic heart disease [77–80]. The schematic diagrams depicted in Figure 8.3 and Figure 8.4 identify mechanisms and potential targets for engineered nanoparticles in circulation.

In vitro Studies

Cardiovascular disease is a complex and dynamic process with many contributing factors and therefore a single *in vitro* model to study the adverse effects of engineered nanomaterials on cardiovascular system is unlikely to emerge. However, platelets, platelet–monocyte aggregation and the vascular endothelium have emerged as useful *in vitro* systems given their key role in the pathogenesis of coronary and ischemic heart disease [81,82].

Compatibility of nanoparticles with blood components is important and a number of studies have employed assays of hemolysis [83], platelet aggregation and activation [84] and platelet–leukocyte aggregation [77]. Direct and indirect toxicity to the cardiovascular system can be determined using cell lines with assays of oxidative stress, inflammation and cell death [85]. Whether observations from these *in vitro* models apply *in vivo* or in the clinic is uncertain and, in the absence of any standards for the assessment of the cardiovascular toxicity of engineered nanomaterials, the role of *in vitro* assessments remains controversial.

Recent studies have shown that engineered nanoparticles can interact with platelets causing activation, aggregation [77,84,86] and hyperreactivity (unpublished observations). Surface charge is thought to be a critical determinant in platelet–nanoparticle activation [77].

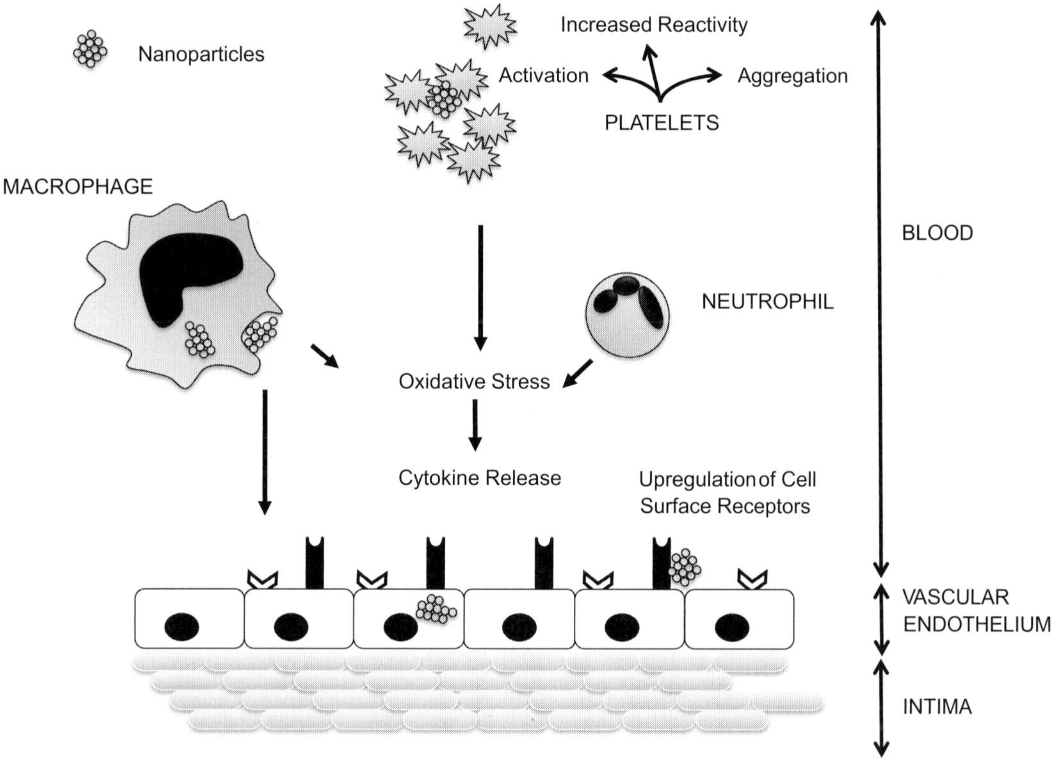

FIGURE 8.3 Overview of interactions of blood-borne engineered nanoparticles.

FIGURE 8.4 Pathways of nanoparticle-mediated cardiovascular morbidity and death.

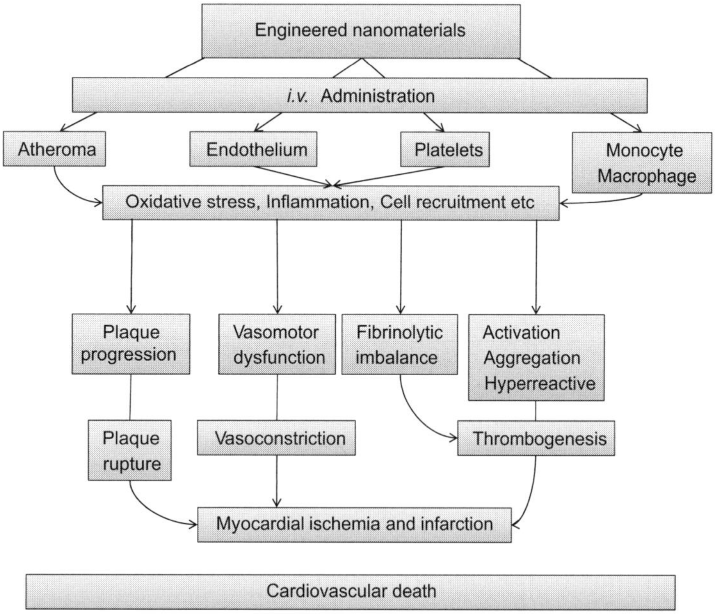

When studying the impact of engineered nanomaterials on a complex multisystem disease, it is important to take into consideration potential downstream, indirect effects as well as direct toxicity. Shaw et al. [87] demonstrated that CDNP induce moderate toxicity and inflammatory effects on endothelial cells following direct incubation, these effects were amplified when endothelial cells were exposed to conditioned supernatant from monocyte-derived macrophages exposed to CDNP due to the release of cytokines from CDNP-laden macrophages. This approach utilizing co-culture and other techniques that model the secondary effects of CDNP [88–90] could be very useful in studying the cardiovascular effects of engineered nanomaterials. Nanomaterials are increasingly being used in cell tracking to monitor disease progression [43,91,92]. It is no longer sufficient to assess the potential toxicity of engineered nanoparticles on the cell carriers alone; it is essential that these cells do not become "time bombs" capable of releasing vast quantities of pro-inflammatory mediators in a vulnerable atherosclerotic plaque. This mechanism may be equally important in the occupational setting where inhalation of nanomaterials has been shown to have systemic effects on the cardiovascular system [93–97].

In vivo Studies

In vivo studies carried out in susceptible animals have contributed significantly to our understanding of cardiovascular disease. In the ApoE knock-out mouse, CDNP instilled into the lungs increases atherogenesis [10,98,99]. Indeed, engineered nanomaterials such as carbon nanotubes and titanium dioxide can accelerate plaque progression in susceptible animal models [74]. Mills et al. exposed healthy human volunteers to dilute diesel exhaust, pure carbon nanoparticles, filtered diesel exhaust, or filtered air, in a randomized double blind cross-over study demonstrating that the adverse vascular effects of diesel exhaust could be attributed mainly to the nanoparticulate component and not to the associated gaseous pollutants [100]. Observations *in vivo* were replicated *in vitro* using wire-myography providing a rationale for testing the effects of engineered nanomaterials on vascular function before introduction to the clinic.

Ex vivo Studies

Ex vivo studies have been very informative in modeling the effects of CDNP on cardiovascular disease and have potential use in determining the effects of engineered nanomaterials on cardiovascular disease. They allow measurements to be taken in tissues or artificial environments outside the organism but with as little as possible deviation from natural conditions with no risk to the patient. *Ex vivo* techniques have not been used widely to study the effects of engineered nanomaterials on the cardiovascular system due to the need for specialist equipment, training, and the high cost involved, however, given the potential to generate meaningful data in the near patient setting, these systems are worth consideration. *Ex vivo* techniques also side-step possible ethical issues associated with exposing healthy or disease populations to the unknown effects of engineered nanoparticles. Traditional *ex vivo* models have assessed the direct effects of CDNP in isolated vessels to assess vasomotor dysfunction. This chapter does not intend to review previous studies as they have largely focused on CDNP, however, this topic is very well covered by many comprehensive reviews (see Møller et al. [74]) and the findings could easily be applied to the study of the adverse effects of engineered nanomaterial [74].

A recent *ex vivo* study used the Langendorff heart as a model system to study the cardiovascular effects of engineered nanoparticles [101]. The isolated heart is a well-established method of modeling cardiac dysfunction and is used extensively to monitor electrophysiological parameters after exposure to investigational products. This model assesses the direct effects of engineered nanoparticles on the intact heart without the contribution from other systemic effects [74]. The authors report that silicon dioxide and titanium dioxide nanoparticles, as well as engineered nanoparticles made of flame soot, caused a significant dose-dependent increase in heart rate and promoted cardiac arrhythmia [101]. They hypothesize that the nanoparticle-induced increase in heart rate might be caused by the release of catecholamines from the sympathetic nerve endings in the heart, which was supported by studies using β1-receptor antagonism post-excision and by depleting catecholamine stores in the neurons pre-excision.

As many medical nanoparticles in development are for use in populations considered particularly vulnerable, it would be desirable to model this *ex vivo*. The Badimon chamber is an *ex vivo* model of deep arterial injury and thrombus formation used previously to study the effects of diesel exhaust inhalation on thrombosis in humans [10,102]. The chamber permits an assessment of thrombus formation in whole native blood without the need for anticoagulants or exposure of the patient to the compounds being tested. While principally used to assess the impact of novel antithrombotic agents, it can just as easily be used to assess the effects of engineered nanomaterials on thrombosis. A pump is used to draw blood from an ante-cubital vein through the chambers maintained at 37°C in a water bath. The rheological conditions in the first chamber simulate those of patent coronary arteries (low-shear rate), whereas those in the second and third chambers simulate those of stenosed coronary artery (high-shear rate) [10]. The thrombus formed in the low shear chamber is macrophage rich whereas the thrombus formed in high shear conditions is platelet rich [10] (Figure 8.5). The chamber is therefore potentially able to assess nanoparticle interactions with specific

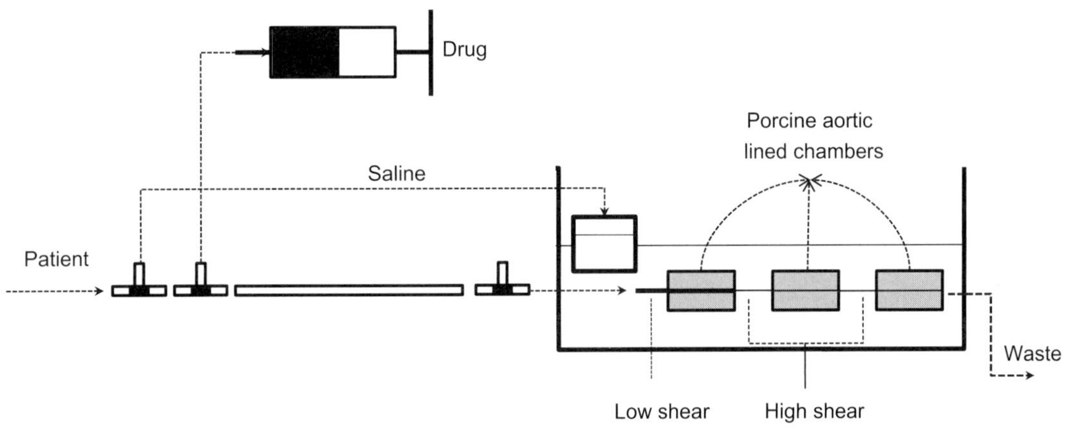

FIGURE 8.5 Schematic of Badimon chamber set-up.

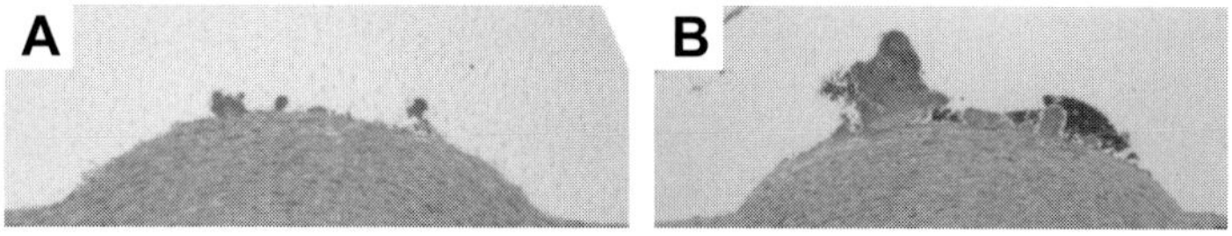

FIGURE 8.6 Porcine aortic strips from the Badimon chamber. Strip A represents a saline control run. Strip B represents a run with PEGylated quantum dots. Magnification: ×100.

cells and types of thrombus. Our preliminary investigations suggest that engineered nanomaterials may increase thrombus formation and that surface charge may be an important determinant (Figure 8.6).

TAKE-HOME MESSAGES

Engineered nanomaterials are likely to have a big impact on diagnostic assays, medical imaging and targeted therapeutics in the near future. Characteristics of these nanoparticles, such as size, surface area, charge and surface chemistry, as well as the method by which they are delivered, will determine their distribution, efficacy and potential toxicity. Nanoparticles designed for medical use are usually derived from materials deemed biologically safe, but it remains unclear how we define a "biologically safe"' material and, in particular, what investigations should be mandatory in the development of these materials. Unfortunately, the behavior of nanoparticles cannot be determined from their bulk properties; when relatively inert or safe materials in their bulk form are presented in the nano-scale they may exert previously unseen harmful effects on cells and systems. Several potential adverse effects on the cardiovascular system are indeed possible, as suggested by available *in vitro*, *in vivo* and *ex vivo* studies. If we are to maximize the potential of nanotechnologies in the field of medicine, we must ensure that novel nanomaterials are rigorously tested before introduction to the clinic.

References

[1] Medina C, Santos-Martinez MJ, Radomski A, Corrigan OI, Radomski MW. Nanoparticles: Pharmacological and toxicological significance. Br J Pharmacol 2007;150:552–8.

[2] Linkov I, Satterstrom FK, Corey LM. Nanotoxicology and nanomedicine: Making hard decisions. Nanomedicine 2008;4:167–71.

[3] Mills NL, Donaldson K, Hadoke PW, et al. Adverse cardiovascular effects of air pollution. Nat Clin Pract Cardiovasc Med 2008;6:36–44.

[4] Mills NL, Robinson SD, Fokkens PH, et al. Exposure to concentrated ambient particles does not affect vascular function in patients with coronary heart disease. Environ Health Perspect 2008;116:709–15.

[5] Donaldson K, Brown DM, Mitchell C, et al. Free radical activity of PM10: Iron-mediated generation of hydroxyl radicals. Environ Health Perspect 1997;105:1285–9.

[6] Renwick LC, Brown D, Clouter A, Donaldson K. Increased inflammation and altered macrophage chemotactic responses caused by two ultrafine particle types. Occup Environ Med 2004;61:442–7.

[7] Barlow PG, Clouter-Baker A, Donaldson K, Maccallum J, Stone V. Carbon black nanoparticles induce type II epithelial cells to release chemotaxins for alveolar macrophages. Part Fibre Toxicol 2005;2:11.

[8] Barath S, Mills NL, Lundback M, et al. Impaired vascular function after exposure to diesel exhaust generated at urban transient running conditions. Part Fibre Toxicol 2010;7:19.

[9] Mills NL, Tornqvist H, Gonzalez MC, et al. Ischemic and thrombotic effects of dilute diesel-exhaust inhalation in men with coronary heart disease. N Engl J Med 2007;357:1075–82.
[10] Lucking AJ, Lundback M, Mills NL, et al. Diesel exhaust inhalation increases thrombus formation in man. Eur Heart J 2008;29:3043–51.
[11] Hoffmann B, Moebus S, Kroger K, et al. Residential exposure to urban air pollution, ankle-brachial index, and peripheral arterial disease. Epidemiology 2009;20:280–8.
[12] Wang TJ. Assessing the role of circulating, genetic, and imaging biomarkers in cardiovascular risk prediction. Circulation 2011;123:551–65.
[13] Viera AJ, Sheridan SL. Global risk of coronary heart disease: Assessment and application. Am Fam Physician 2010;82:265–74.
[14] van Gils JM, Zwaginga JJ, Hordijk PL. Molecular and functional interactions among monocytes, platelets, and endothelial cells and their relevance for cardiovascular diseases. J Leukoc Biol 2009;85:195–204.
[15] Slevin M, Wang Q, Font MA, et al. Atherothrombosis and plaque heterology: Different location or a unique disease? Pathobiology 2008;75:209–25.
[16] Song S, Qin Y, He Y, Huang Q, Fan C, Chen HY. Functional nanoprobes for ultrasensitive detection of biomolecules. Chem Soc Rev 2010;39:4234–43.
[17] Cormode DP, Skajaa T, Fayad ZA, Mulder WJ. Nanotechnology in medical imaging: Probe design and applications. Arterioscler Thromb Vasc Biol 2009;29:992–1000.
[18] Kim BY, Rutka JT, Chan WC. Nanomedicine. N Engl J Med 2010;363:2434–43.
[19] Kim D, Yu MK, Lee TS, Park JJ, Jeong YY, Jon S. Amphiphilic polymer-coated hybrid nanoparticles as CT/MRI dual contrast agents. Nanotechnology 2011;22:155101.
[20] Schladt TD, Schneider K, Schild H, Tremel W. Synthesis and bio-functionalization of magnetic nanoparticles for medical diagnosis and treatment. Dalton Trans 2011;40:6315–43.
[21] Stark DD, Moseley ME, Bacon BR, et al. Magnetic resonance imaging and spectroscopy of hepatic iron overload. Radiology 1985;154:137–42.
[22] Renshaw PF, Owen CS, McLaughlin AC, Frey TG, Leigh Jr JS. Ferromagnetic contrast agents: A new approach. Magn Reson Med 1986;3:217–25.
[23] Uppal R, Caravan P. Targeted probes for cardiovascular MRI. Future Med Chem 2010;2:451–70.
[24] Fukukura Y, Kamiyama T, Takumi K, Shindo T, Higashi R, Nakajo M. Comparison of ferucarbotran-enhanced fluid-attenuated inversion-recovery echo-planar, T2-weighted turbo spin-echo, T2*-weighted gradient-echo, and diffusion-weighted echo-planar imaging for detection of malignant liver lesions. J Magn Reson Imaging 2010;31:607–16.
[25] Beduneau A, Ma Z, Grotepas CB, et al. Facilitated monocyte-macrophage uptake and tissue distribution of superparmagnetic iron-oxide nanoparticles. PLoS One 2009;4:e4343.
[26] Weidner AM, van Lin EN, Dinter DJ, et al. Ferumoxtran-10 MR lymphography for target definition and follow-up in a patient undergoing image-guided, dose-escalated radiotherapy of lymph nodes upon PSA relapse. Strahlenther Onkol 2011;187:206–12.
[27] Krombach GA, Wendland MF, Higgins CB, Saeed M. MR imaging of spatial extent of microvascular injury in reperfused ischemically injured rat myocardium: Value of blood pool ultrasmall superparamagnetic particles of iron oxide. Radiology 2002;225:479–86.
[28] Harisinghani MG, Barentsz J, Hahn PF, et al. Noninvasive detection of clinically occult lymph-node metastases in prostate cancer. N Engl J Med 2003;348:2491–9.
[29] Harisinghani MG, Saini S, Slater GJ, Schnall MD, Rifkin MD. MR imaging of pelvic lymph nodes in primary pelvic carcinoma with ultrasmall superparamagnetic iron oxide (Combidex): Preliminary observations. J Magn Reson Imaging 1997;7:161–3.
[30] Klug G, Bauer L, Bauer WR. Patterns of USPIO deposition in murine atherosclerosis. Arterioscler Thromb Vasc Biol 2008;28:E157–9.
[31] Klug K, Gert G, Thomas K, et al. Murine atherosclerotic plaque imaging with the USPIO Ferumoxtran-10. Front Biosci 2009;14:2546–52.
[32] Sigovan M, Boussel L, Sulaiman A, et al. Rapid-clearance iron nanoparticles for inflammation imaging of atherosclerotic plaque: Initial experience in animal model. Radiology 2009;252:401–9.
[33] Briley-Saebo KC, Mani V, Hyafil F, Cornily JC, Fayad ZA. Fractionated Feridex and positive contrast: *In vivo* MR imaging of atherosclerosis. Magn Reson Med 2008;59:721–30.

[34] Morishige K, Kacher DF, Libby P, et al. High-resolution magnetic resonance imaging enhanced with superparamagnetic nanoparticles measures macrophage burden in atherosclerosis. Circulation 2010;122:1707–15.
[35] Richards JM, Semple SI, Macgillivray TJ, et al. Abdominal aortic aneurysm growth predicted by uptake of ultrasmall superparamagnetic particles of iron oxide: A pilot study. Circ Cardiovasc Imaging 2011;4:274–81.
[36] Tang TY, Howarth SP, Miller SR, et al. Correlation of carotid atheromatous plaque inflammation using USPIO-enhanced MR imaging with degree of luminal stenosis. Stroke 2008;39:2144–7.
[37] Yilmaz A, Rosch S, Klingel K, et al. Magnetic resonance imaging (MRI) of inflamed myocardium using iron oxide nanoparticles in patients with acute myocardial infarction—Preliminary results. Int J Cardiol 2011 (Epub ahead of print).
[38] Trivedi RA, King-Im JM, Graves MJ, et al. *In vivo* detection of macrophages in human carotid atheroma: Temporal dependence of ultrasmall superparamagnetic particles of iron oxide-enhanced MRI. Stroke 2004;35:1631–5.
[39] Kooi ME, Cappendijk VC, Cleutjens KB, et al. Accumulation of ultrasmall superparamagnetic particles of iron oxide in human atherosclerotic plaques can be detected by *in vivo* magnetic resonance imaging 2. Circulation 2003;107:2453–8.
[40] Schmitz SA, Taupitz M, Wagner S, Wolf KJ, Beyersdorff D, Hamm B. Magnetic resonance imaging of atherosclerotic plaques using superparamagnetic iron oxide particles. J Magn Reson Imaging 2001;14:355–61.
[41] Trivedi R, King-Im J, Gillard J. Accumulation of ultrasmall superparamagnetic particles of iron oxide in human atherosclerotic plaque. Circulation 2003;108:e140.
[42] Ruehm SG, Goyen M, Barkhausen J, et al. Rapid magnetic resonance angiography for detection of atherosclerosis. Lancet 2001;357:1086–91.
[43] Jing XH, Yang L, Duan XJ, et al. *In vivo* MR imaging tracking of magnetic iron oxide nanoparticle labeled, engineered, autologous bone marrow mesenchymal stem cells following intra-articular injection. Joint Bone Spine 2008;75:432–8.
[44] Eck W, Nicholson AI, Zentgraf H, Semmler W, Bartling S. Anti-CD4-targeted gold nanoparticles induce specific contrast enhancement of peripheral lymph nodes in X-ray computed tomography of live mice. Nano Lett 2010;10:2318–22.
[45] Thu MS, Najbauer J, Kendall SE, et al. Iron labeling and pre-clinical MRI visualization of therapeutic human neural stem cells in a murine glioma model. PLoS One 2009;4:e7218.
[46] Zhao Y, Zhao L, Zhou L, et al. Quantum dot conjugates for targeted silencing of bcr/abl gene by RNA interference in human myelogenous leukemia K562 cells. J Nanosci Nanotechnol 2010;10:5137–43.
[47] Lodhia J, Mandarano G, Ferris N, Eu P, Cowell S. Development and use of iron oxide nanoparticles (Part 1): Synthesis of iron oxide nanoparticles for MRI. Biomed Imaging Interv J 2010;6:e12.
[48] Liu Z, Lammers T, Ehling J, et al. Iron oxide nanoparticle-containing microbubble composites as contrast agents for MR and ultrasound dual-modality imaging. Biomaterials 2011;32:6155–63.
[49] Briley-Saebo KC, Cho YS, Shaw PX, et al. Targeted iron oxide particles for *in vivo* magnetic resonance detection of atherosclerotic lesions with antibodies directed to oxidation-specific epitopes. J Am Coll Cardiol 2011;57:337–47.
[50] Yang F, Li L, Li Y, Chen Z, Wu J, Gu N. Superparamagnetic nanoparticle-inclusion microbubbles for ultrasound contrast agents. Phys Med Biol 2008;53:6129–41.
[51] Mulder WJ, Strijkers GJ, Briley-Saboe KC, et al. Molecular imaging of macrophages in atherosclerotic plaques using bimodal PEG-micelles. Magn Reson Med 2007;58:1164–70.
[52] Amirbekian V, Lipinski MJ, Briley-Saebo KC, et al. Detecting and assessing macrophages *in vivo* to evaluate atherosclerosis noninvasively using molecular MRI. Proc Natl Acad Sci USA 2007;104:961–6.
[53] Khlebtsov N, Dykman L. Biodistribution and toxicity of engineered gold nanoparticles: A review of *in vitro* and *in vivo* studies. Chem Soc Rev 2011;40:1647–71.
[54] Pan D, Pramanik M, Senpan A, Wickline SA, Wang LV, Lanza GM. A facile synthesis of novel self-assembled gold nanorods designed for near-infrared imaging. J Nanosci Nanotechnol 2010;10:8118–23.
[55] van Schooneveld MM, Cormode DP, Koole R, et al. A fluorescent, paramagnetic and PEGylated gold/silica nanoparticle for MRI, CT and fluorescence imaging. Contrast Media Mol Imaging 2010;5:231–6.
[56] Al-Jamal WT, Al-Jamal KT, Cakebread A, Halket JM, Kostarelos K. Blood circulation and tissue biodistribution of lipid–quantum dot (L-QD) hybrid vesicles intravenously administered in mice. Bioconjug Chem 2009;20:1696–702.

[57] Li SD, Huang L. Pharmacokinetics and biodistribution of nanoparticles. Mol Pharm 2008;5:496–504.
[58] van Tilborg GA, Strijkers GJ, Pouget EM, et al. Kinetics of avidin-induced clearance of biotinylated bimodal liposomes for improved MR molecular imaging. Magn Reson Med 2008;60:1444–56.
[59] Jayagopal A, Russ PK, Haselton FR. Surface engineering of quantum dots for *in vivo* vascular imaging. Bioconjug Chem 2007;18:1424–33.
[60] De Jong WH, Hagens WI, Krystek P, Burger MC, Sips AJ, Geertsma RE. Particle size-dependent organ distribution of gold nanoparticles after intravenous administration. Biomaterials 2008;29:1912–9.
[61] Hirn S, Semmler-Behnke M, Schleh C, et al. Particle size-dependent and surface charge-dependent biodistribution of gold nanoparticles after intravenous administration. Eur J Pharm Biopharm 2011;77:407–16.
[62] Mills NL, Robinson SD, Fokkens PH, et al. Exposure to concentrated ambient particles does not affect vascular function in patients with coronary heart disease. Environ Health Perspect 2008;116:709–15.
[63] Geiser M, Rothen-Rutishauser B, Kapp N, et al. Ultrafine particles cross cellular membranes by nonphagocytic mechanisms in lungs and in cultured cells. Environ Health Perspect 2005;113:1555–60.
[64] Moller W, Brown DM, Kreyling WG, Stone V. Ultrafine particles cause cytoskeletal dysfunctions in macrophages: Role of intracellular calcium. Part Fibre Toxicol 2005;2:7.
[65] Brown LM, Collings N, Harrison RM, Maynard AD, Maynard RL. Ultrafine particles in the atmosphere. London: Imperial College Press; 2003.
[66] Brook RD, Brook JR, Urch B, Vincent R, Rajagopalan S, Silverman F. Inhalation of fine particulate air pollution and ozone causes acute arterial vasoconstriction in healthy adults. Circulation 2002;105:1534–6.
[67] Delfino RJ, Sioutas C, Malik S. Potential role of ultrafine particles in associations between airborne particle mass and cardiovascular health. Environ Health Perspect 2005;113:934–46.
[68] Mills NL, Miller JJ, Anand A, et al. Increased arterial stiffness in patients with chronic obstructive pulmonary disease; A mechanism for increased cardiovascular risk. Thorax 2007;63:306–11.
[69] Mills NL, Donaldson K, Hadoke PW, et al. Adverse cardiovascular effects of air pollution. Nat Clin Pract Cardiovasc Med 2009;6:36–44.
[70] Langrish JP, Mills NL, Newby DE. Air pollution: the new cardiovascular risk factor. Intern Med J 2008;38: 875–8.
[71] Donaldson K. Resolving the nanoparticles paradox. Nanomedicine 2006;1:229–34.
[72] Donaldson K, Seaton A. The Janus faces of nanoparticles. J Nanosci Nanotechnol 2007;7:4607–11.
[73] Donaldson K, Tran L, Jimenez L, et al. Combustion-derived nanoparticles: A review of their toxicology following inhalation exposure 1. Part Fibre Toxicol 2005;2:10.
[74] Møller P, Mikkelsen L, Vesterdal LK, et al. Hazard identification of particulate matter on vasomotor dysfunction and progression of atherosclerosis. Crit Rev Toxicol 2011;41:339–68.
[75] Eisen EA, Costello S, Chevrier J, Picciotto S. Epidemiologic challenges for studies of occupational exposure to engineered nanoparticles; a commentary. J Occup Environ Med 2011;53:S57–61.
[76] Castranova V. Overview of current toxicological knowledge of engineered nanoparticles. J Occup Environ Med 2011;53:S14–7.
[77] McGuinnes C, Duffin R, Brown S, et al. Surface derivatization state of polystyrene latex nanoparticles determines both their potency and their mechanism of causing human platelet aggregation *in vitro*. Toxicol Sci 2011;119:359–68.
[78] Zhu MT, Wang B, Wang Y, et al. Endothelial dysfunction and inflammation induced by iron oxide nanoparticle exposure: Risk factors for early atherosclerosis. Toxicol Lett 2011;203:162–71.
[79] Zhu MT, Wang Y, Feng WY, et al. Oxidative stress and apoptosis induced by iron oxide nanoparticles in cultured human umbilical endothelial cells. J Nanosci Nanotechnol 2010;10:8584–90.
[80] Sun J, Wang S, Zhao D, Hun FH, Weng L, Liu H. Cytotoxicity, permeability, and inflammation of metal oxide nanoparticles in human cardiac microvascular endothelial cells: Cytotoxicity, permeability, and inflammation of metal oxide nanoparticles. Cell Biol Toxicol 2011;27:333–42.
[81] Neun BW, Dobrovolskaia MA. Method for *in vitro* analysis of nanoparticle thrombogenic properties. Meth Mol Biol 2011;697:225–35.
[82] Bihari P, Holzer M, Praetner M, et al. Single-walled carbon nanotubes activate platelets and accelerate thrombus formation in the microcirculation. Toxicology 2010;269:148–54.
[83] Lu S, Duffin R, Poland C, et al. Efficacy of simple short-term *in vitro* assays for predicting the potential of metal oxide nanoparticles to cause pulmonary inflammation. Environ Health Perspect 2009;117:241–7.

[84] Radomski A, Jurasz P, onso-Escolano D, et al. Nanoparticle-induced platelet aggregation and vascular thrombosis 3. Br J Pharmacol 2005;146:882–93.
[85] Dick CA, Brown DM, Donaldson K, Stone V. The role of free radicals in the toxic and inflammatory effects of four different ultrafine particle types. Inhal Toxicol 2003;15:39–52.
[86] Miller VM, Hunter LW, Chu K, et al. Biologic nanoparticles and platelet reactivity. Nanomedicine (Lond) 2009;4:725–33.
[87] Shaw CA, Robertson S, Miller MR, et al. Diesel exhaust particulate-exposed macrophages cause marked endothelial cell activation. Am J Respir Cell Mol Biol 2011;44:840–51.
[88] Inoue K, Takano H. Another systemic impact of inhaled diesel exhaust particles. Toxicol Sci 2010;115:607–8.
[89] Ishii H, Hayashi S, Hogg JC, et al. Alveolar macrophage–epithelial cell interaction following exposure to atmospheric particles induces the release of mediators involved in monocyte mobilization and recruitment. Respir Res 2005;6:87.
[90] Porter M, Karp M, Killedar S, et al. Diesel-enriched particulate matter functionally activates human dendritic cells. Am J Respir Cell Mol Biol 2007;37:706–19.
[91] Jendelova P, Herynek V, Urdzikova L, et al. Magnetic resonance tracking of transplanted bone marrow and embryonic stem cells labeled by iron oxide nanoparticles in rat brain and spinal cord. J Neurosci Res 2004;76:232–43.
[92] Voura EB, Jaiswal JK, Mattoussi H, Simon SM. Tracking metastatic tumor cell extravasation with quantum dot nanocrystals and fluorescence emission-scanning microscopy. Nat Med 2004;10:993–8.
[93] Adamkiewicz G, Ebelt S, Syring M, et al. Association between air pollution exposure and exhaled nitric oxide in an elderly population. Thorax 2004;59:204–9.
[94] Antonini JM. Health effects of welding. Crit Rev Toxicol 2003;33:61–103.
[95] Behndig AF, Mudway IS, Brown JL, et al. Airway antioxidant and inflammatory responses to diesel exhaust exposure in healthy humans 11. Eur Respir J 2006;27:359–65.
[96] Brook RD, Brook JR, Rajagopalan S. Air pollution: The "heart" of the problem. Curr Hypertens Rep 2003;5:32–9.
[97] faro-Moreno E, Nawrot TS, Nemmar A, Nemery B. Particulate matter in the environment: Pulmonary and cardiovascular effects. Curr Opin Pulm Med 2007;13:98–106.
[98] Tornqvist H, Mills NL, Gonzalez M, et al. Persistent endothelial dysfunction in humans after diesel exhaust inhalation. Am J Respir Crit Care Med 2007;176:395–400.
[99] Jacobsen NR, Moller P, Jensen KA, et al. Lung inflammation and genotoxicity following pulmonary exposure to nanoparticles in ApoE-/- mice. Part Fibre Toxicol 2009;6:2.
[100] Mills NL, Miller MR, Lucking AJ, et al. Combustion-derived nanoparticulate induces the adverse vascular effects of diesel exhaust inhalation. Eur Heart J 2011;32:2660–71.
[101] Stampfl A, Maier M, Radykewicz R, Reitmeir P, Gottlicher M, Niessner R. Langendorff heart: A model system to study cardiovascular effects of engineered nanoparticles. ACS Nano 2011;5:5345–53.
[102] Lundback M, Mills NL, Lucking A, et al. Experimental exposure to diesel exhaust increases arterial stiffness in man. Part Fibre Toxicol 2009;6:7.

CHAPTER

9

Neurological System

Sandra Ceccatelli[1], Giuseppe Bardi[2]

[1]Department of Neuroscience, Karolinska Institutet, Stockholm, Sweden

[2]Center for Nanotechnology Innovation @ NEST, Istituto Italiano di Tecnologia, Pisa, Italy

OUTLINE

INTRODUCTION

There is a growing interest in the possible application of engineered nanomaterials (ENM) for treatment of pathological conditions of the central nervous system (CNS). The size of ENM, at least in one or two dimensions (i.e. nanotubes), is such that they can go through very small vessels and, of relevance for the CNS, an "ad hoc" chemical modification could allow them to cross the blood–brain barrier (BBB), an anatomical structure that in the mature healthy brain prevents the entry of many substances, making difficult the delivery of drugs for therapeutical use [1]. ENM are certainly promising carriers for delivery of drugs targeting the brain. However, there is evidence that they may exert neurotoxicity [2] and this adverse potential needs to be carefully evaluated to understand the mechanisms behind the toxic effects in order to avoid any dangerous exposure and to develop safe ENM for novel clinical treatments.

Adverse Effects of Engineered Nanomaterials
DOI: 10.1016/B978-0-12-386940-1.00009-X

IN VIVO AND *IN VITRO* EXPERIMENTAL MODELS FOR NEUROTOXICOLOGICAL STUDIES ON ENM

Because of the complexity of the nervous system, a precise assessment of neurotoxic effects has to be based on multiple testing strategies that include *in vitro* and *in vivo* experimental models. Both approaches have advantages and limitations, which make necessary an integrated view to interpret the data also in light of possible discrepancies. From *in vivo* experimental studies, we know that nanoparticles (NPs) or their components can translocate into the brain: MnO_2 [3], C_{60} [4], fluorescent magnetic [5], Fe_2O_3 [6], lectin-conjugated PEG-PLA [7] and methoxy-PEG-PLA [8]. NPs have been detected after inhalation or intranasal administration. After intravenous (i.v.), intraperitoneal (i.p.), or gastrointestinal administration, which all lead to blood circulating NPs, the presence of silica-coated quantum dots [9] or magnetic NPs [10], TiO_2 [11], PEGylated PHDCA [12], and polysorbate-80 coated NPs [13,14] was detected in several brain regions.

Signs of neurotoxicity, including cellular morphological and functional changes, inflammation, and oxidative stress were observed after *in vivo* exposure to Al_2O_3 [15], TiO_2 [11,16], MnO_2 [3], quantum dots (QDs) [17] and Zn [18] NPs. Sharma et al. [19] showed that i.p. and i.v. administration of Ag, Cu or Al NPs disrupted the BBB of rats and mice. This event could lead to the unexpected entrance of activated immune cells which, in turn, would exacerbate NP-mediated inflammatory reactions.

Useful indications on the mechanisms involved in NPs' neurotoxicity shown by *in vivo* studies have been provided by *in vitro* experiments based on neural cell culture models. A limitation of *in vitro* studies is given by the fact that ENM directly face cells that are often situated in tissues with shielding biological environments (i.e. neurons and glia surrounded by the extracellular matrix). Nevertheless, the collected data from *in vitro* experiments are crucial to understand the mechanism of interaction between ENM and cells. Of relevance are the data obtained with the PC12 cell line showing increased production of reactive oxygen species (ROS) and changes in the dopaminergic system induced by Mn oxide [20,21]. Dopaminergic alterations were also induced by Ag or Cu NPs [21]. Degussa P25, a TiO_2 nanomaterial, induced activation of the microglia cell line BV2, with subsequent release of ROS. Conversely, no direct toxicity was observed in N27 dopaminergic neurons, and altogether the data suggest that neurotoxicity may occur via ROS generated by microglia [22,23].

Oxidative stress seems to be a common mechanism in neurotoxicity induced by NPs [24]. Translocation of NPs into mitochondria followed by functional alterations may also play a critical role in the onset of neurotoxic effects induced by small size particles [25]. Cytotoxicity has also been observed in primary neurons and glial cells exposed to carbon nanotubes [26]. However, their toxic potential is influenced by chemical functionalization associated with changes in solubility and tendency to form agglomerates [27]. *In vitro* data (obtained in solution, not from cell experiments) also indicate that carbon nanotubes, as well as TiO_2, copolymer, cerium oxide and QD NPs promote formation of fibrils [28,29] suggesting that exposure to NPs may enhance the formation of protein fibrils associated with neurodegenerative processes. On the contrary, other types of ENMs like dihydrolipoic acid- (DHLA) capped CdSe/ZnS quantum dots can reduce the fibrillation process [30], making it difficult to find general rules on the neurotoxic effects of nanomaterials.

It should be noted that most of the *in vivo* or *in vitro* mentioned studies have used massive doses of NPs, which raise questions on the actual risk associated with exposure to reasonable doses. It is also not clear whether the entire particles enter the CNS or metallic ions dissociate from the NPs and cross the BBB by transport protein mediated transcytosis.

Incongruence among the available observations may be associated to the type of ENM and also to the cell culture models employed [26, 31,32]. The results may vary depending on the cell type used for the testing (neuronal cell line versus primary cells), the area and animal species used for the preparation of primary cell cultures, as well as the presence or absence of glial cells in the cultures. All these aspects should be considered for a precise evaluation and reproducibility of the results. Most of the information on toxicity profiles of ENM comes from *in vitro* experiments using cell culture systems [26,33–36]. Even if an *in vitro* system can provide valuable information, data from *in vivo* models are required to recognize the safety of ENM, and standard procedures should be adopted to evaluate precisely their potential harmfulness on the nervous system. (For a further discussion on *in vitro* versus *in vivo* assays for nanotoxicity assessment, see Chapter 4.)

BIOCOMPATIBILITY CONCERNS FOR NANOMEDICINE APPLICATION OF ENM IN THE CNS

Specific types of ENM are intentionally engineered to reach the brain for drug delivery or imaging applications. Having a diameter from 1 to 100 nm, NPs may interact with different organelles, channels and protein complexes present on the cell surface, or with many other subcellular structures that are in the range of nanometer scale. These cell–NP interactions can result in very diverse biological effects, including inauspicious ones.

Together with the size, the shape of NPs must also be considered for potential toxicity. For example, different graphitic nanomaterials, such as graphene or carbon nanotubes (often described as rolled graphene sheets) can show dissimilar toxicity in the same cell type, albeit they are made of the same material but with different shapes [37].

The chemical surface of the nanoparticle is probably the most critical aspect to be considered in order to verify the toxicity of a drug carrier. Normally, NPs are chemically modified on their surface (functionalization) to improve the biological goal they have been prepared for. As a simple example, to maximize the interaction with nucleic acids (negatively charged), the NP surface can be covered by positive charges (i.e. NH_3^+) in order to exploit the electrostatic forces [38]. The same surface can also be modified with active molecule (e.g. peptides) able to target specific cells bearing their cognate receptors [39].

It becomes obvious, therefore, that nanoparticles can be tailored differently for a specific function and that cytotoxicity could accordingly vary with the many different possible chemical modifications of their surface. Solubility, binding properties, chemical reactivity, and all the characteristic features of a particle could change by modifying its surface. Moreover, due to their surface/mass ratio, low mass concentrations of NPs can expose many reactive groups on the large surface.

To be sure that a drug carrier is safe, scientists should not overlook the stability of the nanoparticle–chemical functionalization system. This means that any type of NP, carrying every possible modification, must be stable in a biological environment, where incredible amounts of chemical reactions happen in a salty medium. The issue of physicochemical properties of NPs in relation to their biologic effects is extensively discussed in Chapter 1.

In this scenario, NPs aimed at delivering active molecules into the CNS parenchyma in a non-invasive way have to pass the protective BBB. To solve the problem of brain localization and BBB transcytosis, many approaches have been considered, including coating with surfactants, conjugation with ligands for BBB-localized receptors, or antibodies against these receptors [40]. The latter strategy has been proved advantageous over two decades for brain targeted drug delivery [41—45]. The chosen targets are usually receptors overexpressed in the luminal side of brain microvessel endothelial cells, like transferring receptor or insulin receptor [46,47]. This strategy seems successful when applied to ENM with improved drug loading aimed at brain targets [48—50].

Polymeric NPs possess many advantages as nanocarriers for CNS drug delivery systems due to their well-established synthesis, drug-loading capacity and biodegradability of the chosen polymers [51]. Polymers are also commonly used to functionalize inorganic NPs to increase their solubility, stability in biological systems or to allow the binding with appropriate molecules for targeted specific binding [52]. Unfortunately, many polymers or surfactants used to coat the NP surfaces show a certain degree of toxicity, raising numerous criticisms concerning aspects of the formulations. It is possible that some polymers or their degradation products destabilize the integrity and permeability of the BBB, impairing its protective action [53]. This event may increase the space between the endothelial cells, normally joined by tight junctions (TJs), allowing the entry of active molecule and immune cells which, in turn, may start unexpected inflammatory processes within the CNS.

The importance of NP—surfactant stability when applied to primary neurons is emphasized by observations on multiwalled carbon nanotubes (MWCNTs) coated with Pluronic F-127 (PF-127) [54].

On the other hand, PF-127 biocompatibility studies revealed that concentration of the surfactant as low as 0.01% in the buffer/medium suspension can induce apoptosis of mouse primary cortical neurons *in vitro* within 24 h (Figure 9.1). When PF-127 coated MWCNTs were released in primary mixed neural cultures, toxicity was significantly reduced, suggesting that the mechanism triggering the apoptotic program was not dependent on MWCNTs and their presence was tolerated by the primary cells (Figure 9.2). So, as mentioned above, core NP/surface chemical modification stability should be carefully evaluated to avoid possible toxicity induced by the dissociation of a certain surface chemical modification. For an additional discussion on synthesis and modification of NPs, see the Annex.

No damage to the overall organization of mouse brain cortical layers in the area surrounding the lesion after 3 days were present when PF-127/MWCNT suspension was injected *in vivo* and the injury volume induced by the MWCNT suspension was not different from the one induced by the injection of CNT-free controls. Moreover, in a long-term observation 18 days after the injection, MWCNT-treated mice presented the formation of a glial scar in the cortex localized at the injection site without major alterations of the cortical structure. From these observations, it seems that brain cortex could tolerate MWCNT/PF-127 injection *in vivo*, even though one cannot exclude possible cytotoxic events induced by the release of the surfactant showing up at a later stage.

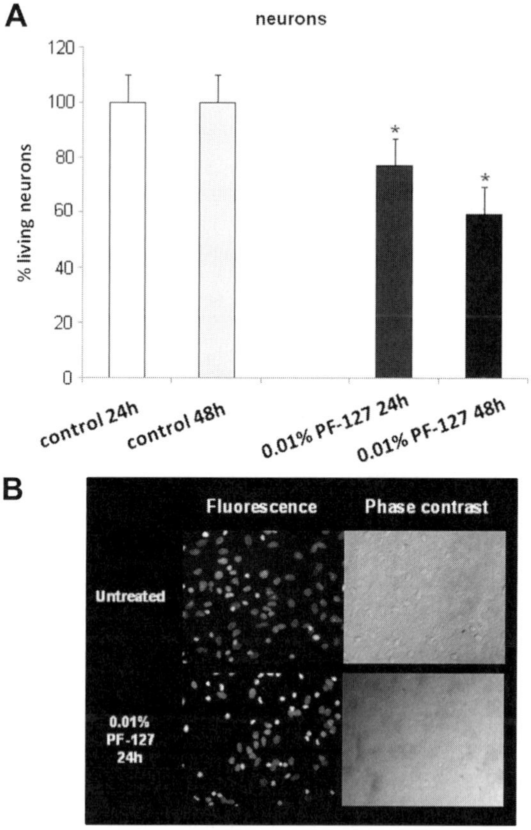

FIGURE 9.1 Pluronic F-127 induces apoptosis of mouse primary neurons *in vitro*. (A) Percentage of cultured primary neurons after treatments with 0.01% PF-127 after 24 h and 48 h at 37°C (mean ± SEM of treatments versus control, $^*P<0.01$). (B) Fluorescence (nuclei previously stained with 1 μg/mL Hoechst 33342) and phase contrast (whole cell body) pictures of mouse cortical neurons and glia untreated and treated for 24 h at 37 C with 0.01% PF-127 are shown. (*Modified from [54] with permission from* Nanomedicine.)

Underlying the potential of these types of ENM, the same authors have used carbon nanotube-mediated RNA interference against caspase-3 (siRNA-CAS 3) in an endothelin-1 induced stroke model *in vivo* [55]. Perilesional administration of siRNA-CAS 3 delivered by amino-functionalized carbon nanotubes reduced neurodegeneration and promoted functional preservation before and after focal ischemic damage of the rodent motor cortex. These observations illustrate the opportunity offered by carbon nanotube-mediated drug or gene delivery in the neuronal tissue in a variety of different neuropathological conditions where intervention at well-localized brain foci may offer therapeutic and functional benefits.

Nevertheless, it has to be mentioned that physiologic and morphologic alterations of neurons have been observed in the presence of other types of functionalized carbon nanotubes [56]. For example, polyethylene glycol/SWCNT induced disruption of the regulated,

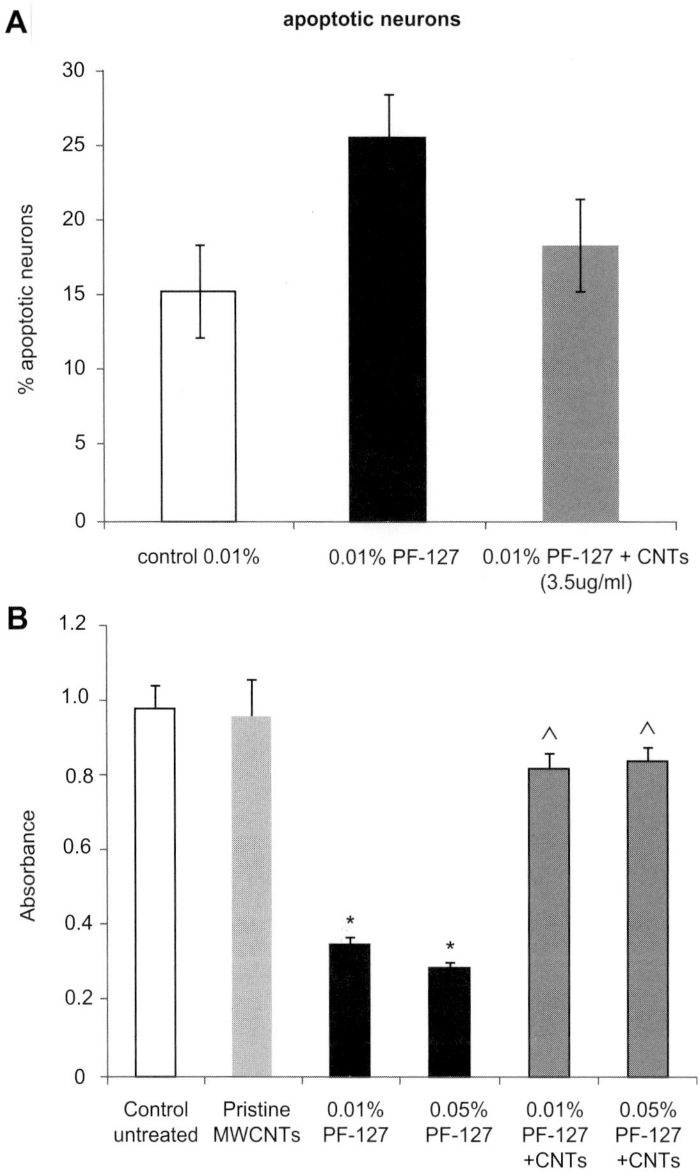

FIGURE 9.2 Adverse effects of Pluronic F-127 are reduced in presence of carbon nanotubes. (A) Apoptotic neurons increase in the presence of PF-127 (black bar) but not in the presence of PF-127/MWCNTs complexes (gray bar). (B) The bar graph shows MTT results after 24 h treatments without (white bar) or with 0.01% and 0.05% PF-127 (black bars); 0.01% and 0.05% PF-127 in presence of 3.5 μg/mL and 17.5 μg/mL MWCNTs, respectively (dark gray bars); Pristine MWCNTs added to the cell medium (light gray bar). Absorbance of MTT of samples are shown compared to control as mean ± SEM (treatments versus control, *$P<0.001$), (PF-127+CNTs vs PF-127 treatment alone, ^$P<0.001$). (*Modified from [54] with permission from* Biomaterials.)

not constitutive, depolarization-dependent plasma membrane/vesicular recycling (endocytotic mechanism) of neurons, which has been suggested to underlie the alteration of morphology and neurite growth of these cells in the presence of single walled carbon nanotubes.

A valuable application of NPs that needs to be stated is the preparation of stable fluorescent probes for bioimaging [57]. Even though the safety principle preserves its importance, the development of nano-probes could provide a precious tool for diagnostic or research purposes. The use of these "real-time visible" NPs can be very helpful to understand not only the nanoparticle–cell interaction and even its possible induced toxicity mechanism, but also the biological cell mechanism undergoing normal cell physiology [58].

Silica NPs or silica shells surrounding fluorescent particles have been frequently associated with major damage [59,60]. However, "silica nanoparticle induced toxicity" does not have a generalizable meaning. Careful characterization of the synthesis, size and surface chemistry must be performed. Choi et al., for instance, showed that low doses of 150–200 nm silica NPs with a negative surface charge (−23 mV) induce ROS production in microglia [60]. On the other hand, we described that 25 and 50 nm positively-charged (+11 mV) QD doped silica NPs were not toxic for primary neurons in a mixed culture *in vitro* model up to a concentration of 10 µg/mL (Figure 9.3), nor did they induce apoptotic cell death *in vivo* when injected at a concentration of 1 mg/mL into the cortical parenchyma [38]. Therefore, investigators must pay attention to all the NP characteristics before defining toxicity induced by a certain type of NP.

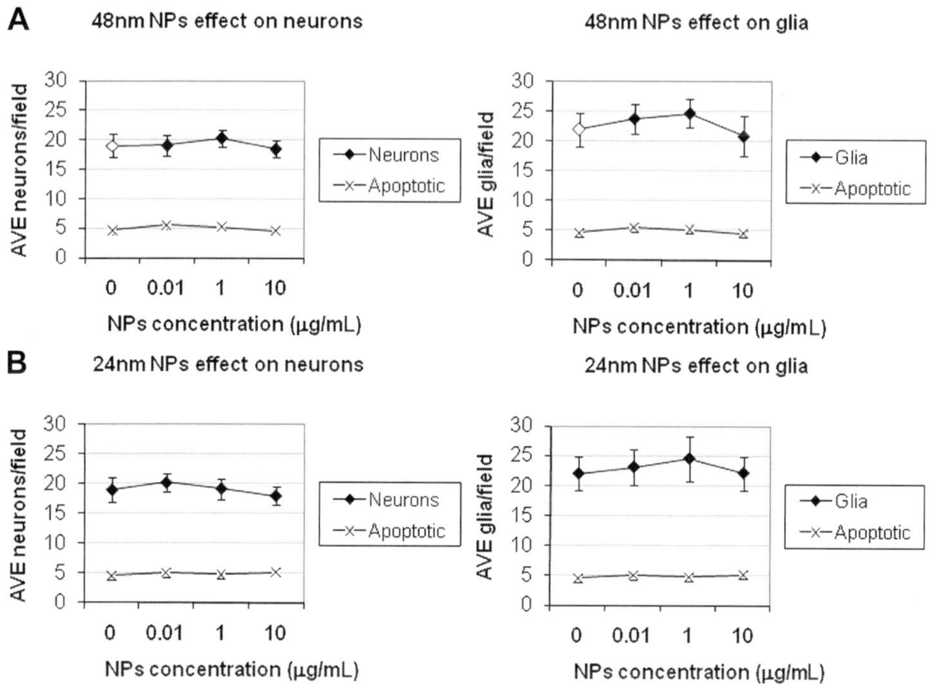

FIGURE 9.3 CdSe/ZnS QDs-doped SiO_2 NPs *in vitro* biocompatibility. Quantification of cell number and apoptotic nuclei after 24 h at 37°C in presence or absence of increasing concentration of SiO_2 NPs 50 nm and 25 nm (from 0.1 µg/mL to 10 µg/mL). Graphs show the number of living neurons or glia (filled diamonds) after 24 h of 50 nm (A) and 25 nm (B) NPs. Numbers of apoptotic cells present in the cultures shown by "-x-" graphs. (*Modified from [38] with permission from* Biomaterials.)

ENVIRONMENTAL EXPOSURE TO NANOPARTICLES AND POSSIBLE ADVERSE EFFECTS ON THE CNS

Although remote, the possibility that NPs released in the environment may translocate to the brain parenchyma does exist. Most of the available information on toxicology of NPs present in the environment is related to the ultrafine particles (UFP) dispersed in air pollution. UFP have been described as "ambient and laboratory-generated NPs that are not produced in a controlled, engineered way" [61].

People exposed to severe air pollution have been reported suffering adverse effects including inflammatory reactions in different areas of the CNS [62]. Experiments on mice have also linked ultrafine particulate matter in polluted air to CNS inflammatory reactions [63,64]. These data on UFP in air pollution, however, may mislead inattentive readers to a direct cause—effect ruling: "NPs present in the air induce brain inflammation". More information is needed to support this interpretation. A role of UFP may not be excluded, but the particle by itself may not have a primary toxic effect, nevertheless, cooperating to a synergistic effect that depends on its nano-structure. For example, due to their high surface-to-volume ratio NPs could aggregate different toxic pollutants, then increasing their concentration in the respiratory tract. Alternatively, a synergistic effect of several pollutants may also induce systemic inflammation, which would affect the stability of the endothelium, including the BBB. This event would likely create a feed forward mechanism of inflammatory state certainly inducing adverse effects on the brain [65]. A major problem of difficult solution in these studies is the finding of the right controls. Epidemiology of long-term ultrafine particle exposure in ambient air gives a basic contribution to nanotoxicology, however, to understand the effects of engineered NPs on the CNS and define a proper risk assessment, NP-specific and laboratory controlled experiments are required [2,66].

NANOPARTICLE INSTABILITY CAN RESULT IN ADVERSE EFFECTS

In laboratory-controlled studies, the best described pathway for NPs to undergo CNS translocation is from nasal epithelium through the olfactory bulb using 30 nm MnO_2 NPs [3]. Elder et al. [3] found Mn in different areas of the rat brain, including striatum and frontal cortex. However, in a review on NP safety concern, Stern and McNeil [67] did not exclude the possibility that the dissociation of MnO_2 particles could have occurred instead of the neuronal translocation of the entire nanoparticle. Indeed, the high level of Mn present in many regions of the CNS after inhalation was partially attributed to the circulatory transport of Mn. The event of MnO_2 dissociation is also suggested by the detection of Mn in the brain after inhalation of micron-size MnO_2 NPs, larger than the diameter of rat olfactory neuronal axons ranging from 100 to 200 nm [68].

The results obtained with inhaled MnO_2 NPs underlines, once more, the importance of nanoparticle stability. If engineered NPs released in a biological environment to accomplish a specific function were accidentally leaking atoms, ions or molecules resulting in increased toxicity or inflammatory reactions, the beneficial effect coming from its supposed primary function (e.g. drug delivery) would be compromised. One of the dangerous consequences of accumulation of manganese in the basal ganglia (a group of structures deep in the brain connected with several brain areas including cerebral cortex, and associated with motor control, cognition and emotion) is a neurological syndrome with movement defects and cognitive problems, very similar to Parkinson's disease [69].

POSSIBLE IMPLICATIONS FOR PATHOLOGICAL CONDITIONS OF THE NERVOUS SYSTEM: TAKE-HOME MESSAGES

Altogether, the experimental data indicate that NPs have the potential of exerting neurotoxicity via major mechanisms involving ROS formation, inflammation, implying the activation of microglial cells [70] and damage to the BBB. Mitochondrial damage and accelerated fibril formation may also have a role in inducing or worsening CNS damage. It is relevant to note that all these alterations have been found associated with CNS pathological conditions, including Parkinson's and Alzheimer's diseases. To be noted are also the alterations induced by Mn NPs on the dopaminergic system.

Increased cell autophagic vacuolization has been observed in the presence of NPs and this phenomenon has suggested an interesting speculation related to possible NP-induced brain damage [71]. Evidence from epidemiological studies shows an association between exposure to combustion-derived ambient particles (which are predominantly nanoscale) and neurodegenerative pathologies similar to Alzheimer's and Parkinson's diseases. Autophagy may be an appropriate cell response to clear NPs, however, the interaction of ENM with the autophagy pathway could have negative consequences that may influence the pathogenesis of neurodegenerative diseases [72,73].

Information coming from UFPs or viability tests is not sufficient to characterize the impact of exposure to ENM. The CNS is highly regulated by the presence of different cell types releasing several diverse active molecules, such as neurotransmitters or hormones. Possible ENM-induced release or ENM-induced suppression of those molecules in specific areas of the CNS could, in principle, have devastating impacts and lead to neurodegenerative diseases.

Activation of the CNS resident immune system, mainly represented by microglia, has also to be considered. Possible ENM-mediated inflammatory reaction could not be excluded. However, CNS access is limited by specific barriers and *in vivo* data on the actual dose of ENM penetration in the nervous system are missing. This lack of crucial nanotoxicological details and the diversity of the several ENMs render it impossible to quantify thresholds above which ENM-mediated neuroinflammation could start.

Certainly, definitive proof of NP-induced neurologic diseases is not available and speculations on the specific type of degenerative diseases are very difficult. Further investigation with specific models is required to identify and prevent any adverse effects that nanomaterials might have on the CNS.

References

[1] Neuwelt EA, Bauer B, Fahlke C, et al. Engaging neuroscience to advance translational research in brain barrier biology. Nat Rev Neurosci 2011;12:169–82.

[2] Simko M, Mattsson MO. Risks from accidental exposures to engineered nanoparticles and neurological health effects: A critical review. Part Fibre Toxicol 2010;7:42.

[3] Elder A, Gelein R, Silva V, et al. Translocation of inhaled ultrafine manganese oxide particles to the central nervous system. Environ Health Perspect 2006;114:1172–8.
[4] Oberdorster E. Manufactured nanomaterials (fullerenes, C60) induce oxidative stress in the brain of juvenile largemouth bass. Environ Health Perspect 2004;112:1058–62.
[5] Kwon J-T, Hwang S-K, Jin H, et al. Body distribution of inhaled fluorescent magnetic nanoparticles in the mice. J Occup Health 2008;50:1–6.
[6] Wang B, Feng WY, Wang M, et al. Transport of intranasally instilled fine Fe2O3 particles into the brain: Micro-distribution, chemical states, and histopathological observation. Biol Trace Elem Res 2007;118:233–43.
[7] Gao X, Tao W, Lu W, et al. Lectin-conjugated PEG-PLA nanoparticles: Preparation and brain delivery after intranasal administration. Biomaterials 2006;27:3482–90.
[8] Zhang Q-Z, Zha L-S, Zhang Y, et al. The brain targeting efficiency following nasally applied MPEG-PLA nanoparticles in rats. J Drug Targeting 2006;14:281–90.
[9] Chen Z, Chen H, Meng H, et al. Bio-distribution and metabolic paths of silica coated CdSeS quantum dots. Toxicol Appl Pharmacol 2008;230:364–71.
[10] Kim JS, Yoon T-J, Yu KN, et al. Toxicity and tissue distribution of magnetic nanoparticles in mice. Toxicol Sci 2006;89:338–47.
[11] Wang J, Zhou G, Chen C, et al. Acute toxicity and biodistribution of different sized titanium dioxide particles in mice after oral administration. Toxicol Lett 2007;168:176–85.
[12] Calvo P, Gouritin B, Chacun H, et al. Long-circulating PEGylated polycyanoacrylate nanoparticles as new drug carrier for brain delivery. Pharmaceut Res 2001;18:1157–66.
[13] Kreuter J, Alyautdin RN, Kharkevich DA, Ivanov AA. Passage of peptides through the blood–brain barrier with colloidal polymer particles (nanoparticles). Brain Res 1995;674:171–4.
[14] Schroeder U, Sommerfeld P, Sabel BA. Efficacy of oral dalargin-loaded nanoparticle delivery across the blood–brain barrier. Peptides 1998;19:777–80.
[15] Chen L, Yokel RA, Hennig B, and Toborek M. Manufactured aluminum oxide nanoparticles decrease expression of tight junction proteins in brain vasculature. J Neuroimmune Pharmacol 2008;3:286–95.
[16] Ma L, Liu J, Li N, et al. Oxidative stress in the brain of mice caused by translocated nanoparticulate TiO_2 delivered to the abdominal cavity. Biomaterials 2010;31:99–105.
[17] Tang M, Li Z, Chen L, et al. The effect of quantum dots on synaptic transmission and plasticity in the hippocampal dentate gyrus area of anesthetized rats. Biomaterials 2009;30:4948–55.
[18] Zhao J, Xu L, Zhang T, Ren G, Yang Z. Influences of nanoparticle zinc oxide on acutely isolated rat hippocampal CA3 pyramidal neurons. Neurotoxicology 2009;30:220–30.
[19] Sharma HS. Nanoneuroscience: emerging concepts on nanoneurotoxicity and nanoneuroprotection. Nanomedicine 2007;2:753–58.
[20] Hussain SM, Javorina AK, Schrand AM, Duhart HM, Ali SF, Schlager JJ. The interaction of manganese nanoparticles with PC-12 cells induces dopamine depletion. Toxicol Sci 2006;92:456–63.
[21] Wang J, Rahman MF, Duhart HM, et al. Expression changes of dopaminergic system-related genes in PC12 cells induced by manganese, silver, or copper nanoparticles. Neurotoxicology 2009;30:926–33.
[22] Long TC, Saleh N, Tilton RD, Lowry GV, Veronesi B. Titanium dioxide (P25) produces reactive oxygen species in immortalized brain microglia (BV2): Implications for nanoparticle neurotoxicity. Environ Sci Technol 2006;40:4346–52.
[23] Long TC, Tajuba J, Sama P, et al. Nanosize titanium dioxide stimulates reactive oxygen species in brain microglia and damages neurons *in vitro*. Environ Health Perspect 2007;115:1631–7.
[24] Donaldson K. Resolving the nanoparticles paradox. Nanomedicine (Lond) 2006;1:229–34.
[25] Xia T, Kovochich M, Liong M, Zink JI, Nel AE. Cationic polystyrene nanosphere toxicity depends on cell-specific endocytic and mitochondrial injury pathways. ACS Nano 2008;2:85–96.
[26] Belyanskaya L, Weigel S, Hirsch C, Tobler U, Krug HF, Wick P. Effects of carbon nanotubes on primary neurons and glial cells. Neurotoxicology 2009;30:702–11.
[27] Coccini T, Roda E, Sarigiannis DA, et al. Effects of water-soluble functionalized multi-walled carbon nanotubes examined by different cytotoxicity methods in human astrocyte D384 and lung A549 cells. Toxicology 2010;269:41–53.
[28] Linse S, Cabaleiro-Lago C, Xue W-F, et al. Nucleation of protein fibrillation by nanoparticles. Proc Natl Acad Sci USA 2007;104:8691–6.

[29] Wu W-H, Sun X, Yu Y-P, et al. TiO2 nanoparticles promote beta-amyloid fibrillation *in vitro*. Biochem Biophys Res Commun 2008;373:315–8.
[30] Thakur G, Micic M, Yang Y, et al. Conjugated quantum dots inhibit the amyloid beta (1-42) fibrillation process. Int J Alzheimers Dis 2011;2011:502386.
[31] Liopo AV, Stewart MP, Hudson J, Tour JM, Pappas TC. Biocompatibility of native and functionalized single-walled carbon nanotubes for neuronal interface. J Nanosci Nanotechnol 2006;6:1365–74.
[32] McKenzie JL, Waid MC, Shi R, Webster TJ. Decreased functions of astrocytes on carbon nanofiber materials. Biomaterials 2004;25:1309–17.
[33] Hasadsri L, Kreuter J, Hattori H, Iwasaki T, George JM. Functional protein delivery into neurons using polymeric nanoparticles. J Biol Chem 2009;284:6972–81.
[34] Hu Z, Liu S, Wei Y, Tong E, Cao F, Guan W. Synthesis of glutathione C60 derivative and its protective effect on hydrogen peroxide-induced apoptosis in rat pheochromocytoma cells. Neurosci Lett 2007;429: 81–6.
[35] Liu S, Xu L, Zhang T, Ren G, Yang Z. Oxidative stress and apoptosis induced by nanosized titanium dioxide in PC12 cells. Toxicology 2010;267:172–7.
[36] Sayes CM, Gobin AM, Ausman KD, Mendez J, West JL, Colvin VL. Nano-C60 cytotoxicity is due to lipid peroxidation. Biomaterials 2005;26:7587–95.
[37] Zhang Y, Ali SF, Dervishi E, et al. Cytotoxicity effects of graphene and single-wall carbon nanotubes in neural phaeochromocytoma-derived PC12 cells. ACS Nano 2010;4:3181–6.
[38] Bardi G, Malvindi MA, Gherardini L, et al. The biocompatibility of amino functionalized CdSe/ZnS quantum-dot-doped SiO2 nanoparticles with primary neural cells and their gene carrying performance. Biomaterials 2010;31:6555–66.
[39] Sundaram S, Roy SK, Ambati BK, Kompella UB. Surface-functionalized nanoparticles for targeted gene delivery across nasal respiratory epithelium. FASEB J 2009;23:3752–65.
[40] Kabanov AV, Batrakova EV. New technologies for drug delivery across the blood brain barrier. Curr Pharm Des 2004;10:1355–63.
[41] Friden PM, Walus LR, Musso GF, Taylor MA, Malfroy B, Starzyk RM. Anti-transferrin receptor antibody and antibody-drug conjugates cross the blood–brain barrier. Proc Natl Acad Sci USA 1991;88:4771–5.
[42] Newton HB. Advances in strategies to improve drug delivery to brain tumors. Expert Rev Neurother 2006;6:1495–509.
[43] Qian ZM, Li H, Sun H, Ho K. Targeted drug delivery via the transferrin receptor-mediated endocytosis pathway. Pharmacol Rev 2002;54:561–87.
[44] Shin SU, Friden P, Moran M, et al. Transferrin-antibody fusion proteins are effective in brain targeting. Proc Natl Acad Sci USA 1995;92:2820–4.
[45] Vyas SP, Singh A, Sihorkar V. Ligand-receptor-mediated drug delivery: an emerging paradigm in cellular drug targeting. Crit Rev Ther Drug Carrier Syst 2001;18:1–76.
[46] Jefferies WA, Brandon MR, Hunt SV, Williams AF, Gatter KC, Mason DY. Transferrin receptor on endothelium of brain capillaries. Nature 1984;312:162–3.
[47] Frank HJ, Pardridge WM. Insulin binding to brain microvessels. Adv Metab Disord 1983;10:291–302.
[48] Beduneau A, Saulnier P, Hindre F, Clavreul A, Leroux JC, Benoit JP. Design of targeted lipid nanocapsules by conjugation of whole antibodies and antibody Fab′ fragments. Biomaterials 2007;28:4978–90.
[49] Mishra V, Mahor S, Rawat A, et al. Targeted brain delivery of AZT via transferrin anchored pegylated albumin nanoparticles. J Drug Target 2006;14:45–53.
[50] Aktas Y, Yemisci M, Andrieux K, et al. Development and brain delivery of chitosan-PEG nanoparticles functionalized with the monoclonal antibody OX26. Bioconjug Chem 2005;16:1503–11.
[51] Tosi G, Costantino L, Ruozi B, Forni F, Vandelli MA. Polymeric nanoparticles for the drug delivery to the central nervous system. Expert Opin Drug Deliv 2008;5:155–74.
[52] Sperling RA, Parak WJ. Surface modification, functionalization and bioconjugation of colloidal inorganic nanoparticles. Phil Transact A Math Phys Eng Sci 2010;368:1333–83.
[53] Sun W, Xie C, Wang H, Hu Y. Specific role of polysorbate 80 coating on the targeting of nanoparticles to the brain. Biomaterials 2004;25:3065–71.
[54] Bardi G, Tognini P, Ciofani G, Raffa V, Costa M, Pizzorusso T. Pluronic-coated carbon nanotubes do not induce degeneration of cortical neurons *in vivo* and *in vitro*. Nanomedicine 2009;5:96–104.

[55] Al-Jamal KT, Gherardini L, Bardi G, et al. Functional motor recovery from brain ischemic insult by carbon nanotube-mediated siRNA silencing. Proc Natl Acad Sci USA 2011;108:10952–7.
[56] Malarkey EB, Reyes RC, Zhao B, Haddon RC, Parpura V. Water soluble single-walled carbon nanotubes inhibit stimulated endocytosis in neurons. Nano Lett 2008;8:3538–42.
[57] Sharma P, Brown S, Walter G, Santra S, Moudgil B. Nanoparticles for bioimaging. Adv Colloid Interface Sci 2006;123-126:471–85.
[58] Fichter KM, Flajolet M, Greengard P, Vu TQ. Kinetics of G-protein-coupled receptor endosomal trafficking pathways revealed by single quantum dots. Proc Natl Acad Sci USA 2010;107:18658–63.
[59] Yamashita K, Yoshioka Y, Higashisaka K, et al. Silica and titanium dioxide nanoparticles cause pregnancy complications in mice. Nat Nanotechnol 2011;6:321–8.
[60] Choi J, Zheng Q, Katz HE, Guilarte TR. Silica-based nanoparticle uptake and cellular response by primary microglia. Environ Health Perspect 2010;118:589–95.
[61] Oberdorster G, Oberdorster E, Oberdorster J. Nanotoxicology: An emerging discipline evolving from studies of ultrafine particles. Environ Health Perspect 2005;113:823–39.
[62] Calderon-Garciduenas L, Franco-Lira M, Henriquez-Roldan C, et al. Urban air pollution: Influences on olfactory function and pathology in exposed children and young adults. Exp Toxicol Pathol 2010;62:91–102.
[63] Campbell A, Oldham M, Becaria A, et al. Particulate matter in polluted air may increase biomarkers of inflammation in mouse brain. Neurotoxicology 2005;26:133–40.
[64] Kleinman MT, Araujo JA, Nel A, et al. Inhaled ultrafine particulate matter affects CNS inflammatory processes and may act via MAP kinase signaling pathways. Toxicol Lett 2008;178:127–30.
[65] Perry VH. The influence of systemic inflammation on inflammation in the brain: Implications for chronic neurodegenerative disease. Brain Behav Immun 2004;18:407–13.
[66] Oberdorster G, Elder A, Rinderknecht A. Nanoparticles and the brain: Cause for concern? J Nanosci Nanotechnol 2009;9:4996–5007.
[67] Stern ST, McNeil SE. Nanotechnology safety concerns revisited. Toxicol Sci 2008;101:4–21.
[68] Fechter LD, Johnson DL, Lynch RA. The relationship of particle size to olfactory nerve uptake of a non-soluble form of manganese into brain. Neurotoxicology 2002;23:177–83.
[69] Guilarte TR. Manganese and Parkinson's disease: A critical review and new findings. Environ Health Perspect 2010;118:1071–80.
[70] Gonzalez-Scarano F, Baltuch G. Microglia as mediators of inflammatory and degenerative diseases. Annu Rev Neurosci 1999;22:219–40.
[71] Stern ST, Johnson DN. Role for nanomaterial-autophagy interaction in neurodegenerative disease. Autophagy 2008;4:1097–100.
[72] Boland B, Kumar A, Lee S, et al. Autophagy induction and autophagosome clearance in neurons: Relationship to autophagic pathology in Alzheimer's disease. J Neurosci 2008;28:6926–37.
[73] Cuervo AM, Stefanis L, Fredenburg R, Lansbury PT, Sulzer D. Impaired degradation of mutant alpha-synuclein by chaperone-mediated autophagy. Science 2004;305:1292–5.

CHAPTER

10

Immune System

Diana Boraschi[1], Albert Duschl[2]

[1]Institute of Biomedical Technologies of the National Research Council, Pisa, Italy
[2]University of Salzburg, Department of Molecular Biology, Salzburg, Austria

OUTLINE

INTRODUCTION

This chapter deals with pathogenic effects of nanomaterials on the immune system. It will not focus on either lung diseases or on genotoxicity leading to neoplastic diseases (see Chapters 7 and 14, respectively). Here we describe both general and specific effects of nanoparticles on the immune system. The immune system is a diffuse organ, present in the entire body, which therefore encompasses essentially all tissues. However, particular attention will be paid to immune cells present in blood. The blood is the environment in which many cells of the immune system exist in their resting states and from which they are transported to sites of infection. A focus on blood can be justified also on a much simpler level: for human subjects this is the main source of immune cells available for studies, much more accessible

Adverse Effects of Engineered Nanomaterials
DOI: 10.1016/B978-0-12-386940-1.00010-6

than solid primary tissues, so for blood cells there is a particularly extensive literature. We will nevertheless cover also immune functions occurring in solid tissue, but for those situations there are more frequent occasions where important data are lacking.

Giving special attention to the immune system is well justified in the context of nanosafety. Immune cells are enriched close to body surfaces (like skin, airways, gastrointestinal tract), where pathogenic microorganisms, but also nanomaterials, are most likely to enter the body. Immunity is specialized to recognize non-self entities and to distinguish whether or not they pose a threat to the integrity of the body. Evaluating the risk posed by foreign particles is, in a nutshell, the core task of nanosafety research as well.

Immune cells are, due to their location and function, particularly well suited to act as sentinels for possible nanothreats. Specific cell types actively try to engage and process non-self materials, and they respond with reactions that can be readily measured. Examples include the secretion of soluble immune hormones (cytokines and chemokines), the expression of activation markers on the cell's surface, the production of reactive oxygen species (ROS) and other signals associated with cellular stress, like activation of NF-κB. Analyzing these endpoints gives a good insight into whether or not immune cells perceive a threat in the presence of specific nanomaterials. Of course, even the professional cells of the immune system make errors that may result in inappropriate outcomes, as in the case of allergies or in autoimmune diseases. Nevertheless, it is instructive to see to what extent the body's primary defense system considers specific nanoparticles as a potential threat or as harmless agents.

We are aware of immunity mostly when the immune system battles a disease, but actually the most common outcome of encounters with non-self entities is tolerance. Most entities surrounding us, even most bacteria, are indeed harmless and ignoring them is the correct response. The immune system is specialized to fight off pathogenic viruses, microorganisms and parasites. If the immune system recognizes entities that are not part of pathogens, as in the case of allergies and of autoimmunity, diseases may ensue. Since engineered nanoparticles are not associated with dangerous pathogens, most immune mechanisms do not come into play as they are not the appropriate response. Activation of cellular stress mechanisms may occur if nanoparticles do act as stressors, but full-blown immune responses like those involving inflammation, complement activation or antibody production are not useful and may even be detrimental to the organism itself, so the decision for tolerance is usually the most appropriate. Tolerance is not the same as failure to recognize non-self. The decision for tolerance is an active one and is reinforced by immune mechanisms preventing future response, like making reactive T cells permanently unresponsive (anergic) [1] or inducing regulatory T cells (Treg) which maintain tolerance by secreting immunosuppressive cytokines [2].

THE OFF SWITCH: IMMUNOSUPPRESSION BY NANOMATERIALS

The most direct cause of immunosuppression caused by nanomaterials is toxicity with immune cell death. Cell death can occur by two main mechanisms, necrosis and apoptosis [3]. Larger microparticulated materials mostly cause necrosis, i.e. mechanical or chemical membrane damage that causes "violent" cell death and consequent induction of a defensive inflammatory reaction [4]. In the case of nanoparticles, besides chemical composition, also size and shape are key elements for determining their direct toxicity. In carbon nanotubes, for instance, the length of the materials makes them much more toxic as compared to spherical or short nanoparticles, as reviewed in [5]. Immune cells, and in particular phagocytes, are professional defense cells and their paramount role is to patrol the body looking for potential dangers. In this perspective, it is likely that immune cells meet and handle nanomaterials more frequently than many other cell types (with the exception of non-professional defense cells such as skin and mucosal epithelial cells). Phagocytic immune cells are also often the first to come in contact with nanomaterials, either in blood (for instance with nanomedicines), at the body surfaces (in particular respiratory mucosa, gastrointestinal tract, skin) and in internal tissues (e.g. in the case of wounds or after extravasation). There are two main types of phagocytic cells, polymorphonuclear leukocytes (PMN) and macrophages. In the blood, both cell types are present in a quiescent state, with PMN representing about 60–70% of the white blood cells, and monocytes (the blood form of macrophages), representing about 3–5%. PMN are short-living cells that usually die rapidly upon encounter with the dangerous agent (indeed, their death is a key element in pathogen destruction). Monocytes, on the other hand, are long-living professional phagocytes and, after meeting with nanoparticles, they take them up abundantly and either degrade them or, if this is impossible, just keep them segregated [6,7]. In tissues, macrophages do the same, i.e. they efficiently take up the particles to clear the tissue. In many cases, that uptake of nanoparticles does not affect monocyte/macrophage viability, even at high doses, except when inducing direct inflammatory activation [8]. Also morphological and functional characteristics (shape, adherence, migration, phagocytosis of other particles) do not seem to be significantly affected by the simple physical presence of loads of nanoparticles. Of course, as in the case of uptake of other kinds of foreign materials, the intracellular release of poisonous ions may cause death of the phagocytic cell. Due to their physical characteristics, nanoparticles (in particular small spherical ones) may enter cells by non-phagocytic mechanisms, thus they can be found, although to a lesser extent, in all types of immune cells, including T and B lymphocytes.

Even in the absence of overt toxicity, nanoparticle-loaded immune cells may be less fit for exerting their defensive and scavenging functions [9]. Although not many data have been published in this sense, nanoparticle-loaded immune cells may undergo apoptosis sooner than normal, or be subjected to anomalous autophagic mechanisms, affecting immunological fitness. Even more relevant is the possibility that nanoparticle-loaded immune cells may be less able to tackle incoming danger in an adequate fashion. For example, when a pathogenic virus enters the body, the immune system recognizes the virus as dangerous and initiates a complex reaction that leads to viral clearance and re-establishment of the normal healthy conditions. If immune cells are loaded with nanoparticles, will they be able to initiate and develop the defensive response, to produce interferons, kill virus-infected cells, and later to repair the damages produced to the infected tissue? This issue will be further developed in the section below dedicated to inflammation, but it needs to be emphasized that immunosuppression can rest on much subtler mechanisms than on the killing of immune cells, and that nanomaterials may have unforeseeable modulatory effects in the absence of any overt sign of toxicity.

Eventually, it should be considered that nanomaterials may affect immune responses by abnormally activating immunosuppressive mechanisms. These mechanisms usually come into play during the final phases of defensive responses, i.e. when the infection/danger has been eliminated and the immune activation must be reconducted to baseline [10]. Among downregulatory mechanisms of innate immune reactions, a series of inflammatory cytokine inhibitors are abundantly expressed and released in the resolution phase of the response (e.g. IL-1Ra, IL-18BP, sIL-1RII, TIR8, TGFβ, IL-10). Other molecules act as anti-inflammatory transcriptional regulators (IL-33, IL-37). Besides directly inhibiting the innate inflammatory response, several of these factors can also create the microenvironmental conditions for the development of regulatory T cells (Treg) that are responsible for the downregulation of specific T-cell responses [11]. Concomitantly, the same factors/microenvironments promote tissue remodeling for re-establishing tissue integrity. The balance between these factors, in terms of amount and timing, is very delicate and, although redundancy usually ensures some stability, variations not only in the quantity of factors produced (more or less of one of the inhibitors) but also in the time of its production (too late or too early) or in its duration (produced for too long or too short a time) can cause significant changes in the outcome of the reaction and be the basis for pathological derangements. Examples relating to the development of autoimmune diseases are given in a subsequent section. What the role of nanomaterials could be in unbalancing the immune regulation and causing diseases is presently unknown.

INNOCENT BYSTANDERS? DENDRITIC CELLS, THEIR INTERACTION WITH T CELLS AND EFFECTS OF NANOMATERIALS

There is extensive literature on dendritic cells and nanoparticles, most of it from the field of nanomedicine. Dendritic cells are the primary sentinels of the immune system. Typically, these cells are in the resting state located close to body surfaces, where they become activated upon contact with non-self materials perceived as dangerous, migrate to local lymph nodes and interact there with specific T cells. This is the normal mechanism of immune activation and it immediately suggests an application for nanoparticles: they are non-self and can even be modified to increase immune responses, so they may be useful for vaccination where a strong immune response is desired. Nanoparticles coated with either protein antigens or with antigenic DNA have the potential to act as adjuvants, enhancing the immune response against the specific antigen and, since objects on the nanoscale are known to enter cells, including dendritic cells, they can also serve as delivery vehicles which facilitate efficient uptake of the vaccine antigens. Cancer antigens have been a particular focus of the field since it is usually difficult to raise an efficient immune response against them. For recent reviews on the application of nanoparticles for vaccine delivery see [12–14].

It is quite clear that nanomedicine requires a thorough understanding of toxicological properties of nanoparticles used for diagnosis or therapy [15]. A striking example of important insights for nanosafety questions derived from medicine is the finding that nano-sized alum particles act by stimulating the Nalp3 inflammasome, probably through intracellular induction of reactive oxygen species [16,17] (Figure 10.1). The mechanism of the long known and medically used adjuvant alum was found to depend on a specific pro-inflammatory signaling pathway in dendritic cells. Furthermore, nanoparticles were shown to rupture lysosomal compartments and act as "danger signals" for the immune system [18], which allows transferring medical experience to research on nanorisks. The main benefit is a better

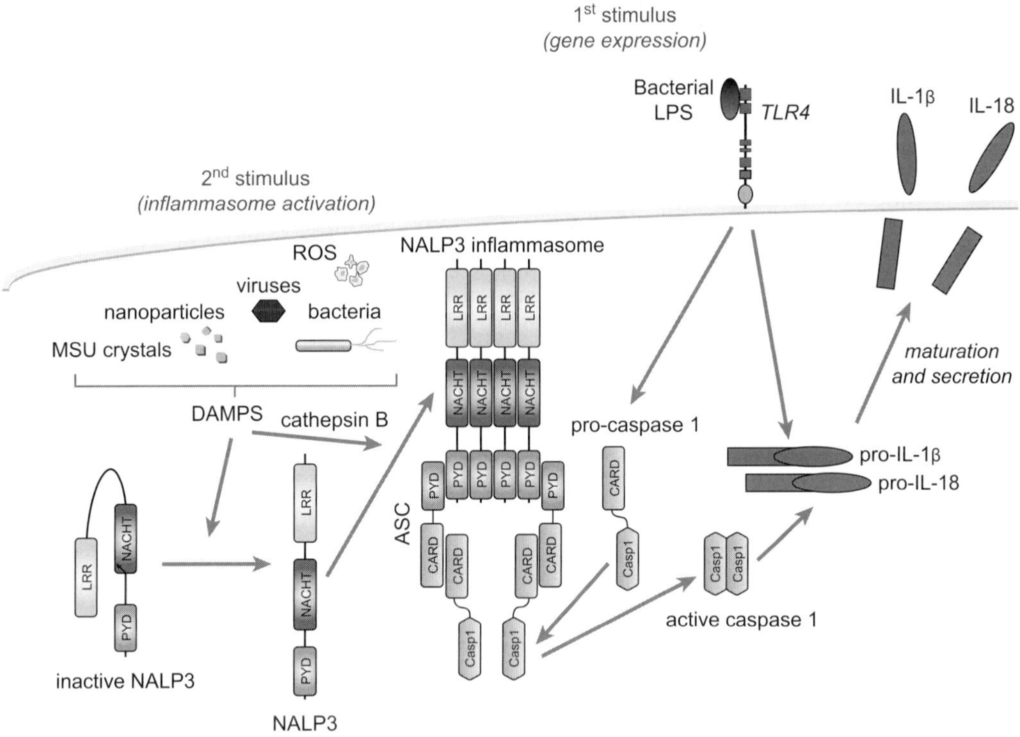

FIGURE 10.1 Nanoparticles and inflammasome activation. Release of active inflammatory cytokines IL-1β and IL-18 is a consequence of a first stimulus (usually TLR activation) that upregulates gene expression and synthesis of the pro-forms of cytokines and caspase-1. Inflammasome activation depends on different stimuli, including intracellular nanoparticles, that can activate the inactive form of an NLR (here the NALP3 molecule) leading to the formation of the inflammasome that matures and activates pro-caspase 1. Active caspase 1 cleaves inactive pro-IL-1β and pro-IL-18 and promotes the extracellular release of the active cytokines.

understanding of nanotoxicity mechanisms, which facilitates the selection of appropriate endpoints. Choice of endpoints is, in a complex system like immunity, a major issue, since there is no single parameter that can describe and reflect the status of the immune system. Indeed, the "metrology" of immunity is one of the most burning issues in immunological research and, without it, there is little possibility of establishing thresholds between what is immunologically safe and what is not.

The main function of the inflammasome is the activation of caspase-1, a protease that cleaves the immature pro-forms of the inflammatory cytokines IL-1β and IL-18, which leads to the secretion of the mature forms and to the induction of inflammation [19,20]. Inflammatory caspases and inflammasomes have been described as the master switches of inflammation [21]. The inflammasome-mediated cleavage of the immature cytokine pro-forms requires signals both from Toll-like receptors (TLR), which are localized in the plasma membrane or in endosomal/lysosomal membranes, and NOD-like receptors (NLR), which are localized in the cytoplasm. Both receptor classes recognize different bacterial compounds, but TLRs have a wider range of ligands and recognize also some self-compounds, like heat shock proteins and defensins. In addition to TLR and NLR-mediated signals, also reactive oxygen species may mediate inflammasome activation [22]. It is known that these activation pathways can be triggered by naturally occurring particles produced within the body, like cholesterol complexes or uric acid crystals. Thus, mechanisms which sense danger in nano-sized uric acid crystals (leading to gout) may be able to respond to non-natural nano-sized particles as well. Several mechanisms can be considered. Rupture of the lysosome, as shown for alum and silica particles, leads to release of the protease cathepsin B, which has been proposed to mediate inflammasome activation [18]. Recognition of nanoparticles by TLR is possible and may not even require uptake of nanoparticles into cells, since several members of the TLR family are localized in the plasma membrane and recognize extracellular stimuli. However, since more than one signal is required for inflammasome activation, binding of nanoparticles to TLR in the extracellular space would by itself not be sufficient to induce an inflammatory response.

The most crucial function of dendritic cells is to take up protein antigens, process them, and present to T cells, bound to their MHC II molecules, the peptides derived from these non-self proteins. The interaction between MHC/peptide complexes and the T-cell receptor can stimulate the activation of the T cell, leading to an antigen-specific immune response. The uptake, processing and presentation of proteins may in principle be subjected to interference by nanoparticles present on the surface of or within the dendritic cell, but there is another more subtle possibility, which will have to be addressed by immunologists in the future. Proteins readily attach to nanoparticles, and the impact of this coating on nanoparticle fate and toxicity has received significant interest. It is known that, in *in vitro* experiments, composition of the culture medium determines adsorption of different amounts and types of proteins to nanoparticles [23,24] and that the protein coating evolves over time from a "soft" layer which can be removed by centrifugation to a "hard" shell which is quite firmly attached [25,26]. Cells will in practice not be exposed to a "naked" nanoparticle—as technically produced—but rather to a nanoparticle–protein complex, with profound consequences both for the properties of the nanomaterials and for the cellular responses towards them (see also Chapter 1 for further information on protein coating of nanoparticles).

However, the consequences for the proteins involved are much less clear. The formation of a "hard" protein shell depends on nanoparticle–protein and on protein–protein interaction, which contribute to a dense packing of proteins in a single layer [25,27]. Bovine serum albumin bound to dextran-poly(isobutylcyanoacrylate) nanoparticles was shown to exist in a specific configuration [28]. Proteins bound to nanoparticles may therefore be perceived by dendritic cells in a form and possibly also a shape that differs from those of proteins in solution. This can influence uptake and processing of the proteins and could ultimately lead to quantitative and qualitative differences in the presentation of peptides on the surface of the dendritic cells. Small post-translational modifications of proteins can have a strong impact on the peptides presented [29–32], since the machinery involved in recognition is quite sensitive even to minor changes. It has been shown that fibrinogen, when bound to negatively-charged poly(acrylic acid)-conjugated gold nanoparticles, becomes partially unfolded and interacts in this form with the integrin receptor, Mac-1, which mediates pro-inflammatory signals [33].

The composition of the protein coat is highly influenced by the serum concentration, as shown in a recent study in which nanoparticles were exposed to concentrations of serum proteins between 3% and 80% [23]. The protein coat composition was quite different at different serum concentrations, both for silica and for polystyrene nanoparticles. Cell culture medium contains usually 5–10% serum, so the identities of the proteins attaching to nanoparticles in culture conditions may be different from the proteins attaching in the human body, even when considering intravenous delivery. This issue will have to be addressed in future investigations.

INFLAMMATION

Innate immunity or inflammation (the two terms are almost synonyms) is the earliest defense mechanism of the body. At variance with adaptive immunity, which develops only after a first encounter with the pathogen, inflammation is an immediate response that is always ready to come into action and does not need to "learn". The advantages of innate reactions are in their rapidity and potency, while the disadvantage is in their limited specificity. Thus, innate immunity/inflammation is a key element in the overall defense strategy of the body but, because of its potency and its lack of specificity, can cause unwanted damage to the body itself. Indeed, the inflammatory reaction, while eliminating dangers, has collateral effects by causing concomitant damage to the body's tissues. The concluding steps of the inflammatory response, once the danger is eliminated, encompass a complex series of reactions that restore the physical and functional integrity of the damaged tissue (tissue repair and remodeling, with cell proliferation and angiogenesis) (Figure 10.2).

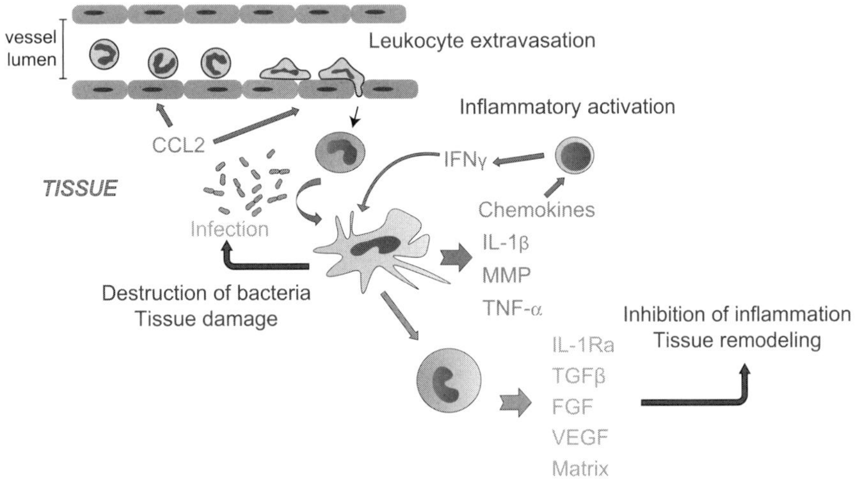

FIGURE 10.2 The development of the innate/inflammatory defense reaction. A tissue injury such as an infection leads to the release of chemotactic factors (e.g. CCL2) that recruit monocytes from the circulation to the tissues, where they meet the infectious agents and become activated to produce inflammatory factors. These are both involved in pathogen destruction and in recruiting other cells that amplify the defensive response (e.g. T lymphocytes that produce IFN-γ). Upon elimination of danger, monocytes stop producing inflammatory factors and contribute to tissue repair and remodeling by producing angiogenic and proliferation-inducing factors.

Each phase of the inflammatory response is an integral part of the overall defense reaction, and is finely regulated in order to avoid pathological derangements. Variations in the timing, amplitude, and duration of the various phases may transform the defensive response into a pathological reaction that damages the body. This can happen, for example, in the case of persistence of the inflammatory stimuli and consequent persistence of the defense reaction, which can cause the unremitting destruction of the surrounding tissues. Nanoparticles may have these characteristics, i.e. being persistent in the body and cause permanent inflammatory activation when they are made of materials that cannot be digested by the scavenger cells. The inflammatory effects of nanoparticles have been described in several instances, both *in vivo* and *in vitro* (reviewed in [34,35]). However, since the size of nanoparticles is such that they may easily escape the particle-clearing defensive mechanisms (e.g. phagocytosis), it is probable that, in many cases, no direct inflammatory response is triggered by nanomaterials [36]. Inflammatory effects that have been observed by many researchers may have been caused by concurring events. For instance, the different state of aggregation or agglomeration of nanoparticles upon contact with the biological environment may change the size of the entities coming in contact with the immune cells, which tend to recognize and uptake better the larger particles ($>0.5\ \mu m$) and readily clear or sequester them, with less pathological/toxic consequences as compared to the smaller particles [37,38]. In addition, inflammation may be due to bioactive molecules adsorbed on the nanomaterial surface (e.g. bacterial moieties; see below) rather than by the nanomaterial itself.

As mentioned above, nanoparticles entering the body get coated with proteins and other molecules, which can be modified by such interactions. In addition, the particulate shape of an antigen (as in the case of microorganisms or complex antigens as opposed to haptens) allows much better recognition by dendritic cells, which can then process and present it, thereby initiating an autoimmune response. Thus, the alteration which may be caused to self-proteins by the interaction with nanoparticles and the persistence of these new entities may provide several of the conditions for the initiation of chronic and autoimmune pathologies, i.e. recognition of self-antigens and persistent stimulation.

Another factor contributing to the development of autoimmunity is the defective clearance of apoptotic cells by scavenger phagocytes [39]. Although there is no direct evidence of a role for nanoparticles in the autoimmune pathogenesis, it should be noted that nanoparticle-loaded human macrophages have impaired capacity of clearing apoptotic cells *in vitro* [9].

While these considerations do not by any means propose that nanoparticles can cause autoimmunity, autoimmune diseases are complex diseases to the pathogenesis of which multiple risk factors can concur (genetic, environmental, nutritional, stress, inflammation, infection). That nanoparticles or nanoparticle-induced inflammation may be among the risk factors should be considered as a possibility and addressed with targeted studies.

ALLERGY

The rise in allergic diseases which has been observed in association with a "Western lifestyle" during the last century [40,41] has prompted an intensive search for causes. One factor suspected to contribute to the dramatic rise in allergies is environmental pollution, since combustion-derived particles and ultrafine particles are associated with inhaling lung diseases, including allergy and asthma [42,43]. This history has led to two lasting legacies in the field of nanosafety research: (1) due to the overlap in size, combustion-derived ultrafine particles and engineered nanoparticles are often discussed together, since at least size-dependent effects may be similar in both types of materials [44–46]; and (2) the paradigm of toxicity mediated by oxidative stress, originally developed for combustion-generated particles, has become an important consideration for engineered nanoparticles as well [46,47].

There are few indications that engineered nanoparticles can misdirect immunity specifically towards T_H2 responses, which are associated with type 1 allergies like hay fever, animal hair allergy, house dust mite allergy or allergic asthma. Some studies in mouse models, however, have reported an enhancement of allergic parameters induced by nanomaterials. Nano-sized TiO_2 increased allergic lung inflammation, T_H2 cytokine production and IgE levels in mice sensitized to ovalbumin [48,49], and enhanced allergy to the same allergen was also reported following exposure to single- and multiwalled carbon nanotubes [50]. On the other hand, loading the major birch pollen allergen Bet v 1 onto nano-sized poly (D,L-lactic-co-glycolic acid) (PLGA) reduced allergy parameters in previously Bet v 1-sensitized mice, as compared to exposure to Bet v 1 alone [51]. All these studies on allergic mice were done using the BALB/c strain, which readily develops allergy, and for this reason the most widely used allergy model is the BALB/c mouse allergic reaction to ovalbumin. However, BALB/c mice are so extremely prone to (allergic) T_H2 responses that an increased response may simply indicate increased immune activation, which in another mouse might manifest in a different form, in particular as T_H1 inflammation.

Anti-allergic effects of silver nanoparticles have been described after treatment of ovalbumin-allergic C57BL/6 mice with nano-sized silver particles [52], and calcium carbonate or calcium phosphate nanoparticles were reported to reduce nickel allergy by binding nickel ions [53]. However, nickel allergy is based on completely different mechanisms than most common allergies, and neither T_H2 cytokines nor IgE antibodies are involved. Overall, the literature does not support a contribution of engineered nanoparticles to the allergy epidemics, and comparisons to the much more complex combustion-derived ultrafine particles have to be made with care. Given increasing exposure and intense consumer awareness, specific assays for detecting allergy-promoting effects will nevertheless have to be considered for products where this issue is particularly relevant, as in the cosmetics industry [54,55].

It is now evident that climate changes influence vegetation, including prolonged periods of pollen production, which affect seasonal allergies [56]. Since particulate matter contributes to climate changes, there is a potential for engineered nanoparticles to act via this rather indirect pathway [46]. However, for the foreseeable future, combustion-generated ultrafine particles will remain far more relevant in this perspective compared to engineered nanoparticles of the same size class.

COMPLEMENT ACTIVATION

The complement system encompasses a group of serum proteins which cascade-interact with each other, as well as with antibodies and with immune cells. Complement is an important part of the immune system and can "tag" pathogens in a process called opsonization. Opsonized pathogens are quickly attacked by antibodies and eliminated by phagocytic immune cells. In addition, complement can induce immune attack on pathologically deviant cells of the body (e.g. virus-infected cells and tumor cells), and it can also by itself destroy body cells. Complement is activated via relatively unspecific mechanisms and it has the ability to damage severely body tissues. For this reason, complement activities need to be tightly controlled and any effect of nanoparticles on complement needs to be ruled out.

There are two main ways how microorganisms can stimulate complement activation (the "classical" and the "alternative" pathways), both of which were activated by single-walled and multiwalled carbon nanotubes [57]. It is highly problematic that the nanomaterial is in this case apparently recognized and treated like a pathogenic microorganism. A number of nanoparticle types has by now been reported to activate complement, and charged nanomaterials appear to be more efficient activators of the complement system compared to their neutral counterparts, an effect shown for a number of different materials (see reviews [58,59]).

Appropriate coating seems to be the method of choice for eliminating unwanted complement activation properties of nanoparticles, even though it may be difficult to produce nanomaterials with sufficient solubility in blood without incurring the risks of recognition by complement factors [60]. Lipid nanocapsules displayed a very low capacity of complement activation when they contained polyethylene glycol at high density [61,62]. Apart from the material used for coating, its density may have surprising effects. Block copolymer polyaxamine 908 can lead to reduced complement activation when used for coating [63], but more recent investigations by the same group showed that, in fact, different complement activation pathways were triggered depending on the density of the polymers [64]. The need to minimize complement activation is particularly relevant for nanomedicine, where delivery of large amounts of nanoparticles into the blood of patients poses a clear risk of adverse reactions mediated by complement. It was indeed shown that hypersensitivity reactions towards Doxil (a doxorubicin-formulation encapsulated in nanoliposomes used in cancer chemotherapy) were due to complement activation [65]. The quick and rather unspecific responses mediated by the complement system are quite difficult to control. Excluding activation of complement will therefore be an important issue for all nanomaterials which may be taken up by the human body.

DEFINITELY NOT INNOCENT BYSTANDERS: THE CASE OF BACTERIAL CONTAMINANTS

An issue of major relevance when assessing the possible effects of nanoparticles on the immune system is the fact that nanoparticles are not "naked". We have already discussed the adsorption of proteins and other endogenous molecules on the nanoparticle surface that occurs when the particles come in contact with the body. However, nanoparticles can interact with other biologically active entities in the environment, most frequently microorganisms and molecules derived thereof. In these circumstances, it is difficult to attribute a possible effect to the nanoparticle rather than to the associated contaminating molecules. For immunological effects this observation is particularly relevant because the possible contaminating agents (microorganisms) are major inducers of inflammation and immune responses. Thus, the activity displayed by nanoparticles in *in vivo* or *in vitro* assays of immune response could be due to contaminating bioactive bacterial products rather than to the nanoparticles themselves.

A clear example is that of bacterial lipopolysaccharide (LPS, or endotoxin). LPS is a cell wall component of Gram-negative bacteria and is one of the most potent activators of innate immunity by interacting with TLR4 receptors on the surface of monocytes/macrophages and other cells [66]. Even minute amounts of LPS can fully activate the inflammatory reaction of human monocytes, e.g. by upregulating the expression of the inflammatory cytokine IL-1β [67,68]. Trace amounts of LPS are found practically everywhere, even in the absence of living bacteria, due to its molecular structure that makes it very resistant to cleaning and sterilization procedures (solvents, heat, etc.) and to its "stickiness" to surfaces. Thus, all glassware and reagents must be specially treated and accurately tested as "LPS-free" before being used for running immune assays. When testing the immunosafety of nanoparticles, both *in vitro* and *in vivo*, it is therefore of key importance that the test nanomaterial is completely free of endotoxin contamination, to avoid the risk of testing the effects of LPS on the immune responses rather than those of the nanoparticles [69]. An issue that may hamper appropriate evaluation of the presence of LPS in nanomaterial preparations is that nanoparticles can physically interfere with standard methods for endotoxin evaluation, thus risking providing misleading results if not accurately validated [70,71]. Two types of approaches can be followed in order to evaluate the immunosafety of nanomaterials avoiding the biological interference by LPS. The first, which can be applied to engineered nanoparticles, is to perform the entire synthetic process in endotoxin-free conditions, a procedure expensive in terms of facilities and equipment but leading to excellent results [72]. The second, suitable for all kinds of nanoparticles, is that of devising assays in which LPS cannot interfere (LPS-resistant mice for *in vivo* experiments, deliberate activation with LPS *in vitro*). This latter experimental setting, i.e. an assay in which cells are deliberately activated so as to simulate the reaction to an infectious agent, has the advantage of allowing us to use nanoparticles as they are, including their environmentally-induced changes and contaminations, a situation closer to real life (Boraschi et al., unpublished observatons).

LEARNING FROM THE IMMUNE SYSTEM

As already mentioned, the immune system has evolved to recognize whether non-self entities pose a threat to the body. Immunity may thus be able to recognize threats that are indirect and not immediately obvious, as opposed to those, for instance, which induce cell death. Activation of the immune system is therefore a warning sign that can be exploited for safer nanoparticle design. Investigation of immunosafety holds specific advantages for the development of nanosafety strategies, but also of nanotechnology at large:

- Immune cells are specialized for recognizing dangerous non-self materials, and they are concentrated close to body surfaces where pathogens, but also engineered nanoparticles, are most likely to intrude. *The sentinel function of immunity can be used to investigate health risks of nanomaterials.* Powerful complement activation or a strong pro-inflammatory response would clearly label a nanomaterial as a health threat. On the other hand, if no immune modulation is found the chances are good that the material is quite harmless, since it is treated as harmless by the immune system.
- There is an extensive body of knowledge on *how immune cells interact with non-self entities.* This is useful to elucidate the mechanisms of interaction between nanomaterials and

biological systems at large. Findings from immune cells may provide paradigms relevant for other cell types as well.

- Immune activation follows various pathways, but in a rather foreseeable manner. Processes like cellular stress, inflammation or allergy can be followed using a limited number of markers. Since these processes arrive, independent of the initiating event, at specified end-points, it is possible to *define general end-points that characterize the activation process*, that can be applied to whatever type of nanomaterial, route of delivery, species, etc. For example, cellular stress will involve NF-κB activation, inflammation will involve inflammasome activation and IL-1 secretion, and type 1 allergy will involve production of IL-4. End-points of such a general nature are well suited to address immunosafety, independently of specific material or system.
- While immune activation processes are part of our defense system, the end-points described above are not markers of pathology, but simply markers of immune activation. Thus, the context in which such activation occurs and its characteristics (in terms of extent and duration) makes the difference between the normal defensive response, which resolves, and a pathological reaction that persists and causes tissue damage. A *kinetical evaluation of the immune response* in the presence of nanoparticles is therefore needed.
- Redundancy is a major feature of the immune system, where many molecules and molecular pathways may be used to reach the same result, which ensures an efficacious response to danger even in the case of significant molecular damage. However, this also means that different organisms are likely to use different mechanisms. For the mouse, the most frequently used animal model for studying immunity, it is evident that many defense processes, including regulation of inflammation, are different from those preferentially active in humans (see for instance IL-37, which has an important role in downregulating human inflammation, but which is absent in the mouse) [73]. Thus, immunosafety assays and models should be designed in such a way that *pathways and mechanisms relevant to human immunity* are considered.
- For safety evaluation, models and assays are generally designed having as the target the healthy organism. This is not really the case in real life, since many if not most people are not in perfect health. For example, many workers have chronic lung conditions, like allergic asthma or chronic obstructive pulmonary disease (COPD). Other people are old, with an immune system less able to react to many challenges, or babies/young children, with a still immature immune system. Do these conditions matter if these people inhale or ingest nanoparticles? Probably yes. Immunotoxicology has excellent tools to simulate disease states, and design *models representing different conditions of immunological frailty.*

TURNING THE TABLES

Nanomaterials have the potential to suppress or to activate immune functions. In the absence of other toxic effects (cytotoxicity, genotoxicity, etc.), a nanomaterial that displays immunomodulatory effects may be an interesting lead substance for medical application. Activation of the immune system is highly desired in cancer, or in cases of chronic microbial infections that cannot be cleared by normal mechanisms. On the other hand, suppression of immune responses is key to treatment of autoimmune diseases and allergies. The field of nanomedicine would therefore strongly benefit from a deeper interaction with nanosafety studies. In particular in the area of immunosafety, the thorough characterization of the immunomodulatory effects of a nanomaterial would open the way for its exploitation in the treatment of many diseases, in which the therapeutic indication includes immunostimulation (infections, tumors) or immunosuppression (allergies, autoimmunity, chronic and degenerative diseases).

TAKE-HOME MESSAGES

Various components of the immune system can recognize nanoparticles. Mostly this does not result in detrimental reactions, so the nanoparticles involved can be considered as harmless. Some nanomaterials, however, either stimulate inflammatory responses or impair normal immune functions, resulting in immunotoxicity. These effects may occur at much lower concentrations than found for overall toxicity, so immunosafety needs to be taken into account for safety assessement. The high sensitivity of immune cells makes them particularly suitable for testing risks associated with nanomaterials.

References

[1] Chappert P, Schwartz RH. Induction of T cell anergy: Integration of environmental cues and infectious tolerance. Curr Opin Immunol 2010;22:552–9.

[2] Sakaguchi S. Regulatory T cells: History and perspective. Methods Mol Biol 2011;707:3–17.

[3] Fink SL, Cookson BT. Apoptosis, pyroptosis, and necrosis: Mechanistic description of dead and dying eukaryotic cells. Infect Immun 2005;73:1907–16.

[4] Aust AE, Cook PM, Dodson RF. Morphological and chemical mechanisms of elongated mineral particle toxicities. J Toxicol Environ Health B Crit Rev 2011;14:40–75.

[5] Johnston HJ, Hutchison GR, Christensen FM, et al. A critical review of the biological mechanisms underlying the *in vivo* and *in vitro* toxicity of carbon nanotubes: The contribution of physico-chemical characteristics. Nanotoxicology 2010;4:207–46.

[6] Alexis F, Pridgen E, Molnar LK, Farokhzad OC. Factors affecting the clearance and biodistribution of polymeric nanoparticles. Mol Pharm 2008;5:505–15.

[7] Yoo JW, Chambers E, Mitragotri S. Factors that control the circulation time of nanoparticles in blood: Challenges, solutions and future prospects. Curr Pharm Des 2010;16:2298–307.

[8] Reisetter AC, Stebounova LV, Baltrusaitis J, et al. Induction of inflammasome dependent pyroptosis by carbon black nanoparticles. J Biol Chem 2011;286:21844–52.

[9] Witasp E, Shvedova AA, Kagan VE, Fadeel B. Single-walled carbon nanotubes impair human macrophage engulfment of apoptotic cell corpses. Inhal Toxicol 2009;21(Suppl. 1):131–6.

[10] Valledor AF, Comalada M, Santamaria-Babi LF, Lloberas J, Celada A. Macrophage proinflammatory activation and deactivation: A question of balance. Adv Immunol 2010;108:1–20.

[11] Shevach EM. Mechanisms of foxp3+ T regulatory cell-mediated suppression. Immunity 2009;30:636–45.

[12] De Koker S, Lambrecht BN, Willart MA, et al. Designing polymeric particles for antigen delivery. Chem Soc Rev 2011;40:320–39.

[13] Xiang SD, Selomulya C, Ho J, Apostolopoulos V, Plebanski M. Delivery of DNA vaccines: An overview on the use of biodegradable polymeric and magnetic nanoparticles. Wiley Interdiscip Rev Nanomed Nanobiotechnol 2010;2:205–18.

[14] Rice-Ficht AC, Arenas-Gamboa AM, Kahl-McDonagh MM, Ficht TA. Polymeric particles in vaccine delivery. Curr Opin Microbiol 2010;13:106–12.

[15] Fadeel B, Garcia-Bennett AE. Better safe than sorry: Understanding the toxicological properties of inorganic nanoparticles manufactured for biomedical applications. Adv Drug Deliv Rev 2010; 62:362–74.
[16] Eisenbarth SC, Colegio OR, O'Connor W, Sutterwala FS, Flavell RA. Crucial role for the Nalp3 inflammasome in the immunostimulatory properties of aluminium adjuvants. Nature 2008;453:1122–6.
[17] Li H, Willingham SB, Ting JP, Re F. Cutting edge: Inflammasome activation by alum and alum's adjuvant effect are mediated by NLRP3. J Immunol 2008;181:17–21.
[18] Hornung V, Bauernfeind F, Halle A, et al. Silica crystals and aluminum salts activate the NALP3 inflammasome through phagosomal destabilization. Nat Immunol 2008;9:847–56.
[19] Sutterwala FS, Ogura Y, Flavell RA. The inflammasome in pathogen recognition and inflammation. J Leukoc Biol 2007;82:259–64.
[20] Franchi L, Eigenbrod T, Nunez G. Cutting edge: TNF-alpha mediates sensitization to ATP and silica via the NLRP3 inflammasome in the absence of microbial stimulation. J Immunol 2009;183:792–6.
[21] Martinon F, Tschopp J. Inflammatory caspases and inflammasomes: Master switches of inflammation. Cell Death Differ 2007;14:10–22.
[22] Martinon F. Signaling by ROS drives inflammasome activation. Eur J Immunol 2010;40:616–9.
[23] Monopoli MP, Walczyk D, Campbell A, et al. Physical-chemical aspects of protein corona: Relevance to *in vitro* and *in vivo* biological impacts of nanoparticles. J Am Chem Soc 2011;133:2525–34.
[24] Maiorano G, Sabella S, Sorce B, et al. Effects of cell culture media on the dynamic formation of protein-nanoparticle complexes and influence on the cellular response. ACS Nano 2010;4:7481–91.
[25] Casals E, Pfaller T, Duschl A, Oostingh GJ, Puntes V. Time evolution of the nanoparticle protein corona. ACS Nano 2010;4:3623–32.
[26] Dell'Orco D, Lundqvist M, Oslakovic C, Cedervall T, Linse S. Modeling the time evolution of the nanoparticle-protein corona in a body fluid. PLoS One 2010;5:e10949.
[27] Rocker C, Potzl M, Zhang F, Parak WJ, Nienhaus GU. A quantitative fluorescence study of protein monolayer formation on colloidal nanoparticles. Nat Nanotechnol 2009;4:577–80.
[28] Vauthier C, Lindner P, Cabane B. Configuration of bovine serum albumin adsorbed on polymer particles with grafted dextran corona. Colloids Surf B Biointerfaces 2009;69:207–15.
[29] Wallberg M, Bergquist J, Achour A, Breij E, Harris RA. Malondialdehyde modification of myelin oligodendrocyte glycoprotein leads to increased immunogenicity and encephalitogenicity. Eur J Immunol 2007;37:1986–95.
[30] Qiao SW, Bergseng E, Molberg O, Jung G, Fleckenstein B, Sollid LM. Refining the rules of gliadin T cell epitope binding to the disease-associated DQ2 molecule in celiac disease: Importance of proline spacing and glutamine deamidation. J Immunol 2005;175:254–61.
[31] Hill JA, Southwood S, Sette A, Jevnikar AM, Bell DA, Cairns E. Cutting edge: The conversion of arginine to citrulline allows for a high-affinity peptide interaction with the rheumatoid arthritis-associated HLA-DRB1*0401 MHC class II molecule. J Immunol 2003;171:538–41.
[32] Moss CX, Matthews SP, Lamont DJ, Watts C. Asparagine deamidation perturbs antigen presentation on class II major histocompatibility complex molecules. J Biol Chem 2005;280:18498–503.
[33] Deng ZJ, Liang M, Monteiro M, Toth I, Minchin RF. Nanoparticle-induced unfolding of fibrinogen promotes Mac-1 receptor activation and inflammation. Nat Nanotechnol 2011;6:39–44.
[34] Chang C, Gershwin ME. Drugs and autoimmunity—a contemporary review and mechanistic approach. J Autoimmun 2010;34:J266–275.
[35] Nagai H, Toyokuni S. Biopersistent fiber-induced inflammation and carcinogenesis: Lessons learned from asbestos toward safety of fibrous nanomaterials. Arch Biochem Biophys 2010;502:1–7.
[36] Kunzmann A, Andersson B, Vogt C, et al. Efficient internalization of silica-coated iron oxide nanoparticles of different sizes by primary human macrophages and dendritic cells. Toxicol Appl Pharmacol 2011;253:81–93.
[37] Geiser M. Update on macrophage clearance of inhaled micro- and nanoparticles. J Aerosol Med Pulm Drug Deliv 2010;23:207–17.
[38] Groves E, Dart AE, Covarelli V, Caron E. Molecular mechanisms of phagocytic uptake in mammalian cells. Cell Mol Life Sci 2008;65:1957–76.
[39] Munoz LE, Lauber K, Schiller M, Manfredi AA, Herrmann M. The role of defective clearance of apoptotic cells in systemic autoimmunity. Nat Rev Rheumatol 2010;6:280–9.

[40] Heinrich J. Influence of indoor factors in dwellings on the development of childhood asthma. Int J Hyg Environ Health 2011;214:1–25.

[41] Holgate ST, Lack G. Improving the management of atopic disease. Arch Dis Child 2005;90:826–31.

[42] Nel AE, Diaz-Sanchez D, Ng D, Hiura T, Saxon A. Enhancement of allergic inflammation by the interaction between diesel exhaust particles and the immune system. J Allergy Clin Immunol 1998;102:539–54.

[43] van Zijverden M, Granum B. Adjuvant activity of particulate pollutants in different mouse models. Toxicology 2000;152:69–77.

[44] Oberdorster G, Oberdorster E, Oberdorster J. Nanotoxicology: An emerging discipline evolving from studies of ultrafine particles. Environ Health Perspect 2005;113:823–39.

[45] Duschl A. Nanoparticle effects on the immune system. In: Zhang YN, editor. Nanotoxicology. Interactions of nanomaterials with biological systems. Stevenson Ranch: American Scientific Publishers; 2007. p. 144–65.

[46] Chang C. The immune effects of naturally occurring and synthetic nanoparticles. J Autoimmun 2010;34:J234–246.

[47] Li N, Xia T, Nel AE. The role of oxidative stress in ambient particulate matter-induced lung diseases and its implications in the toxicity of engineered nanoparticles. Free Radic Biol Med 2008;44:1689–99.

[48] Larsen ST, Roursgaard M, Jensen KA, Nielsen GD. Nano titanium dioxide particles promote allergic sensitization and lung inflammation in mice. Basic Clin Pharmacol Toxicol 2010;106:114–7.

[49] Rossi EM, Pylkkanen L, Koivisto AJ, et al. Inhalation exposure to nanosized and fine TiO_2 particles inhibits features of allergic asthma in a murine model. Part Fibre Toxicol 2010;7:35.

[50] Nygaard UC, Hansen JS, Samuelsen M, Alberg T, Marioara CD, Lovik M. Single-walled and multi-walled carbon nanotubes promote allergic immune responses in mice. Toxicol Sci 2009;109:113–23.

[51] Scholl I, Weissenbock A, Forster-Waldl E, et al. Allergen-loaded biodegradable poly(D, L-lactic-co-glycolic) acid nanoparticles down-regulate an ongoing Th2 response in the BALB/c mouse model. Clin Exp Allergy 2004;34:315–21.

[52] Park HS, Kim KH, Jang S, et al. Attenuation of allergic airway inflammation and hyperresponsiveness in a murine model of asthma by silver nanoparticles. Int J Nanomed 2010;5:505–15.

[53] Vemula PK, Anderson RR, Karp JM. Nanoparticles reduce nickel allergy by capturing metal ions. Nat Nanotechnol 2011;6:291–5.

[54] Robertson TA, Sanchez WY, Roberts MS. Are commercially available nanoparticles safe when applied to the skin? J Biomed Nanotechnol 2010;6:452–68.

[55] Wiechers JW, Musee N. Engineered inorganic nanoparticles and cosmetics: Facts, issues, knowledge gaps and challenges. J Biomed Nanotechnol 2010;6:408–31.

[56] Shea KM, Truckner RT, Weber RW, Peden DB. Climate change and allergic disease. J Allergy Clin Immunol 2008;122:443–53. quiz 454–455.

[57] Salvador-Morales C, Flahaut E, Sim E, Sloan J, Green ML, Sim RB. Complement activation and protein adsorption by carbon nanotubes. Mol Immunol 2006;43:193–201.

[58] Dobrovolskaia MA, Patri AK, Zheng J, et al. Interaction of colloidal gold nanoparticles with human blood: Effects on particle size and analysis of plasma protein binding profiles. Nanomedicine 2009;5:106–17.

[59] Dobrovolskaia MA, Aggarwal P, Hall JB, McNeil SE. Preclinical studies to understand nanoparticle interaction with the immune system and its potential effects on nanoparticle biodistribution. Mol Pharm 2008;5:487–95.

[60] Sim RB, Wallis R. Surface properties: Immune attack on nanoparticles. Nat Nanotechnol 2011;6:80–1.

[61] Mosqueira VC, Legrand P, Gulik A, et al. Relationship between complement activation, cellular uptake and surface physicochemical aspects of novel PEG-modified nanocapsules. Biomaterials 2001;22:2967–79.

[62] Vonarbourg A, Passirani C, Saulnier P, Simard P, Leroux JC, Benoit JP. Evaluation of pegylated lipid nanocapsules versus complement system activation and macrophage uptake. J Biomed Mater Res A 2006;78:620–8.

[63] Al-Hanbali O, Rutt KJ, Sarker DK, Hunter AC, Moghimi SM. Concentration dependent structural ordering of poloxamine 908 on polystyrene nanoparticles and their modulatory role on complement consumption. J Nanosci Nanotechnol 2006;6:3126–33.

[64] Hamad I, Al-Hanbali O, Hunter AC, Rutt KJ, Andresen TL, Moghimi SM. Distinct polymer architecture mediates switching of complement activation pathways at the nanosphere-serum interface: Implications for stealth nanoparticle engineering. ACS Nano 2010;4:6629–38.

[65] Chanan-Khan A, Szebeni J, Savay S, et al. Complement activation following first exposure to pegylated liposomal doxorubicin (Doxil): Possible role in hypersensitivity reactions. Ann Oncol 2003;14:1430–7.

[66] Jerala R. Structural biology of the LPS recognition. Int J Med Microbiol 2007;297:353–63.
[67] Guha M, Mackman N. LPS induction of gene expression in human monocytes. Cell Signal 2001;13:85–94.
[68] Oostingh GJ, Casals E, Italiani P, et al. Problems and challenges in the development and validation of human cell-based assays to determine nanoparticle-induced immunomodulatory effects. Part Fibre Toxicol 2011;8:8.
[69] Shi Y, Yadav S, Wang F, Wang H. Endotoxin promotes adverse effects of amorphous silica nanoparticles on lung epithelial cells *in vitro*. J Toxicol Environ Health A 2010;73:748–56.
[70] Dobrovolskaia MA, Neun BW, Clogston JD, Ding H, Ljubimova J, McNeil SE. Ambiguities in applying traditional Limulus amebocyte lysate tests to quantify endotoxin in nanoparticle formulations. Nanomedicine (Lond) 2010;5:555–62.
[71] Neun BW, Dobrovolskaia MA. Detection and quantitative evaluation of endotoxin contamination in nanoparticle formulations by LAL-based assays. Meth Mol Biol 2011;697:121–30.
[72] Vallhov H, Qin J, Johansson SM, et al. The importance of an endotoxin-free environment during the production of nanoparticles used in medical applications. Nano Lett 2006;6:1682–6.
[73] Nold MF, Nold-Petry CA, Zepp JA, Palmer BE, Bufler P, Dinarello CA. IL-37 is a fundamental inhibitor of innate immunity. Nat Immunol 2010;11:1014–22.

CHAPTER

11

Skin

Nancy A. Monteiro-Riviere[1], *Francesca Larese Filon*[2]

[1]North Carolina State University, Center for Chemical Toxicology Research and Pharmacokinetics, Raleigh, NC, USA

[2]Unit of Occupational Medicine, University of Trieste, Trieste, Italy

OUTLINE

Adverse Effects of Engineered Nanomaterials
DOI: 10.1016/B978-0-12-386940-1.00011-8

SKIN ANATOMY AND FUNCTIONS

Skin is one of the largest organs of the body and is the primary barrier to absorption from the environment. Due to its accessibility and continuous exposure to a myriad of chemicals, both accidental and deliberate (e.g. cosmetics), the skin is a common route of exposure and target for nanomaterials and toxic chemicals. There are properties of a chemical/nanomaterial that will determine its propensity to cause dermatotoxicity which is the ability to penetrate the skin and subsequent interactions with the biological components of skin that could elicit a toxicological response. Due to the recent advances in nanotechnology, this chapter will review nanomaterials used in biological applications, consumer products as well as occupational exposure and discuss their toxicity, mutagenicity, carcinogenicity and irritation potential as well as penetration in intact skin or skin cells (keratinocytes). (For a further discussion on genotoxicity of nanomaterials, see Chapter 14.) In addition, basic anatomical factors that contribute to the barrier properties of skin that may affect the processes of absorption or determine targets for toxicity, as well as how these factors determine whether a compound can be absorbed or can penetrate in skin to elicit a toxicological response will be presented.

Functions of Skin

Skin can act as an environmental barrier to protect major organs, as a diffusion barrier that minimizes insensible water loss that results in dehydration, and as a metabolic barrier that can metabolize a compound to eliminate more easily products after absorption has occurred. Skin can function in temperature regulation, as an immunological effector axis by having Langerhans cells processing antigens and as an effector axis by initiating an inflammatory response to a foreign insult. Skin has a well-developed stroma, which supports all other organs and has neurosensory properties where receptors sense the modalities of touch, pain and heat. Skin can act as an endocrine organ by synthesizing vitamin D, is a target for androgens that regulate sebum production and a target for insulin that regulates carbohydrate and lipid metabolism. In addition, it contains several sebaceous glands that secrete sebum, a complex mixture of lipids that function as antibacterial agents, and apocrine and eccrine sweat glands that produce a secretion that contains scent and functions in territorial demarcation. Skin also plays a role in the biosynthesis of keratin, collagen, melanin, lipids, carbohydrates, as well as in respiration, and biotransformation of xenobiotics and vitamin D. Anatomically, skin can be divided into two principal and distinct components: a stratified, avascular, outer cellular epidermis, and an underlying dermis consisting of connective tissue with numerous cell types and special adnexial structures. The epidermis consists of stratified squamous keratinized epithelium that undergoes a programmed proliferation and differentiation that eventually results in the formation of the major barrier to chemical penetration: the stratum corneum. Two primary cell types, the keratinocytes (stratum basale, stratum spinosum, stratum granulosum, stratum lucidum, and stratum corneum) and the non-keratinocytes (melanocytes, Merkel cells [tactile epithelioid cells] and Langerhans cells) are present. However, the stratum corneum is the primary anatomical barrier responsible for modulating chemical, drug and particle absorption into the skin [1–3].

Skin Morphology

The epidermis consists of a keratinized stratified squamous epithelium and undergoes an orderly pattern of proliferation, differentiation and keratinization. The stratum corneum is the outermost superficial epidermal layer that serves as the first line of defense between the body and the environment, and the lower part of this layer is crucial in preventing the penetration of irritants and allergens [4]. The stratum corneum can vary in thickness depending on the regions of the body and the species [5]. Each stratum corneum cell (corneocyte) is 30 μm in diameter and 0.5 to 0.8 μm in thickness [6]. These cells are highly organized and stacked upon one another to form vertical interlocking columns having a flattened tetrakaidecahedron shape [7,8]. This spatial arrangement helps facilitate the maintenance of the skin's efficient barrier function. Lipids between the stratum corneum are derived from lamellar granules (Odland bodies) of the stratum granulosum to form the intercellular lipid component of the stratum corneum barrier. These lipids are arranged into lamellar sheets that constitute the epidermal permeability barrier. The number of lamellae may vary within the same area of the stratum corneum and when viewed by transmission electron microscopy (TEM) consists of a pattern of alternating electron-dense and electron-lucent bands which represent paired bilayers formed from fused lamellar granule disks [9–13]. These lipids consist of ceramides, cholesterol, fatty acids, and small amounts of cholesterol esters, as well as hydrolytic enzymes such as acid phosphates, proteases, lipases and glycosidases. Each corneocyte is embedded in the lipid matrix produced by the lamellar granules. Cells are bathed in the lipid matrix to form the so-called "brick and mortar" structure where the non-viable corneocytes represent the "bricks" and the intercellular lipids represent the "mortar" [14,15]. Most chemicals are absorbed through skin via this intercellular pathway, with partitioning into and diffusion through the lipid predictive of compound absorption. In atopic dermatitis, the skin barrier is impaired and allergens can penetrate through the skin causing IgE sensitization. Skin exposure to irritants in an occupational setting or environmental scenario may cause a disruption of the stratum corneum layers by protein denaturation agents such as detergents or by extraction of lipids from stratum corneum with solvents [16–20]. Lipid extraction studies conducted in the abdominal, inguinal and back regions of pigs demonstrated relative proportions of individual lipids (ceramides 1–6, fatty acids, cholesterol, triglycerides, and cholesterol esters) were similar across all body regions [18]. Skin is permeable to some chemicals but the permeability may increase four to 100 times in atopic subjects with damaged skin [21,22], thereby increasing the risk for penetration, irritation and sensitization. Regional and species differences in skin thickness, hair follicle density, blood flow, age, and disease states may all influence barrier function [1,19].

NANOPARTICLE (NP) PENETRATION

People are in daily contact with many different substances and the dermal absorption of these substances needs to be considered for risk evaluation studies [23–25]. There is much information on skin absorption in relation to drugs, cosmetics, toxins and metal salts and powders [25], and many of these toxic substances are labeled with "skin notation" by the American Conference of Governmental Industrial Hygienists (ACGIH) which warns that cutaneous exposure to these compounds may significantly contribute to total systemic exposure [25]. However, there is very little information regarding nanoparticles and their ability to enter through skin. Recently, the transdermal route has been investigated with a variety of results due to the fact there are so many different types of nanomaterials [26]. European projects such as Nanoderm stated that both titanium oxide and zinc oxide nanoparticles (NP) are unable to permeate skin, yet many investigators had shown permeation of NP through skin [27–29].

There are four potential pathways in which compounds can penetrate through healthy skin, depending on their physicochemical properties: intercellular (between cells), transcellular (through cells), through hair follicles, and through sweat glands [1,30,31]. Most dermal absorption literature is based on chemicals and drugs that are very small (<500 Daltons) and soluble in dosing formulations that move via passive diffusion across the stratum corneum intercellular lipids [32,33]. Also, there are a variety of factors that can influence the extent of uptake in the skin including barrier integrity, a contaminated surface, the anatomical structure, and the presence of skin diseases such as allergic and irritant contact dermatitis, atopic eczema, and psoriasis [19]. Moreover, mechanical flexion [34,35], irritant detergents and some chemicals [36,37] may also increase skin absorption.

It should be noted that NP may inhibit skin penetration of other components. In fact, a thin layer of glycerin containing NP of either calcium carbonate or calcium phosphate applied to isolated *in vitro* pig skin and to mice *in vivo* prevented the penetration of nickel ions into the skin. It is thought that the NP captured nickel ions by cation exchange which remained on the surface of the skin, thereby making them easily removed by simple washing techniques. Particles smaller than 500 nm added to topical creams may be an effective way to limit the exposure of metal ions that may cause skin irritation [38].

Nanoparticle (NP) Penetration—Is it Feasible?

The penetration of NP in the skin is a controversial subject [39–47], partly because many factors that influence absorption were not studied and the opinions of these investigators were generalized based on limited studies with only a few types of NP. Many of these studies were conducted on very large particles that were not of nanosize (<100 nm). Also, these discrepancies may be related to differences in NP composition, surface chemistry, vehicles or solvents, techniques and methods of exposure, and analytical analysis, laboratory conditions and duration of the experiment. The respiratory route of exposure has been studied extensively [48–59] (and see Chapter 7), but skin exposure has been neglected being considered less permeable with a lower perception of risk. For a more complete discussion on the aspects of skin penetration with NP see the review by Monteiro-Riviere and Baroli [26] and Monteiro-Riviere and Riviere [32].

Nanomaterials have widespread applications in biomedical applications, but also in occupational and consumer scenarios. Consumers or workers may have compromised skin with an altered barrier from conditions such as atopic dermatitis, psoriasis or eczema on their hands or other parts of their body which could influence the rate of absorption or penetration of nanomaterials through skin. The transdermal flux of nanomaterials has obvious implications to toxicology, therapeutics and drug delivery, but there is a need to know the depth and mechanism of nanoparticle penetration through the rate-limiting barrier of the stratum corneum. Several studies have evaluated the hypothesis that NP may become lodged in the lipid matrix of the skin [60–62]. The importance of NP size, shape, charge, surface properties, and vehicle as well as animal species will determine penetration and should be considered when designing studies to assess whether or not nanoparticle penetration occurs. There is limited information on the *in vivo* dermal penetration of nanomaterials in humans; however, there are several published studies in various *in vitro* model systems and on excised animal and human skins.

Barrier Modifications

The effects of solvents and vehicles on skin absorption are well known. The barrier properties may be altered either by mechanical action, ultraviolet radiation, or by chemical agents, after occupational or accidental spillage, or topical creams such as sunscreens, cosmetics, insect repellents, or drugs. Many solvents can cause a range of structural alterations to the intercellular lipid bilayers of the stratum corneum by increasing the fluidization, and delipidization of the lipid, thereby changing the polarity, and phase separation causing vesicles within the lipid layers. Damage to desmosomes may create fissures within the lipid bilayer and damage to the fibrillar keratin of the stratum corneum may create vacuoles. All of these solvent actions will influence the rate of absorption for chemicals [18,20,33,63,64] but with nanomaterials only a few studies have investigated these effects.

Mechanical Action

Mechanical stressors may affect how NP can traverse through the skin. In an occupational setting, repetitive motions such as constant hand movements could accentuate biomechanical forces that may alter the structural organization of skin thereby leading to an increase in the permeability of the skin. The effects of flexing for 30 min on normal skin with FITC-conjugated dextran beads with a diameter of 0.5 μm showed penetration through the stratum corneum layer and into the viable epidermis [34]. Studies in which flow-through diffusion cells were used showed penetration of a very small fullerene amino acid-derivatized peptide nanoparticle of 3.5 nm topically applied to dermatomed porcine skin flexed at a 45 degree angle at a frequency of 20 flexes/min for 60 min or 90 min. After 8 h and 24 h, TEM analysis revealed NP localization within the intercellular spaces of the stratum granulosum, suggesting that penetration of the derivatized fullerene occurs through intercellular lipid moieties [35,60]. The effects of applied strain on quantum dot (QD) cellular uptake showed that application of physiological load conditions may increase the permeability of the cell membrane allowing for an increase in QD uptake [65]. However, QD applied to rat skin flexed for 60 min showed that larger nanoparticles QD655-COOH (18 nm) and QD565-COOH (14 nm) did not penetrate at 8 and 24 h [61].

QD penetration was assessed in intact, tape stripped, and abraded skins which are methods used to mimic barrier perturbation. Studies were conducted with two different sizes of QD with carboxylic acid coatings on rat skin in flow-through diffusion cells for 8 and 24 h. Tape-stripped skin showed QD remained on the surface of the viable epidermis. In abraded skin, QD penetrated into the viable dermis at both 8 and 24 h, while intact skin depicted QD only in the stratum corneum [61]. QD565-COOH topically applied to human skin that was

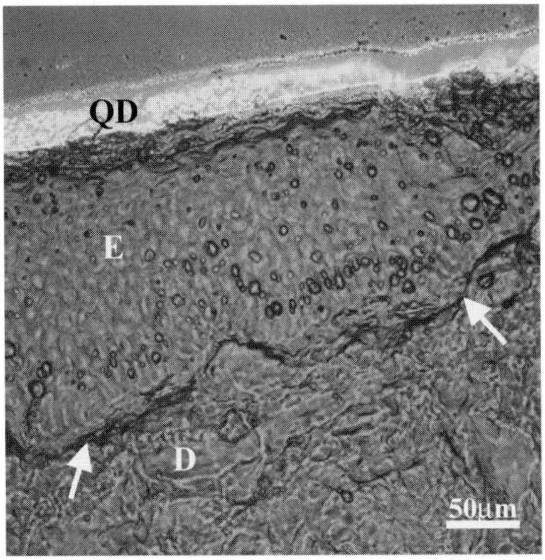

FIGURE 11.1 Laser confocal microscopy of taped-stripped human skin exposed to QD565-COOH for 8 h in flow-through diffusion cells. The arrows depict the interface of the epidermis (E) and dermis (D) of the skin. QD remained on the surface of the epidermis.

tape-stripped 30 times to remove the stratum corneum in flow-through diffusion cells for 8 h and was detected by laser confocal microscopy on the surface of the viable epidermal layer of the skin [66] (Figure 11.1).

UVB-Damaged Skin

Studies with sunscreens containing zinc isotopes (^{68}Zn) topically applied in humans for 5 days have been conducted on sunburned damaged (UVB) (ultraviolet B radiation) skin and showed higher levels of zinc in the blood and urine of individuals receiving the nano-sunscreen. However, whether ^{68}Zn was absorbed as ZnO particles or soluble Zn is unknown [67]. Sunscreens containing titanium dioxide (TiO_2) and zinc oxide (ZnO) NP in different water and oil formulations were studied on *in vivo* and *in vitro* UVB-damaged porcine skin and TiO_2 showed deep penetration into the stratum corneum (Figure 11.2) while ZnO remained primarily on the surface (Figure 11.3) (more detail can be found under the TiO_2 and ZnO section). However, with more sophisticated microscopic techniques, such as time-of-flight-secondary-ion-mass-spectrometry (TOF-SIMS), Ti was detected within the epidermis and superficial dermis and Zn was limited to the stratum corneum. This study showed TiO_2 and ZnO sunscreen formulations were slightly enhanced in UVB-damaged skin, but transdermal absorption was not detected [68].

Solvent Effects

The impact of solvents on skin penetration was studied with pristine fullerenes (C_{60}) in toluene, cyclohexane, chloroform and mineral oil in weanling pigs. The fullerene in the

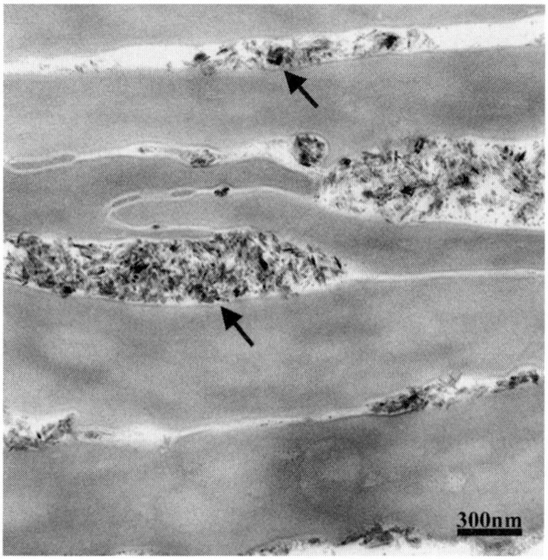

FIGURE 11.2 Transmission electron micrograph of TiO_2 NP (arrows) within the intercellular lipid space of the stratum corneum layers of *in vivo* skin treated for 48 h with a sunscreen formulation.

solvent was applied to the skin daily for 4 days to simulate occupational exposures. The dose sites were tape-stripped 26 times for quantitative analysis by simplified liquid–liquid extraction along with skin biopsies. The C_{60} dose in toluene, cyclohexane or chloroform showed that the pristine fullerenes penetrated deeper into the stratum corneum layers, but

FIGURE 11.3 Transmission electron micrograph of ZnO NP on the surface of stratum corneum of UVB-exposed skin treated with ZnO sunscreen formulation for 48 h.

more C_{60} was found in the stratum corneum when dosed in chloroform. No C_{60} was detected in the skin with the mineral oil. Similar studies were conducted with *in vitro* flow-through diffusion cells which also showed no fullerenes were present in the receptor fluid [69].

Terpenes are highly lipophilic aliphatic compounds in essential oils that are generally regarded as safe and have been found to change the stratum corneum lipid organization from orthorhombic to hexagonal [20,70,71]. Human skin was pretreated for 1 h with the terpenes (menthol, eucalyptol, limonene), or ethanol vehicle and dosed with different charged QD of two different sizes in the *in vitro* flow-through diffusion cells for 8 h. Laser scanning confocal microscopy showed no penetration of QD into the viable layers of the skin, but detected QD in the upper stratum corneum layers. TEM showed slightly deeper QD penetration in the stratum corneum in the terpene-pretreated skin, but QD were not detected in the perfusate. These studies showed that terpenes caused only a slight enhancement of QD penetration into the stratum corneum but no absorption occurred through skin [72].

Skin Penetration Studies with Carbon-Based Nanomaterials

Carbon-based nanomaterial absorption through skin may not be similar to chemical absorption since these nanomaterials are engineered differently and of various sizes, and therefore are not homogeneous like chemicals [73]. Carbon nanomaterials are one of the most important classes of NP because of their large variety of applications. They include: (1) carbon nanotubes, single walled carbon nanotubes (SWCNT) and multiwalled carbon nanotubes (MWCNT), with diameters ranging from a few to hundreds of nanometers and length up to a few micrometers, and (2) fullerenes with a size less than 100 nm. Carbon nanotubes are quite large compared to carbon fullerene NP and will not penetrate through the skin. Therefore, most studies have used fullerenes due to their small size to study the penetration through skin. A study in which C_{60} with a fluorescently tagged nuclear localization signal (C_{60}-NLS-FITC) was topically applied at 4.4 μg and 1.1 $\mu g/cm^2$ to porcine skin in flow-through diffusion cells for 24 h showed penetration of the NP through all epidermal layers by confocal microscopy. TEM showed that the C_{60}-NLS-FITC localized within the intercellular bilipid layers of the stratum corneum cell layers [35]. The addition of 1% Pluronic F127 surfactant greatly enhanced the permeability for all treatments [74].

Quantum Dot Penetration

Several investigators have studied the penetration of quantum dots (QD) in skin because they are easy to track by fluorescent methods due to their intense and photostable fluorescence and their commercial availability with different surface coatings. QD of different shapes, sizes and different charges were studied. QD565 and QD655 contain a cadmium/selenium (CdSe) core with a zinc sulfide (ZnS) shell. QD565 are spherical with a diameter of 4.6 nm, while QD655 are ellipsoid with a size of 6 nm (minor axis) × 12 nm (major axis). The hydrodynamic diameters for QD565 are 35 nm for the polyethylene glycol (PEG), uncharged, 14 nm for the carboxylic acid (COOH), negatively charged, and 15 nm for the PEG-amine (NH_2), positively charged. The hydrodynamic diameters of QD655 were 45 nm (PEG), 18 nm (COOH), and 20 nm (NH_2) [28]. In comparison, QD621-PEG have a CdSe core and CdS shell coated with PEG polymer coils, and are nail shaped as seen by TEM, with the mean width of 5.78±0.97 nm and length of 8.40±1.9 nm and a hydrodynamic size of 39±1 nm as evaluated by size-exclusion chromatography [75]. Penetration of QD565 and QD655 with diverse physicochemical properties was studied in porcine flow-through diffusion cells at an occupationally relevant dose over a work day. QD565 with PEG, PEG-NH_2, or COOH coatings showed penetration into the stratum corneum and localization within the epidermal and dermal layers by 8 h. PEG and PEG-NH_2 coated QD655 were localized within the epidermal layers by 8 h. The penetration of QD655-COOH was evident only at 24 h in the epidermal layers [28]. Skin absorption of QD565 with three surface modifications was evaluated in human skin with *in vitro* flow-through and static diffusion cells at various pHs. Laser scanning confocal microscopy detected penetration into intact viable epidermis of skin for QD-PEG at pH 8.3, but not at pH 7.0. QD did not penetrate beyond the stratum corneum of intact skin but, after tape stripping 30 times to remove most of the stratum corneum, were found in the viable epidermal layers [76]. QD621 topically applied to dermatomed porcine skin in flow-through diffusion cells were located primarily in the normal stratum corneum, although a small amount of fluorescence was detected in the upper epidermal layers. Occasionally, QD621 were seen in the outer root sheath of the hair follicle [62]. By TEM, the QD were limited to the upper intercellular spaces of the stratum corneum of porcine skin (Figure 11.4). Neutral and charged QD621 were infused into the arterial vascular supply of the isolated perfused skin flap model to mimic systemic distribution after exposure by any route. Biodistribution studies showed that QD621 can migrate out of the capillaries into the surrounding tissue [77]. QD621 that were intradermally injected into SKH-1 hairless mice migrated from the injection site to regional lymph nodes through the lymphatic duct system and then to the liver and other organs [78].

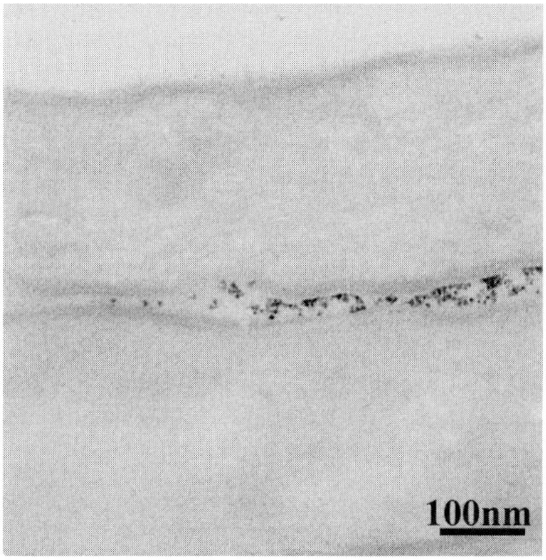

FIGURE 11.4 Transmission electron micrograph showing the localization of QD621 within the intercellular lipid bilayers of the upper stratum corneum.

Titanium Dioxide and Zinc Oxide Penetration

The global production of TiO_2 and ZnO for use in skincare products was 1000 tons in 2003 and 2004 [79]. The Australian Therapeutic Goods Administration estimated that, in 2005, 70% of all sunscreens containing TiO_2 and 30% containing ZnO were formulated with NP [80]. Sunscreen use is highly recommended to prevent skin cancer, sunburn, photoaging, and skin wrinkles but traditional sunblocks may cause irritation. Nanoscale titanium dioxide (TiO_2) and micronized ZnO nanoparticles (NP) less than 100 nm have been incorporated into sunscreens to provide effective broad spectrum physical sunblockers and this transparent form is more aesthetically acceptable to the consumers. Several *in vivo* and *in vitro* studies, including the European NANODERM project, have shown negligible or no penetration of intact skin with different formulations and concentrations of TiO_2 NP under various experimental conditions [81–95]. The EU Scientific Committee on Consumer Products in 2007 stated, that a safety assessment of nanosized TiO_2 in cosmetics should also take into account abnormal skin conditions and the impact of mechanical effects on skin penetration [47].

Some investigators have suggested that penetration can occur when the skin barrier is damaged. Humans exposed for 5 days outdoors to sunscreens containing ZnO NP showed small increases in the Zn tracer in the blood and urine by very sensitive methods but it is not known whether ^{68}Zn was absorbed as ZnO or soluble Zn [67]. Recently, *in vitro* and *in vivo* studies conducted with TiO_2 and ZnO NP to UVB damaged porcine skin with four different sunscreen formulations showed slightly enhanced TiO_2 or ZnO NP penetration with TOF-SIMS analysis compared to normal skin (see Figures 11.2 and 11.3), in the absence of any transdermal absorption [68].

Gold (Au) NP Penetration

In vitro skin permeation of Au NP was first studied by Sonavane [96] in rat skin Franz static diffusion cells with three sizes of Au particles (15 nm, 102 nm and 19 nm). Au NP showed size-dependent permeation through rat skin and TEM revealed accumulation of small NP within the deep layer of skin but large particles were observed mainly in the epidermis and dermis. This is not surprising because the larger particles were not of nanosize (<100 nm). Other authors confirmed that Au NP of 12.6 nm can permeate intact and damaged human skin in Franz static diffusion cells in a dose-dependent manner (Figure 11.5). Also, when greater amounts of Au were applied (three times), there was a parallel increase in permeation (three times). Evaluation of the Au into skin revealed that Au concentration decreased from the superficial to deeper skin layers and a significant amount was found in damaged skin compared to intact skin. The abraded skin had an increase in only the amount of Au that remained in the skin but not in the receptor compartment. This may be due to a strong interaction between NP and the skin cells and extracellular matrix that might hinder the permeation of particle migration, and the short 24-h period of time [22].

Silver (Ag) NP Penetration

Ag NP absorption through intact skin is low (<1 ppm) since much of the free ion is precipitated as Ag sulfide in the superficial layers of the stratum corneum. An *in vivo* daily

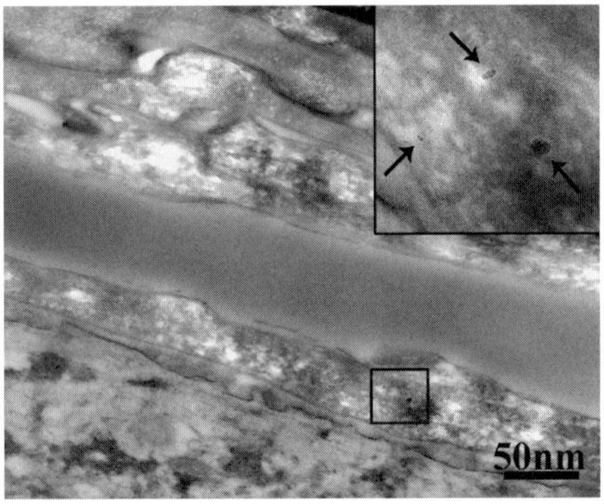

FIGURE 11.5 Transmission electron micrograph of 13 nm Au NP (arrows) localized in the stratum corneum of human skin.

repetitive skin study for 14 days was conducted using 25 nm and 35 nm carbon-coated Ag NP, as well as freshly synthesized or thoroughly washed 20 nm, 50 nm and 80 nm Ag NP topically applied to the backs of weanling pigs. Ag NP penetration was observed only in the superficial layers of the stratum corneum by TEM and confirmed by EDS which detected the localization of Ag [97]. Skin penetration of Ag NP of 25 nm was demonstrated in intact and damaged *in vitro* human skin in static diffusion cells [21] but with low flux compared to the absorption rates for metal powders [98]. TEM confirmed the presence of Ag NP in the stratum corneum and the outermost surface of the epidermis but not in the dermis. These results suggest that a small fraction of the NP were dissolved and diffused through the skin layers as elemental Ag.

TOXICITY STUDIES

Irritation

There is little information whether NP on the surface of the skin can cause a local irritation due to mechanical actions [99]. In particular, local irritation of carbon nanotubes have been reported on the conjunctiva of the eye of animals but no effect was noted on the skin [100]. Carbon fiber dermatitis has been reported in humans after exposure to carbon nanofibers (hyperkeratosis) [101].

Sensitization

The clinical manifestation of skin sensitization is allergic contact dermatitis, but the sensitization potential of nanomaterials is unknown. NP that penetrate into the epidermis may cause an allergic reaction due to the presence of sensitization substances such as metals (e.g. nickel, cobalt and chromium) that can induce an allergic contact dermatitis and to their large surface area allowing them a larger contact with biological surfaces in comparison to bulk material. To date, however, there are no data on their allergenicity, and future studies are required to confirm this hypothesis. On the other hand, it has been suggested that fullerenes may play a leading role in the inhibition of the *in vitro* and *in vivo* IgE-mediated allergic response, thus blocking histamine release. Fullerenes are also capable of inhibiting mast cells and basophilic allergic responses *in vitro* and anaphylaxis *in vivo*. This biological effect critically depends on the moieties that are added to the carbon cage, and it has been suggested they act on different FcepsilonRI-specific molecules in mast cells and basophils. These fullerene derivatives represent a new class of compounds that interfere with FcepsilonRI signaling pathways to stabilize mast cells and basophils, and thus fullerene-based therapies may be a new approach for treating allergic diseases [102], and perhaps to control diseases largely influenced by these cells, including type I hypersensitivity, anaphylaxis, hay fever, asthma, arthritis, multiple sclerosis, and other autoimmune diseases. Obviously, a great deal of research is required to assess this hypothesis.

C_{60} Fullerene Toxicity

Buckyballs or fullerenes are molecular structures made up of 60 or more carbon atoms. Fullerenes applied topically to mouse skin over 72 h had no effect on DNA synthesis. However, repeated application of fullerenes to mouse skin after initiation with dimethlybenzanthracene (DMBA) for 24 weeks showed promotion with 12-0-tetradecanoylphorbol-13-acetate (TPA) that resulted in benign skin tumors [103].

Cytotoxicity Effects on Keratinocytes with Carbon-Based Nanomaterials

A major concern in conducting toxicity studies with carbon nanomaterial is the nature of the controls (reference material of known toxicity) employed. Carbon black (CB), a nanoscale particulate, has often been recommended to serve as a negative (non-toxic) control. The use of CB has grown out of its previous role in serving as a control in inhalational toxicology studies. However, caution must be taken when utilizing CB because of its ability to adsorb colorimetric chemical markers used in biochemical assays, thus leading to unreliable findings [73]. Therefore, multiple approaches including TEM, mRNA assay, and protein expression assays should be used to assess the toxicological effects of these NP [73]. Ten different viability assays were used to assess cytotoxicity in HEK with five different nanomaterials (SWCNT, C_{60}, CB, nC_{60}, and QD). Nanomaterials tested were carbon or non-carbon based and soluble or non-soluble in aqueous media. Although these assays provide accurate viability data for small molecule cytotoxicity studies, they were less reliable when assessing nanomaterials or NP [104].

Due to the poor solubility of fullerenes, limited research has been conducted in biological systems. Functionalization of C_{60} with the addition of amino acids, hydroxyl groups, or carboxyl groups can increase the ability of the NP to interact with the biological environment [105–107]. ^{14}C-labeled underivatized C_{60} exposure to immortalized HEK demonstrated no effect on proliferation and fibroblasts but showed that the radiolabel was taken up; it was unclear whether NP actually entered the cell or were associated with the cell membrane [108]. HEK exposed to nC_{60} ranging from 0.0000025–0.0005 mg/mL and $C_{60}(OH)_{24}$ ranging from 0.000005–0.001 mg/mL for 24 and 48 h showed a significant decrease in viability by 24 h. Interleukin (IL)-8 and IL-6 were significantly greater than controls but IL-10, IL-1β, and TNF-α were below detectable limits. The $C_{60}(OH)_{24}$ showed no toxicity at all concentrations. These results showed that derivatized C_{60} $(OH)_{24}$ is non-toxic, and nC_{60} is toxic at concentrations as low as 0.00025 mg/mL in HEK [109]. HEK exposed to a functionalized C_{60}-based amino acid (Baa) fullerene at various concentrations depicted a statistically significant decrease in viability for concentrations of 0.4 and 0.04 mg/mL and a significant increase in IL-8 for the 0.04 mg/mL dose by 8, 12, and 24 h. IL-6 and IL-1β were greater at 24 and 48 h, but no significant TNF-α or IL-10 expression was noted. TEM depicted this Baa in vacuoles of HEK [110]. HEK exposed to 0.4 to 0.001 mg/mL of C_{60}-NLS-FITC for 24 h and 48 h showed a decrease in viability that was statistically significant at 0.4 mg/mL at both 24 and 48 h. Cytokines IL-8 and IL-6 showed a statistically significant increase at 0.2 mg/mL and 0.4 mg/mL. TEM of the C_{60}-NLS-FITC showed localization in vacuoles in close proximity to the nucleus (Figure 11.6) [74]. The endocytic pathways of a water-soluble fullerene-substituted phenylalanine (Bucky amino acid, Baa) poly-lysine derivative (Baa-Lys(FITC)-$(Lys)_8$-OH) with a positive surface charge and an FITC label was studied in HEK. The cellular uptake pathway was mediated by caveolae/lipid rafts and cytoskeletal components and the inhibitor for the scavenger receptor completely blocked the uptake [111].

SWCNT Toxicity

Immortalized non-tumorigenic human epidermal (HaCaT) cells exposed to unrefined SWCNT showed the formation of free radicals and a decrease in cell viability upon exposure for 18 h, however, inflammatory markers of irritation or inflammation were not evaluated [112]. HEK dosed with 6-aminohexanoic acid-derivatized single-walled carbon

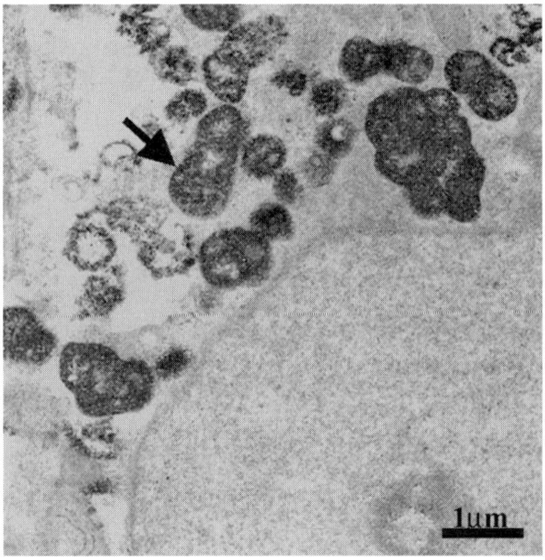

FIGURE 11.6 Transmission electron micrograph of the bioaccumulation of C_{60}-NLS-FITC NP localized in the cytoplasm but in close proximity to the nucleus of the HEK.

nanotubes (AHA-SWCNT) at various concentrations exhibited a significant decrease in viability ($P<0.05$) from 0.00005 mg/mL to 0.05 mg/mL after 24 h. Biomarkers of inflammation such as IL-6, increased in HEK treated with 0.05 mg/mL from 1 to 48 h, while IL-8 showed a significant increase at 24 and 48 h. Expression of other cytokines such as TNF-α, IL-10 and IL-1β was not significant. TEM showed that AHA-SWCNT localized within intracytoplasmic vacuoles [113]. Studies with 1% Pluronic®F127 surfactant used to disperse the AHA-SWCNT aggregates in the cell culture medium caused less toxicity [113]. Another study showed that SWCNT can activate NF-κB in a dose-dependent manner in HEK [114]. Granulomas similar to those induced by foreign body reaction were seen 1 week after carbon nanotubes resembling stacked hats were implanted in the subcutaneous tissue of rats. Nanotubes were present in macrophages. No severe inflammatory responses were noted after 4 weeks [115].

MWCNT Toxicity

HEK exposed to unfunctionalized MWCNT at 0.1, 0.2 and 0.4 mg/mL concentrations for 1, 2, 4, 8, 12, 24 and 48 h showed a dose-dependent decrease in viability and a dose-dependent increase in the proinflammatory mediator IL-8. MWCNT were localized within cytoplasmic vacuoles (59% at 24 h, and 84% at 48 h) (Figure 11.7) [116]. When Pluronic F127 surfactant was added to the media it dispersed the MWCNT and reduced aggregation and was non-toxic to the HEK. MWCNT alone without surfactants were equally toxic but caused an increase in IL-8 expression. But MWCNT with surfactants caused a decrease in IL-8 suppression than when exposed to medium alone. This suggests that surfactant-nanotube interactions alter the surface properties, thus modulating their interaction with viable cells similar to derivatized particles have altered cytotoxicity [117]. Studies with lectin inhibitors that normally inhibit melanosome transfer pathways were studied to determine whether they play a role in reducing the cellular uptake of MWCNT. The lectin cocktails decreased MWCNT at the cell surface and reduced cellular uptake [118]. MWCNT have been shown to alter protein expression associated with metabolism, cell signaling, stress and expression of cytoskeletal elements and vesicular trafficking [119,120].

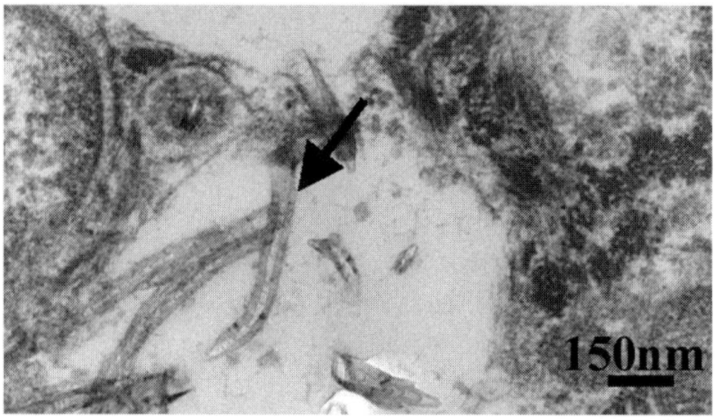

FIGURE 11.7 Transmission electron micrograph depicting MWCNT (arrow) within an intracellular cytoplasmic vacuole of an HEK.

QD Cytotoxicity

The cytotoxicity and pro-inflammatory potential of QD with different surface coatings were studied to determine if they cause differential cytotoxic effects on HEK and cytokine release. Toxicity was observed for PEG-amine and QD-COOH at 48 h with slight toxicity with PEG alone. The QD-COOH caused an increase in IL-1β, IL-6, and IL-8. Therefore, this study showed that the surface coatings of QD can play a role in cytotoxicity and immunotoxicity [121]. Endocytic pathways of untargeted QD were studied in HEK with three different surface coatings to determine the time-course effects of surface coating, temperature, culture medium or inhibitors of the cell cycle and endocytosis. The time-course studies showed an increase in cellular interactions with time and media supplements, but decreasing the temperature decreased the magnitude of interactions of all three surface coatings. The biochemical mediators able to inhibit endocytosis were influenced by these variables [122]. Cytotoxicity of QD621 showed a dose-dependent decrease in viability from 0.3125 nM to 10 nM dose that was statistically significant at 1.25 nM–10 nM compared to the controls. The cytotoxic and inflammatory responses of IL-8 and IL-6 were dose and time dependent. TEM verified that QD621 also were localized within the cells and along the periphery of the cell membrane after 24 h [62].

Silver (Ag) NP Toxicity

Incorporation of Ag NP is commonly used in the professional and personal healthcare fields and in consumer products such as surgical and food handling tools, clothing, cosmetics, contact lens cases, catheters, and even children's toys. Skin exposure to Ag NP is particularly high due to their increasing inclusion into textiles, burn creams, jewelry, acupuncture needles, and catheters. In fact, according to the database of the Nanotechnology Project of the Woodrow Wilson International Center for Scholars, Ag NP are the most commonly used material in all nanotechnology consumer products [123]. Release of Ag ions from wound dressings can be triggered by wound fluids and tissue exudates. The Ag content of dressings varies from $<0.1\ mg/cm^2$ to more than $1\ mg/cm^2$ with the amount of Ag ions released ranging from <1 ppm to >70 ppm. Several studies have evaluated the toxicity of Ag from Ag NP-containing wound dressings and have shown elevated serum enzyme levels (including liver enzymes) as well as argyria [124]. Argyria is a benign skin condition characterized by the bluish-graying of the skin that occurs through the preferential Ag deposition in the basal lamina of skin.

Ag in its bulk metallic form is inert and exerts no biocidal action but ionizes in the presence of water or tissue fluids to release antimicrobial Ag ions. The speciation of Ag does not play a large part in its toxicity, unlike other heavy metals. The toxicity of Ag^{3+} or Ag^{2+} is unlikely due to the transient nature of these forms of Ag, particularly in an aqueous environment. Furthermore, there is no discernable difference between Ag ions released from NP from those generated by other sources such as electrolysis or dissolution of ions from Ag salts. However, the rate of Ag ion production, as well as the distribution of Ag in tissue, may differ substantially between Ag NP and other sources of silver ions due to the ability of Ag NP to release ions into specific tissues, cell membranes, or cells. Additionally, once NP are located within a site, they have the ability for protracted release of Ag ions, thus increasing the potential toxic impact. Furthermore, agglomeration of multiple Ag NP may change the rate of Ag ion offloading from their surface.

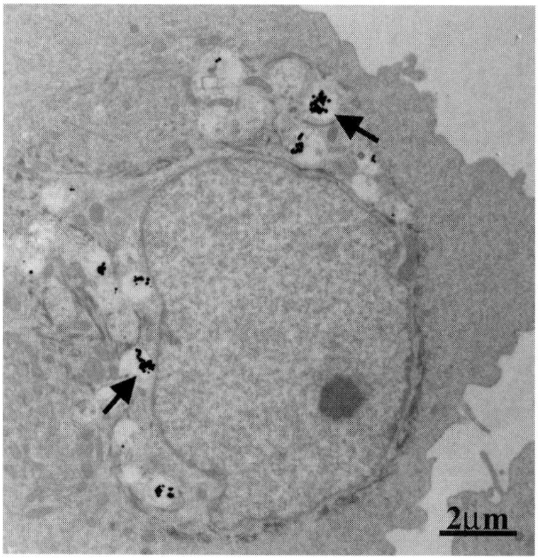

FIGURE 11.8 Transmission electron micrograph of 20 nm Ag NP (arrows) within the intracytoplasmic vacuoles of an HEK.

The cytotoxicity and inflammatory potential of Ag NP were evaluated in HEK following 24 h exposure to 25 nm and 35 nm carbon-coated Ag NP, as well as to either freshly synthesized or thoroughly washed 20 nm, 50 nm and 80 nm Ag NP at concentrations ranging from 0.00054 ppm to 1.7 ppm [97]. Although cytotoxicity was only observed with freshly synthesized Ag NP, small agglomerates of each of the Ag NP were observed within the intracytoplasmic vacuoles of HEK (Figure 11.8). These same Ag NP of different sizes, surface conditions, and synthesis methods were evaluated in *Escherichia coli*, Ag-resistant *Escherichia coli*, *Staphylococcus aureus*, methicillin-resistant *Staphylococcus aureus* (MRSA) and *Salmonella* spp. This study yielded a range of minimal inhibitory concentrations, and was based on whether the Ag NP solutions were contaminated by particle synthesis reducing agents. While freshly synthesized Ag NP were toxic to all bacterial strains at concentrations between 3.0 and 8.0 ppm, the washed Ag NP and the carbon-coated Ag NP were toxic to all bacterial strains at concentrations between 64.0 and 1024.0 ppm [125]. HEK and fibroblasts treated with Ag NP dissolved out of commercially bought Ag-based wound dressings showed cytotoxicity at approximately 15 μg/ml [126]. Interactions of Ag NP ranging from 7 to 20 nm in size were studied in human skin carcinoma cells and human fibrosarcoma cells and showed the onset of apoptosis at concentrations of 0.78 μg/mL and 1.56 μg/mL respectively [127].

Other Metals

Studies with 6.9 nm maghemite NP coated with tetramethylammonium hydroxide (TMAOH) and dispersed in a TMAOH aqueous solution, and iron NP (51%: 4.9 nm by TEM) coated with sodium bis(2-ethylexyl) sulfosuccinate (AOT) and dispersed in an AOT-rich aqueous solution were applied to *in vitro* human skin for 24 h. NP were located at the

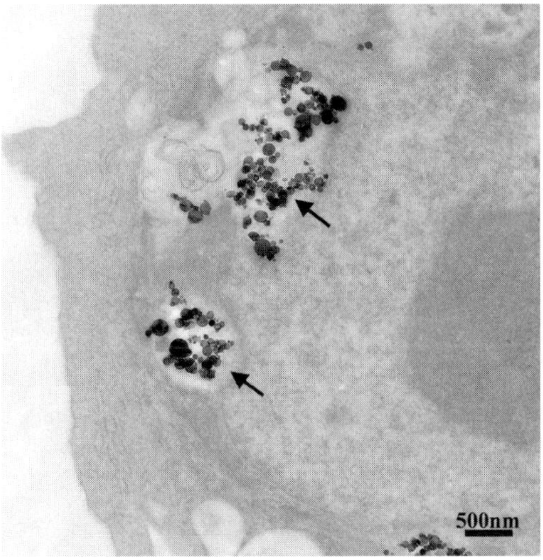

FIGURE 11.9 Transmission electron micrograph of 80 nm Al NP localized within the intracytoplasmic vacuoles of an HEK.

stratum corneum—stratum granulosum interface, in hair follicles, and in the viable epidermis by TEM and EDS-SEM. NP were not found in the receptor fluid by ICP-OES [29]. Aluminum (Al) is one of the most ubiquitous elements in our environment and is covered by an oxide film that generally inhibits corrosion. Al is found in explosives, artillery surface coatings and on rocket fuels. The cytotoxicity of Al NP of 30 nm did not cause mortality in immortalized mouse keratinocytes [128]. HEK exposed to 50 nm and 80 nm of Al NP in various concentrations showed no toxicity with four different viability assays. Al NP were also evaluated at different times (0 h and 24 h) and at different temperatures (25°C and 37°C) in both water and cell culture medium to determine what factors influenced cellular uptake in HEK. Al NP were observed in HEK (Figure 11.9) in all conditions. Al NP agglomerated in deionized water and increased in size in the suspension at both 25°C and 37°C. The hydrodynamic size of the NP was much greater in the cell culture medium, probably due to an increase in agglomeration that would reduce the effective dose of the NP to the HEK [129]. Others have suggested that Al found in antiperspirants may be linked to breast cancer [130].

TAKE-HOME MESSAGES

1. Current literature has shown minimal penetration of NP into the skin.
2. NP physicochemical properties are important determinants to modulate skin penetration.
3. Long-term *in vivo* studies under different conditions are needed to assess the true potential for NP penetration.
4. No firm conclusion on possible adverse effects of NP on skin is possible on the basis of available data. Once again, long-term studies are needed.

References

[1] Monteiro-Riviere NA. Comparative anatomy, physiology, and biochemistry of mammalian skin. In: Hobson DW, editor. Dermal and ocular toxicology fundamentals and methods. Boca Raton: CRC Press; 1991. p. 3–71.

[2] Monteiro-Riviere NA. The integument. In: Eurell J, Frappier B, editors. Dellmann's textbook of veterinary histology. 6th ed. Ames: Blackwell Publishing; 2006. p. 320–49.

[3] Monteiro-Riviere NA. Structure and function of skin. In: Monteiro-Riviere NA, editor. Toxicology of the skin—target organ series, Vol. 29. New York: Informa Healthcare; 2010. p. 1–18.

[4] Cork MJ, Robinson DA, Vasilopoulos A, et al. New perspectives on epidermal barrier dysfunction in atopic dermatitis: Gene–environment interactions. J Allergy Clin Immunol 2006;118:3–21.

[5] Monteiro-Riviere NA, Bristol DG, Manning TO, Riviere JE. Interspecies and interregional analysis of the comparative histological thickness and laser Doppler blood flow measurements at five cutaneous sites in nine species. J Invest Dermatol 1990;95:582–6.

[6] Holbrook K, Odland G. Regional differences in the thickness (cell layers) of the human stratum corneum: An ultrastructural analysis. J Invest Dermatol 1974;62:415–22.

[7] Menten DA. A minimum surface mechanism to account for the organization of cells into columns in the mammalian epidermis. Am J Anat 1976;145:1–22.

[8] Menten DA. Liquid film model of tetrakaidecahedral packing to account for the establishment of epidermal cell columns. J Invest Dermatol 1976;66:283–91.

[9] Landmann L. Epidermal permeability barrier: Transformation of lamellar granule-disks into intercellular sheets by a membrane-fusion process, a freeze-fracture study. J Invest Dermatol 1986;87:202–9.

[10] Swartzendruber DC, Wertz PW, Madison KC, Downing DT. Evidence that the corneocyte has a chemically bound lipid envelope. J Invest Dermatol 1987;88:709–13.

[11] Swartzendruber DC, Wertz PW, Kitko DJ, Madison KC, Downing DT. Molecular models of the intercellular lipid lamellae in mammalian stratum corneum. J Invest Dermatol 1989;92:251–7.

[12] Madison KC, Swartzendruber DC, Wertz PW. Presence of intact intercellular lipid lamellae in the upper layers of the stratum corneum. J Invest Dermatol 1987;88:714–8.

[13] Menon GK, Feinfold KR, Elias PM. The lamellar secretory response to barrier disruption. J Invest Dermatol 1992;98:279–89.

[14] Elias PM. Epidermal lipids, barrier function and desquamation. J Invest Dermatol 1983;80:44s–9s.

[15] Elias PM. Defensive functions of the stratum corneum: Intergrative aspects. In: Monteiro-Riviere NA, editor. Toxicology of the skin—target organ series, Vol. 29. New York: Informa Healthcare; 2010. p. 19–28.

[16] Elsner P. Irritant dermatitis in the workplace. Dermatol Clin 1994;12:461–7.

[17] Fluhr JW, Dickel H, Kuss O, Weyher I, Diepgen TL, Berardesca EI. Impact of anatomical location on barrier recovery, surface pH and stratum corneum hydration after acute barrier disruption. Br J Dermatol 2002;146:770–6.

[18] Monteiro-Riviere NA, Inman AO, Mak V, Wertz P, Riviere JE. Effect of selective lipid extraction from different body regions on epidermal barrier function. Pharm Res 2001;18:992–8.

[19] Monteiro-Riviere NA. Anatomical factors that affect barrier function. In: Zhai H, Wilhelm KP, Maibach HI, editors. Dermatotoxicology. 7th ed. New York: CRC Press; 2008. p. 39–50.

[20] Roberts MS, Gierden A, Riviere JE, Monteiro-Riviere NA. Solvents and vehicle effects on the skin. In: Roberts MS, Walters KA, editors. Dermal absorption and toxicity assessment. 2nd ed. New York: Informa Healthcare; 2008. p. 433–47.

[21] Larese Filon F, D'Agostin F, Bovenzi M, et al. Human skin penetration of silver nanoparticles through intact and damaged skin. Toxicology 2009;255:33–7.

[22] Larese Filon F, Crosera M, Adami G, Bovenzi M, Rossi F, Maina G. Human skin penetration of gold nanoparticles through intact and damaged skin. Nanotoxicology 2011. online 1-9 DOI 10.3109/ 1743390. 2010.551428.

[23] Fiserova-Bergerova V, Pierce JT, Droz PO. Dermal absorption potential of industrial chemicals: Criteria for skin notation. Am J Ind Med 1990;17:617–35.

[24] Nielsen JB, Grandjean P. Criteria for skin notation in different countries. Am J Ind Med 2004;45:275–80.

[25] Sartorelli P, Ahlers HW, Alanko K, et al. How to improve skin notation. Position paper from a workshop. Regul Toxicol Pharmacol 2007;49:301–7.

[26] Monteiro-Riviere NA, Baroli B. Nanomaterial penetration. In: Monteiro-Riviere NA, editor. Toxicology of the skin—target organ series, Vol. 29. New York: Informa Healthcare; 2010. p. 333–46.
[27] Lansdown AB, Taylor A. Zinc and titanium oxides: Promising UV-absorbers but what influence do they have on the intact skin? Int J Cosmet Sci 1997;19:167–72.
[28] Ryman-Rasmussen JP, Riviere JE, Monteiro-Riviere NA. Penetration of intact skin by quantum dots with diverse physicochemical properties. Toxicol Sci 2006;91:159–65.
[29] Baroli B, Ennas MG, Loffredo F, Isola M, Pinna R, Lopez-Quintela A. Penetration of metallic nanoparticles in human full-thickness skin. J Invest Dermatol 2007;127:1701–12.
[30] Scheuplein RJ. Mechanisms of percutaneous absorption. II. Transient diffusion and the relative importance of various routes of skin penetration. J Invest Dermatol 1967;48:79–88.
[31] Riviere JE. Absorption and distribution. In: Hodgson E, Levi PR, editors. Introduction to biochemical toxicology. 2nd ed. Appleton & Lange Publishers, CT; 1994. p. 11–49.
[32] Monteiro-Riviere NA, Riviere JE. Interaction of nanomaterials with skin: Aspects of absorption and biodistribution. Nanotoxicology 2009;3:188–93.
[33] Riviere JE, Monteiro-Riviere NA. Toxicokinetics: Dermal exposure and absorption of chemicals and nanomaterials. In: McQueen C, editor. Comprehensive toxicology. 2nd ed. Vol. 1. New York: Elsevier; 2010. p. 111–22.
[34] Tinkle SS, Antonini JM, Rich BA, et al. Skin as a route of exposure and sensitization in chronic beryllium disease. Environ Health Perspect 2003;111:1202–8.
[35] Rouse JG, Yang J, Ryman-Rasmussen JP, Barron AR, Monteiro-Riviere NA. Effects of mechanical flexion on the penetration of fullerene amino acid-derivatized peptide nanoparticles through skin. Nano Lett 2007;7:155–60.
[36] Filon FL, Boeninger M, Maina G, Adami G, Spinelli P, Damian A. Skin absorption of inorganic lead and the effects of skin cleansers. J Occup Environ Med 2006;48:692–9.
[37] Nielsen JB, Nielsen F, Sørensen JA. Defense against dermal exposures is only skin deep: Significantly increased penetration through slightly damaged skin. Arch Dermatol Res 2007;299:423–31.
[38] Vermula PK, Anderson RR, Karp JM. Nanoparticles reduce nickel allergy by capturing metal ions. Nat Nanotechnol 2011;5:291–5.
[39] Bronaugh RL. Skin penetration of nanoparticles. Proceedings of PPP2008, Perspectives in percutaneous penetration, La Grande Motte; 25-29th of March, 2008. p. 1.
[40] Chen X, Schluesener HJ. Nanosilver: A nanoproduct in medical application. Toxicol Lett 2008;176:1–12.
[41] EPA. Nanotechnology White Paper. Prepared for the US Environmental Protection Agency by members of the Nanotechnology Workgroup, a group of EPA's Science Policy Council. Washington, DC: US Environmental Protection Agency; 2007.
[42] NIOSH, National Institute for Occupational Safety and Health. Progress Toward Safe Nanotechnology in the Workplace—A Report from the NIOSH Nanotechnology Research Center, DHHS (NIOSH). Publication No. 2007–123, available on line, www.cdc.gov/niosh; 2007.
[43] Roco MC. Environmentally responsible development of nanotechnology. Environ Sci Technol 2005;39:106A–12A.
[44] Gwinn MR, Vallyathan V. Nanoparticles: health effects—pros and cons. Environ Health Perspect 2006;114:1818–25.
[45] Kielhorn J, Melching-Kollmuß S, Mangelsdorf I. World Health Organization (WHO), environmental health criteria 235 dermal absorption. 2006.
[46] Oberdörster G, Oberdörster E, Oberdörster J. Nanotoxicology: An emerging discipline evolving from studies of ultrafine particles. Environ Health Perspect 2005;113:823–39.
[47] SCCP—Scientific Committee on Consumer Products. Preliminary opinion on safety of nanomaterials in cosmetic products. Brussels, Belgium: European Commission, http://ec.europa.eu/health/archive/ph_risk/committees/04_sccp/docs/sccp_o_123.pdf; 2007.
[48] Donaldson K, Aitken R, Tran L, et al. Carbon nanotubes: A review of their properties in relation to pulmonary toxicology and workplace safety. Toxicol Sci 2006;92:5–22.
[49] Geiser M, Schurch S, Gehr P. Influence of surface chemistry and topography of particles on their immersion into the lung's surface-lining layer. J Appl Physiol 2003;94:1793–801.
[50] Geys J, Coenegrachts L, Vercammen J, et al. *In vitro* study of the pulmonary translocation of nanoparticles: a preliminary study. Toxicol Lett 2006;160:218–26.

[51] Lam CW, James JT, McCluskey R, Hunter RL. Pulmonary toxicity of single-wall carbon nanotubes in mice 7 and 90 days after intratracheal instillation. Toxicol Sci 2004;77:126–34.
[52] Limbach LK, Wick P, Manser P, Grass RN, Bruinink A, Stark WJ. Exposure of engineered nanoparticles to human lung epithelial cells: Influence of chemical composition and catalytic activity on oxidative stress. Environ Sci Technol 2007;41:4158–63.
[53] Magrez A, Kasas S, Salicio V, et al. Cellular toxicity of carbon-based nanomaterials. Nano Lett 2006;6:1121–5.
[54] Muller J, Huaux F, Moreau N, et al. Respiratory toxicity of multi-wall carbon nanotubes. Toxicol Appl Pharmacol 2005;207:221–31.
[55] Nel A, Xia T, Mädler L, Li N. Toxic potential of materials at the nanolevel. Science 2006;311:622–7.
[56] Oberdörster G, Maynard A, Donaldson K, et al, and ILSI Research Foundation/Risk Science Institute Nanomaterial Toxicity Screening Working Group. Principles for characterizing the potential human health effects from exposure to nanomaterials: Elements of a screening strategy. Part Fibre Toxicol 2005;2:8.
[57] Rotoli BM, Bussolati O, Bianchi MG, et al. Non-functionalized multi-walled carbon nanotubes alter the paracellular permeability of human airway epithelial cells. Toxicol Lett 2008;178:95–102.
[58] Shimada A, Kawamura N, Okajima M, Kaewamatawong T, Inoue H, Morita T. Translocation pathway of the intratracheally instilled ultrafine particles from the lung into the blood circulation in the mouse. Toxicol Pathol 2006;34:949–57.
[59] Shvedova AA, Kisin ER, Mercer R, et al. Unusual inflammatory and fibrogenic pulmonary responses to single-walled carbon nanotubes in mice. Am J Physiol Lung Cell Mol Physiol 2005;289:L698–708.
[60] Monteiro-Riviere NA, Inman AO, Ryman-Rasmussen JP. Dermal effects of nanomaterials. In: Monteiro-Riviere NA, Tran CL, editors. Nanotoxicology: Characterization, Dosing, and Health Effects. New York: Informa Healthcare; 2007. p. 317–38.
[61] Zhang LW, Monteiro-Riviere NA. Assessment of quantum dot penetration into intact, tape stripped, abraded and flexed rat skin. Skin Pharmacol Physiol 2008;21:166–80.
[62] Zhang LW, Yu WW, Colvin VL, Monteiro-Riviere NA. Biological interactions of quantum dot nanoparticles in skin and in human epidermal keratinocytes. Toxicol Appl Pharmacol 2008;228:200–11.
[63] Riviere JE, Brooks JD. Predicting skin permeability from complex chemical mixtures. Toxicol Appl Pharmacol 2005;208:99–110.
[64] Riviere JE, Brooks JD. Predicting skin permeability from complicated chemical mixtures: Dependency of quantitative structure permeability relationships (QPSR) on biology of skin models used. Toxicol Sci 2011;119:224–32.
[65] Rouse JG, Haslauer CM, Loboa EG, Monteiro-Riviere NA. Cyclic tensile strain increases interactions between human epidermal keratinocytes and quantum dot nanoparticles. Toxicol In Vitro 2008;22:491–7.
[66] Monteiro-Riviere NA, Inman AO. Evaluation of quantum dot nanoparticle penetration in human skin. Toxicologist 2008;102(Suppl. 1):1029.
[67] Gulson B, McCall M, Korsch M, et al. Small amounts of zinc from zinc oxide particles in sunscreens applied outdoors are absorbed through human skin. Toxicol Sci 2010;118:140–9.
[68] Monteiro-Riviere NA, Wiench K, Landsiedel R, Schulte S, Inman AO, Riviere JE. Safety evaluation of sunscreen formulations containing titanium dioxide and zinc oxide nanoparticles in UVB sunburned skin: An *in vitro* and *in vivo* study. Toxicol Sci 2011;123:264–80.
[69] Xia XR, Monteiro-Riviere NA, Riviere JE. Skin penetration and kinetics of pristine fullerenes (C_{60}) topically exposed in industrial organic solvents. Toxicol Appl Pharmacol 2010;242:29–37.
[70] Sapra B, Jain S, Tiwary AK. Percutaneous permeation enhancement by terpenes: Mechanistic view. AAPS J 2008;10:120–32.
[71] Dos Anjos JLV, Alonso A. Terpenes increase the partitioning and molecular dynamics of an amphipathic spin label in stratum corneum membranes. Int J Pharm 2008;350:103–12.
[72] Monteiro-Riviere NA, Inman AO, Erdmann D, Xia X, Riviere JE. Terpene effects on penetration of nanoparticles in human skin. Toxicologist 2011;120(Suppl. 2):2154.
[73] Monteiro-Riviere NA, Inman AO. Challenges for assessing carbon nanomaterial toxicity to the skin. Carbon 2006;44:1070–8.
[74] Monteiro-Riviere NA, Yang J, Inman AO, Ryman-Rasmussen J, Barron AR, Riviere JE. Skin penetration of fullerene substituted amino acids and their interactions with human epidermal keratinocytes. Toxicol Sci 2006;90(Suppl. 1):168.

[75] Yu WW, Chang E, Falkner JC, et al. Forming biocompatible and nonaggregated nanocrystals in water using amphiphilic polymers. J Am Chem Soc 2007;129:2871–9.
[76] Prow TW, Monteiro-Riviere NA, Inman AO, et al. Quantum dot penetration into viable human skin. Nanotoxicology 2011; DOI-10.3109/17435390.2011.569092.
[77] Lee HA, Imran M, Monteiro-Riviere NA, Colvin VL, Yu WW, Riviere JE. Biodistribution of quantum dot nanoparticles in perfused skin: Evidence of coating dependency and periodicity in arterial extraction. Nano Lett 2007;9:2865–70.
[78] Gopee NV, Roberts DW, Webb P, et al. Migration of intradermally injected quantum dots to sentinel organs in mice. Toxicol Sci 2007;98:249–57.
[79] Borm PJA, Robbins D, Haubold S, et al. The potential risks of nanomaterials: A review carried out for ECETOC. Part Fibre Toxicol 2006;3:11.
[80] Faunce T, Murray K, Nasu H, Bowman D. Sunscreen safety: The precautionary principle, the Australian Therapeutic Goods Administration and nanoparticles in sunscreens. Nanoethics 2008;2:231–40.
[81] Dussert AS, Gooris E. Characterization of the mineral content of a physical sunscreen emulsion and its distribution onto human stratum corneum. Int J Cosmet Sci 1997;19:119–29.
[82] Gamer AO, Leibold E, van Ravenzwaay B. The *in vitro* absorption of microfine zinc oxide and titanium dioxide through porcine skin. Toxicol In Vitro 2006;20:301–7.
[83] Tan MH, Commens CA, Burnett L, Snitch PJ. A pilot sudy on the percutaneous absorption of microfine titanium dioxide from sunscreens. Australas J Dermatol 1996;37:185–7.
[84] Lademann J, Weigmann HJ, Rickmeyer C, et al. Penetration of titanium dioxide in sunscreen formulation into the horny layer and the follicular orifice. Skin Pharmacol Appl Skin Physiol 1999;12:247–56.
[85] Pflücker F, Wendel V, Hohenberg H, et al. The human stratum corneum layer: An effective barrier against dermal uptake of different forms of topically applied micronized titanium dioxide. Skin Pharmacol Appl Skin Physiol 2001;14(S1):92–7.
[86] Schulz J, Hohenberg H, Pflücker F, et al. Distribution of sunscreens on skin. Adv Drug Deliv Rev 2002;54(S1):S157–63.
[87] Mavon A, Miquel C, Lejeune O, Payre B, Moretto P. *In vitro* percutaneous absorption and *in vivo* stratum corneum distribution of an organic and mineral sunscreen. Skin Pharmacol Physiol 2007;20:10–20.
[88] Bennat C, Müller-Goymann CC. Skin penetration and stabilization of formulations containing microfine titanium dioxide as physical UV filter. Int J Cosmet Sci 2000;22:271–83.
[89] Cross SE, Innes B, Roberts M, Tsuzuki T, Robertson TA, McCormick P. Human skin penetration of sunscreen nanoparticles: *In-vitro* assessment of a novel micronized zinc oxide formulation. Skin Pharmacol Physiol 2007;20:148–54.
[90] Sadrieh N, Wokovich AM, Gopee NV, et al. Lack of significant dermal penetration of titanium dioxide from sunscreen formulations containing nano- and submicron-size TiO_2 particles. Toxicol Sci 2010;115:156–66.
[91] Kertész Zs, Szikszai Z, Gontier E, et al. Nuclear microprobe study of TiO_2-penetration in the epidermis of human skin xenografts. Nucl Inst Meth Phys Res B Beam Interact Mater Atoms 2005;231:280–5.
[92] Menzel F, Reinert T, Vogt J, Butz T. Investigations of percutaneous uptake of ultrafine TiO_2 particles at the high energy ion nanoprobe LIPSION. Nucl Inst Meth Phys Res B Beam Interact Mater Atoms 2004;219-220:82–6.
[93] Kiss B, Biró T, Czifra G, et al. Investigation of micronizided titanium dioxide penetration in human skin xenografts and its effect on cellular functions of human skin-derived cells. Exp Dermatol 2008;17:659–67.
[94] NANODERM. Quality of skin as a barrier to ultra-fine particles. Final Report. (Project Number: QLK4-CT-2002–02678), www.uni-leipzig.de/~nanoderm/; 2007.
[95] SCCNFP. Opinion of the scientific committee on cosmetic products and non-food products intended for consumers concerning titanium dioxide. Brussels, Belgium: European Commission; 2000. Colipa No S 75.
[96] Sonavane G, Tomoda K, Sano A, Ohshima H, Terada H, Makino K. *In vitro* permeation of gold nanoparticles through rat skin and rat intestine: Effect of particle size. Colloids Surf B 2008;65:1–10.
[97] Samberg ME, Oldenburg SJ, Monteiro-Riviere NA. Evaluation of silver nanoparticle toxicity in skin *in vivo* and keratinocytes in vitro. Environ Health Perspect 2010;118:407–13.
[98] Larese Filon F, Adami G, Venier M, et al. *In vitro* percutaneous absorption of cobalt. Int Arch Environ Health 2004;77:85–9.
[99] Crosera M, Bovenzi M, Maina G, et al. Nanoparticle dermal absorption and toxicity: A review of the literature. Int Arch Occup Environ Health 2009;82:1043–55.

[100] Kishore AS, Surekha P, Murthy PB. Assessment of the dermal and ocular irritation potential of multi-walled carbon nanotubes by using *in vitro* and *in vivo* methods. Toxicol Lett 2009;191:268–74.
[101] Eedy DJ. Carbon-fibre-induced airborne irritant contact dermatitis. Contact Dermatit 1996;35:362–3.
[102] Norton SK, Dellinger Z, Zhou Z, et al. A new class of human mast cell and peripheral blood basophil stabilizers that differentially control allergic mediator release. Clin Transl Sci 2010;3:158–69.
[103] Nelson MA, Frederick ED, Bowden GT, Hooser SB, Fernando Q, Carter DE. Effects of acute and subchronic exposure of topically applied fullerene extracts on the mouse skin. Toxicol Ind Health 1993;9:623–30.
[104] Monteiro-Riviere NA, Inman AO, Zhang LW. Limitations and relative utility of screening assays to assess engineered nanoparticle toxicity in a human cell line. Toxicol Appl Pharmacol 2009;234:222–35.
[105] Yang J, Barron AR. A new route to fullerene substituted phenylalanine derivatives. Chem Commun (Cambridge) 2004;24:2884–5.
[106] Colvin VL. The potential environmental impact of engineered nanomaterials. Nature Biotech 2003;21:1166–70.
[107] Sayes CM, Fortner JD, Guo W, et al. The differential cytotoxicity of water-soluble fullerenes. Nano Lett 2004;4:1881–7.
[108] Scrivens WA, Tour JM, Creek KE, Pirisi L. Synthesis of ^{14}C-labeled C_{60}, its suspension in water, and its uptake by human keratinocytes. J Am Chem Soc 1994;116:4517–8.
[109] Inman AO, Sayes CM, Colvin VL, Monteiro-Riviere NA. Nano-C_{60} and derivatized C_{60} toxicity in human epidermal keratinocytes. Toxicol Sci 2006;90(S-1):167.
[110] Rouse JG, Yang J, Barron AR, Monteiro-Riviere NA. Fullerene based amino acid nanoparticle interactions with human epidermal keratinocytes. Toxicol In Vitro 2006;20:1313–20.
[111] Zhang LW, Yang J, Barron AR, Monteiro-Riviere NA. Endocytic mechanisms and toxicity of a functionalized fullerene in human cells. Toxicol Lett 2009;191:149–57.
[112] Shvedova AA, Castranova V, Kisin E, et al. Exposure to carbon nanotube material: Assessment of cytotoxicity using human keratinocytes cells. J Toxicol Environ Health 2003;66:1909–26.
[113] Zhang LW, Zeng L, Barron AR, Monteiro-Riviere NA. Biological interactions of functionalized single-wall carbon nanotubes in human epidermal keratinocytes. Int J Toxicol 2007;26:103–13.
[114] Manna SK, Sarkar S, Barr J, et al. Single-walled carbon nanotube induces oxidative stress and activates nuclear transcription factor-kB in human keratinocytes. Nano Lett 2005;5:1676–84.
[115] Yokoyama A, Sato Y, Nodasaka Y, et al. Biological behavior of hat-stacked carbon nanofibers in the subcutaneous tissue in rats. Nano Lett 2005;5:157–61.
[116] Monteiro-Riviere NA, Nemanich RJ, Inman AO, Wang YY, Riviere JE. Multi-walled carbon nanotube interactions with human epidermal keratinocytes. Toxicol Lett 2005;155:377–84.
[117] Monteiro-Riviere NA, Inman AO, Wang YY, Nemanich RJ. Surfactant effects on carbon nanotube interactions with human epidermal keratinocytes. Nanomed Nanotechnol Biol Med 2005;1:293–9.
[118] Zhang LW, Monteiro-Riviere NA. Lectins modulate multi-walled carbon nanotubes cellular uptake in human epidermal keratinocytes. Toxicol In Vitro 2010;24:546–51.
[119] Witzmann FA, Monteiro-Riviere NA. Multi-walled carbon nanotube exposure alters protein expression in human keratinocytes. Nanomed Nanotechnol Biol Med 2006;2:158–68.
[120] Witzmann FA, Monteiro-Riviere NA. Effect of carbon nanotube exposure on keratinocyte protein expression. In: Monteiro-Riviere NA, Tran CL, editors. Nanotoxicology: Characterization, dosing, and health effects. New York: Informa Healthcare; 2007. p. 197–224.
[121] Ryman-Rasmussen JP, Riviere JE, Monteiro-Riviere NA. Surface coatings determine cytotoxicity and irritation potential of quantum dot nanoparticles in epidermal keratinocytes. J Invest Dermatol 2007;127:143–53.
[122] Ryman-Rasmussen JP, Riviere JE, Monteiro-Riviere NA. Variables influencing interactions of untargeted quantum dot nanoparticles with skin cells and identification of biochemical modulators. Nano Lett 2007;7:1344–8.
[123] Woodrow Wilson International Center for Scholars. A nanotechnology consumer products inventory, www.nanotechproject.org/consumerproducts; 2011.
[124] Wijnhoven SWP, Peijnenburg WJGM, Herberts CA, et al. Nano-silver: A review of available data and knowledge gaps in human and environmental risk assessment. Nanotoxicology 2009;3:109–38.
[125] Samberg ME, Orndorff PE, Monteiro-Riviere NA. Antibacterial efficacy of silver nanoparticles of different sizes, surface conditions and synthesis methods. Nanotoxicology 2011;5:244–53.

[126] Burd A, Kwok CH, Hung SC, et al. A comparative study of the cytotoxicity of silver-based dressings in monolayer cell, tissue explant, and animal models. Wound Repair Regen 2007;15:94–104.
[127] Arora S, Jain J, Rajwade JM, Paknikar KM. Cellular responses induced by silver nanoparticles: *In vitro* studies. Toxicol Lett 2008;179:93–100.
[128] Rooney AD, Jones RL, Mattie DR, Schlager JJ, Hussain SM. *In vitro* toxicity of nanoparticles in mouse keratinocytes and endothelial cells. United States Air Force Research Laboratory 2004. AFRL-HE-WP-TR-2005-0091.
[129] Monteiro-Riviere NA, Oldenburg SJ, Inman AO. Interactions of aluminum nanoparticles in human epidermal keratinocytes. J Appl Toxicol 2010;30:276–85.
[130] Exley C, Charles LM, Barr L, Martin C, Polwart A, Darbre PD. Aluminium in human breast tissue. J Inorg Biochem 2007;101:1344–6.

CHAPTER

12

Gastrointestinal Tract

Mark A. Jepson

School of Biochemistry, University of Bristol, Bristol, UK

OUTLINE

INTRODUCTION

The rapid expansion in nanotechnology brings an inevitable risk of exposure to engineered nanomaterials (ENM) and a consequent need to assess the potential impact of such exposure. Despite substantial ongoing research effort in the nanotoxicology field, relatively little attention has been given to the impact, behavior and interaction of ENM in the gastrointestinal tract (GIT). The widespread use of NM such as silver nanoparticles (Ag NP) in food packaging, toothpaste and nanomedicine, and of zinc oxide (ZnO) and titanium dioxide (TiO_2) NM in food, toothpaste, cosmetics and suncream carry a clear risk of ingestion for a large proportion of the population. Furthermore, the more widely studied inhalation route of ENM will also result in secondary exposure of GIT following clearance from the respiratory tract. The extent of exposure to ENM is currently poorly quantified and there is a need to understand the potential impact NM may have within the GIT itself and the extent to which

Adverse Effects of Engineered Nanomaterials
DOI: 10.1016/B978-0-12-386940-1.00012-X

they may be translocated from GIT to the circulation and peripheral organs (primarily the liver), where they have the potential to exert toxic effects. One area that has been studied more extensively is the uptake of "nano-carriers" developed as vehicles for oral drug and vaccine delivery, which depend on their ability to interact with and, in some cases, to be transported by, GIT epithelia. While data on the uptake of such particulate delivery devices identify potential routes for uptake of other ENM, they also highlight the complexity of particle interactions in the GIT which currently prevents accurate prediction of the behavior of individual ENM.

In this chapter, I will discuss the different routes for uptake and transport of NM that exist in the GIT and review available literature characterizing these routes and their relevance for uptake of NM into the body. Much of the available data on NP uptake and transport from both *in vivo* and *in vitro* studies have come more from the pharmaceutical research field where development and characterization of particulate delivery vehicles for drugs and/or vaccines have been hot topics for many years. Unfortunately, here we encounter a risk of confusion because, in pharmaceutical research, the term nanoparticle is generally used to describe any particles of submicron size rather than the stricter definition of less than 100 nm that has become established in the nanotoxicology and nanotechnology fields. Particulate delivery vehicles are more often formulated in the 200–500 nm size range, which increases encapsulation efficiency, but novel nano-sized carriers are also being developed. Although information on the efficacy of these preparations is useful for predicting potential uptake of other particulates, including ENM, limited data are currently available on uptake and transport of most classes of ENM, including those that may arrive in the GIT inadvertently. Uptake and retention of at least some true ENM has been demonstrated following oral delivery in small animals [1,2] and, since it is already known that many larger particulates are absorbed in the gut [3–5], it can be assumed that at least some ENM will access transport routes available to larger particles and probably also other uptake mechanisms. It is important to understand the potential routes for uptake and the factors that may limit or promote such uptake in order to estimate the risk to the organism of ENM transport from the GIT lumen to peripheral tissues.

The chapter will also outline what is known about the toxicity of NM in the gut, and also highlight where information is lacking. Although it has been hypothesized that particulates such as TiO_2 might play a role in the pathogenesis of inflammatory bowel disease [6,7], direct evidence of a causal link between NM and gut pathology is lacking. The relatively sparse data on NM toxicity in intestinal cells have been largely derived from studies on Caco-2 cells, a human adenocarcinoma cell line that, when grown on permeable supports, differentiates to form a polarized epithelial layer with well-developed microvilli and tight junctions that closely models the epithelium of small intestine and is widely used to study drug absorption among other aspects of gut function [8]. Detailed discussion of the potential toxicity mechanisms of ENM in GIT cells is beyond the scope of this chapter and to a large extent these mirror the mechanisms described in other cell types. However, it should also be borne in mind that differentiation into fully polarized epithelia with microvilli that more closely resemble *in vivo* GIT lining may reduce NM uptake [9,10] and some studies have suggested that inflammatory responses to ENM may be less prominent in GIT compared with other cells [11,12]. The specialized morphology and function of differentiated GIT epithelial cells also needs to be carefully considered in the context of nanotoxicity assessment because effects of ENM on epithelia may severely compromise GIT function in ways that would not be apparent from traditional toxicity assays.

CONDITIONS AND BARRIERS IN THE GASTROINTESTINAL TRACT

Ingested NM encounter conditions within the GIT that are very different from those encountered when contacting the organism via other routes and it is not yet clear how this may affect their potential toxicity. For example, the low pH of the stomach may critically affect ENM and the likelihood that they dissociate to yield products with differing toxicity profiles. The extreme pH shifts encountered during gut transit and the presence of exogenous molecules co-ingested with ENM and the proteins, mucus and bile acids secreted within the gut may also affect the physicochemical and aggregation properties of ENM. In comparison with some other exposure routes, such as in the alveoli, residence time within GIT may be much shorter due to normal gut transit, reducing cellular exposure to ingested ENM unless they are mucoadhesive or bioadhesive.

There are outstanding questions about the quantity and status of ENM arriving in the gut. It is also important to consider the cellular composition of the GIT epithelium and additional barriers that limit ENM exposure. The bulk of the epithelial lining in small intestine is made up of enterocytes, columnar epithelial cells with well-developed microvillus brush-border and sealed by tight junctions that act as a barrier to passage of most materials. Interspersed between the enterocytes are the goblet cells which, by secreting mucus into the gut lumen, provide an additional barrier to the diffusion of particulates towards the epithelium [13]. The way particulate material interacts with secreted mucus depends to a large extent on particle size and surface properties, smaller particles penetrating better and positively-charged particles being mucoadhesive [13]. The cell surface glycocalyx, which is composed of glycosylated proteins within the membrane, constitutes a further barrier to the close approach of particulates to the apical membrane of enterocytes. The glycocalyx has been shown to reduce interaction of even nano-sized material with receptors on host cells [14]. With certain exceptions, the villus epithelium is largely refractory to microparticles, but potential routes for nano-sized material exist which will be discussed below. Although there are some anatomical and functional distinctions, the epithelial lining of the large intestine can be considered to be similar to that of the small intestine with regards to its function as a barrier to the passage of particles.

A minor part of the GIT lining is composed of specialized epithelium that plays a prominent role in particle uptake. It has been recognized since the early 1970s that Peyer's patches and analogous gut-associated lymphoid tissues (GALT) in the large intestine are able to accumulate particulate material from the gut lumen, including some in the nanoparticle size-range such as India ink and ferritin [15]. This phenomenon is a consequence of morphological and functional specialization of the epithelium overlying the lymphoid follicles (the follicle-associated epithelium or FAE), including the presence of specialized antigen-sampling M cells [15,16] which exhibit a remarkable capacity to transport a range of material, from proteins to bacteria, and deliver them to the underlying lymphoid cells. A fuller discussion of the role of M cells in translocation of particulates will be provided below, after initial discussion of other uptake routes.

PARTICLE UPTAKE IN THE GASTROINTESTINAL TRACT

Uptake of ENM in the GIT is potentially a major exposure route with important implications for their toxicity, in peripheral organs as well as locally within the GIT. Particle uptake in the GIT is sometimes assumed to be an exclusive feature of M cells but several lines of evidence suggest that significant uptake of particulate material over a broad size range occurs in non-lymphoid tissues. For example, a range of particulate carriers including poly(D,L-lactide) and chitosan particles were not selectively taken up by rodent Peyer's patches compared to villi [13,17]. Similarly, while uptake of polystyrene particles in the GIT of orally dosed rats was greater into lymphoid tissues, a significant proportion was also absorbed by apparently normal villi [18,19]. Uptake also increased as size decreased from micron to 100 nm and below [18,19] suggesting that nano-sized particles may access additional uptake routes to those available to larger particles. These data indicate that the lower efficiency of ENM uptake in villus epithelium might be offset to a significant extent by its vastly greater surface area compared with the specialized FAE. Other studies have demonstrated significant uptake of gold NP following oral delivery to rats [1] and suggested that the GIT is a significant route of silver NP uptake in fish [20,21].

Although these studies have not precisely located sites of uptake at the cellular level, it is frequently assumed that uptake is transcellular due to the apparent integrity of the villus epithelium. Furthermore, experiments with cultured GIT epithelial cells have also demonstrated their capacity to internalize ENM [21,22], as discussed below. While uptake of larger particulates can only be achieved through phagocytosis or macropinocytosis which are not considered to be major uptake routes in fully differentiated GIT epithelia, except for the specialized antigen-sampling M cells, particles in the nano-size range may access additional uptake routes such as clathrin- and caveolin-dependent endocytosis which operate in polarized epithelia, although brush-border development may also limit these forms of endocytosis [9,23].

One potential route for uptake of particles of various sizes is the phenomenon of "persorption" which was described in the 1960s by Volkheimer who described GIT translocation of particles up to several microns in diameter (e.g. yeast and starch granules) following oral dosing of experimental animals and humans [24]. Although this phenomenon is not generally considered to be a very significant route of transport, limited uptake of luminal material including ENM is an inevitable consequence of the infrequent breaches in the epithelial lining occurring even in healthy GIT due to sloughing of aged enterocytes and localized damage. The concept that normal cell turnover may play a role in particle absorption was revived by that the demonstration of nano-gold uptake through dying enterocytes [25].

The tight junctions between GIT epithelial cells are required for normal barrier and absorptive functions due in part to their blockage of the diffusion of molecules and ions between cells. This is, nevertheless, a highly regulated barrier and, under some conditions, significant tight junction leakage occurs, leading to paracellular transport of molecules and of particulate material including bacteria [26]. There is some experimental evidence that NM of 1–2 nm might access the paracellular transport route, even in healthy GIT [1]. In addition, polycationic molecules can trigger selective and reversible opening of tight junctions [27], a property that is being investigated for its potential to increase drug absorption from particulate delivery vehicles [28]. It has also recently been reported that polymer-coated gold NP of 5–20 nm diameter and varying surface charge can transiently open tight junctions of polarized Caco-2 cell layers, a property which may contribute to their own transport and that of bystander molecules [29]. Similarly, nano-sized insulin complexes based on amine-modified comb-like polyesters reversibly decrease transepithelial electrical resistance (TEER) and enhance uptake into Caco-2 cell layers [30]. In general, it seems unlikely that paracellular transport will be a widespread route for ENM penetration of the healthy GIT wall apart from those ENM that are small enough, or have the necessary surface properties to mediate tight junction opening. However, it is also important to bear in mind that this strict limitation on paracellular transport may not apply to all individuals since bacterial translocation can be enhanced by conditions such as trauma, inflammation, stroke and chronic alcohol use [26,31] and this leakage will most likely be mirrored by increased uptake of other material. Support for this suggestion arises from the findings that alcohol treatment and irradiation enhance penetration of 2 μm polystyrene particles by causing tight junction leakage in cultured Caco-2 cells [32,33].

There is limited information available from *in vivo* studies to address questions regarding the mechanisms of particle uptake in GIT. Mechanistic studies are more readily performed in cell culture models which are amenable to experimental manipulation. The most relevant *in vitro* studies of ENM transport and toxicity have largely focused on Caco-2 cells, a human adenocarcinoma cell line widely used in pharmaceutical research [8]. Caco-2 cells are also becoming a popular cell model in the broader nanotoxicology area since they can provide a physiologically relevant *in vitro* model for initial testing, with consequent limitation of the need to utilize animal models. Unfortunately, many studies on uptake of ENM and their effects on Caco-2 cells have focused only on non-polarized cells grown on solid supports which are of more limited value because they do not represent the physiological properties of the GIT epithelium. For example, it is known that brush-border differentiation is incomplete in Caco-2 cells grown for short periods and that uptake mechanisms are downregulated with brush-border differentiation [9].

A diverse range of ENM are transported across fully differentiated Caco-2 cells. For example, the polarized Caco-2 cell model has revealed transport of TiO_2, CeO_2 and silver nanoparticles, seemingly by a transcellular route [21,22]. Furthermore, uptake of PLGA nanoparticles of average diameter 95 nm was detected in polarized Caco-2, although this was significantly lower than in other epithelial cells lines [10]. Despite the availability of this model, there is still no clear consensus as to which specific uptake route is more important and it remains likely that these will depend on the specific properties of individual NM. For example, dendrimers, which are <10 nm NP with potential as drug carriers [34], were shown in one study to be transported across Caco-2 cell layers with negligible toxicity [35], whereas another report suggested the paracellular transport with concomitant loss of transepithelial electrical resistance (TEER), which might indicate toxicity and that was reduced by PEGylation [36].

ACCESSING THE HIGH CAPACITY PARTICLE UPTAKE ROUTE IN GUT-ASSOCIATED LYMPHOID TISSUES

Gut-associated lymphoid tissue (GALT) such as Peyer's patches and appendix serve as antigen sampling and inductive sites of the mucosal immune system. It has been recognized for many years that these sites exhibit a high capacity for accumulating particulate material. The capacity of these regions to sample luminal contents is associated with morphological and functional specialization of the overlying epithelium, the FAE, which has been described in detail elsewhere [3,37,38]. Although there is some variation in morphology between species and GALT sites [3,39,40], certain common features characterize the FAE. The enterocytes present within this epithelium generally have reduced expression of absorptive enzymes and less well-developed microvilli, and goblet cells are reduced in number in FAE compared with neighbouring villi [41,42]. These features may promote interaction of material with the FAE but the most prominent anatomical feature responsible for uptake of material is the presence of specialized antigen-sampling M cells.

M cells were first described in GALT FAE and analogous cells have subsequently been described in the epithelium overlying the lymphoid follicles of other mucosae [43] and more rarely in villi, especially those close to GALT sites [44]. The first identification of M cells came from electron microscopy studies that revealed morphological characteristics of irregular microvilli, numerous endosomes and a basolateral membrane invagination that creates a leukocyte-containing pocket [3,16,37]. The morphological specialization of M cells underlies remarkable functional specialization for antigen-sampling in that these cells can internalize and transcytose a wide range of soluble molecules, inert particulates [5,45–47] and bacteria [41,48] (Table 12.1). In addition to their morphological features which may facilitate uptake and transport of particulate material, M cells also have a poorly developed glycocalyx compared to other enterocytes [14]. Direct evidence of the potential for the glycocalyx to limit NM interaction with the apical membrane came from elegant studies carried out in the 1990s on interaction of material with rabbit Peyer's patch tissue. These revealed that the glycocalyx forms a size-selective barrier to particle interaction with glycolipids in the apical membrane [14,73]. While 1.13 μm and 120 nm latex particles coated with cholera toxin B subunit (CtxB) did not bind either M cells or enterocytes, despite their expression of CtxB receptors, the thinner glycocalyx of M cells allowed CtxB-coated 29 nm diameter gold NP to bind to M cells only [14,73].

Much of the recent focus on GIT uptake of particulates has been on the use of delivery vehicles where antigen or drugs are encapsulated or attached to carriers of various sorts to protect them from degradation and/or facilitate uptake. The extensive range of different carriers that has been developed is beyond the scope of this chapter and has been reviewed elsewhere [74,75] so discussion will be limited to a few specific examples, some of which are included in Table 12.1. Since M cells offer an efficient route for uptake [76] that is directly channeled towards generation of a mucosal immune response, these cells are a logical target for particulate vaccine delivery systems. Despite the broad range of particles and macromolecules that are known to be transported by M cells, this is not an entirely promiscuous process since particle interaction with M cells is greatly influenced by size and surface properties [45,52,77]. For example, studies on particulate polymer delivery vehicles have found

TABLE 12.1 Examples of Particulate Material and Microorganisms Selectively Absorbed by Peyer's Patch Follicle-associated Epithelium (FAE)/M Cells

		In vivo FAE/M Cells (species)	*In vitro* Human M Cell Model	Reference(s)
Particles	Polystyrene (50 nm–2 μm)	Rabbit, mouse etc	Yes	[5,46,49,50,51]
	Poly(lactide co-glycolide) (100–500 nm)	Pig, mouse, rabbit, mouse	Yes	[45,52–55]
	Liposomes (160–500 nm)	Rat, mouse	N/A	[56,57]
	Poly(lactic acid) (200–300 nm)	Mouse	N/A	[58]
	Chitosan derivatives (100–300 nm)	Mouse	Yes	[59,60]
Bacteria	*Salmonella enterica*	Mouse	Yes	[48,61,62]
	Yersinia species	Mouse, rabbit	Yes	[63–66]
	Shigella flexneri	Rabbit, mouse		[67,68]
	Vibrio cholera	Rabbit	Yes	[50,69]
Viruses	HIV-1	Rabbit	Yes	[70,71]
	Polio virus		Yes	[72]

that reducing particle size to 100 nm can increase their uptake across the gut epithelium [52], so it can be hypothesized that similar size dependency might extend to other particle types. As we have seen, the physicochemical properties of ENM influence their ability to bypass the mucus barrier to contact the GIT epithelium and it also clear that binding to M cells is material dependent. For example, 0.5 μm polystyrene microparticles selectively targeted, and were efficiently endocytosed by rabbit Peyer's patch M cells, but PLGA microparticles adhered in much lower numbers and without clear M cell-specificity [5,45]. Similarly, uptake of several particulate carriers, including PLGA microparticles and polymerized liposomes is relatively inefficient in mice [4,78]. The limited uptake of native particulate delivery vehicles and their lack of targeting to M cells have stimulated attempts to improve their efficacy using M cell targeting ligands [38,79].

Potential strategies for M cell targeting have arisen from the discovery of selective lectin binding to M cells [39,80–82], selective expression of receptors targeted by pathogens on M cells [63,83] and additional M cell specific proteins identified from expression profiling studies [84]. Each of these strategies has offered some promise. For example, the M cell selective lectin UEA-1 increased uptake of polystyrene particles and liposomes by murine M cells [49,78]. The UEA-1 targeting approach was subsequently supported by enhancement of immune responses [56]. Further studies with a UEA-1 mimetic, tetragalloyl lysine dendrimer, to avoid the stability and toxicity problems associated with native lectins, further validated this approach by demonstrating effective targeting to primate M cells [85]. β1 integrins, which are atypically expressed on the apical surface of M cells, act as receptors for invasin, a surface protein expressed by *Yersinia* species [63] and coating 0.5 μm polystyrene particles with a fragment of invasin increased their uptake across rat intestinal epithelium [86]. Although the cellular location of particle uptake was not examined in this study, they were most abundant in the rat distal ileum, where Peyer's patches are concentrated [86]. The ability of bacterial adhesins to improve the efficacy of live vaccines has been confirmed in recent studies which characterized the role of type 1 pili in adherence to glycoprotein 2 on M cells [83,87]. Lastly, incorporation of a peptide targeting the M cell surface expressed protein claudin-4 enhanced uptake of submicron-sized PLGA particles by murine intestinal M cells [88].

The experimental data summarized above demonstrate that, although M cells have a high capacity for uptake of particles, it cannot be assumed they will be a major route for uptake for all classes of ENM. Indeed, the results of vaccine delivery experiments indicate that interactions with GIT wall, M cells and luminal materials will be influenced by their surface properties. As already discussed, GIT uptake mechanisms are not easily characterized by *in vivo* studies, so the development of a reliable cultured cell model representing the FAE and M cells has provided a major advance in this field [3,50,89].

An *in vitro* M cell model based on co-culture of Caco-2 cells and mouse Peyer's patch lymphocytes [50] or the Raji human B lymphocyte [89] is being used to examine mechanisms of interaction between particulates and these specialized cells. Methodology and key features of this model have been described extensively elsewhere [89–91] so will not be included here. Although some doubts remain regarding how closely this model recapitulates the native FAE, many aspects of the morphology and behavior of the co-cultured cells do resemble true M cells. For example, the brush border loses microvilli, sucrose isomaltase and villin expression [50] but gains expression of other FAE markers [84]. Appropriate functional specialization was supported by the fact that the M cell model transports polysytrene beads (200 and 500 nm) and some, but not all, bacteria to a greater extent than Caco-2 monocultures, reflecting the *in vivo* specificity of M cells [41,61,89,92]. Subsequent studies with this model have revealed that it recapitulates at least some of the selective gene expression characteristics of native M cells and its use has confirmed the efficacy of M cell targeting using a lectin mimetic [85] and β1 integrin binding motifs [53]. Size and temperature dependence has been clearly demonstrated for latex microspheres [51] consistent with particle transport being predominantly transcellular and energy dependent. The optimum size for latex particle uptake by the co-culture appears to be less than 1 μm in diameter [51] as already observed *in vivo* [52]. Inhibitor studies have suggested that uptake of inert particles and some bacteria into the M cell model occurs via a macropinocytosis-like process [91]. Transport of bacteria across the M cell model reveals species-specificity and dependence on virulence factors that reflect *in vivo* infection characteristics [61,62,64]. The M cell model has been widely used to test the efficacy of particulate carriers, for example demonstrating the effectiveness of M cell targeting ligands as discussed above revealing increased transport by PEGylation of PLGA [91,93] and revealing optimal size and pH for transport of chitosan-DNA NP [59].

It is clear from the *in vivo* and *in vitro* studies summarized here that multiple routes exist for ENM to enter the body via ingestion and passage across the GIT wall. The relative importance of these routes has been characterized for some ENM and some mechanistic details are emerging. However, we remain some way from being able to predict accurately the extent or even the mechanisms of uptake of individual ENM since these are influenced by multiple, poorly characterized, environmental and cellular factors in addition to the physicochemical properties of the ENM themselves. (For a further discussion of some of these factors, see Chapter 1.)

POTENTIAL TOXICITY OF ENGINEERED NANOMATERIALS IN THE GASTROINTESTINAL TRACT

Several recent reports and reviews have highlighted the lack of data on the potential toxicity of food-borne particulates and the need to prioritize research in this area [94]. The relative lack of solid data on ENM toxicity in GIT is compounded by the complexity of the environment, shortcomings of *in vitro* models, and limited information about exposure levels. A summary of some recent ENM toxicity studies highlights where data are emerging and also the lack of data in key areas.

As we have seen, a number of studies have recently investigated the extent to which ENM ingestion results in uptake from the gut lumen, and the sites of this uptake. Some studies have provided evidence of toxicity of metal NP (e.g. copper, silver and gold) at relatively high doses in mice or rats [95–98]. The highly acidic environment encountered in the stomach has been proposed to play a major role in the toxicity of copper NP by triggering release of ions [99]. This and other studies highlight the potential influence of the pH shifts encountered during GIT transit in both stability and dispersal/aggregation of ENM, potentially influencing local toxicity as well as their ability to penetrate the GIT barriers [59,100]. The possibility that gastric fluid potentiates ENM toxicity has also been highlighted by *in vitro* data suggesting low pH-mediated release of cadmium ions from CdSe quantum dots exacerbates toxicity of these NP in GIT cells [100].

While most studies to date have focused on gross responses of animals to the administered ENM, for example lack of weight gain and mortality, it is also important to consider local effects on GIT pathophysiology, and the potential for indirect and chronic effects on GIT health. One under-explored area of relevance here is the potential effect of ENM on the normal bacterial microflora which is now known to play key roles in GIT development and regulation of inflammation [101,102]. For example, concerns have been raised about the potential impact that the known antibacterial effects of silver NP may have on the microflora, and its potential impact on GIT function. The extent to which any such effects of ENM influence GIT health is currently unclear.

Since dysregulation of inflammation is a common cause of GIT disease, the well-documented inflammatory responses induced by ENM in airways for example, give rise to concern that ENM might trigger inflammatory bowel disease. This possibility received further attention due to the identification of pigmented cells in human Peyer's patches which result from macrophages accumulating mineral material such as TiO_2 absorbed from the gut lumen by Peyer's patch FAE and the presence of similar material in inflammatory aggregates of Crohn's disease patients [103]. The potential role of dietary particles including ENM in inflammatory bowel disease has been further characterized and discussed elsewhere [11,104] and, although subsequent studies have failed to identify a clear causal link with gut pathology, the possibility remains that dietary particulates may regulate inflammatory responses, perhaps in concert with bacterial antigens [7,105]. Although some evidence has been presented that inflammatory responses to ENM are less marked in GIT [12,105], it is not possible to exclude the possibility of potentially damaging inflammatory responses being triggered by ENM, especially those that elicit such responses elsewhere. It is also important to bear in mind that inflammatory bowel disease is known to be multifactorial and that some individuals are predisposed to development of disease. It is therefore plausible that ENM responses might vary dramatically between individuals and the presence of other causative factors.

As with the characterization of ENM uptake, most *in vitro* assessment of ENM toxicity in GIT cells has focused on the Caco-2 adenocarcinoma cell-line. For example, a range of food-related ENM, including TiO_2, SiO_2 and ZnO, induced significant cytotoxicity and oxidative DNA damage in cultured Caco-2 cells [106], but the significance of these findings to real GIT exposure remains uncertain due to limitation of these studies to non-polarized cells, which poorly reflect the GIT epithelium, and the continuing uncertainty over realistic exposure levels. Other studies have shown pure cobalt and cobalt-doped tungsten carbide NP exert toxicity in non-polarized Caco-2, and suggest that these cells are unusually sensitive to cobalt in NP form [107,108]. In contrast, other studies have pointed to a relative lack of toxicity of many particulates in Caco-2 cells, including delivery vehicles, TiO_2 and ferrous oxide NP, although ZnO NP exhibited significant toxicity in this cell model [109].

It must be emphasized that most studies measure toxicity by rather crude analysis of cell viability and metabolic activity and have focused on non-polarized Caco-2 cells. Extrapolation of such studies to *in vivo* GIT must be approached with caution since true GIT may be less responsive as we have seen, but is also prone to specific damage mechanisms that cannot be modeled in crude cell models. Subtler effects of ENM on GIT cells that potentially represent profound effects on GIT function may be revealed by studying GIT cells such as Caco-2 in their fully differentiated and polarized form, since this more accurately recapitulates GIT function. For example, when grown as fully polarized layers on permeable supports for 2–3 weeks, Caco-2 cells develop the prominent microvillous brush border required for their function in nutrient absorption, and tight junctions that underpin all GIT epithelial function. Monitoring TEER can give an indication of effects on tight junction function, which may be reversible if in the absence of effects on cellular viability. Similarly, the development of microvilli provides an additional index of the effects of ENM on GIT function, and some ENM have indeed been shown to reduce microvillus length and/or density, effects that might lead to malabsorption if similar changes are prominent *in vivo*. For example, titanium dioxide NP [22], hematite NP [110] and quantum dots [111] have all been shown to disrupt TEER and/or microvillus architecture in polarized Caco-2 cells. In the case of TiO_2 NP, altered microvillar organization on Caco-2 cells coincided with a rise in intracellular-free calcium, providing a potential mechanism for the morphological changes seen [22]. Other studies have suggested that polarized Caco-2 cell layers may reveal unexpected toxic effects that are absent in non-polarized cells. While polystyrene NP have mainly been considered to be biologically inert, a recent study suggested that they can induce widespread apoptosis in Caco-2 layers that is propagated to bystander cells via generation of hydrogen peroxide [112].

Together, the studies carried out to date on ENM suggest that GIT epithelia may be prone to damage that is not always readily modeled *in vitro*, at least without the relatively laborious and expensive use of fully differentiated GIT cell models. However, the relative paucity of data regarding toxicity mechanisms and considerable uncertainty regarding realistic exposure levels and behavior of ENM in the complex GIT environment make it impossible to predict the likely health effects of ENM.

CONCLUSIONS

Widespread exposure to ENM via ingestion is an inevitable consequence of the expanding use of ENM in food and other consumer products and is a cause for concern. However, the extent of ingestion and the potential risks this exposure imposes remain poorly defined. The studies discussed in this chapter go some way towards defining uptake mechanisms and suggest potential toxic mechanisms that warrant further investigation, but fall short of answering the most fundamental questions regarding whether our exposure to ENM via ingestion is safe or not.

It is clear that multiple routes exist for ENM to cross the GIT epithelial barrier and to be disseminated throughout the body, raising the possibility that ingested ENM may have adverse effects in peripheral organs. Significant uptake of some ENM has been demonstrated in animal models following oral administration but these studies have largely focused on relatively high doses. Animal studies and cell culture models have highlighted the potential high capacity uptake route provided by the specialized M cells and also demonstrated that these specialized cells, which are massively outnumbered by other enterocytes, are not the only portal for ENM penetration of the GIT wall. It is not yet clear to what extent uptake routes differ between ENM types and the extent to which additional barriers (e.g. mucus and glycocalyx) and the changing GIT environment affect/influence ENM uptake. The prevailing data do, nevertheless, suggest that there is some commonality but also distinct differences in the behavior of ENM in GIT.

It is clear that at high doses some ENM, e.g. metal and metal oxide NP, and quantum dots, can exhibit severe toxicity in animal and cell culture models. In addition to cytotoxic effects, additional, more subtle effects have been identified with appropriate GIT culture models that might affect GIT function. However, it is not yet clear to what extent ENM will have adverse effects at more realistic doses and the extent to which some of the *in vitro* data reflect the *in vivo* situation. As with other areas of nanotoxicology, attention must be given to the stability of ENM, with release of toxic ions potentially exacerbating toxicity of some ENM and it is noteworthy that some recent studies have demonstrated the potential for the acidic gastric environment to increase ENM toxicity. No clear evidence yet exists for ENM playing a role in inflammatory bowel disease or other immune-related pathologies (see also Chapter 10, on immune effects of ENM). Nevertheless, the delicate balance of immune regulation in the GIT, and the key role played by commensal bacteria in this balance are additional causes of concern since the known effects of some ENM suggests they have potential to contribute to disease.

TAKE-HOME MESSAGES

In conclusion, significant questions remain about the potential for ENM to have adverse effects in GIT or elsewhere, via GIT uptake. Studies have to date identified several potential uptake and toxicity mechanisms that will require further research to characterize fully.

Additional, poorly characterized interactions peculiar to GIT epithelia and the luminal environment might also have significant implications for GIT health in at least some individuals ingesting substantial levels of ENM.

Acknowledgments

The work on M cells and particle interactions with epithelia has been supported mainly by BBSRC, MRC, NERC and the Wellcome Trust. I would like to acknowledge many colleagues past and present who have contributed to my studies on particle uptake and whose insights have helped formulate some of the concepts discussed here. Thanks go especially to Barry Hirst, Ann Clark, Nick Simmons, Caroline Mason, Neil Foster, Alan Baird, David Brayden, Margaret Saunders, Caroline Sands, Isa Martinez-Argudo, Andrea Buda, Birgit Gaiser, Vicki Stone, Teresa Fernandez, Charles Tyler, Jamie Lead and Gary Hutchison.

References

[1] Schleh C, Semmler-Behnke M, Lipka J, et al. Size and surface charge of gold nanoparticles determine absorption across intestinal barriers and accumulation in secondary target organs after oral administration. Nanotoxicology 2011.

[2] Johnston BD, Scown TM, Moger J, et al. Bioavailability of nanoscale metal oxides TiO(2), CeO(2), and ZnO to fish. Environ Sci Technol 2010;44:1144–51.

[3] Brayden DJ, Jepson MA, Baird AW. Keynote review: Intestinal Peyer's patch M cells and oral vaccine targeting. Drug Discov Today 2005;10:1145–57.

[4] Jepson MA, Clark MA, Foster N, et al. Targeting to intestinal M cells. J Anat 1996;189:507–16.

[5] Jepson MA, Simmons NL, Savidge TC, James PS, Hirst BH. Selective binding and transcytosis of latex microspheres by rabbit intestinal M cells. Cell Tissue Res 1993;271:399–405.

[6] Lomer MC, Thompson RP, Powell JJ. Fine and ultrafine particles of the diet: Influence on the mucosal immune response and association with Crohn's disease. Proc Nutr Soc 2002;61:123–30.

[7] Butler M, Boyle JJ, Powell JJ, Playford RJ, Ghosh S. Dietary microparticles implicated in Crohn's disease can impair macrophage phagocytic activity and act as adjuvants in the presence of bacterial stimuli. Inflamm Res 2007;56:353–61.

[8] Artursson P, Borchardt RT. Intestinal drug absorption and metabolism in cell cultures: Caco-2 and beyond. Pharm Res 1997;14:1655–8.

[9] Hansen GH, Rasmussen K, Niels-Christiansen LL, Danielsen EM. Endocytic trafficking from the small intestinal brush border probed with FM dye. Am J Physiol Gastrointest Liver Physiol 2009;297:G708–15.

[10] Cartiera MS, Johnson KM, Rajendran V, Caplan MJ, Saltzman WM. The uptake and intracellular fate of PLGA nanoparticles in epithelial cells. Biomaterials 2009;30:2790–8.

[11] Powell JJ, Harvey RS, Ashwood P, Wolstencroft R, Gershwin ME, Thompson RP. Immune potentiation of ultrafine dietary particles in normal subjects and patients with inflammatory bowel disease. J Autoimmun 2000;14:99–105.

[12] Powell JJ, Faria N, Thomas-McKay E, Pele LC. Origin and fate of dietary nanoparticles and microparticles in the gastrointestinal tract. J Autoimmun 2010;34:J226–33.

[13] Behrens I, Pena AI, Alonso MJ, Kissel T. Comparative uptake studies of bioadhesive and non-bioadhesive nanoparticles in human intestinal cell lines and rats: The effect of mucus on particle adsorption and transport. Pharm Res 2002;19:1185–93.

[14] Frey A, Giannasca KT, Weltzin R, et al. Role of the glycocalyx in regulating access of microparticles to apical plasma membranes of intestinal epithelial cells: Implications for microbial attachment and oral vaccine targeting. J Exp Med 1996;184:1045–59.

[15] Bockman DE, Cooper MD. Pinocytosis by epithelium associated with lymphoid follicles in the bursa of Fabricius, appendix, and Peyer's patches. An electron microscopic study. Am J Anat 1973;136:455–77.

[16] Owen RL, Jones AL. Epithelial cell specialization within human Peyer's patches: An ultrastructural study of intestinal lymphoid follicles. Gastroenterology 1974;66:189–203.

[17] McClean S, Prosser E, Meehan E, et al. Binding and uptake of biodegradable poly-DL-lactide micro- and nanoparticles in intestinal epithelia. Eur J Pharm Sci 1998;6:153–63.
[18] Hillery AM, Jani PU, Florence AT. Comparative, quantitative study of lymphoid and non-lymphoid uptake of 60 nm polystyrene particles. J Drug Target 1994;2:151–6.
[19] Jani P, Halbert GW, Langridge J, Florence AT. Nanoparticle uptake by the rat gastrointestinal mucosa: Quantitation and particle size dependency. J Pharm Pharmacol 1990;42:821–6.
[20] Scown TM, Santos EM, Johnston BD, et al. Effects of aqueous exposure to silver nanoparticles of different sizes in rainbow trout. Toxicol Sci 2010;115:521–34.
[21] Gaiser BK, Fernandes TF, Jepson M, Lead JR, Tyler CR, Stone V. Assessing exposure, uptake and toxicity of silver and cerium dioxide nanoparticles from contaminated environments. Environ Health 2009;8(Suppl. 1):S2.
[22] Koeneman BA, Zhang Y, Westerhoff P, Chen Y, Crittenden JC, Capco DG. Toxicity and cellular responses of intestinal cells exposed to titanium dioxide. Cell Biol Toxicol 2010;26:225–38.
[23] Danielsen EM, Hansen GH. Lipid raft organization and function in the small intestinal brush border. J Physiol Biochem 2008;64:377–82.
[24] Volkheimer G. Persorption of particles: Physiology and pharmacology. Adv Pharmacol Chemother 1977;14:163–87.
[25] Hillyer JF, Albrecht RM. Gastrointestinal persorption and tissue distribution of differently sized colloidal gold nanoparticles. J Pharm Sci 2001;90:1927–36.
[26] Berg RD. Bacterial translocation from the gastrointestinal tract. Adv Exp Med Biol 1999;473:11–30.
[27] McEwan GT, Jepson MA, Hirst BH, Simmons NL. Polycation-induced enhancement of epithelial paracellular permeability is independent of tight junctional characteristics. Biochim Biophys Acta 1993;1148:51–60.
[28] Makhlof A, Werle M, Tozuka Y, Takeuchi H. A mucoadhesive nanoparticulate system for the simultaneous delivery of macromolecules and permeation enhancers to the intestinal mucosa. J Control Release 2011;149: 81–8.
[29] Lin IC, Liang M, Liu TY, Ziora ZM, Monteiro MJ, Toth I. Interaction of densely polymer-coated gold nanoparticles with epithelial Caco-2 monolayers. Biomacromolecules 2011;12:1339–48.
[30] Simon M, Behrens I, Dailey LA, Wittmar M, Kissel T. Nanosized insulin-complexes based on biodegradable amine-modified graft polyesters poly[vinyl-3-(diethylamino)-propylcarbamate-co-(vinyl acetate)-co-(vinyl alcohol)]-graft-poly(L-lactic acid): Protection against enzymatic degradation, interaction with Caco-2 cell monolayers, peptide transport and cytotoxicity. Eur J Pharm Biopharm 2007;66:165–72.
[31] Berg RD. Bacterial translocation from the gastrointestinal tract. J Med 1992;23:217–44.
[32] Moyes SM, Morris JF, Carr KE. Culture conditions and treatments affect Caco-2 characteristics and particle uptake. Int J Pharm 2010;387:7–18.
[33] Moyes SM, Killick EM, Morris JF, Kadhim MA, Hill MA, Carr KE. Changes produced by external radiation in parameters influencing intestinal permeability and microparticle uptake *in vitro*. Int J Radiat Biol 2008;84:467–86.
[34] Duncan R, Izzo L. Dendrimer biocompatibility and toxicity. Adv Drug Deliv Rev 2005;57:2215–37.
[35] El-Sayed M, Ginski M, Rhodes C, Ghandehari H. Transepithelial transport of poly(amidoamine) dendrimers across Caco-2 cell monolayers. J Control Release 2002;81:355–65.
[36] Sweet DM, Kolhatkar RB, Ray A, Swaan P, Ghandehari H. Transepithelial transport of PEGylated anionic poly(amidoamine) dendrimers: Implications for oral drug delivery. J Control Release 2009;138:78–85.
[37] Neutra MR, Frey A, Kraehenbuhl JP. Epithelial M cells: Gateways for mucosal infection and immunization. Cell 1996;86:345–8.
[38] Jepson MA, Clark MA, Hirst BH. M cell targeting by lectins: A strategy for mucosal vaccination and drug delivery. Adv Drug Deliv Re 2004;56:511–25.
[39] Jepson MA, Mason CM, Clark MA, Simmons NL, Hirst BH. Variations in lectin binding properties of intestinal M cells. J Drug Target 1995;3:75–7.
[40] Clark MA, Jepson MA, Simmons NL, Hirst BH. Differential surface characteristics of M cells from mouse intestinal Peyer's and caecal patches. Histochem J 1994;26:271–80.
[41] Clark MA, Jepson MA. Intestinal M cells and their role in bacterial infection. Int J Med Microbiol 2003;293:17–39.
[42] Smith MW. Selective expression of brush border hydrolases by mouse Peyer's patch and jejunal villus enterocytes. J Cell Physiol 1985;124:219–25.
[43] Gebert A, Pabst R. M cells at locations outside the gut. Semin Immunol 1999;11:165–70.

[44] Jang MH, Kweon MN, Iwatani K, et al. Intestinal villous M cells: An antigen entry site in the mucosal epithelium. Proc Natl Acad Sci USA 2004;101:6110–5.
[45] Jepson MA, Simmons NL, O'Hagan DT, Hirst BH. Comparison of poly(DL-lactide-co-glycolide) and polystyrene microsphere targeting to intestinal M cells. J Drug Target 1993;1:245–9.
[46] Pappo J, Ermak TH. Uptake and translocation of fluorescent latex particles by rabbit Peyer's patch follicle epithelium: A quantitative model for M cell uptake. Clin Exp Immunol 1989;76:144–8.
[47] Neutra MR, Phillips TL, Mayer EL, Fishkind DJ. Transport of membrane-bound macromolecules by M cells in follicle-associated epithelium of rabbit Peyer's patch. Cell Tissue Res 1987;247:537–46.
[48] Clark MA, Jepson MA, Simmons NL, Hirst BH. Preferential interaction of *Salmonella typhimurium* with mouse Peyer's patch M cells. Res Microbiol 1994;145:543–52.
[49] Foster N, Clark MA, Jepson MA, Hirst BH. Ulex europaeus 1 lectin targets microspheres to mouse Peyer's patch M-cells *in vivo*. Vaccine 1998;16:536–41.
[50] Kerneis S, Bogdanova A, Kraehenbuhl JP, Pringault E. Conversion by Peyer's patch lymphocytes of human enterocytes into M cells that transport bacteria. Science 1997;277:949–52.
[51] des Rieux A, Ragnarsson EG, Gullberg E, Preat V, Schneider YJ, Artursson P. Transport of nanoparticles across an *in vitro* model of the human intestinal follicle associated epithelium. Eur J Pharm Sci 2005;25:455–65.
[52] Desai MP, Labhasetwar V, Amidon GL, Levy RJ. Gastrointestinal uptake of biodegradable microparticles: Effect of particle size. Pharm Res 1996;13:1838–45.
[53] Garinot M, Fievez V, Pourcelle V, et al. PEGylated PLGA-based nanoparticles targeting M cells for oral vaccination. J Contr Release 2007;120:195–204.
[54] Gupta PN, Khatri K, Goyal AK, Mishra N, Vyas SP. M-cell targeted biodegradable PLGA nanoparticles for oral immunization against hepatitis B. J Drug Target 2007;15:701–13.
[55] Torche AM, Jouan H, Le Corre P, et al. *Ex vivo* and *in situ* PLGA microspheres uptake by pig ileal Peyer's patch segment. Int J Pharm 2000;201:15–27.
[56] Aramaki Y, Tomizawa H, Hara T, Yachi K, Kikuchi H, Tsuchiya S. Stability of liposomes *in vitro* and their uptake by rat Peyer's patches following oral administration. Pharm Res 1993;10:1228–31.
[57] Foster N, Hirst BH. Exploiting receptor biology for oral vaccination with biodegradable particulates. Adv Drug Deliv Rev 2005;57:431–50.
[58] Primard C, Rochereau N, Luciani E, et al. Traffic of poly(lactic acid) nanoparticulate vaccine vehicle from intestinal mucus to sub-epithelial immune competent cells. Biomaterials 2010;31:6060–8.
[59] Slutter B, Plapied L, Fievez V, et al. Mechanistic study of the adjuvant effect of biodegradable nanoparticles in mucosal vaccination. J Contr Release 2009;138:113–21.
[60] Kadiyala I, Loo Y, Roy K, Rice J, Leong KW. Transport of chitosan-DNA nanoparticles in human intestinal M-cell model versus normal intestinal enterocytes. Eur J Pharm Sci 2010;39:103–9.
[61] Martinez-Argudo I, Jepson MA. *Salmonella* translocates across an *in vitro* M cell model independently of SPI-1 and SPI-2. Microbiology 2008;154:3887–94.
[62] Clark MA, Reed KA, Lodge J, Stephen J, Hirst BH, Jepson MA. Invasion of murine intestinal M cells by *Salmonella typhimurium* inv mutants severely deficient for invasion of cultured cells. Infect Immun 1996;64: 4363–8.
[63] Jepson MA, Clark MA. Studying M cells and their role in infection. Trends Microbiol 1998;6:359–65.
[64] Clark MA, Hirst BH, Jepson MA. M-cell surface beta1 integrin expression and invasin-mediated targeting of *Yersinia pseudotuberculosis* to mouse Peyer's patch M cells. Infect Immun 1998;66:1237–43.
[65] Schulte R, Kerneis S, Klinke S, et al. Translocation of *Yersinia entrocolitica* across reconstituted intestinal epithelial monolayers is triggered by *Yersinia* invasin binding to beta1 integrins apically expressed on M-like cells. Cell Microbiol 2000;2:173–85.
[66] Autenrieth IB, Firsching R. Penetration of M cells and destruction of Peyer's patches by *Yersinia enterocolitica*: An ultrastructural and histological study. J Med Microbiol 1996;44:285–94.
[67] Sansonetti PJ, Phalipon A. M cells as ports of entry for enteroinvasive pathogens: Mechanisms of interaction, consequences for the disease process. Semin Immunol 1999;11:193–203.
[68] Wassef JS, Keren DF, Mailloux JL. Role of M cells in initial antigen uptake and in ulcer formation in the rabbit intestinal loop model of shigellosis. Infect Immun 1989;57:858–63.
[69] Owen RL, Cray Jr WC, Ermak TH, Pierce NF. Bacterial characteristics and follicle surface structure: Their roles in Peyer's patch uptake and transport of *Vibrio cholerae*. Adv Exp Med Biol 1988;237:705–15.

[70] Neutra MR. Interactions of viruses and microparticles with apical plasma membranes of M cells: Implications for human immunodeficiency virus transmission. J Infect Dis 1999;179(Suppl. 3):S441–3.
[71] Fotopoulos G, Harari A, Michetti P, Trono D, Pantaleo G, Kraehenbuhl JP. Transepithelial transport of HIV-1 by M cells is receptor-mediated. Proc Natl Acad Sci USA 2002;99:9410–4.
[72] Ouzilou L, Caliot E, Pelletier I, Prevost MC, Pringault E, Colbere-Garapin F. Poliovirus transcytosis through M-like cells. J Gen Virol 2002;83:2177–82.
[73] Mantis NJ, Frey A, Neutra MR. Accessibility of glycolipid and oligosaccharide epitopes on rabbit villus and follicle-associated epithelium. Am J Physiol Gastrointest Liver Physiol 2000;278:G915–23.
[74] Singh M, Chakrapani A, O'Hagan D. Nanoparticles and microparticles as vaccine-delivery systems. Expert Rev Vaccines 2007;6:797–808.
[75] Petkar KC, Chavhan SS, Agatonovik-Kustrin S, Sawant KK. Nanostructured materials in drug and gene delivery: A review of the state of the art. Crit Rev Ther Drug Carrier Syst 2011;28:101–64.
[76] Clark MA, Jepson MA, Hirst BH. Exploiting M cells for drug and vaccine delivery. Adv Drug Deliv Rev 2001;50:81–106.
[77] Clark MA, Hirst BH, Jepson MA. Lectin-mediated mucosal delivery of drugs and microparticles. Adv Drug Deliv Rev 2000;43:207–23.
[78] Clark MA, Blair H, Liang L, Brey RN, Brayden D, Hirst BH. Targeting polymerised liposome vaccine carriers to intestinal M cells. Vaccine 2001;20:208–17.
[79] des Rieux A, Fievez V, Garinot M, Schneider YJ, Preat V. Nanoparticles as potential oral delivery systems of proteins and vaccines: A mechanistic approach. J Contr Release 2006;116:1–27.
[80] Clark MA, Jepson MA, Hirst BH. Lectin binding defines and differentiates M-cells in mouse small intestine and caecum. Histochem Cell Biol 1995;104:161–8.
[81] Clark MA, Jepson MA, Simmons NL, Booth TA, Hirst BH. Differential expression of lectin-binding sites defines mouse intestinal M-cells. J Histochem Cytochem 1993;41:1679–87.
[82] Gebert A, Hach G. Differential binding of lectins to M cells and enterocytes in the rabbit cecum. Gastroenterology 1993;105:1350–61.
[83] Hase K, Kawano K, Nochi T, et al. Uptake through glycoprotein 2 of FimH(+) bacteria by M cells initiates mucosal immune response. Nature 2009;462:226–30.
[84] Lo D, Tynan W, Dickerson J, et al. Cell culture modeling of specialized tissue: Identification of genes expressed specifically by follicle-associated epithelium of Peyer's patch by expression profiling of Caco-2/Raji co-cultures. Int Immunol 2004;16:91–9.
[85] Misumi S, Masuyama M, Takamune N, et al. Targeted delivery of immunogen to primate m cells with tetragalloyl lysine dendrimer. J Immunol 2009;182:6061–70.
[86] Hussain N, Florence AT. Utilizing bacterial mechanisms of epithelial cell entry: Invasin-induced oral uptake of latex nanoparticles. Pharm Res 1998;15:153–6.
[87] Martinez-Argudo I, Jepson MA. Identification of adhesin-receptor interactions driving bacterial translocation through M cells. Future Microbiol 2010;5:549–53.
[88] Rajapaksa TE, Stover-Hamer M, Fernandez X, Eckelhoefer HA, Lo DD. Claudin 4-targeted protein incorporated into PLGA nanoparticles can mediate M cell targeted delivery. J Control Release 2010;142:196–205.
[89] Gullberg E, Leonard M, Karlsson J, et al. Expression of specific markers and particle transport in a new human intestinal M-cell model. Biochem Biophys Res Commun 2000;279:808–13.
[90] Kerneis S, Caliot E, Stubbe H, Bogdanova A, Kraehenbuhl J, Pringault E. Molecular studies of the intestinal mucosal barrier physiopathology using cocultures of epithelial and immune cells: A technical update. Microbes Infect 2000;2:1119–24.
[91] des Rieux A, Fievez V, Theate I, Mast J, Preat V, Schneider YJ. An improved *in vitro* model of human intestinal follicle-associated epithelium to study nanoparticle transport by M cells. Eur J Pharm Sci 2007;30: 380–91.
[92] Martinez-Argudo I, Sands C, Jepson MA. Translocation of enteropathogenic Escherichia coli across an *in vitro* M cell model is regulated by its type III secretion system. Cell Microbiol 2007;9:1538–46.
[93] des Rieux A, Fievez V, Momtaz M, et al. Helodermin-loaded nanoparticles: Characterization and transport across an *in vitro* model of the follicle-associated epithelium. J Contr Release 2007;118:294–302.
[94] Card JW, Jonaitis TS, Tafazoli S, Magnuson BA. An appraisal of the published literature on the safety and toxicity of food-related nanomaterials. Crit Rev Toxicol 2011;41:22–49.

[95] Park EJ, Bae E, Yi J, et al. Repeated-dose toxicity and inflammatory responses in mice by oral administration of silver nanoparticles. Environ Toxicol Pharmacol 2010;30:162–8.

[96] Kim YS, Song MY, Park JD, et al. Subchronic oral toxicity of silver nanoparticles. Part Fibre Toxicol 2010;7:20.

[97] Chen Z, Meng H, Xing G, et al. Acute toxicological effects of copper nanoparticles *in vivo*. Toxicol Lett 2006;163:109–20.

[98] Zhang XD, Wu HY, Wu D, et al. Toxicologic effects of gold nanoparticles *in vivo* by different administration routes. Int J Nanomed 2010;5:771–81.

[99] Meng H, Chen Z, Xing G, et al. Ultrahigh reactivity provokes nanotoxicity: Explanation of oral toxicity of nano-copper particles. Toxicol Lett 2007;175:102–10.

[100] Wang L, Nagesha DK, Selvarasah S, Dokmeci MR, Carrier RL. Toxicity of CdSe nanoparticles in Caco-2 cell cultures. J Nanobiotechnol 2008;6:11.

[101] Reiff C, Kelly D. Inflammatory bowel disease, gut bacteria and probiotic therapy. Int J Med Microbiol 2010;300:25–33.

[102] Stecher B, Hardt WD. Mechanisms controlling pathogen colonization of the gut. Curr Opin Microbiol 2011;14:82–91.

[103] Shepherd NA, Crocker PR, Smith AP, Levison DA. Exogenous pigment in Peyer's patches. Hum Pathol 1987;18:50–4.

[104] Powell JJ, Ainley CC, Harvey RS, et al. Characterisation of inorganic microparticles in pigment cells of human gut associated lymphoid tissue. Gut 1996;38:390–5.

[105] Lomer MC, Hutchinson C, Volkert S, et al. Dietary sources of inorganic microparticles and their intake in healthy subjects and patients with Crohn's disease. Br J Nutr 2004;92:947–55.

[106] Gerloff K, Albrecht C, Boots AW, Forster I, Schins RPF. Cytotoxicity and oxidative DNA damage by nanoparticles in human intestinal Caco-2 cells. Nanotoxicol 2009;3:355–64.

[107] Bastian S, Busch W, Kuhnel D, et al. Toxicity of tungsten carbide and cobalt-doped tungsten carbide nanoparticles in mammalian cells *in vitro*. Environ Health Perspect 2009;117:530–6.

[108] Horev-Azaria L, Kirkpatrick CJ, Korenstein R, et al. Predictive toxicology of cobalt nanoparticles and ions: Comparative *in vitro* study of different cellular models using methods of knowledge discovery from data. Toxicol Sci 2011;122:489–501.

[109] Semete B, Booysen L, Lemmer Y, et al. *In vivo* evaluation of the biodistribution and safety of PLGA nanoparticles as drug delivery systems. Nanomedicine 2010;6:662–71.

[110] Zhang W, Kalive M, Capco DG, Chen Y. Adsorption of hematite nanoparticles onto Caco-2 cells and the cellular impairments: Effect of particle size. Nanotechnology 2010;21:355103.

[111] Koeneman BA, Zhang Y, Hristovski K, et al. Experimental approach for an *in vitro* toxicity assay with non-aggregated quantum dots. Toxicol In Vitro 2009;23:955–62.

[112] Thubagere A, Reinhard BM. Nanoparticle-induced apoptosis propagates through hydrogen-peroxide-mediated bystander killing: Insights from a human intestinal epithelium *in vitro* model. ACS Nano 2010;4:3611–22.

CHAPTER

13

Reproductive Toxicity

Karin Sørig Hougaard[1], *Luisa Campagnolo*[2]

[1]National Research Center for the Working Environment, Copenhagen, Denmark

[2]Department of Public Health and Cell Biology, University of Tor Vergata, Rome, Italy

OUTLINE

INTRODUCTION

The sensitivity of reproductive and developmental outcomes to toxic substances causes significant public concern. Unfortunately, there is at present a large discrepancy between the number of chemicals in commerce and the number that have been evaluated for reproductive toxicity [1]. For nanomaterials this aspect is even graver. Thus, toxicity of the bulk material might not be adequately assessed in the first place, and downsizing and engineering might convey physical and chemical qualities that differ from that of the corresponding bulk material. Exposure to nanomaterials might have detrimental consequences for fertility, pregnancy and development of the offspring. However, very

Adverse Effects of Engineered Nanomaterials
DOI: 10.1016/B978-0-12-386940-1.00013-1

few studies have investigated this aspect as of 2011, even if the usage of engineered nanomaterials supposedly will grow exponentially during the next few years. At the same time, legislative measures aimed to control the exposure to engineered nanoparticles (ENP) are currently lacking for pregnant women, as well as for women and men of childbearing age. The present chapter aims to provide an overview on reproductive and developmental toxicity of engineered nanomaterials, based on published investigations of the most commonly studied ENP.

FERTILITY

Fertility includes the processes that are necessary for initiation of pregnancy. Production of the male and female germ cells (gametogenesis) is a highly regulated process that requires balance between proliferation, differentiation and apoptosis. Also several less visible biological processes need to function correctly, e.g. correct timing of hormonal activity or physiological parameters impacting the ability to fertilize, as well as libido, sexual behavior and development of the fertilized ovum up to and including implantation [1,2]. Pollutants (among which nanoparticles [NPs] can be included) may adversely affect human reproductive health by directly influencing reproductive tissues or indirectly, by behaving as endocrine disrupters or interfering with the hormonal regulation of germ cell differentiation and maturation [3,4]. Perturbation of reproductive processes by chemicals might lead to DNA mutations in somatic or in germ cells [5]. Environmental and occupational issues are considered by some as a new threat to fertility. Studying the impact of nanotechnology on the reproductive health of both sexes is of major importance, since, for example, mutations in reproductive organs and cells represent not only a risk for the individual, but also for his/her progeny. In this respect, a differential sensitivity to mutation-inducing agents is expected between males and females as the male process of spermiogenesis is continuous and takes place throughout the male fertile age. In comparison, female oogenesis is cyclic in nature and relies on the ovarian pool of oocytes that was established during fetal life.

Several reports demonstrate that ultrafine particles, the non-manufactured counterparts to ENP, can adversely affect male reproductive function [6,7]. Due to the often carbonaceous nature of ultrafine particles, e.g. diesel exhaust particles, these are difficult to trace in biological tissues. In contrast, the chemical constitution of most ENP allows such tracing. Of relevance for the occupational setting, mice exposed by nose-only inhalation to iron-containing NPs (4.89 and 9.34×10^5 particles/cm^3, 4 h/day, 5 days/week, for 4 weeks) showed particles in various organs, including specific regions of the testes and the ovaries [8]. Other kinds of NP have been shown to reach the reproductive organs in adult animals, and pass the blood–testes barrier, and some of the most representative studies will be summarized in the following sections. Relatively little information is available as to the specific localization inside reproductive organs, although some preference for the interstitial regions of the testes seems possible [8,9]. Furthermore, findings in one study indicate that smaller particles access the testes whereas larger particles might not. In rats, only 10 nm gold NPs accessed the testes 24 hours after an intravenous injection (0.2% of the dose), whereas 50 and 100 nm nanoparticles did not [10].

Male Fertility

Whether engineered particles interfere with male reproductive function has been addressed *in vivo* in a few studies. The distribution to the testes of 50 nm magnetic particles administered intraperitoneally (25, 50 and 100 mg/kg) to mice daily for 4 weeks was not associated with histological changes in this organ at any dose level [9]. For carbon black NPs, fertility parameters were investigated after intratracheal instillation of particles of different sizes (15, 56 and 95 nm) to sexually mature male mice (Figure 13.1). A dose of 0.1 mg/mouse, was administered once weekly for 10 weeks. No changes in body or reproductive organ weights were observed. Seminiferous tubules appeared vacuolated and daily sperm production was decreased in all treated groups. Plasma levels of testosterone were increased, although only significantly so in the 14 and 56 nm particle groups. Additional males were exposed to the 14 nm particle but at particle numbers corresponding to the 56 nm particle group (weekly dose of 1.56 μg CB 14 nm/mouse). In this group, daily sperm production was significantly decreased [11].

Repeated intravenous administration of multiwalled carbon nanotubes (MWCNT; 5 mg/kg every 3 days for 15 days) altered the normal architecture of the seminiferous tubules, reducing thickness of the germ layer and the number of Sertoli and spermatogonial cells, 15 days after the first dose. Increased testicular level of malondialdehyde indicated increased level of oxidative stress, returning to control levels at 60 and 90 days. Histological alterations were only sporadically observed at 60 and 90 days, suggesting an ability of the seminiferous epithelium to repair. No changes in plasma levels of sex hormones, DNA damage in sperm, total sperm concentration, motility, acrosome integrity, or mating behavior were observed. Effects did not depend on functionalization by carboxylation (negatively charged) or amination (positively charged). At all time points, minimal change was observed after a single

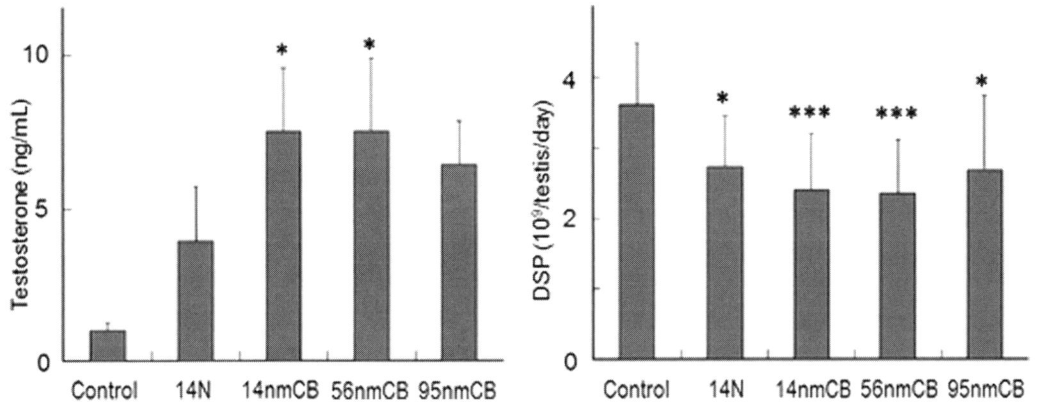

FIGURE 13.1 Effects on male reproductive parameters after exposure to carbon black particles of different sizes. Male mice were exposed to 0.1 mg carbon black/mouse by intratracheal instillation, once per week for 10 weeks. Particles were of three different sizes, i.e. 14, 56 and 96 nm. Additional males were exposed to the 14 nm particle but at particle numbers corresponding to the 56 nm particle group (14N; weekly dose of 1.56 μg CB 14 nm/mouse). Mean±SD. *$P<0.05$, *** $P<0.001$. Left: Serum testosterone concentration, determined by ELISA (n=15–16). Right: The efficiency of spermatogenesis, calculated as Daily Sperm Production (n=15–16). *(Courtesy of* International Journal of Andrology *[11].)*

intravenous injection [12]. In contrast, a protective role on the male reproductive apparatus has been proposed for the hydroxylated forms of fullerenes (fullerols or fullerenols). Administration of hydrated C_{60} fullerene in drinking water reduced reproductive alterations observed in male rats with streptozotocin-induced diabetes [13]. Similarly, pretreatment of male rats with fullerenol ameliorated the nephrotesticular phenotype induced by doxorubicin, probably through antioxidant capability or ability to bind free iron [14].

Some *in vitro* studies have investigated the sensitivity of isolated cell types from the testis to NPs. Several metal NPs interfered dose dependently with cell metabolic activity and viability of a mouse spermatogonial stem cell line. The effects occurred independently of surface modifications, with silver NPs showing the highest toxicity [15]. The inhibition by silver NPs of cell division seemed to depend on interaction with an intracellular kinase. Smaller particles exerted the greatest effect [16]. In the testes, Leydig cells synthesize testosterone, which is critical for the male phenotype and spermatogenesis. In the mouse TM3 Leydig cell line, titanium dioxide (TiO_2, 25–70 nm) and carbon black (CB, 14 nm) NPs were both internalized by the cells, and TiO_2 particles, at a concentration of 10 and 100 μg/mL, transiently impaired cell proliferation and viability. The ability of 10 and 30 μg/mL of CB NPs to alter testosterone production was demonstrated by quantitative PCR of the steroidogenic acute regulatory (StAR) gene, which is an important molecule in the process of testosterone synthesis [17]. Gold NPs (9 nm) were added to a single donor sample of human sperm cells. Particles penetrated into sperm heads and tails and 25% of the sperm cells were immotile, compared to 5% in the control sample. Motility was the only end-point assessed [18]. When bovine sperm cells were incubated for 2 hours in an aqueous colloid solution of Fe_3O_4 NPs, coated with poly(vinyl alcohol), particles penetrated the cells, but did not interfere with motility or ability to undergo the acrosome reaction, both fundamental processes for fertilization [19].

Female Fertility

Reports on the action of NPs on the female reproductive process are scant. After intraperitoneal administration of three different doses (100, 50, and 25 mg/kg) of 50 nm magnetic particles, particles were observed in the uterus of mice, but without associated histological changes [9]. Lowered pregnancy rates were observed after intratracheal instillation, once a day for 5 days, with 7.5 mg/kg of either zinc oxide or composite carbon and zinc oxide NPs during gestation. Control pregnancy rate was 100%, compared to 57.2% and 66.7% in the two treated groups [20]. *In vitro*, granulosa cells from pre-ovulatory rat follicles were incubated with 10 nm gold particles from 1 to 24 h. Particles were internalized shortly after addition to media, and transmission electron microscopy showed that particles interacted with mitochondria. Estrogen accumulation and secretion changed in a time-dependent manner, initially increasing but decreasing after 24 h [21]. Similarly, calcium phosphate NPs were internalized in human granulosa cells. Apoptosis increased, but hormonal production was unaltered [22].

NANOPARTICLES AND DEVELOPMENTAL TOXICITY

Developmental toxicity embraces any effect interfering with normal development, before, as well as after birth. Embryotoxicity might manifest as reduced body weight, growth and developmental retardation, organ toxicity, structural defects, death and abortion. Impaired postnatal development might appear as delayed physical development. Functional deficits are developmental outcomes that do not have obvious structural correlates. They might appear before birth, but will, in many cases, not appear until after birth, e.g. impaired cognitive function or enhanced propensity to develop allergy. Generally, developmental effects might appear up to and including normal pubertal development [2,23]. Whereas in most cases adverse effects from exposure to chemicals in the adult organism might be reversible, a similar exposure but of the developing organism might lead to permanent effects [23]. Furthermore, the developing fetus is characterized by intense cellular division and differentiation, an environment that may represent a different milieu, than that of an established tissue or even monocultures of dividing cells *in vitro*. In a model system for neuronal differentiation, PC12 cells, silver NPs produced no or little oxidative stress in undifferentiated, dividing cells. This changed when the cells were induced to differentiate by addition of nerve growth factor. Then silver NPs were associated with robust oxidative stress and, at the same time, impairment of differentiation into the determined phenotype [24]. Furthermore, in mice, pregnancy has been shown to increase the lung inflammatory response to NPs administered through the airways [25,26]. For developmental toxicity, the placenta presents at least a potential barrier between the maternal blood and the fetal compartment [27]. For some types of particles, this barrier seems almost complete [28] whereas for others some access to fetal tissues has been reported [29,30]. Reported below are the nanoparticles whose effects on developmental toxicity have been investigated.

Metal-Based Nanoparticles

Titanium Dioxide (TiO_2)

TiO_2 particles have been investigated for their effects on offspring male reproductive function, after exposure of pregnant mice to 0.1 mg/mouse 25–70 nm TiO_2 anatase particles by subcutaneous injection, at gestation days 3, 7, 10 and 14. Four days or 6 weeks after birth, aggregated particles of TiO_2 were identified in brain and testicular tissue from male offspring. Particles were detected in spermatids and Leydig and Sertoli cells alongside abnormal testicular morphology and reduced number of Sertoli cells and daily sperm production. Sperm morphology was comparable between groups. In the brain, numerous cells in the olfactory bulb stained positive for caspase-3, a biomarker of apoptosis [31]. Whether developmental exposure also influences function of the nervous system function has been assessed in a couple of studies. Exposure of lactating females to 25 nm anatase TiO_2 by the intragastric route caused neurological alterations in the offspring, evidenced by changes in electrophysiological recordings *in vivo*. After a similar exposure, but throughout gestation, less evident changes were observed. Observation of changed function of the hippocampus, which is pivotal for higher order neural function, suggests developmental exposure to TiO_2 might interfere with cognitive function in the offspring [32]. This was specifically studied by means of a neurobehavioral test battery in offspring of time-mated mice exposed by inhalation on gestation days 8–18 to 21 nm rutile TiO_2 modified with Al, Si and Zr, and polyalcohols. Prenatally exposed offspring tended to avoid the central zone of the open field, and exposed female offspring displayed enhanced prepulse inhibition. These data suggest that prenatal exposure somewhat influenced development of the nervous system, although learning and memory function were unaffected. This study also investigated functional reproductive implications by cross-mating the mature male and female F1 offspring to non-exposed mature mice. Time-to-delivery of the first F2 litter was somewhat (although non-significantly) delayed for exposed male offspring (Figure 13.2). Litter size, sex distribution and birth weights of F2 litters were comparable to those of controls [34].

Offspring propensity to develop allergy was investigated after intranasal insufflation of pregnant mice with a single dose of 50 μg TiO_2 at gestation day 14. When offspring were sensitized and exposed to the aerosolized model allergen of ovalbumin, neonates of exposed mothers developed a more pronounced asthmatic phenotype than did sham exposed control offspring [25].

Gold Nanoparticles

Gold NPs have already been extensively used for applications in biology and medicine, and are also used in consumer products such as cosmetics. *In vivo* experiments have primarily been carried out with the zebra fish embryo as a model. A "random walk" of gold NPs in embryonic tissues has been proposed. When cleavage-stage zebra fish

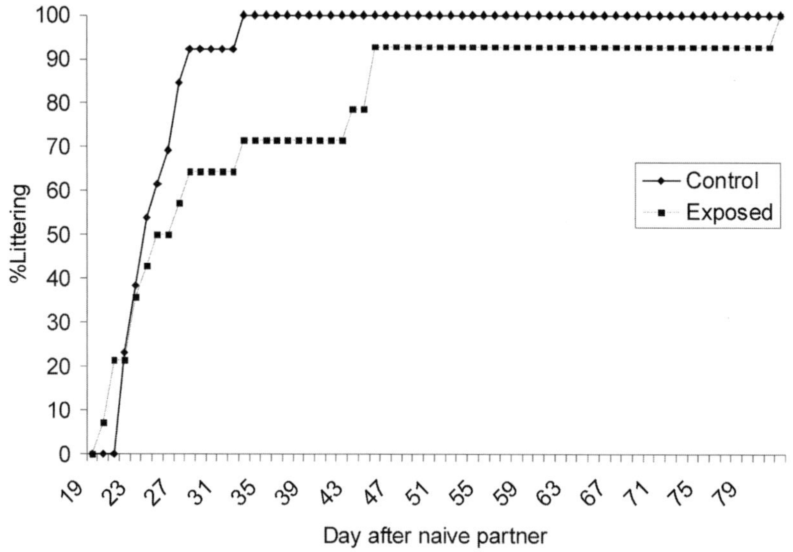

FIGURE 13.2 Time-to-first-delivery of F2 litter. The littering curve shows the time it took for cohabiting male and female mouse couples to deliver their first litter for male offspring of sham and TiO_2-exposed pregnant mice (exposed by inhalation 1h/day to 42 mg/m^3 aerosolized TiO_2-powder (1.7×10^6 particles/cm^3; peak-size: 97 nm) on gestation days 8–18, respectively. As adults, these male offspring were cross-mated to mature, naïve female mice and time-to-first-delivery of F2 litter was recorded. Time to delivery of first F2 litter was delayed in exposed compared to control male offspring, however non-significantly so (P=0.12). Other studies report statistically significant effects on male reproductive measures after prenatal exposure to ENP [31,33]. This endpoint warrants further investigations in future studies. (*Courtesy of* Particle and Fibre Toxicology, in [34].)

embryos were incubated for 120 h with stable, mono-dispersed (16.6±0.9 nm), highly purified gold NPs, in concentrations ranging from 0.025 nM to 1.2 nM, particles diffused through the chorionic pores and into the embryos via Brownian motion. A small number of dead or abnormal embryos was observed, with no clear dose–response dependence, probably due to a random distribution of NPs within the embryonic tissues, that might lead to stochastic effects on embryonic development [35]. In a further experiment using the same model, embryonic uptake was confirmed but low developmental toxicity was observed after exposure to 30–100 nm gold NPs, even at the highest concentration (250 μM). Similar findings were found for triphenylphosphine monosulfate-functionalized gold nanoparticles [36]. When mercaptobenzoic acid-coupled gold NPs of 25 and 40 nm in diameter were injected directly into zebra fish at the one-cell embryonic stage, particles distributed to all cells and tissues during the subsequent embryo development but, again, in the absence of appreciable toxicity [37]. Altogether, gold NPs seem of low toxicity for embryonic development of the zebra fish *in vivo*, independently of size (30 to 100 nm) and tested surface modifications, even at high concentrations (250 μM). *In vivo* data on mammalian development are needed, as to date only one study has investigated the embryotoxic potential of gold NPs in mammalian cells, *in vitro*. Using a modified version of the embryonic stem cell test, gold NPs covered by hyaluronic acid were classified as weakly embryotoxic, while silane and gold coated cobalt ferrite NPs appeared non-embryotoxic [38].

Placental crossing has been studied for gold NPs. In an *ex vivo* model of placental perfusion, 10 and 15 nm gold NPs accumulated in the maternal layer of the placenta, but no particles reached the fetal side of the organ [39]. Contradictory results have been obtained in animal models *in vivo*. Some transplacental passage of 5 and 30 nm radiolabeled gold NPs has been demonstrated in rats [40], whereas placental impermeability to 2 or 40 nm has been reported in mice [28].

Silver Nanoparticles

Silver NPs are being widely used in many consumer products due to their antibacterial properties. The only data reporting on the effect of silver NPs on mammalian embryonic development were obtained by culturing mouse pre-implantation embryos *in vitro* using relatively high concentrations of 25 and 50 μM [41]. Incubation for 24 h of morula stage embryos with silver NPs impaired their ability to reach the blastocyst stage, increased the percentage of apoptotic cells and reduced ability to undergo *in vitro* outgrowth. In addition, when blastocysts pre-incubated with 50 μM silver NPs were transferred to the uterus of pseudo-pregnant females, implantation success was significantly decreased compared to non-exposed blastocysts. Also, the percentage of implanted embryos that failed to develop normally was markedly higher in the treated group. Comparison of same-size gold and silver nanoparticle embryotoxicity in the zebra fish model demonstrated that silver is more toxic than gold, suggesting that nanoparticle effects depend more on surface chemical composition than surface area. In addition, a time and dose dependence of the embryotoxic response was observed [36]. Twenty-four hours post-fertilization, smaller NPs induced the highest toxicity and, after 96 h, all embryos were dead, independent of size thus indicating a size-independent toxic effect at longer term. Several data are available on the effect of silver NPs on embryonic development of other aquatic species [42–44]. Apparently, embryonic development did not appear impaired over a wide range of low concentrations but was completely blocked at a threshold concentration. In comparison, the adults responded to the particles in a dose-dependent manner [44]. These results suggest that silver NPs possess embryotoxic potential, also in non-mammalian species. In contrast with gold, silver NPs seem to have a pronounced developmental toxicity.

Carbonaceous Particles

Carbon-based NPs present as a heterogeneous class of particles with very different structure whose common feature is a regular organization of carbon atoms.

Carbon Nanotubes

Despite being considered among the most promising ENP for several applications (from industry to biomedicine), very scant information on carbon nanotube embryotoxic potential is available. Recently, it has been demonstrated that single walled carbon nanotubes (SWCNTs) can interfere with embryonic development in mice [45]. When either pristine or oxidized nanotubes were intravenously administered to pregnant mice soon after implantation (day 5.5 post coitum), early miscarriages and fetal malformations were observed, with the oxidized material having the highest abortive or teratogenic effect, suggesting that the presence of oxygenated functionalities was partly responsible for the outcome. Developmental alterations were observed at concentrations as low as 100 ng/mouse, and consisted of growth retardation, head and abdominal wall deformities, limb hypoplasia, and abnormal torsion of the trunk. These characteristics were associated with extensive placental vascular lesions and increased production of reactive oxygen species (ROS). In the same study, SWCNT embryotoxicity was evaluated in parallel *in vitro* experiments using the embryonic stem cell test (EST), a validated assay developed for predicting embryotoxicity of soluble chemical compounds [38]. The EST predicted the *in vivo* data, identifying oxidized SWCNTs as the more toxic compound. A summary of the results is provided in Figure 13.3. Intravenous exposure of male mice to multiwalled carbon nanotubes (MWCNTs) (5 mg/kg once every 3 days for 15 days) followed by pairing with naïve female mice, 15 or 60 days after the first injection, did not affect delivery, litter size and offspring viability compared to pregnancies fathered by control mice. MWCNTs were therefore not associated with male-mediated developmental toxicity [12]. The effect of carbon nanotubes on development has also been studied in the zebra fish model. When chorion-intact zebra fish embryos were incubated with pristine SWCNTs, no sign of toxicity was observed [46]. Ultra-microscopy revealed that SWCNT were agglomerated and unable to cross the chorion [35]. In embryos manually dechorionated, SWCNTs were observed on the inner layer of the chorion, suggesting the chorion forms an efficient barrier to these ENP. The only effect of exposure was a delay in hatching, which did not influence embryo development. No hatching delay was observed when embryos were incubated with carbon black as control [46]. Injection of fluorescent-labeled MWCNTs in embryos at the one cell stage or in embryos 72 h post fertilization, however, induced acute and long-term adverse effects [47]. The effect of carbon nanotubes on mammalian embryonic cells was also investigated *in vitro* using mouse embryonic stem cells [48]. Exposure of embryonic cells to MWCNTs reduced expression of stem cell markers

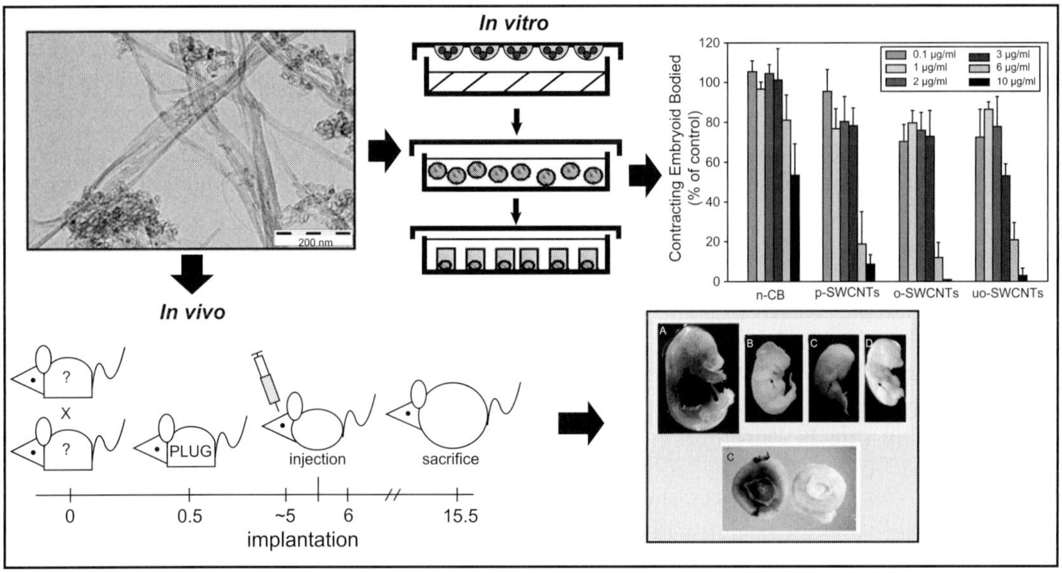

FIGURE 13.3 The figure summarizes the experimental approach and shows the results obtained after administration of SWCNTs to pregnant mice and to mouse embryonic stem cells. *In vitro* and *in vivo* experiments were carried out in parallel to assess the embryotoxic potential of pristine (p-SWCNTs) and oxidized (o- and uo-SWCNTs) carbon nanotubes. *(Reprinted with permission from [45] and courtesy of American Chemical Society.)*

after 2, 4 and 24 h, while the NPs entered the cells. In addition, cell damage, including DNA damage, was observed [47].

Fullerenes and Fulleroles

Fullerenes are a class of NPs with a peculiar spherical cage structure usually composed of 60 carbon atoms (C_{60}). Similarly to carbon nanotubes, also for fullerenes not much information about their embryotoxic potential is available. A pioneering study investigated the effect of fullerenes on mouse development [49]. Fullerene C_{60} solubilized in poly(vinylpyrrolidone) was intraperitoneally injected at day 10 of gestation, in doses ranging from 25 to 137 mg/kg. Eighteen hours later, C_{60} was found to distribute in the embryo, and to induce severe structural abnormalities leading to death at the highest dose. At lower dosages, C_{60} caused abnormalities especially around the head region and tail, the lowest effective dose being 25 mg/kg. Analysis of the yolk sac revealed narrowing of blood vessels which might account for the observed alterations. Limitations to this study appear, as the doses used in these experiments were extremely high and only two mice per group were included [49]. Most of the data on fullerenes and embryonic development have been obtained in aquatic species to investigate their eco-toxicological potential on ground water pollution, and very little information on mammalian development is available. For example, fullerenes have been demonstrated to cause hatching delay, decreased heartbeat, pericardial edema and ultimately death in zebra fish embryos at two different concentrations (1.5 mg/L and 200 µg/L), while much higher concentrations of fullerols appeared ineffective [48,51].

Carbon Black

Carbon black (CB) is a common denomination for particles with a carbonaceous core, produced by incomplete combustion of heavy petroleum products and used as a black pigment. It has been used extensively as a model for diesel emission particles without adhering hydrocarbon and metals, and thus some mammalian studies exist for developmental toxicity. Effects of maternal pulmonary exposure to 14 nm carbon black on gestation, lactation and genetic damage were evaluated. Time mated mice were exposed by: (1) inhalation of 42 mg/m^3 for 1 h/day on gestation days 8–18; or (2) intratracheal instillations, on gestation days 7, 10, 15 and 18, with total doses of 11, 54 and 268 μg/animal. Neither inhalation nor instillation affected gestational and lactational measures. Inhalation exposure to particles increased levels of DNA strand breaks in the liver of mothers and offspring, whereas intratracheal instillation did not [52]. In another study, pregnant mice were exposed to 14 nm carbon particles by intratracheal instillation at day 7 and 14 of gestation. Each dose was 0.2 mg. No effects were observed for gestational parameters, e.g. length of gestation, litter size and birth weight. Reproductive end-points were evaluated in male offspring at the ages of 5, 10 and 15 weeks. The testes from exposed offspring displayed significantly more seminiferous tubule damage and decreased daily sperm production compared to controls. Sperm morphology and serum testosterone levels were comparable to control values [33]. The potential involvement of maternal gestational exposure to respirable-sized carbon black in development of allergy was studied in the same animal model as described above for TiO_2, although intratracheal instillation dose for the pregnant mice was five times higher, i.e. 250 μg CB. As for TiO_2, the exposed offspring developed a more pronounced asthmatic phenotype. A similar phenotype was observed after maternal exposure to 50 μg diesel exhaust particles [25].

Polystyrene Nanoparticles

Among nanostructured polymeric materials, polystyrene has attracted much interest for its specific characteristics such as stiffness, transparency and easy processability. Polystyrene beads have been proposed for drug delivery, protein adsorption, and DNA transfection and are considered to be among the safest particles for experimental use. A recent report [53] demonstrated that mixed-size (between 40 and 120 nm) fluorescent polystyrene-based NPs were internalized by 2-cell mouse embryos. Initially, particles were localized in the cytoplasm but, at later stages, they translocated to the nuclei. Up to the blastocyst stage (the latest stage investigated), no impairment in the development was observed and, although a preferential localization of the NPs to the trophoblast cells was recorded, no significant difference in the ability of blastocysts to implant was evident between exposed and non-exposed groups [53]. Developmental toxicity of polystyrene NPs has also been investigated in post-implantation mouse embryos. Embryos were collected 7.5 days post coitum, i.e. the beginning of mouse gastrulation, and injected via the extra-embryonic tissues with different sizes (20, 100, 200 and 500 nm) of amine- and carboxyl-functionalized fluorescent polystyrene NPs. After 12 h of incubation, smaller carboxyl-functionalized NPs (20 nm) were distributed in all three germ layers of both the extra-embryonic tissues and the embryo proper, while bigger particles ($\geq$100 nm) distributed exclusively in the extra-embryonic compartment. In contrast, 200 nm amine-modified beads were able to access the embryonic tissues and induced growth retardation, thus again indicating that surface chemistry plays a primary role for NP behavior and toxicity [54]. In an *in vitro* study, mouse embryonic stem cells were incubated with two different concentrations (0.4 and 12.2 million/mL) of mixed-size (40–120 nm) fluorescently-labeled particles for 6 days, under conditions that would lead to differentiation of the cells. After one day, a slight reduction in cell viability was recorded. At 6 days, nanoparticle influence in the type of differentiation was evident, since a shift from a squamous to a fibroblast and columnar phenotype was observed. Cytoskeletal reorganization was suggested to be responsible for this shift [55]. Overall, data indicate the possibility that polystyrene NPs might interfere with proper differentiation of tissues. As the ability of 50–500 nm fluorescent polystyrene NPs to cross the blood–placental barrier has been demonstrated in the *ex vivo* model of human placental transfusion [29], possible interference of polystyrene NPs with embryonic development *in vivo* needs to be properly assessed.

GENERAL DISCUSSION

ENP of various types are able to influence reproduction and development in several ways, as it appears from the above synthesis. Regarding reproductive function, experimental data suggest that ENP are able to escape the blood–testis barrier, and ENP have been observed in female reproductive organs. Exposure to some ENP interferes adversely with important aspects of reproductive function, including spermatogenesis, establishment of pregnancy, and endocrine homeostasis—also after exposure through the airways. Effects might be transient, as one study in males indicates that the seminiferous epithelium is able to repair damage induced by carbon nanotubes when followed up [12]. Effects of ENPs on central regulation of sex hormones that might indirectly interfere with spermatogenesis and oogenesis have not been reported so far. However, polyelectrolyte multilayer coated gold NPs (a potential nano-drug) have been reported to accumulate among others in the hypothalamus [56]. Interaction of ENP with the hypothalamus–hypophysis–gonad axes might therefore warrant investigation.

Most ENP have been shown to interfere with embryonic development, irrespective of size and chemical composition. Prenatal exposure to TiO_2 NPs influenced development of the offspring nervous system [34] and propensity to develop allergy [25]. Silver NPs appear to seriously impair embryonic development both in mammalian and non-mammalian species [36,41,44]. For carbonaceous nanoparticles, SWCNTs interfered adversely with mouse embryonic development, fullerenes induced structural and functional abnormalities in mouse and zebra fish embryos [49,51] and inhalation of carbon black affected DNA in mouse offspring [52]. Gold NPs appear to be of low embryotoxic potential, as functionalized and non-functionalized particles only slightly affected zebra fish development [35,36]. The sensitivity of male reproductive organs and function to maternal particulate exposure has been investigated for both TiO_2 and CB in mice. It is of major concern that this organ system shows sensitivity over a range of particle sizes as well as different routes of maternal exposure [31,33].

Several of the *in vivo* studies on developmental toxicity used the zebra fish model. In this model, large numbers of embryos can be generated rapidly at low cost and the entire embryo development takes only 4 days. Development occurs outside the mother, allowing direct visualization of morphological changes in the offspring. In addition, genes involved in zebra fish development have human orthologs and genetic pathways between the two classes are often conserved. Although development and embryo morphology might not compare directly to that of mammals, zebra fish might represent a valuable tool to screen for NP embryotoxicity and give useful indications for risk assessment of NPs in humans. The lack of a placenta in this model precludes its exploitation for studying transfer of NP from mother to fetus, a central aspect in environmental and occupational risk assessment of developmental toxicity. The *ex vivo* placental perfusion system might overcome this limitation. Two reports have already exploited such a system to study the mother–fetus transfer of gold and polystyrene NPs, showing that the ability of NPs to cross the blood–placental barrier was not size dependent, as polystyrene NPs up to 240 nm were able to reach the fetal side of the organ, while much smaller gold NPs (30 nm) were confined to the maternal side [29,39].

The concept of engineering of nanomaterials is too recent for focused epidemiological studies to exist. However, the association between exposure to ambient air (and thus ultrafine) particles and adverse effects on especially fetal development has been investigated. Shah and Balkhair systematically reviewed the association between air pollutants and birth outcomes based on 41 studies with well-defined maternal exposures and birth outcomes. Fine particulate matter ($\leq 2.5\,\mu m$) was associated with fetotoxicity, i.e. low birth weight, preterm birth and being born small for gestational age [57]. As such, the experimental studies on prenatal exposure to ENP do not suggest fetotoxic potential of ENP *per se*, at least not for maternal airway exposure [33,34,52]. The basis for this conclusion is faint, however. Even if easily available, most mammalian studies on ENP do not provide information on common maternal and gestational measures (e.g. maternal weight gain, litter size, birth weights, length of gestation). In animals, rather many studies have investigated the effects of prenatal exposure to diesel exhaust and reported reduced body weight, altered immunological measures, increased frequency of DNA deletions, changes in sexual differentiation and sperm production in the offspring [58]. Most studies used whole diesel exhaust, i.e. a mixture of particulates and gases. A few studies resuspended diesel exhaust particles, generally corroborating previous findings [58–60]. However, the relative roles of diesel exhaust particles, associated chemicals and exhaust gases in developmental toxicity remain to be clarified.

Mechanistically, exposure to ENP may potentially affect reproduction and development through several modes of action. Toxicity can arise due to toxicological properties related to specific chemical constituents of the particles, as ENP are often enriched with functional groups, or due to contaminants secondary to their production. Dissociation of these or of molecules from the particulate core might lead to toxicity, e.g. after placental transfer. As an example, the embryotoxic effect of silver NPs has been proposed to be a consequence of silver ions released secondarily to *in vivo* nanoparticle destabilization [36]. The placenta acts as a barrier for some but not all NPs, and particles have been identified in offspring tissues several weeks after birth ([28,29,31] and summarized in [30]). Direct effects of particles on fetal development following maternal exposure are therefore possible, although the mere physical presence of particles does not necessarily interfere with development [36,61]. However, many ENP induce oxidative stress which poses a potential threat to germ cell, fetal and placental DNA integrity. Some NPs have been demonstrated to interact with proteins involved in DNA mismatch repair [62] and altered DNA mismatch-repair machinery has been reported to cause germ line mutations [63]. Placental crossing of ENP or even presence of particles in maternal blood might not be necessary in order for developmental toxicity to occur. NPs have been shown to induce DNA damage without direct contact with the affected tissue. The underlying mechanism of such external exposure is suggested to involve alterations in signaling across tissue barriers [64]. Finally, lung exposure to particulates might induce airway inflammation (see Chapter 7) with the resulting mediators representing potential adverse agents for reproductive and fetal development [34]. Interestingly, the lung inflammatory response to particles seems enhanced in pregnancy [25,26].

Published reports present a great diversity of study designs, e.g. with respect to choice of ENP and surface modifications, model systems and animal species, dose levels, routes of exposure and end-points. This diversity makes it difficult to deduce general rules as to the reproductive and developmental toxicity of ENP. Thus, changes in hippocampal electrophysiological signals suggests that developmental exposure to particulate TiO_2 in rats interferes with brain development that would be predicted to manifest as changes in cognitive function in offspring [32]. However, when cognitive function was assessed in offspring of mothers inhaling TiO_2 during gestation, exposed offspring performed as well as control offspring [34]. This discrepancy might occur due to differences in exposure, with respect to both dose levels and route of exposure [32]. Also, lactational exposure seemed to have a larger impact on hippocampal recordings than did gestational exposure. This could be due to a relatively increased transfer of the compound from mother to offspring during lactation or to increased sensitivity of the rat developing nervous system in lactation compared to the prenatal period. The latter is very important when translating these results to humans. Rodents are relatively delayed in CNS development at birth compared to humans and, at birth, a rat compares to a human fetus approximately two trimesters into pregnancy [65]. Increased sensitivity during lactation in the rat would therefore indicate that the human fetus would display increased sensitivity during the last gestational trimester and shortly after birth.

Overall, available data on the effect of ENP on male and female fertility as well as on developmental toxicity suggest that ENP posses the potential to interfere with fertility and development. The diversity in applied study designs and investigated ENP hampers deduction of general rules as to reproductive and developmental toxicity of ENP. For ENP in general, NP size or surface area is considered pivotal for the adverse response [66]. However, findings in several experimental studies indicate that response does not depend strictly on size. Furthermore, surface chemistry seems important in some [54], but not other cases [12,36]. Also, the core material might be the primary determinator of toxicity, e.g. for silver particles [36]. As reported above, in the *ex vivo* model of placental transfer, 20–25 nm gold particles coated with polyethyleneglycol did not pass the placenta [39], whereas much larger polystyrene beads passed with almost no hindrance [29]. Finally, one study suggests that the route of exposure is crucial for developmental toxicity [52]. This will definitely be the case for reproductive and developmental toxicity suggested to occur due to particle-induced lung inflammation [34]. At present, all nanomaterials developed for industrial or biomedical purposes should therefore in principle be tested for adverse effects on reproduction and development. Accidental exposures should be avoided, and use should occur consciously under controlled conditions, as data suggest several NPs interfere with the reproductive apparatus even after airway exposure.

In protection against airborne exposure to chemicals in the workplace, occupational exposure limits are pivotal. Awaiting the full implementation of Registration, Evaluation, Authorization and Restriction of Chemicals (REACH) in the European Union (see Chapter 6), limit values are in the EU set such that "exposure, even when repeated on a regular basis throughout a working life, will not lead to adverse effects on the health of exposed persons and/or their progeny at any time" [2]. In principle, limit values therefore also protect the unborn child and future generations. However, limit values are "predicted from the contemporary state of knowledge", which evidently is scarce for ENP and reproductive and developmental toxicity. Even if only limited data are available, "The absence of relevant data [on reproductive and developmental toxicity] will not normally be a factor in determining the size of an Uncertainty Factor" [2]. Furthermore, nanomaterials are at present regulated similarly to the bulk material. Occupational exposure limits therefore provide little protection against potential reproductive and developmental toxicity of ENP. Once pregnancy is a fact and "official", however, employers are obliged to assess risks and to remove, avoid, or reduce risks for pregnancy and the unborn child. These policies ignore that men and women might be exposed to reproductive toxicants while aiming to establish pregnancy, as well as the lag of time from conception to acknowledgement of pregnancy [67]. ENP that are non-biodegradable and resistant to excretion might present a special case, if particles accumulate in the body during chronic exposure. This might translate into a slow release during pregnancy, even if exposure during pregnancy is avoided. Although speculative, it has been suggested for nervous tissue, that long-term and low-dose exposure might lead to chronic inflammation-like conditions [68] (see also Chapters 1, 4 and 9), which might prove detrimental for some reproductive processes.

FUTURE RESEARCH NEEDS

At present, most research within reproductive and developmental toxicity of ENP must be categorized as hypothesis generating rather than directed towards risk assessment. Future research ought to address more systematically reproductive and developmental toxicity of nanomaterials. Essential goals of research in reproductive and developmental toxicity of ENP must be to: (1) identify most sensitive end-points; and (2) to deduce general rules as to the relative role of, for example, particle size and surface area, surface chemistry, core material, and route of exposure. Better knowledge of potential mechanisms might prove important in this task. Whether the mere presence of particles interferes with reproductive tissue and tissues undergoing development is especially important to clarify. Specifically, there is a lack of *in vivo* studies investigating several dose levels and comparing nano-sized with micro-sized particles.

Reproductive function is one of the main topics that needs to be addressed when assessing the toxic potential of ENP. Very little work has been done on particle effects on female reproductive capacity, and most of the information has been obtained *in vitro*. For the male reproductive system, concerns relate to the ability to bypass the blood–testis barrier and accumulate within the seminiferous epithelium and affect spermatogenesis, either directly or indirectly. It is also imperative to establish if effects are permanent or can be reverted with time, and if toxicants affect germ line cells through inheritable mutations or subtle epigenetic modifications (e.g. DNA methylation) [5].

Finally, important data could be provided without much effort if reporting included information on common maternal and gestational measures (e.g. maternal weight gain, litter size, birth weights), and if study designs adhered to commonly accepted practices in reproductive and developmental toxicity, e.g. by consulting guidelines and established protocols before initiating experiments [23,69].

TAKE-HOME MESSAGES

The possibility that ENP might adversely affect embryonic development appears real, although in most cases information has been obtained indirectly from *in vitro* experiments or *in vivo* from non-mammalian species. Several studies suggest fertility is sensitive to nanomaterials. Systematic investigation of the reproductive and developmental toxicity of ENP is therefore warranted and urgently needed.

References

[1] Lawson CC, Schnorr TM, Daston GP, et al. An occupational reproductive research agenda for the third millennium. Environ Health Perspect 2003;111:584–92.

[2] Scientific Committee on Occupational Exposure Limits (SCOEL). Methodology for the derivation of occupational exposure limits: Key documentation (version 6). Bruxelles: European Commission; 2009. p 1–39.

[3] Pryor JL, Hughes C, Foster W, Hales BF, Robaire B. Critical windows of exposure for children's health: The reproductive system in animals and humans. Environ Health Perspect 2000;108(Suppl. 3):491–503.

[4] Lucas B, Fields C, Hofmann MC. Signaling pathways in spermatogonial stem cells and their disruption by toxicants. Birth Defects Res Part C 2009;87:35–42.

[5] Singer TM, Yauk CL. Germ cell mutagens: Risk assessment challenges in the 21st century. Environ Mol Mutagen 2010;51:919–28.

[6] Yoshida S, Sagai M, Oshio S, et al. Exposure to diesel exhaust affects the male reproductive system of mice. Int J Androl 1999;22:307–15.

[7] Yauk C, Polyzos A, Rowan-Carroll A, et al. Germ-line mutations, DNA damage, and global hypermethylation in mice exposed to particulate air pollution in an urban/industrial location. Proc Natl Acad Sci USA 2008;105:605–10.

[8] Kwon JT, Hwang SK, Jin H, et al. Body distribution of inhaled fluorescent magnetic nanoparticles in the mice. J Occup Health 2008;50:1–6.

[9] Kim JS, Yoon TJ, Yu KN, et al. Toxicity and tissue distribution of magnetic nanoparticles in mice. Toxicol Sci 2006;89:338–47.

[10] de Jong WH, Hagens WI, Krystek P, Burger MC, Sips AJ, Geertsma RE. Particle size-dependent organ distribution of gold nanoparticles after intravenous administration. Biomaterials 2008;29:1912–9.

[11] Yoshida S, Hiyoshi K, Ichinose T, et al. Effect of nanoparticles on the male reproductive system of mice. Int J Androl 2009;32:337–42.

[12] Bai Y, Zhang Y, Zhang J, et al. Repeated administrations of carbon nanotubes in male mice cause reversible testis damage without affecting fertility. Nat Nanotechnol 2010;5:683–9.

[13] Bal R, Turk G, Tuzcu M, et al. Protective effects of nanostructures of hydrated C(60) fullerene on reproductive function in streptozotocin-diabetic male rats. Toxicology 2011;282:69–81.

[14] Srdjenovic B, Milic-Torres V, Grujic N, Stankov K, Djordjevic A, Vasovic V. Antioxidant properties of fullerenol C60(OH)24 in rat kidneys, testes, and lungs treated with doxorubicin. Toxicol Mech Methods 2010;20:298–305.

[15] Braydich-Stolle L, Hussain S, Schlager JJ, Hofmann MC. *In vitro* cytotoxicity of nanoparticles in mammalian germline stem cells. Toxicol Sci 2005;88:412–9.

[16] Braydich-Stolle LK, Lucas B, Schrand A, et al. Silver nanoparticles disrupt GDNF/Fyn kinase signaling in spermatogonial stem cells. Toxicol Sci 2010;116:577–89.

[17] Komatsu T, Tabata M, Kubo-Irie M, et al. The effects of nanoparticles on mouse testis Leydig cells *in vitro*. Toxicol In Vitro 2008;22:1825–31.
[18] Wiwanitkit V, Sereemaspun A, Rojanathanes R. Effect of gold nanoparticles on spermatozoa: The first world report. Fertil Steril 2009;91:e7–8.
[19] Makhluf SB, bu-Mukh R, Rubinstein S, Breitbart H, Gedanken A. Modified PVA-Fe3O4 nanoparticles as protein carriers into sperm cells. Small 2008;4:1453–8.
[20] Zhang H, Yang D, Yang H, Liu H. Effect on conception and offspring development in female parenatal rats following intratracheal instillation of nano-C/ZnO and C-ZnO composite nanoparticles. Weisheng Yanjiu 2008;37:654–6. (Article in Chinese, abstract in English).
[21] Stelzer R, Hutz RJ. Gold nanoparticles enter rat ovarian granulosa cells and subcellular organelles, and alter *in-vitro* estrogen accumulation. J Reprod Dev 2009;55:685–90.
[22] Liu X, Qin D, Cui Y, et al. The effect of calcium phosphate nanoparticles on hormone production and apoptosis in human granulosa cells. Reprod Biol Endocrinol 2010;8:32.
[23] United States Environmental Protection Agency. Guidelines for developmental toxicity risk assessment. Fed Regist 1991;56:63798–826.
[24] Powers CM, Badireddy AR, Ryde IT, Seidler FJ, Slotkin TA. Silver nanoparticles compromise neurodevelopment in PC12 cells: Critical contributions of silver ion, particle size, coating, and composition. Environ Health Perspect 2011;119:37–44.
[25] Fedulov AV, Leme A, Yang Z, et al. Pulmonary exposure to particles during pregnancy causes increased neonatal asthma susceptibility. Am J Respir Cell Mol Biol 2008;38:57–67.
[26] Lamoureux DP, Kobzik L, Fedulov AV. Customized PCR-array analysis informed by gene-chip microarray and biological hypothesis reveals pathways involved in lung inflammatory response to titanium dioxide in pregnancy. J Toxicol Environ Health A 2010;73:596–606.
[27] Saunders M. Transplacental transport of nanomaterials. Wiley Interdiscip Rev Nanomed Nanobiotechnol 2009;1:671–84.
[28] Sadauskas E, Wallin H, Stoltenberg M, et al. Kupffer cells are central in the removal of nanoparticles from the organism. Part Fibre Toxicol 2007;4:10.
[29] Wick P, Malek A, Manser P, et al. Barrier capacity of human placenta for nanosized materials. Environ Health Perspect 2010;118:432–6.
[30] Hougaard KS, Fadeel B, Gulumian M, Kagan VE, Savolainen K. Developmental toxicity of engineered nanoparticles. In: Gupta R, editor. Reproductive and developmental toxicology. San Diego: Academic Press/Elsevier; 2011. p. 269–90.
[31] Takeda K, Suzuki K, Ishihara A, et al. Nanoparticles transferred from pregnant mice to their offspring can damage the genital and cranial nerve systems. J Health Sci 2009;55:95–102.
[32] Gao X, Yin S, Tang M, et al. Effects of developmental exposure to TiO(2) nanoparticles on synaptic plasticity in hippocampal dentate gyrus area: An *in vivo* study in anesthetized rats. Biol Trace Elem Res 2011.
[33] Yoshida S, Hiyoshi K, Oshio S, Takano H, Takeda K, Ichinose T. Effects of fetal exposure to carbon nanoparticles on reproductive function in male offspring. Fertil Steril 2009;93:1695–9.
[34] Hougaard KS, Jackson P, Jensen KA, et al. Effects of prenatal exposure to surface-coated nanosized titanium dioxide (UV-Titan). A study in mice. Part Fibre Toxicol 2010;7:16.
[35] Browning LM, Lee KJ, Huang T, Nallathamby PD, Lowman JE, Xu XH. Random walk of single gold nanoparticles in zebrafish embryos leading to stochastic toxic effects on embryonic developments. Nanoscale 2009;1:138–52.
[36] Bar-Ilan O, Albrecht RM, Fako VE, Furgeson DY. Toxicity assessments of multisized gold and silver nanoparticles in zebrafish embryos. Small 2009;5:1897–910.
[37] Wang Y, Seebald JL, Szeto DP, Irudayaraj J. Biocompatibility and biodistribution of surface-enhanced Raman scattering nanoprobes in zebrafish embryos: *In vivo* and multiplex imaging. ACS Nano 2010;4:4039–53.
[38] Di GC, Lopez DR, De Lapuente J, Mallafre JM, Suarez MB. Embryotoxicity of cobalt ferrite and gold nanoparticles: A first *in vitro* approach. Reprod Toxicol 2010;30:271–6.
[39] Myllynen PK, Loughran MJ, Howard CV, Sormunen R, Walsh AA, Vahakangas KH. Kinetics of gold nanoparticles in the human placenta. Reprod Toxicol 2008;26:130–7.
[40] Takahashi S, Matsuoka O. Cross placental transfer of 198Au-colloid in near term rats. J Radiat Res 1981;22:242–9.

[41] Li PW, Kuo TH, Chang JH, Yeh JM, Chan WH. Induction of cytotoxicity and apoptosis in mouse blastocysts by silver nanoparticles. Toxicol Lett 2010;197:82–7.
[42] Wu Y, Zhou Q, Li H, Liu W, Wang T, Jiang G. Effects of silver nanoparticles on the development and histopathology biomarkers of Japanese medaka (*Oryzias latipes*) using the partial-life test. Aquat Toxicol 2010;100:160–7.
[43] Laban G, Nies LF, Turco RF, Bickham JW, Sepulveda MS. The effects of silver nanoparticles on fathead minnow (*Pimephales promelas*) embryos. Ecotoxicology 2010;19:185–95.
[44] Ringwood AH, Levi-Polyachenko N, Carroll DL. Fullerene exposures with oysters: Embryonic, adult, and cellular responses. Environ Sci Technol 2009;43:7136–41.
[45] Pietroiusti A, Massimiani M, Fenoglio I, et al. Low doses of pristine and oxidized single-wall carbon nanotubes affect mammalian embryonic development. ACS Nano 2011;5:4624–33.
[46] Cheng J, Flahaut E, Cheng SH. Effect of carbon nanotubes on developing zebrafish (*Danio rerio*) embryos. Environ Toxicol Chem 2007;26:708–16.
[47] Cheng J, Chan CM, Veca LM, et al. Acute and long-term effects after single loading of functionalized multi-walled carbon nanotubes into zebrafish (*Danio rerio*). Toxicol Appl Pharmacol 2009;235:216–25.
[48] Zhu X, Zhu L, Li Y, Duan Z, Chen W, Alvarez PJ. Developmental toxicity in zebrafish (*Danio rerio*) embryos after exposure to manufactured nanomaterials: Buckminsterfullerene aggregates (nC60) and fullerol. Environ Toxicol Chem 2007;26:976–9.
[49] Tsuchiya T, Oguri I, Yamakoshi YN, Miyata N. Novel harmful effects of [60]fullerene on mouse embryos *in vitro* and *in vivo*. FEBS Lett 1996;393:139–45.
[50] Yadav BC, Kumar R. Structure, properties and applications of fullerenes. Int J Nanotech Appl 2008; 2:15–24.
[51] Usenko CY, Harper SL, Tanguay RL. *In vivo* evaluation of carbon fullerene toxicity using embryonic zebrafish. Carbon NY 2007;45:1891–8.
[52] Jackson P, Hougaard KS, Boisen AM, et al. Pulmonary exposure to carbon black by inhalation or instillation in pregnant mice: Effects on liver DNA strand breaks in dams and offspring. Nanotoxicology 2011. in press.
[53] Bosman SJ, Nieto SP, Patton WC, Jacobson JD, Corselli JU, Chan PJ. Development of mammalian embryos exposed to mixed-size nanoparticles. Clin Exp Obstet Gynecol 2005;32:222–4.
[54] Tian F, Razansky D, Estrada GG, et al. Surface modification and size dependence in particle translocation during early embryonic development. Inhal Toxicol 2009;21:92–6.
[55] Tran DN, Ota LC, Jacobson JD, Patton WC, Chan PJ. Influence of nanoparticles on morphological differentiation of mouse embryonic stem cells. Fertil Steril 2007;87:965–70.
[56] Sousa F, Mandal S, Garrovo C, et al. Functionalized gold nanoparticles: A detailed *in vivo* multimodal microscopic brain distribution study. Nanoscale 2010;2:2826–34.
[57] Shah PS, Balkhair T. Air pollution and birth outcomes: A systematic review. Environ Int 2011;37:498–516.
[58] Hougaard KS, Jensen KA, Nordly P, et al. Effects of prenatal exposure to diesel exhaust particles on postnatal development, behavior, genotoxicity and inflammation in mice. Part Fibre Toxicol 2008;5:3–18.
[59] Hemmingsen JG, Hougaard KS, Talsness C, et al. Prenatal exposure to diesel exhaust particles and effect on the male reproductive system in mice. Toxicology 2009;264:61–8.
[60] Ritz C, Ruminski W, Hougaard KS, Wallin H, Vogel U, Yauk CL. Germline mutation rates in mice following *in utero* exposure to diesel exhaust particles by maternal inhalation. Mut Res 2011; 712:55–58.
[61] Park MV, Annema W, Salvati A, et al. *In vitro* developmental toxicity test detects inhibition of stem cell differentiation by silica nanoparticles. Toxicol Appl Pharmacol 2009;240:108–16.
[62] Gupta K, Baweja L, Gurbani D, Pandey AK, Dhawan A. Interaction of C60 fullerene with the proteins involved in DNA mismatch repair pathway. J Biomed Nanotechnol 2011;7:179–80.
[63] Tijsterman M, Pothof J, Plasterk RH. Frequent germline mutations and somatic repeat instability in DNA mismatch-repair-deficient *Caenorhabditis elegans*. Genetics 2002;161:651–60.
[64] Parry MC, Bhabra G, Sood A, et al. Thresholds for indirect DNA damage across cellular barriers for orthopaedic biomaterials. Biomaterials 2010;31:4477–83.
[65] Morgane PJ, Austin-LaFrance RJ, Bronzino JD, Tonkiss J, Galler JR. Malnutrition and the developing central nervous system. In: Isaacson RL, Jensen KF, editors. The vulnerable brain and environmental risks. New York: Plenum Press; 1992. p. 3–44.
[66] Nel A, Xia T, Madler L, Li N. Toxic potential of materials at the nanolevel. Science 2006;311:622–7.

[67] European Agency for Safety and Health at Work. Gender issues in safety and health at work. A review. In: Kauppinen K, Kumpulainen R, Houtman I, Copsey S, editors. Luxemburg: Office for Official Publications of the European Communities; 2003.
[68] Simko M, Mattsson MO. Risks from accidental exposures to engineered nanoparticles and neurological health effects: A critical review. Part Fibre Toxicol 2010;7:42–57.
[69] Parker RM. Testing for reproductive toxicity. In: Hood RD, editor. Developmental and reproductive toxicology. A practical approach. LLC, Boca Raton: Taylor & Francis Group; 2006. p. 425–87.

CHAPTER

14

Genotoxicity and Cancer

Shareen H. Doak[1], *Ying Liu*[2], *Chunying Chen*[2]

[1]Institute of Life Science, College of Medicine, Swansea University, Swansea, Wales, UK

[2]National Centre for Nanoscience and Technology, Beijing, China

OUTLINE

INTRODUCTION

Genetic toxicology is the study of chemical and physical perturbations in DNA following exposure to endogenous or exogenous agents. The damage may consist of small lesions arising at very specific sites within the DNA molecule, such as DNA adducts, DNA cross-links, abasic sites and point mutations. Alternatively, gross abnormalities can arise at the chromosomal level including aneuploidy (alterations in chromosome copy number), clastogenicity (chromosome fragmentation) and structural chromosomal rearrangements. The type of DNA damage induced by a genotoxic substance is entirely dependent upon its mechanism of action. This has been well characterized for many chemical or pharmaceutical products humans may be exposed to, for example S_N2 alkylating agents have a tendency to target highly nucleophilic centers within the DNA macromolecule and therefore largely induce DNA alkyl-adducts at sites such as the N7 position in guanine bases or the N3 position in adenine [1,2]. However, for nanomaterials this level of understanding is lacking at present.

Adverse Effects of Engineered Nanomaterials
DOI: 10.1016/B978-0-12-386940-1.00014-3

The mechanisms likely to be responsible for nanomaterial-induced genotoxicity fall into two main categories, namely primary and secondary mechanisms, the former of which can be further subdivided into direct and indirect acting agents (Figure 14.1).

Primary mechanisms for genotoxicity are those that are imparted by the nanomaterials themselves at the level of the single cell. It requires cellular internalization of the nanomaterials and often their subsequent interaction with biomolecules.

Direct acting agents are those that are DNA reactive, i.e. they come into direct contact with the genetic material causing physical or chemical damage. This might include the formation of DNA lesions at specific sites within the DNA molecule, which can lead to mutagenesis as they may destabilize the nucleotide they damage resulting in its loss and therefore an abasic site which is often subject to error prone repair. Alternatively, during replication the damaged base may not be recognized by the polymerase enzymes resulting in a random selection of nucleotides placed opposite to enable duplication of the genetic material to continue, thus leading to point mutations or strand breakages [1,2]. Direct acting agents may also have the capacity to intercalate into the DNA double helix creating a structure that stalls DNA replication either leading to strand breaks or the insertion of additional nucleotides [3]. Alternatively, these agents may cause physical damage to the DNA in the form of single- or double-strand breaks. Direct DNA damage can therefore only be induced if the

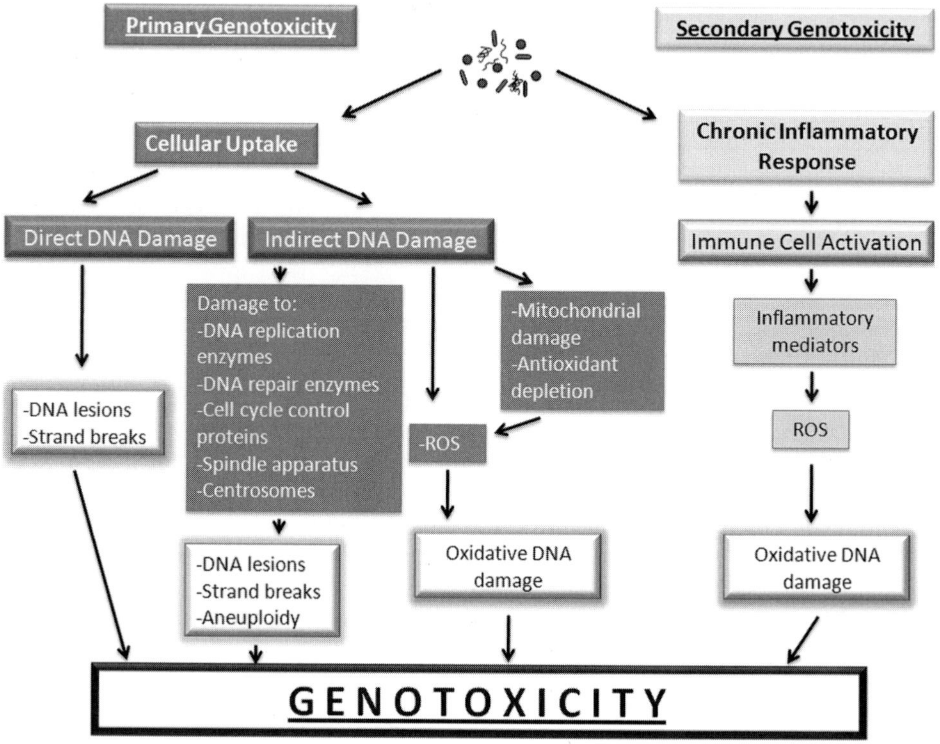

FIGURE 14.1 Mechanisms for nanomaterial-induced DNA damage.

nanomaterials are able to penetrate the cell nucleus, or those free in the cytoplasm (i.e. not contained within membrane bound vesicles) may have the opportunity to come into direct contact with DNA during mitosis when the nuclear membrane breaks down. Specific targeting of nanomaterials to the cell nucleus can be achieved by modifying their surfaces with nuclear localizing signals [4,5]. Nonetheless, most nanomaterials associated with human exposure would not have such surface functionalization and thus their uptake into the nucleus would be substantially limited, but a number of studies have observed that some nanomaterials, including titanium dioxide (TiO_2), zinc oxide (ZnO), quantum dots and silica, are able to enter the nuclear region in the absence of targeting moieties [6–10]. However, it has yet to be established whether or not nanomaterials directly interact with DNA to induce damage.

Indirect acting agents are those which induce genetic damage via intermediate biomolecules that are often components of the cell division machinery. DNA replication and cell division is a complex multifactorial process that involves a large number of proteins, which are potential targets for indirect acting genotoxins. For example, damage may occur to proteins required for the mechanical segregation of chromosomes (including the centrioles, microtubules, kinetochore structures on the chromosomes); enzymes involved in replication of the DNA (such as polymerases, topoisomerases); DNA repair enzymes or signal transduction proteins involved in cell cycle checkpoints that monitor the chromosome replication and segregation process [1]. Consequently, damage to such components often results in structural or numerical chromosomal aberrations. Aside from interfering with the DNA replication process, indirect DNA damage may arise as a result of the induction of oxidative stress. A normal by-product of aerobic metabolism is reactive oxygen species (ROS), which readily oxidize biomolecules such as proteins, lipids and DNA. To prevent oxidative damage, cellular antioxidants neutralize ROS but, if there is an imbalance between the generation of ROS and the level of antioxidants within a cell, then the result is oxidative stress. This imbalance could be caused by enhanced generation of ROS after exposure to an exogenous agent or depletion of cellular antioxidants, the result of which can be the formation of oxidative DNA lesions that are subsequently fixed as mutations [11]. Currently, there is more evidence to support an indirect mode of action for nanomaterials with regards to DNA damage. For example, single-wall carbon nanotubes (SWCNTs) have been found to interfere mechanically with the spindle apparatus, culminating in centrosome fragmentation and multipolar mitotic spindles, which were associated with aneuploidy [12]. Nanosilica is able to induce nuclear protein aggregation, which is of concern as some of these proteins include enzymes important in DNA replication processes, particularly topoisomearase I [8]. However, most studies have demonstrated that oxidative stress is a central cause for indirect DNA damage, with *in vivo* studies detecting oxidative DNA lesions (8-hydroxy-deoxyguanosine; 8-OHdG) in the liver and lungs of rodents following oral administration of nanoparticles [13,14]. Additionally, a recent study examining the consequences of 70 nm diameter silica nanoparticles on human keratinocyte cells detected 8-OHdG adducts and DNA damage that was abrogated when the cells were pretreated with N-acetylcystein (an ROS scavenger). The latter study thus provides direct evidence for the role of oxidative stress in the genotoxic responses induced by these nanoparticles [15].

Secondary mechanism for genotoxicity largely refers to the ability of a substance to induce a (chronic) inflammatory response *in vivo* that results in excessive generation of ROS by macrophage and neutrophil cells recruited to the exposure site. This is a defense mechanism within the body to protect against invading pathogens—acute inflammation often enables the removal of foreign bodies by the immune cells, but if this fails due to biopersistence then the result is chronic inflammation, promoting oxidative stress that is subsequently responsible for damaging surrounding cells within the tissue. As described in the previous section, oxidative stress generates free radicals that damage biomolecules including DNA and consequently chronic inflammation is associated with carcinogenesis [16]. In this model, the nanomaterial itself may exhibit no genotoxic potential within an *in vitro* testing system, but their physicochemical characteristics may be sufficient to promote a chronic immune response *in vivo* that, in turn, generates ROS responsible for inducing genotoxicity within the tissue cells at the nanomaterial deposition site. The classical example of such secondary genotoxicity involves inhalation of asbestos—these fibers are comprised of biopersistent materials that cannot be degraded by immune cells, their dimensions (very long and thin) prevent efficient phagocytosis, and they often contain redox-active iron. Consequently, inhalation of these fibers is associated with a chronic inflammatory response, and the persistent oxidative stress leads to secondary genotoxicity (as evidenced by 8-OHdG lesions with concomitant elevated mutagenicity) culminating in progressive fibrosis and, ultimately, mesothelioma [17–19]. Many parallels have been drawn between asbestos and carbon nanotubes due to their high aspect ratio and biopersistence. Consequently, multiwall carbon nanotubes (MWCNTs) have been reported to induce an asbestos-like inflammatory response and the formation of granulomas *in vivo* following intraperitoneal administration [20]. However, carcinogenicity studies are limited and those in the literature are conflicting with one reporting that MWCNTs promote the development of mesothelioma following the administration of a single intraperitoneal dose in p53 heterzygous mice [21], while another failed to find a carcinogenic response to MWCNTs (both with and without structural defects) in rats after a 2-year period following an intraperitoneal exposure [22]. Although, evidence specifically involving nanomaterials is limited at present, this concept of secondary genotoxicity is important because it explains why non-mutagenic materials *in vitro* are still able to induce carcinogenesis *in vivo*—an example of which is carbon black. *In vitro* studies indicated carbon black did not induce genotoxicity [23], yet *in vivo* investigations indicated inhalation of the dust was able to promote tumor formation [24,25].

While we have an understanding of the potential mechanisms by which nanomaterials may induce DNA damage, it must be noted that there are only a limited number of studies that have experimentally demonstrated the means by which genotoxicity is induced by specific nanomaterials. This area therefore requires more focus to elucidate fully the factors which govern the induction of genotoxicity by nanomaterials.

GENOTOXICITY TO CANCER

Genetic instability involving gross chromosomal abnormalities through to comparatively minute aberrations such as single base pair substitutions is closely associated with carcinogenesis and has long been detected in all malignancies. Thus, it is now widely accepted that cancer is a phenotype—the result of an accumulation of multiple genome-wide mutations that drive tumorigenesis in a multistep fashion, progressively transforming normal cells into pre-malignant foci, then subsequently to localized tumors that develop into invasive and metastatic lesions [26]. Cells are continuously subjected to potential mutagenic assaults through their exposure to environmental carcinogens, or simply as a result of normal metabolism. Genetic alterations are therefore constantly arising within a cell's genome, but stringent monitoring usually enables repair of such damage before it is transmitted to subsequent progeny cells. In extreme cases, where the genetic damage is beyond repair, the cell commits suicide (i.e. apoptosis) to prevent the advent of abnormal cell populations.

The first step in tumorigenesis (initiation) involves a carcinogen interacting with DNA and creating a permanent alteration in the cell's genome. Tumors can be initiated by a single mutation arising in an indispensable gene within an individual cell as a consequence of an exogenous insult (e.g. mediated by carcinogens) or an endogenous strike (caused by metabolic by-products such as ROS). Cells that have inherited a defective gene may then be more susceptible to tumor initiation depending upon the signaling pathway affected. If the initiating genetic abnormalities arise in vital genes involved in pathways required for maintenance of genomic integrity and cell cycle control, a cascade of further mutational events potentially results in malignant transformation.

Several types of mutations exist varying from gross chromosomal abnormalities involving numerical and structural alterations to single base pair changes, all of which have the potential to alter significantly the gene product's function. Not all genetic aberrations involved in carcinogenesis directly affect a gene's nucleotide sequence. Disruptions that alter gene expression or biochemical properties of protein products without altering the gene's DNA sequence are also important factors known as epigenetic mutations. One of the most prominent epigenetic changes that arises in tumorigenesis is hypermethylation of gene control regions that act to silence their expression. The 5′ regions of genes are relatively rich in CpG dinucleotides when compared to the rest of the human genome and these sites are unmethylated in genes that are actively expressed. However, if aberrant methylation patterns arise as a consequence of exposure to an exogenous agent, the transcriptional status of the gene may be altered resulting in a reduction or complete loss of expression [27]. Indeed, quantum dots have been found to induce epigenetic changes through hypoacetylation of histones, altering chromatin condensation, which also has the potential to modify global gene expression patterns [28].

Consequently, nanomaterials that are able to induce genotoxicity have the capacity to initiate carcinogenesis due to the alterations imposed on the genetic material. The ramifications are substantial and highlight the necessity to establish the genotoxic and carcinogenic potential of nanomaterials to which humans may be exposed (Table 14.1).

CARBON-BASED NANOMATERIALS

Carbon-based nanomaterials have become an important focus of the nanotechnology industry with many derivatives such as fullerenes, carbon nanotubes, graphene, carbon black and carbon diamond. Consequently, in recent years, a great deal of attention has been given to their toxicity, genotoxicity and carcinogenesis assessments.

Fullerene (C_{60}) has attracted increasing attention because of its unique structural, physical, chemical and electric properties, with large-scale production of C_{60} representing a milestone

TABLE 14.1 Genotoxicity and Toxicity of Nanomaterials

Nanoparticle			*In vitro/in vivo*	Genotoxicity and Toxicity Assays	Findings	Reference
Carbon-based nanomaterials	C_{60}	$Gd@C_{82}(OH)_{22}$	*In vitro* and *in vivo* HepG2 cells LLC cells MCF7 cells	CCK-8	[Gd@C82(OH)22]n nanoparticles have no toxicity	[29–32]
		C_{60}	*In vitro* and *in vivo*	ROS	C_{60} has the capacity to generate singlet oxygen that induces lipid peroxidation of linoleate which leads to oxidative DNA damage	[33–35]
		C_{60} suspension	*In vitro* *gpt* delta transgenic mouse primary embryo fibroblasts (MEF cells)	MTT mutation analysis	C_{60} induced cytotoxicity in transgenic MEF cells. Increased mutation frequencies at red/gam gene loci in response to C_{60} exposure	[36]
		Aqu/nC_{60} and EthOH/nC_{60} suspensions	*In vivo* (PBL)	Comet assay	Aqu/nC_{60} suspensions elicited higher genotoxic response than EthOH/nC_{60} at the same dose	[37]
		$C_{60}(OH)_{24}$	*In vitro* CHO cells Hela cells HEK293 cells	CCK-8 micronucleus (MN) test	Structural chromosomal damage was increased	[38]
		C_{60}	*In vitro* FE1-Mutatrade mark-Mouse lung epithelial cells	Comet assay FE1-MML mutagenicity analysis	The mutant frequency was slightly decreased; non-cytotoxic concentrations did not result in increased levels of strand breaks	[39]

CNTs	SWCNTs	*In vitro* FE1-Mutatrade mark-Mouse lung epithelial cells	Comet assay FE1-MML mutagenicity analysis	Concentrations below cytotoxicity did not result in increased levels of strand breaks	[39]
		In vitro lung fibroblast V79 cells	Comet assay MN test Salmonella gene mutation assay	No DNA damage in the Ames and Mn tests, but some significance was detected by the comet assay	[40]
		In vitro the mouse macrophage cell line RAW 264.7	ROS both necrosis and apoptosis chromosomal aberrations cellular ultrastructural alteration	SWCNTs are cyto- and genotoxic to RAW 264.7	[41]
	MWCNTs	*In vitro* epithelial cells (Type II pneumocytes (AT-II); MCF-7; RLE cells)	*Ex vivo* and *in vitro* Mn assay	Significant dose-dependent increase in Mn	[42]
		In vitro mouse embryonic stem cells	Double strand break repair protein assay	Cellular apoptosis and activation of p53, and increased mutation frequency	[43]
		In vivo p53 heterozygous mice	/	MWCNT induced mesothelioma in p53 heterozygous mice	[44]
		In vivo male Fischer 344 rats	/	Mesothelioma observed	[45]

(*Continued*)

TABLE 14.1 Genotoxicity and Toxicity of Nanomaterials—cont'd

Nanoparticle			*In vitro/in vivo*	Genotoxicity and Toxicity Assays	Findings	Reference
			In vitro the mouse macrophage cell line RAW 264.7	ROS both necrosis and apoptosis chromosomal aberrations cellular ultrastructural alteration	SWCNTs are cyto- and genotoxic to RAW 264.7	[41]
	others	nanodiamonds	*In vitro* embryonic stem cells	Expression of DNA repair proteins	Oxidized nanodiamonds induced more DNA damage than the pristine/raw forms, showing the surface chemistry specific genotoxicity	[46]
Metallic nanoparticles	QDs	QDs	Plasmid DNA	Gel electrophoreses analyses assay of reactive oxygen intermediates	Photoactivation of QD-conjugated nucleic acids for an extended period of time is not favorable for their stability	[47]
		QDs modified with various molecules	*In vitro*	Comet assay	The properties of QDs are not related to those of QD-core materials but to molecules covering the surface of QDs	[48]
	Au NPs	PDDAC/CTAB/PSS-coated Au NPs	*In vitro and in vivo* MCF-7 -mesenchymal stem cells		PDDAC-coated Au NRs have high cellular uptake efficiency. CTAB-coated Au NRs have distinct effects on	[49–51]

					cell viability of cancer cells while posing negligible impacts on normal and mesenchymal stem cells	
		Au NP m)	*In vitro* human fetal lung fibroblast cell line (MRC-5)	HPLC to measure 8-OHdG Cell proliferation (trypan blue)	Significant oxidative DNA damage; DNA repair genes downregulated (cyclin C, Hus1, BRCAI/BRCC1)	[52]
	Ag-NPs	Ag NP (25 nm)	*In vitro* mouse embryonic stem cells (MES) and fibroblasts (MEF)	Expression of DNA repair proteins H2AX phosphorylation MTT assay	Increased expression of Rad51, p53 and phospho- H2AX proteins; time-dependent decrease in cell viability	[53]
Metal oxides	TiO_2	5 nm 40 nm	*In vitro* *gpt* delta transgenic mouse primary embryo fibroblasts	MTT mutation analysis	TiO_2 induced cytotoxicity in transgenic MEF cells; increased mutation frequencies at red/gam gene loci detected	[36]
		10 * 40 nm <25 nm <5 μm	*In vitro* human bronchial epithelial BEAS 2B cells	Mn test Comet assay	Uncoated nanosized anatase TiO_2 and fine rutile TiO_2 are more efficient than SiO_2-coated nanosized rutile TiO_2 in inducing DNA damage, whereas only nanosized anatase is able to slightly induce micronuclei	[54]

(*Continued*)

TABLE 14.1 Genotoxicity and Toxicity of Nanomaterials—cont'd

Nanoparticle		*In vitro/in vivo*	Genotoxicity and Toxicity Assays	Findings	Reference
	<25 nm <100 nm	*In vitro* human hepatoma HepG2 cells	Comet assay DNA damage responsive genes	The toxic potential of TiO_2 nanoparticles varies not only with particle size but also with crystalline structure	[55]
	25 nm	*In vivo* PBL	Mn test Comet assay	Dose-dependent increase in Mn frequency and ROS generation; time-dependent upregulation of phospho- and total-p53	[56]
	<100 nm	*In vitro* human lymphoblastoid cells	MTT assay Population growth assay Flow cytometry Cytokinesis block micronucleus assay Comet assay Gene mutation assay	TiO_2 can cause genotoxicity and cytotoxicity in cultured human cells	[57]
zinc oxide	40–70 nm	*In vitro and in vivo* PBL human sperm cells	Comet assay Trypan blue staining	Dose-dependent increase in DNA damage	[58]
	100 nm	*In vitro* CHO cells	Chromosomal aberration test	Photo-clastogenicity detected	[59]
SiO_2	6.57, 8.2 and 196.52 nm	*In vitro* human lymphoblastoid cells	Mn assay Comet assay HPRT assay	Mn induced and significant increase in HPRT mutations at 120 mg/mL, but no increases in comet tail detected	[60]

in the development of nanomaterials. The Nobel Prize in Chemistry for 1996 was awarded to Kroto, Curl and Smalley for their discovery of C_{60} in 1985 [61]. C_{60} now has several technological and medicinal applications including cosmetics [62], lubricants, food additives, building materials, fuel cells, microreactors, drug delivery devices [29], and it has demonstrated antitumor effects in several tumor xenograft-bearing mice models [29–32].

Fullerenes have largely been found to induce limited toxicity and, in some cases, demonstrate antioxidant properties, but there is evidence that C_{60} also has the capacity to generate singlet oxygen that induces lipid peroxidation of linoleate which leads to oxidative DNA damage [33]. Indeed Oberdörster found that fullerenes were lipophilic, localized into lipid-rich regions such as cell membranes *in vitro*, and were redox active [34]. Several studies have reported C_{60} and its derivatives generate ROS with some evidence indicating these nanomaterials are able to damage DNA possibly via an oxidative stress-based mechanism [35].

C_{60} and its derivatives can also form complexes with DNA, induce DNA strand breakages, mutagenicity and chromosomal damage. For example, the induction of mutagenesis by C_{60} has been demonstrated in gpt delta transgenic mouse primary embryo fibroblasts [36]; colloidal C_{60} suspensions in ethanol and water induced genotoxicity in human lymphocytes using single-cell gel electrophoresis assay (comet assay) [37]; and long-term exposure (80 days) of the water-soluble $C_{60}(OH)_{24}$ to Chinese hamster ovary cells, human cervical epithelial carcinoma (HeLa) cells and human embryonic kidney (HEK293) cells increased structural chromosomal damage [38]. Despite these studies reporting significant DNA damaging responses and toxicity following exposure to fullerenes, there are conflicting reports where no such damage has been observed. Jacobsen and colleagues found that fullerenes did not induce significant increases in DNA strand breakages as detected by the comet assay [39], whereas Dhawan and colleagues reported positive genotoxic responses using the same test [37]. Additionally, surface derivatization and structural alterations can largely change the toxicity of fullerenes. Sayes and colleagues compared the cytotoxicity of C_{60} and water-soluble species in two different human cell lines and found the lethal dose of fullerene changed over seven orders of magnitude with relatively minor alterations in fullerene structure [63]. Thus, discrepancies in the literature are likely to be due to slight physicochemical variations in the fullerenes studied [64].

Carbon nanotubes (CNTs) are molecular-scale tubes of graphitic carbon with unique characteristics, and among the stiffest and strongest fibers known. The discovery of CNTs is a promising breakthrough in the field of material science and has attracted huge academic and industrial interest. Some investigations reported that CNTs, both SWCNTs and MWCNTs, have minimal or insignificant cellular responses [65,66]. However, others demonstrated that CNTs were cytotoxic to several cell types and, due to the fibrous structure, the toxicity of CNTs is based on a high-aspect ratio and biopersistence, leading to prolonged inflammatory responses. (For a further discussion on the pathogenic fiber paradigm, see Chapter 7.)

A limited number of investigations have considered the genotoxic potential of CNTs, and they have been found to promote structural and numerical chromosomal damage (both aneugenic and clastogenic events) [40,42,43], together with point mutations [19,67], which indicates potential carcinogenicity. SWCNTs not only enter the cytoplasm, but can also become localized in the cell nucleus, causing cell mortality in a dose-dependent manner

[68]. Significant DNA strand breakages are induced by exposure to different SWCNTs and MWCNTs in a variety of model cell lines and *in vivo*. Additionally, there are now two studies that have reported that MWCNTs promote mesothelioma *in vivo* following the administration of an intraperitoneal and intrascrotal dose [21,44]. However, there is a contradictory report that failed to detect a carcinogenic response to MWCNTs 2 years after an intraperitoneal exposure [22].

Thus, it appears that CNTs have genotoxic and carcinogenic potential, but the underlying mechanism requires further study. Most investigations have highlighted the importance of oxidative stress-mediated DNA damage; however, CNTs have a potential to interact with cellular biomolecules, such as the mitotic spindle, as well as the motor proteins that separate the chromosomes during cell division. Indeed, the disruption of centrosomes and mitotic spindles by CNTs has been found to result in monopolar, tripolar and quadrapolar divisions of chromosomes [45]. Furthermore, the physicochemical characteristics of CNTs are important in governing the resultant biological response following exposure, with surface chemistry representing a key parameter that has a substantial effect on genotoxic potential [69].

To date, there are limited reports examining the genotoxicity and carcinogenesis of other carbon-based materials, including graphene, carbon black, carbon nanofibers, nanorods, nanowires and nanodiamonds. These carbon-based materials are all generating great interest in industrial sectors such as energy production and electronics, thus occupational exposure is likely. There is evidence demonstrating carbon black is cytotoxic and genotoxic to RAW 264.7 mouse macrophages [70]. Cellular uptake, generation of oxygen radicals and gross chromosomal damage have also been observed in several cell types following exposure to carbon nanofibers [64]. While nanodiamonds lead to mildly increased expression of DNA repair proteins (p53 and MOGG-1) in embryonic stem cells, surface chemistry was important as oxidized nanodiamonds caused more DNA damage than unmodified nanodiamonds [41]. In contrast, carbon-fiber-reinforced polybutyleneterephthalate (CFRPBT) was non-hemolytic, non-immunogenic and showed no genotoxicity [46]. Evidence for the genotoxicity and carcinogenesis of carbon-based nanomaterials is accumulating with a clear dependency on physicochemical features, but their long-term impact on human health has yet to be definitively established.

METALLIC NANOPARTICLES

A vast array of metallic nanoparticles (NPs) for a myriad of applications has been developed. Their genotoxicity has been reviewed in detail elsewhere [64], thus here we will only summarize the genotoxic potential of those in greatest industrial fabrication.

Quantum dots (QDs) are semiconductor nanocrystals that possess unique optical properties, making them promising tools for bioimaging. QDs can be conjugated with specific receptors to target particular cell types or molecules, such as in the directed labeling of neoplastic cells [71]. However, QDs can transfer energy to nearby oxygen molecules, which could induce the generation of ROS and, in turn, lead to cell damage or death [72]. The main disadvantage of these nanoparticles is that they consist of highly toxic elements such as tellurium (Te), cadmium (Cd), lead (Pb) and selenium (Se), which is likely to hinder their use *in vivo*, as these dangerous materials could be released into the cellular environment, which may induce cyto- or genotoxicity [73]. Indeed, uncoated CdTe QD elevated cellular ROS levels leading to plasma membrane, mitochondria and nuclear damage (shrinkage and chromatin condensation), accompanied by upregulation of the antioxidant enzyme superoxide dismutase 1 (SOD-1) [28,74]. Capping the QDs with zinc sulfide (ZnS) can reduce this effect by protecting cells from the toxic core, but two groups have demonstrated that CdSe/ZnS QDs can cause DNA nicking and fragmentation in cell-free systems [75]. Other methods to reduce QDs toxicity involve coating their surface with biocompatible molecules such as polyethylene glycol (PEG). However, some functional groups or coating layers may also be cyto- or genotoxic, thus careful consideration should be given when selecting these modifications. In addition, purification of the material is an important factor in its toxicity, as crude QD coated with carboxyl groups prepared only by membrane filtration caused greater DNA damage than purified QD-COOH [47].

Gold nanoparticles (GNPs) have been utilized in imaging, biosensing, gene and drug delivery, cancer diagnostics and therapy, due to their unique optical properties and biocompatibility. Their small size enables GNPs to penetrate easily into cells and translocate throughout the intracellular space (including the nucleus) causing a variety of responses. Size, shape, and surface properties (surface charge and functionalization) influence the biological effects of GNPs, particularly their cytotoxicity [48–50]. However, GNP-induced genotoxicity is controversial. In embryonic lung fibroblasts, GNPs induced oxidative DNA damage (8-OHdG adducts) and decreased expression of DNA repair genes and the cell cycle checkpoint genes MAD2, cyclinB1 and cyclinB2, which could lower the cellular DNA damage response pathways leading to genetic instability [51]. Moreover, GNPs could induce DNA damage indirectly via an oxidative stress response in a size- and cell-type-dependent manner. Of some concern, is the observation that exposure to GNPs causes failure of nuclear chromatin decondensation in mouse sperm *in vitro*. Thus, some limited indication of GNPs genotoxicity has been presented but there appears to be some nanoparticle-size and cell-type dependency.

Silver nanoparticles (Ag NPs) are used in various consumer products including cosmetics, textiles, and health-care products owing to their strong antimicrobial activity. However, there is accumulating evidence to suggest that Ag NPs are highly toxic to a variety of cells. As with many nanomaterials, physicochemical properties are important in defining cellular impact and polysaccharaide-coated Ag NPs exhibit a more severe DNA damage response than uncoated Ag NPs [52], although both types were able to increase p53 expression and phosphorylation, upregulate the DNA damage repair protein Rad51 and elevate the phosphorylation of H2AX. Ag NPs have also been found to reduce the ATP content of cells which damages mitochondria, and increase production of ROS and DNA damage, indicating that Ag NPs can disrupt the mitochondrial respiratory chain leading to the interruption of ATP synthesis and generation of oxidative stress, which in turn causes DNA damage [53]. Finally, the level of bulky DNA adducts induced by Ag NPs has not only been correlated with the level of cellular ROS, but is also inhibited by antioxidant pretreatment, suggesting Ag NPs mediate ROS-induced genotoxicity [76].

Metallic NPs are promising nanomaterials in biomedical and commercial applications, but there are limited data on their genotoxic potential. Once inside cells, a related issue is the long-term fate of nanomaterials. One concern that has yet to be fully investigated is that the metallic NPs may themselves not induce DNA damage, but they may corrode over time in biological environments to release metallic ions which could induce genotoxicity [64]. Thus, the long-term genotoxic responses may be significantly different from short-term exposure datasets and further work in this area is required.

METAL OXIDES

To date, TiO_2 nanoparticles have been the focus in the majority of nanogenotoxicity studies due to their current use in sunscreens and cosmetics. These investigations have tested a range of TiO_2 forms with different sizes and crystalline structure, in an array of mammalian cell lines, using a variety of exposure times and assay systems. However, the primary factor that appears to be critical with regards to the genotoxicity of TiO_2 is its crystalline structure as anatase TiO_2 nanoparticles generally induce DNA strand breaks and chromosomal damage, while rutile TiO_2 is largely non-genotoxic *in vitro* [54,55,77]. Only a small number of studies have directly compared rutile to anatase TiO_2 genotoxicity and, in these reports, the rutile nanoparticles are larger than the anatase forms investigated [54,55,77]. Additionally, many earlier studies exist that report significant increases in DNA damage but have either not detailed crystalline structure or used a mixed anatase/rutile sample [56,78,79]. Thus, further detailed investigation into the influence of this physicochemical parameter on TiO_2 genotoxicity is required.

Only two *in vivo* carcinogenicity studies have been performed to date with TiO_2, both of which yielded negative results but one study was performed with the rutile form [57], while the other does not specify the crystalline structure [80]. Xu and colleagues topically applied rutile TiO_2 nanoparticles to the skin of C-Ha-*ras* proto-oncogen transgenic mice (which are more sensitive to chemically induced skin carcinogens than wild type mice), and observed that the nanoparticles were not able to penetrate through the epidermis to reach the dermis where skin tumors arise [57]. Yokohira and colleagues failed to detect any histological lesions in the lungs of male rats that were exposed to TiO_2 nanoparticles via intratracheal instillation [80]. However, mice that were exposed to mixed TiO_2 (75% anatase/25% rutile) in their drinking water were found to have 8-OHdG adducts in their livers, gross chromosomal damage in their peripheral blood cells and the fetuses of the maternally exposed animals demonstrated an increased DNA deletion frequency [14]. Hence, there is clear justification for further carcinogenicity studies focusing on the role of the TiO_2 crystalline structure.

Due to the use of TiO_2 in sunscreens, one concern has been the influence of UV irradiation on the induction of genotoxicity by these nanoparticles, which unfortunately has not been the subject of many studies. However, one investigation has demonstrated an enhanced level of clastogenicity in the presence of UV light *in vitro* when compared to its genotoxicity in the dark [81]. Furthermore, ZnO nanoparticles, which are also utilized as an ingredient in sunscreens, have also demonstrated the same response following simultaneous and pre-irradiation [58,81].

Silica represents the other most rigorously studied metal oxide nanoparticle for genotoxic responses *in vitro*, but the evidence currently available is conflicting even where similar nanoparticles have been studied in the same cell line. For example 34 nm silica induced positive genotoxic responses in 3T3 mouse fibroblasts in one study [59], while a second reported no genotoxicity in these cells [82]. A similar contradiction has also been seen in A549 lung carcinoma cells [83,84]. However, it appears that the conflict is due to the genotoxicity test applied as all those studies utilizing the comet assay to assess silica demonstrate no significant DNA damage [82–84], yet when the micronucleus assay is used the authors report a genotoxic response [59,83]. Furthermore, when both test systems were used in parallel within the same investigation, the same pattern was observed whereby the silica induced significant genotoxicity in the micronucleus assay but no increase in comet tail was detected [83,85]. The comet assay generally detects DNA strand breaks and alkali-labile sites that can be repaired, while the micronucleus assay reports on chromosome fragmentation or whole chromosome loss that are present post-replication and are therefore fixed mutations. The reason for the discrepancy between these two assays when testing silica is currently unclear but it suggests the micronucleus assay may be more appropriate in the genotoxicity testing of silica due to the type of DNA damage induced.

A small number of studies have also been performed on several other metal oxide nanoparticles including iron oxide, aluminum oxide and cerium oxide with all inducing genotoxic changes both *in vitro* and *in vivo* [60,86,87]. However, the limited array of these studies makes it difficult to identify the physicochemical characteristics that influence genotoxic potential. It has therefore been clearly demonstrated that metal oxide nanoparticles have the capacity to induce a range of genotoxic responses, nevertheless, it is yet to be established if these mutagenic alterations are sufficient to induce carcinogenesis *in vivo*.

TAKE-HOME MESSAGES

Our understanding of the genotoxicity induced by nanomaterials is slowly developing. As more nanomaterials are being investigated, there is increasing evidence supporting their ability to induce DNA damage primarily through the promotion of oxidative stress, although some direct interference with DNA and cell division machinery has been observed. However, it must be stressed that the genotoxic potential of nanomaterials is clearly dependent upon their physicochemical characteristics. Unfortunately, the high degree of variation in materials tested and genotoxicity assessment regimens utilized currently makes it very difficult to form firm conclusions on those features that are directly responsible for inducing genotoxic responses. It is therefore important that future studies focus on investigating the genotoxicity of a range of nanomaterials in parallel, assessing several DNA damage endpoints to determine the true impact of specific physicochemical features and elucidate the mechanisms which govern the induction of genotoxicity by nanomaterials. In the case of metal nanoparticles, which often become trapped once they have become internalized into cells, the importance of corrosion in biological environments and thus its long-term effect is also poorly understood and requires further investigation. Finally, many of the studies to date have only been based on acute *in vitro* and *in vivo* genotoxicity assays and extended *in vivo* carcinogenicity investigations are currently lacking. Substances that promote genotoxicity are highly suspect carcinogens, but this needs to be confirmed in appropriate studies. Thus, carcinogenicity assessments are now required to determine if their genotoxic impact is sufficient to initiate tumorigenesis and to establish the long-term health effects following exposure to nanomaterials.

References

[1] Jenkins GJS, Doak SH, Johnson G, Quick E, Waters E, Parry J. Assessment of the mechanistic basis for genotoxic thresholds induced by alkylating agents. Mutagenesis 2005;20:389–98.

[2] Doak SH, Johnson G, Quick E, Jenkins GJS, Parry JM. Mechanistic influences for mutation induction curves after exposure to DNA-reactive carcinogens. Cancer Res 2007;67:3904–11.

[3] Ferguson LR, Denny WA. Genotoxicity of non-covalent interactions: DNA intercalators. Mutat Res 2007;623: 14–23.

[4] Berry CC, de la Fuente JM, Mullin M, Chu SW, Curtis AS. Nuclear localization of HIV-1 tat functionalized gold nanoparticles. IEEE Trans Nanobiosci 2007;6:262–9.

[5] Nativo P, Prior IA, Brust M. Uptake and intracellular fate of surface-modified gold nanoparticles. ACS Nano 2008;2:1639–44.

[6] Hackenberg S, Friehs G, Froelich K, et al. Intracellular distribution, geno- and cytotoxic effects of nanosized titanium dioxide particles in the anatase crystal phase on human nasal mucosa cells. Toxicol Lett 2010;195:9–14.

[7] Hackenberg S, Scherzed A, Technau A, et al. Cytotoxic, genotoxic and pro-inflammatory effects of zinc oxide nanoparticles in human nasal mucosa cells *in vitro*. Toxicol *in Vitro* 2011;25:657–63.

[8] Chen M, von Mikecz A. Formation of nucelopasmic protein aggregates impairs nuclear function in response to SiO_2 nanoparticles. Exp Cell Res 2005;305:51–62.

[9] Nabiev I, Mitchell S, Davies A, et al. Nonfunctionalised nanocrystals can exploit a cell's active transport machinery delivering them to specific nuclear and cytoplasmic compartments. Nano Lett 2007;7:3452–61.

[10] Shukla RK, Sharma V, Pandey AK, Singh S, Sultana S, Dhawan A. ROS-mediated genotoxicity induced by titanium dioxide nanoparticles in human epidermal cells. Toxicol *in Vitro* 2011;25:231–41.

[11] Cooke MS, Evans MD, Dizdaroglu M, Lunec J. Oxidative DNA damage: Mechanism, mutation, and disease. FASEB J 2003;17:1195–214.

[12] Sargent LM, Shvedova AA, Hubbs AF, et al. Induction of aneuploidy by single-walled carbon nanotubes. Environ Mol Mutagen 2009;50:708–17.

[13] Folkmann JK, Risom L, Jacobsen NR, Wallin H, Loft S, Moller P. Oxidatively damaged DNA in rats exposed by oral gavage to C_{60} fullerenes and single-walled carbon nanotubes. Environ Health Perspect 2009;117: 703–8.

[14] Trouiller B, Reliene R, Westbrook A, Solaimani P, Schiestl RH. Titanium dioxide nanoparticles induce DNA damage and genetic instability *in vivo* in mice. Cancer Res 2009;69:8784–9.

[15] Nabeshi H, Yoshikawa T, Matsuyama K, et al. Amorphous nanosilica induce endocytosis-dependent ROS generation and DNA damage in human keratinocytes. Part Fibre Toxicol 2011;8:1.

[16] Federico A, Morgillo F, Tuccillo C, Ciardiello F, Loguercioc C. Chronic inflammation and oxidative stress in human carcinogenesis. Int J Cancer 2007;121:2381–6.

[17] Marczynski B, Kraus T, Rozynek P, Raithel HJ, Baur X. Association between 8-hyrdoxy-2-deoxyguanosine levels in DNA of workers highly exposed to asbestos and their clinical data, occupational and non-occupational confounding factors, and cancer. Mutat Res 2000;468:203–312.

[18] Rihn B, Coulais C, Kauffer E, et al. Inhaled crocidolite mutagenicity in lung DNA. Environ Health Perspect 2000;108:341–6.

[19] Unfried K, Schurkes C, Abel J. Distinct spectrum of mutations induced by crocidolite asbestos: Clues for 8-hydroxydeoxyguanosine-dependent mutagenesis *in vivo*. Cancer Res 2002;62:99–104.

[20] Poland CA, Duffin R, Kinlock I, et al. Carbon nanotubes introduced into the abdominal cavity of mice show asbestos-like pathogenicity in a pilot study. Nat Nanotech 2008;3:423–8.

[21] Takagi A, Hirose A, Nishimura T, et al. Induction of mesothelioma in p53 +/- mouse by intraperitoneal application of multi-wall carbon nanotube. J Toxicol Sci 2008;33:105–16.

[22] Muller J, Delos M, Panin N, Rabolli V, Huaux F, Lison D. Absence of carcinogenic response to multiwall carbon nanotubes in a 2-year bioassay in the peritoneal cavity of the rat. Toxicol Sci 2009;110:442–8.

[23] Zhong BZ, Whong WZ, Ong TM. Detection of mineral-dust-induced DNA damage in two mammalian cell lines using the alkaline single cell gel/comet assay. Mutat Res 1997;393:181–7.

[24] Driscoll KE, Carter JM, Howard BW, et al. Pulmonary inflammatory, chemokine, and mutagenic responses in rats after sub-chronic inhalation of carbon black. Toxicol Appl Pharmacol 1996;136:372–80.

[25] Schins RPF. Mechanisms of genotoxicity of particles and fibres. Inhalat Toxicol 2002;14:57–78.
[26] Bozic I, Antal T, Ohtsuiki H, et al. Accumulation of driver and passenger mutations during tumor progression. Proc Natl Acad Sci USA 2010;43:18545–50.
[27] Sharma S, Kelly TK, Jones PA. Epigenetics in cancer. Carcinogenesis 2010;31:27–36.
[28] Choi AO, Brown SE, Szyf M, Maysinger D. Quantum dot-induced epigenetic and genotoxic changes in human breast cancer cells. J Mol Med 2008;86:291–302.
[29] Liang XJ, Meng H, Wang Y, et al. Metallofullerene nanoparticles circumvent tumor resistance to cisplatin by reactivating endocytosis. Proc Natl Acad Sci USA 2010;107:7449–54.
[30] Chen C, Xing G, Wang J, et al. Multihydroxylated nanoparticles: Antineoplastic activity of high efficiency and low toxicity. Nano Lett 2005;5:2050–7.
[31] Liu Y, Jiao F, Qiu Y, et al. The effect of Gd-$C_{82}(OH)_{22}$ nanoparticles on the release of Th1/Th2 cytokines and induction of TNF-alpha mediated cellular immunity. Biomaterials 2009;30:3934–45.
[32] Liu Y, Jiao F, Qiu Y, et al. Immunostimulatory properties and enhanced TNF- alpha mediated cellular immunity for tumor therapy by $C_{60}(OH)_{20}$ nanoparticles. Nanotechnology 2009;20:415102.
[33] Sera N, Tokiwa H, Miyata N. Mutagenicity of the fullerene C_{60}-generated singlet oxygen dependent formation of lipid peroxides. Carcinogenesis 1996;17:2163–9.
[34] Oberdorster E. Manufactured nanomaterials (fullerenes, C_{60}) induce oxidative stress in the brain of juvenile largemouth bass. Environ Health Perspect 2004;112:1058–62.
[35] Markovic Z, Trajkovic V. Biomedical potential of the reactive oxygen species generation and quenching by fullerenes (C_{60}). Biomaterials 2008;29:3561–73.
[36] Xu A, Chai Y, Nohmi T, Hei TK. Genotoxic responses to titanium dioxide nanoparticles and fullerene in gpt delta transgenic MEF cells. Part Fibre Toxicol 2009;6:3.
[37] Dhawan A, Taurozzi JS, Pandey AK, et al. Stable colloidal dispersions of C_{60} fullerenes in water: Evidence for genotoxicity. Environ Sci Technol 2006;40:7394–401.
[38] Yasuharu N, Naoharu I. Genotoxicity in cell lines induced by chronic exposure to water soluble fullerenes using micronucleus test. Environ Health Prev Med 2006;11:292–7.
[39] Jacobsen NR, White P, Moller P, et al. Genotoxicity, cytotoxicity, and reactive oxygen species induced by singlewalled carbon nanotubes and C(60) fullerenes in the FE1-Mutatrade mark-Mouse lung epithelial cells. Environ Mol Mutagen 2008;49:12.
[40] Kisin ER, Murray AR, Keane MJ, et al. Single-walled carbon nanotubes: Geno- and cytotoxic effects in lung fibroblast V79 cells. J Toxicol Environ Health A 2007;70:2071–9.
[41] Xing Y, Xiong W, Zhu L, Osawa E, Hussin S, Dai L. DNA damage in embryonic stem cells caused by nanodiamonds. ACS Nano 2011.
[42] Muller J, Decordier I, Hoet PH, et al. Clastogenic and aneugenic effects of multi-wall carbon nanotubes in epithelial cells. Carcinogenesis 2008;29:427–33.
[43] Zhu L, Chang DW, Dai L, Hong Y. DNA damage induced by multiwalled carbon nanotubes in mouse embryonic stem cells. Nano Lett 2007;7:3592–7.
[44] Sakamoto Y, Nakae D, Fukumori N, et al. Induction of mesothelioma by a single intrascrotal administration of multi-wall carbon nanotube in intact male Fischer 344 rats. J Toxicol Sci 2009;34:65–76.
[45] Sargent LM, Reynolds SH, Castranova V. Potential pulmonary effects of engineered carbon nanotubes: *In vitro* genotoxic effects. Nanotoxicology 2010;4:396–408.
[46] Brooks RA, Jones E, Storer A, Rushton N. Biological evaluation of carbon-fibre-reinforced polybutyleneterephthalate (CFRPBT) employed in a novel acetabular cup. Biomaterials 2004;25:3429–38.
[47] Hoshino A, Fujioka K, Oku T, et al. Physicochemical properties and cellular toxicity of nanocrystal quantum dots depend on their surface modification. Nano Lett 2004;4:2163–9.
[48] Qiu Y, Liu Y, Wang L, et al. Surface chemistry and aspect ratio mediated cellular uptake of Au nanorods. Biomaterials 2010;31:7606–19.
[49] Wang L, Li YF, Zhou L, et al. Characterization of gold nanorods *in vivo* by integrated analytical techniques: Their uptake, retention, and chemical forms. Anal Bioanal Chem 2010;396:1105–14.
[50] Wang L, Liu Y, Li W, et al. Selective targeting of gold nanorods at the mitochondria of cancer cells: Implications for cancer therapy. Nano Lett 2011;11:772–80.
[51] Li JJ, Zou L, Hartono D, Ong CN, Bay BH, Lanry Yung LY. Gold nanoparticles induce oxidative damage in lung fibroblasts *in vitro*. Adv Mater 2008;20:138–42.

[52] Ahamed M, Karns M, Goodson M, et al. DNA damage response to different surface chemistry of silver nanoparticles in mammalian cells. Toxicol Appl Pharm 2008;233:404–10.
[53] AshaRani PV, Low Kah Mun G, Hande MP, Valiyaveettil S. Cytotoxicity and genotoxicity of silver nanoparticles in human cells. ACS Nano 2009;3:279–90.
[54] Petkovic J, Zegura B, Stevanovic M, et al. DNA damage and alterations in expression of DNA damage responsive genes induced by TiO_2 nanoparticles in human hepatoma HepG2 cells. Nanotoxicology in press 2011. doi:10.3109/17435390.2010.507316.
[55] Gurr JR, Wang AS, Chen CH, Jan KY. Ultrafine titanium dioxide particles in the absence of photoactivation can induce oxidative damage to human bronchial epithelial cells. Toxicology 2005;213:66–73.
[56] Wang JJ, Sanderson BJS, Wang H. Cyto- and genotoxicity of ultrafine TiO_2 particles in cultured human lymphoblastoid cells. Mutat Res 2007;628:99–106.
[57] Xu J, Sagawa Y, Futakuchi M, et al. Lack of promoting effect of titanium dioxide particles on ultraviolet B-initiated skin carcinogenesis in rats. Food Chem Toxicol 2011. in press.
[58] Dufour EK, Kumaravel T, Nohynek GJ, Kirkland D, Toutain H. Clastogenicity, photo-clastogenicity or pseudo-photo-clastogenicity: Genotoxic effects of zinc oxide in the dark, in pre-irradiated or simultaneously irradiated Chinese hamster ovary cells. Mutat Res 2009;607:215–24.
[59] Park MVDZ, Verharen HW, Zwart E, et al. Genotoxicity evaluation of amorphous silica nanoparticles of different sizes using the micronucleus and the plasmid lacZ gene mutation assay. Nanotoxicology in press 2011. doi:10.3109/17435390.2010.506016.
[60] Auffan M, Rose J, Orsiere T, et al. CeO_2 nanoparticles induce DNA damage towards human dermal fibroblasts *in vitro*. Nanotoxicology 2009;3:161–71.
[61] Kroto HW, O'Brien SC, Smalley RE. C_{60}: Buckminsterfullerene. Nature 1985;318:2.
[62] Kato S, Taira H, Aoshima H, Saitoh Y, Miwa N. Clinical evaluation of fullerene-C_{60} dissolved in squalane for anti-wrinkle cosmetics. J Nanosci Nanotechnol 2010;10:6769–74.
[63] Sayes CM, Fortner JD, Guo W, et al. The differential cytotoxicity of water-soluble fullerenes. Nano Lett 2004;4:7.
[64] Singh N, Manshian B, Jenkins GJ, et al. NanoGenotoxicology: The DNA damaging potential of engineered nanomaterials. Biomaterials 2009;30:3891–914.
[65] Davoren M, Herzog E, Casey A, et al. *In vitro* toxicity evaluation of single walled carbon nanotubes on human A549 lung cells. Toxicol *In Vitro* 2007;21:438–48.
[66] Worle-Knirsch JM, Pulskamp K, Krug HF. Oops they did it again! Carbon nanotubes hoax scientists in viability assays. Nano Lett 2006;6:1261–8.
[67] Ge C, Meng L, Xu L, et al. Acute pulmonary and moderate cardiovascular responses of spontaneously hypertensive rats after exposure to single-wall carbon nanotubes. Nanotoxicology 2011. doi:10.3109/17435390.2011.587905.
[68] Porter AE, Gass M, Muller K, Skepper JN, Midgley PA, Welland M. Direct imaging of single-walled carbon nanotubes in cells. Nat Nanotechnol 2007;2:713–7.
[69] Ge C, Lao F, Li W, et al. Quantitative analysis of metal impurities in carbon nanotubes: Efficacy of different pretreatment protocols for ICPMS spectroscopy. Anal Chem 2008;80:9426–34.
[70] Giorgio ML, Di Bucchianico S, Ragnelli AM, Aimola P, Santucci S, Poma A. Effects of single- and multi-walled carbon nanotubes on macrophages: Cytotoxicity, genotoxicity and ultrastructural damage. Mutat Res 2011.
[71] Gao X, Cui Y, Levenson RM, Chung LWK, Nie S. *In vivo* cancer targeting and imaging with semiconductor quantum dots. Nat Biotechnol 2004;22:969–76.
[72] Bakalova R, Ohba H, Zhelev Z, Ishikawa M, Baba Y. Quantum dots as photosensitizers? Nat a-z index 2004;22:1360–1.
[73] Gaponik N, Talapin DV, Rogach AL, et al. Thiol-capping of CdTe nanocrystals: An alternative to organometallic synthetic routes. J Phys Chem B 2002;106:7177–85.
[74] Lovric J, Cho SJ, Winnik FM, Maysinger D. Unmodified cadmium telluride quantum dots induce reactive oxygen species formation leading to multiple organelle damage and cell death. Chem Biol 2005;12: 1227–34.
[75] Anas AA, Akita H, Harashima H, Itoh T, Ishikawa M, Biju V. Photosensitized breakage and damage of DNA by CdSe- ZnS quantum dots. J Phys Chem B 2008;112:10005–11.
[76] Foldbjerg R, Dang D, Autrup H. Cytotoxicity and genotoxicity of silver nanoparticles in the human lung cancer cell line, A549. Arch Toxicol 2010;85:743–50.

[77] Falck GCM, Lindberg HK, Suhonen S, et al. Genotoxic effects of nanosized and fine TiO_2. Hum Exp Toxicol 2009;28:339–52.
[78] Karlsson HL, Cronholm P, Gustafsson J, Moller L. Copper oxide nanoparticles are highly toxic: A comparison between metal oxide nanoparticles and carbon nanotubes. Chem Res Toxicol 2008;21:1726–32.
[79] Kang SJ, Kim BM, Lee YJ, Chung HW. Titanium dioxide nanoparticles trigger p53-mediated damage response in peripheral blood lymphocytes. Environ Mol Mutagen 2008;49:399–405.
[80] Yokohira M, Hashimoto N, Yamakawa K, et al. Lung carcinogenic bioassay of CuO and TiO_2 nanoparticles with intratracheal instillation using F344 male rats. J Toxicol Pathol 2009;22:71–8.
[81] Gopalan RC, Osman IF, Amani A, De Matas M, Anderson D. The affect of zinc oxide and titanium dioxide nanoparticles in the comet assay with UVA photoactivation of human sperm and lymphocytes. Nanotoxicology 2009;3:33–9.
[82] Barnes CA, Elsaesser A, Arkusz J, et al. Reproducible comet assay of amorphous silica nanoparticles detects no genotoxicity. Nano Lett 2008;8:3069–74.
[83] Gonzalez L, Thomassen LCJ, Plas G, et al. Exploring the aneugenic and clastogenic potential in the nanosize range: A549 human lung carcinoma cells and amorphous monodisperse silica nanoparticles as models. Nanotoxicology 2010;4:382–95.
[84] Jin Y, Kannan S, Wu M, Zhao JX. Toxicity of luminescent silica nanoparticles to living cells. Chem Res Toxicol 2007;20:1126–33.
[85] Wang JJ, Sanderson BJS, Wang H. Cytotoxicity and genotoxicity of ultrafine crystalline SiO_2 particulate in cultured human lymphoblastoid cells. Environ Mol Mutagen 2007;48:151–7.
[86] Balasubramanyam A, Sailaja N, Mahboob M, Rahman MF, Hussain SM, Grover P. *In vivo* genotoxicity assessment of aluminium oxide nanomaterials in rat peripheral blood cells using the comet assay and micronucleus test. Mutagenesis 2009;24:245–51.
[87] Singh N, Jenkins GJS, Asadi R, Doak SH. Potential toxicity of superparamagnetic iron oxide nanoparticles (SPION). Nano Rev 2010;1:5358.

SECTION C

BIOMEDICAL APPLICATIONS OF ENGINEERED NANOMATERIALS

CHAPTER

15

Diagnostic Applications

Kai Cheng, Zhen Cheng

Molecular Imaging Program at Stanford (MIPS), Canary Center at Stanford for Cancer Early Detection, Department of Radiology and Bio-X Program, School of Medicine, Stanford University, Stanford, CA, USA

OUTLINE

INTRODUCTION

Cancer is a complex human disease that is characterized by uncontrolled growth and spread of abnormal cells potentially affecting almost any tissue in the body. It is overtaking heart disease as the no. 1 killer in the USA. An estimated 569 490 Americans will die from cancer in 2010 alone, according to the American Cancer Society, and the worldwide incidence of cancer continues to increase [1]. In fact, cancer is not a single disease but a conglomerate of many diseases caused by the accumulation of multiple molecular alterations in the genetic material. In recent years, many cancer patients have benefited from improved treatments, including surgery, chemotherapy, targeted therapies, and radiation therapy. However, success is often marginal and serious side effects are common and the outcome remains

Adverse Effects of Engineered Nanomaterials
DOI: 10.1016/B978-0-12-386940-1.00015-5

disappointing, especially for those patients with advanced cancer. For instance, surgery followed by systemic chemotherapy and radiotherapy is only recommended for the subset of cancer patients who have been fortunate to have their disease completely resected. Although radiotherapy eradicates cancer cells and prevents disease recurrence to a certain degree, it can only be administered to a specific region and its delivery and dose are limited to organs within the radiation field. Chemotherapy kills actively dividing cells including cancer cells, but its administration is limited because of severe toxicity to non-cancer cells. Furthermore, many improvements in survival obtained in certain type of cancers have not been observed in most advanced cancers because they are often diagnosed late and thus systemic therapies are limited. Most patients die from the malignancy within a year of their diagnosis, despite extensive surgical resection and aggressive adjuvant radiotherapy and chemotheray. Thus, it is clear that development of techniques augmenting diagnosis of cancer at an early stage and/or better adjuvant therapies will significantly improve patients survival and help cancer management.

Development of novel imaging techniques has currently improved our ability to detect cancers at earlier stages, quantify invasion or metastasis, or monitor therapeutic responses [2]. In previous decades, structural abnormalities relative to the diseases or treatment efficacy were usually determined by conventional imaging modalities which relied largely on monitoring the change of gross anatomy. Recently, with advances in imaging instrumentations and imaging agents, it has become possible to image or characterize the biological processes at the cellular, subcellular, or even the molecular level in living subjects [2–4]. As a result, a new discipline known as molecular imaging has emerged in the field of biomedical research [4]. Indeed, the approaches developed in molecular imaging provide essential breakthroughs in the fight against cancer. Combining molecular biology, imaging instruments as well as molecular probes, novel imaging techniques not only display high sensitivity for detecting cancer at a molecular level but also provide high spatial resolution.

Many molecular imaging modalities, such as magnetic resonance imaging (MRI), nuclear imaging (i.e. single-photon-emission computed tomograpy [SPECT], single-photon-emission tomography [PET]), x-ray computed tomography (CT), optical imaging and ultrasound imaging, have become indispensable for the clinic [2–7]. Depending on the clinical need, relevant molecular imaging modalities are chosen for detecting both primary and metastatic tumors and monitoring cancer-associated physiological processes in real-time. Importantly, these techniques are all non-invasive and can provide complementary information for cancer diagnosis. For example, due to its excellent soft-tissue contrast properties and high spatial resolution (pixel dimension of 10–100 μm), MRI enables sensitive detection of soft-tissue pathologies, and provides quantitative morphometric information and valuable physiological readouts [2]. As a classical anatomical imaging modality, CT can non-invasively produce images as rich in anatomical information as MRI. Consequently, CT is primarily used to display internal anatomy of living subjects, especially skeletal structures and the lung. However, the sensitivity of both MRI and CT is insufficient to detect abnormalities at the cellular or molecular level. In contrast, nuclear imaging techniques (PET or SPECT), dependent on the availability of radioactive molecular probes, can provide the high detection sensitivity required for monitoring the biomarkers associated with cancers. Optical imaging techniques, such as fluorescence and bioluminescence imaging, provide fast and highly sensitive approaches to map specific molecular targets *in vivo*. However, either nuclear or optical imaging has limited capability in producing anatomical information. Therefore, it is necessary to develop multimodal imaging strategies to validate fully the results of each single molecular imaging technique. Recent advances in cancer molecular imaging were successfully made through combining different imaging modalities (PET-CT, PET-MRI, MRI-optical, etc.) in order to enhance synergistically overall sensitivity and specificity. Table 15.1 provides an overview of imaging systems.

Nanotechnology, as a multidisciplinary reseach field that involves chemistry, engineering, biology, physics and material science, leads to a generation of new diagnostic and therapeutic agents for early cancer detection and personalized treatment. This hopefully will result in dramatically improved outcomes [3,4,8]. In the field of cancer diagnosis, molecular imaging takes advantage of both conventional and newly emerging imaging techniques and sometimes requires the administration of molecular probes to measure cancer biomarkers. However, it is still challenging to design and develop desirable molecular probes that have

TABLE 1 Overview of Imaging Systems

Technique	Resolution	Depth	Time	Representative Imaging Agents	Cost
Magnetic resonance imaging (MRI)	10–100 μm	No limit	Min–hours	Gadolinium-doped NPs, iron oxide NP	$$$
X-ray computed tomography (CT)	50 μm	No limit	Min	Iodine- or gadolinium-doped NPs, Au NPs, Bi_2S_3 NPs, TaO_x NPs	$$
Ultrasound imaging	50 μm	mm	Min	Microbubbles, perfluorocarbon (PFC) emulsion NPs	$$
Positron emission tomography (PET) imaging	1–2 mm	No limit	Min	^{64}Cu-labeled NPs, ^{64}CuS NPs	$$$
Single photon emission tomography (SPECT) imaging	1–2 mm	No limit	Min	^{111}In-labeled NPs	$$
Fluorescence imaging	1–2 mm	<1 cm	Min	NIR fluorochromes, quantum dot	$
Bioluminescence imaging (BLI)	1–2 mm	cm	Min	Luciferin	$
Raman imaging	1–2 mm	cm	Min	Gold rods, gold nanocages, gold nanoshells, carbon nanotubes	$$
Photoacoustic imaging (PAI)	250 μm	cm	Min–hours	Gold rods, gold nanocages, gold nanoshells, carbon nanotubes	$$

Cost of system: $: <100 000; $$: 100–300 000; $$$: >300 000.

the ability specifically to reach the target of interest and display high target to non-target signal ratio *in vivo*. Thus far, most molecular probes available to the clinic are small molecule-based radioactive probes.

Recent advances in nanotechnology have resulted in an unprecedented development of nanoparticle-based molecular probes [4,8]. Engineered nanomaterials have the potential to revolutionize early cancer detection and personalized cancer management by augmenting the sensivity and specificity of the "state of the art" imaging techniques. The engineered nanomaterials exhibit superior physical and chemical properties when their sizes are smaller than several tenths to hundredths of nanometers, arguably due to large relative surface area and dominant quantum confinement effects. Nanomaterials typically have a size range of 1–100 nm, comparable to most biological functional units, such as enzymes, antibodies and receptors. Furthermore, the decrease in the particle size dramatically increases the reactivity of nanomaterials and enhances their ability to couple with various functional moities. As a result, engineered nanomaterials are capable of maintaining the stability and specificity within the biological environment in order to allow unique interaction with biological systems at the molecular level. Moreover, nanomaterials usually have uniform size and shape so that each individual nanoparticle (NP) has nearly identical physical and chemical properties for controlled biodistribution, bioelimination and contrast effects. When nanoparticles are administered in the bloodstream, their size and surface properties favor extravasation through leaky vasculature, followed by deep penetration and retention at the tumor sites, which is a phenomenon known as the enhanced permeability and retention (EPR) effect. Once immobilized with targeting moieties, the nanoparticles can actively accumulate in tumors by the specific binding with tumor-associated biomarkers such as over-expressed receptors, tumor extracellular matrix and enzymes (Figure 15.1).

Typically, nanomaterials suitable for moleuclar imaging include organic nanoparticles (such as liposomes, polymer vesicles) or inorganic nanoparticles (such as metal NPs, carbon nanotubes, metal oxide NPs, quantum dots). Among various nanoparticles, inorganic nanomaterials have been extensively used as molecular probes because of their superior optical, electrical and magnetic properties, leading to significant enhancement in imaging contrast. For example, superparamagnetic iron oxide nanoparticles have been used as contrast agents for MRI. Gold-based nanoparticles, such as gold rods, gold nanoshells and gold nanocages, have been extensively investigated due to their unique surface plasmon resonance (SPR) in optical imaging, Raman spectroscopy and photoacoustic imaging. Quantum dots have recently received immense attention in cancer imaging because of their attractive fluorescent properties.

Compared to conventional molecular platforms used for imaging, the use of nanoparticle-based molecular probes shows several distinctive advantages and has great potential in both non-invasive detection as well as the treatment monitoring of human cancers. First, multifunctionality is a key advantage of nanomaterial-based imaging probes for cancer diagnosis. Through systematic conjugation with different targeting moieties, contrast enhancers, permeation enhancers, or even therapeutic agents, various molecular imaging nanoprobes can bypass biological barriers, achieve the capability to recognize tumor-specific biomarkers, and visualize and quantify tumor growth or regression during therapy with improved imaging sensitivity and specificity. Oftentimes, multiple modalities are involved. Secondly, many nanomaterials possess unique physical properties which can be used as sensors/reporters for molecular imaging applications, and these properties simply

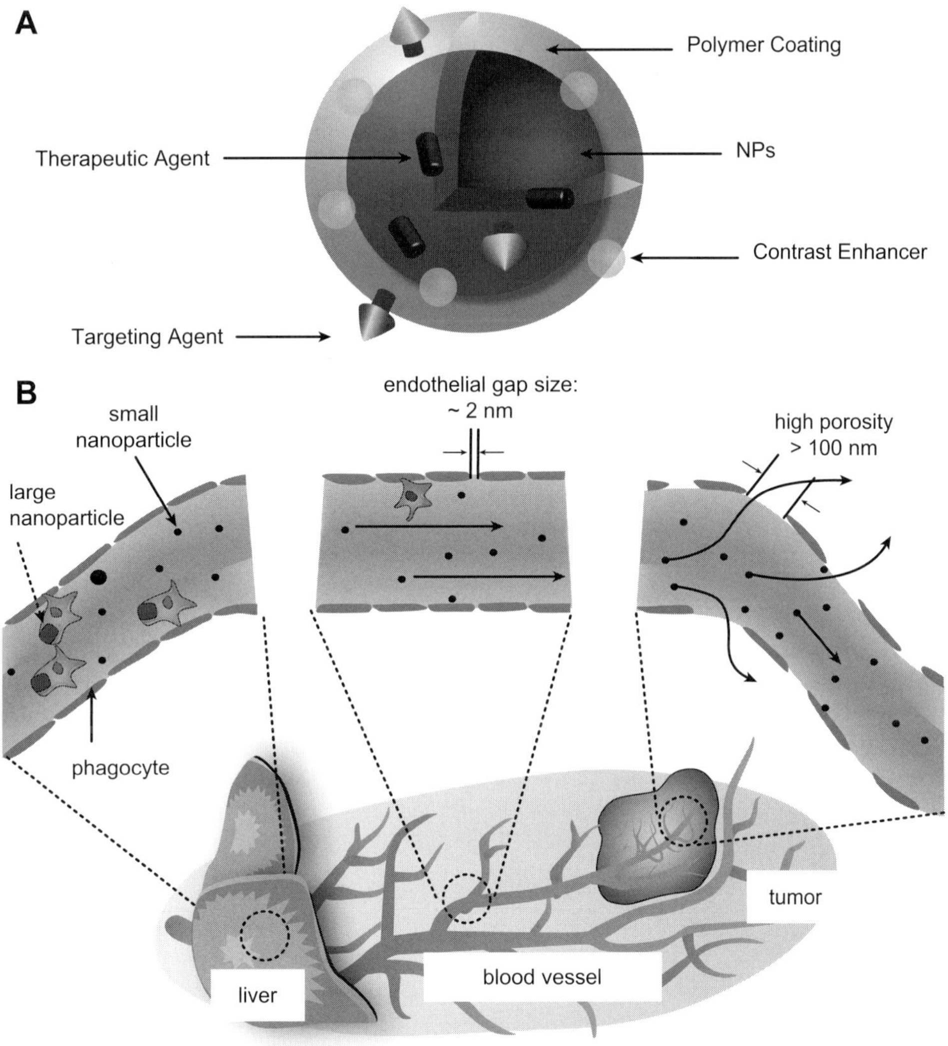

FIGURE 15.1 (A) Multifunctional nanoparticles (NP). Coupled with biological targeting agents and drug molecules, such multifunctional NPs have shown great promise as target-specific delivery vehicles with enhanced imaging sensitivity and therapeutic efficacy. (B) Behavior of nanoparticles in blood vessels in the liver, normal tissue, and tumor tissue regions. The enhanced permeability and retention (EPR) effect of the tailored multifunctional NPs is greatest at cancer cells. See color plate section. *(Reprinted with permission from [3]. Copyright 2005, Nature Publishing Group.)*

cannot be achieved by other molecular platforms. Furthermore, the unique and diverse physical and chemical properties of nanoprobes are derived from the composition, size and shape of relevant nanomaterials. Because each individual nanoprobe has nearly identical physical and chemical properties, they can be designed to have the ability to detect a large number of different biomarkers at the same time to provide sufficient information needed to characterize cancers, which is known as multiplexing. Thirdly, engineered nanomaterials can potentially alter and improve the pharmacokinetics of molecular probes *in vivo*. By coating nanomaterials with biocompatible polymers, the circulation half-life of molecular probes can be extended to have more opportunities to penetrate though vascular endothelium and reach the desired sites while minimizing uptake by the reticuloendothelial system (RES). All these advantages indicate that nanoprobes can play important roles in cancer diagnosis. With the ongoing efforts to enhance their targeting ability and endow more functions, the future of nanomaterial-based molecular imaging probes is highly promising. This chapter summarizes some of the most recent developments of nanomaterials in state-of-the-art molecular imaging and highlights their impact on cancer diagnosis.

MAGNETIC RESONANCE IMAGING (MRI)

MRI is one of the most powerful non-invasive diagnostic techniques applied for whole body imaging. The technology is based on the relaxation of proton spins in a magnetic field [9–11]. The signal intensity in MRI corresponds to the magnetic relaxation rate of water protons. Under a strong static field B_0 (ranging from 1.5 to 3 Tesla (T) in clinic and 7 T or higher in laboratories), unpaired nuclear spins align themselves with the field in either a parallel or an anti-parallel orientation. An electromagnetic radiofrequency pulse is applied to change the alignment of the spins. The flipped spins tend to return to their original state, which is recorded as either longitudinal (T_1 recovery, spin–lattice) or transverse (T_2 decay, spin–spin) relaxations. Water protons in different tissues or organs have different relaxation times, resulting in the endogenous MRI contrast. T_1 recovery leads to a brighter image due to the increase in magnetic signal, but T_2 decay reduces the imaging intensity. Introduction of appropriate exogenous contrast agents will help to accelerate the relaxation time of either T_1 or T_2, thereby further enhancing the local contrast. Commercially available T_1 contrast agents are currently paramagnetic complexes, such as gadolinium (Gd^{3+}) chelates (Gd-DTPA), gadolinium-containing liposomes/micelles. The most representative T_2 contrast agents are iron oxide nanoparticles.

Nanoparticle-Based MRI Contrast Agents

The iron oxide nanoparticles (IONPs), including magnetite (Fe_3O_4) or maghemite (γ-Fe_2O_3), have been extensively studied, marketed and widely used as MRI contrast agents in clinic [10,11]. Among these IONPs, the superparamagnetic iron oxide nanoparticles (SPIOs) are very useful for biomedical applications as they are not subject to strong magnetic interactions in dispersion and are readily stabilized under physiological conditions [9,12,13]. As T_2-weight contrast reagents, SPIOs can be magnetically saturated in the normal range of magnetic field strengths used in MRI scanners (usually 3 Tesla), thereby establishing a substantial locally perturbed dipolar field that leads to a remarkably shortening of proton relaxation of the surrounding solid tissue area [14,15]. Several commercial SPIOs, such as Feridex, Resovist and Combidex, consist of multiple iron oxide cores within a dextran stabilization shell and with sizes in the range of 50–160 nm. These nanoparticles have been widely exploited for liver, spleen and lymph node imaging because they are more likely to be recognized by the RES and subjected to macrophage uptake compared to the nanoparticles with sizes smaller than 50 nm. Such a feature of SPIOs was thus applied for the imaging of liver tumors or metastases because pathologic liver tissues are devoid of macrophages and can be imaged as white contrasts on the T_2 weighted maps, hence increasing the contrast between the diseased tissues and healthy tissues. For instance, Ferumoxides, dextran-coated IONPs with a hydrodynamic size between 120 and 180 nm, were one of the first tested for liver imaging. However, the fast hepatic and splenic clearance of nanoparticles from the blood result in a dramatic decrease in nanoparticles' circulation time, thereby adversely influencing their ability to target the desired sites.

A decrease in overall size can usually reduce RES uptake and increase circulation time of NPs. Recently, ultra-small Fe_3O_4 NPs (10 nm in hydrodynamic diameter) have been prepared for tumor-specific detection [16]. Fe_3O_4 NPs with a 4.5 nm core size were synthesized and *in situ* conjugated with c(RGDyK) peptide through one step Mannich reaction. The targeting ability of c(RGDyK)-Fe_3O_4 NPs to integrin $\alpha_v\beta_3$ *in vivo* was evaluated with mice bearing U87MG tumors. The relaxivity of RGD-IONPs was significantly higher than that of the commercial Feridex NPs with similar core size. After intravenous injection, these particles were found to concentrate selectively in U87MG tumor cells. Although the decrease in IONP size leads to less RES uptake and longer circulation time, by which IONPs have enough time to reach the target sites, the magnetic moments of IONPs gradually reduce from 220 to 78 $mM^{-1}s^{-1}$ as their sizes drop from 12 nm to 4 nm, due to the surface effect.

The magnetism of NPs at a certain magnetic field is dependent on magnetocrystalline anisotropy and the size of the NP. Recently, metal alloy magnetic nanoparticles, such as FeCo and FePt, were found typically to show higher magnetic moments than those of ferromagnetic nanoparticles. Dai and co-workers synthesized FeCo nanoparticles stabilized with single graphitic carbon shells (FeCo/GC) [17]. All the magnetic spins of these NPs can align parallel to the external magnetic field, resulting in ultrahigh saturation magnetization, r_1/r_2 relaxivities. The *in vivo* intravascular T_1 MRI in rabbits successfully demonstrated long-circulating positive contrast enhancement ability of those NPs.

Target-Specific Molecular Imaging

The sensitivity of magnetic nanoparticles for detection and molecular imaging of cancers has been further evaluated. In a recent comprehensive study on spinel metal ferrites, Mn-doped iron oxide NPs ($MnFe_2O_4$) were shown to have a relatively high mass magnetization value of 110 (emu per mass of magnetic atoms) [18]. Upon conjugation with appropriate cancer-targeting molecules, Herceptin, which is an antibody specifically binding to the human epidermal growth factor receptor 2 (HER2)/neu over-expressed on the surface of breast and ovarian cancers, the engineered $MnFe_2O_4$ nanoprobes were utilized for the target-specific molecular imaging of those cancers. As seen in Figure 15.2, the positive tumor treated with the $MnFe_2O_4$-Herceptin displays obvious contrast change in the color-coded MRI. In contrast, those treated with the Fe_3O_4-Herceptin or Herceptin conjugated cross-linked iron oxide (CLIO) at the same dosage do not exhibit apparent change in the color-coded MR images, which is probably attributed to either lower magnetism of both Fe_3O_4 and CLIO or worse extravasation of the large size of CLIO.

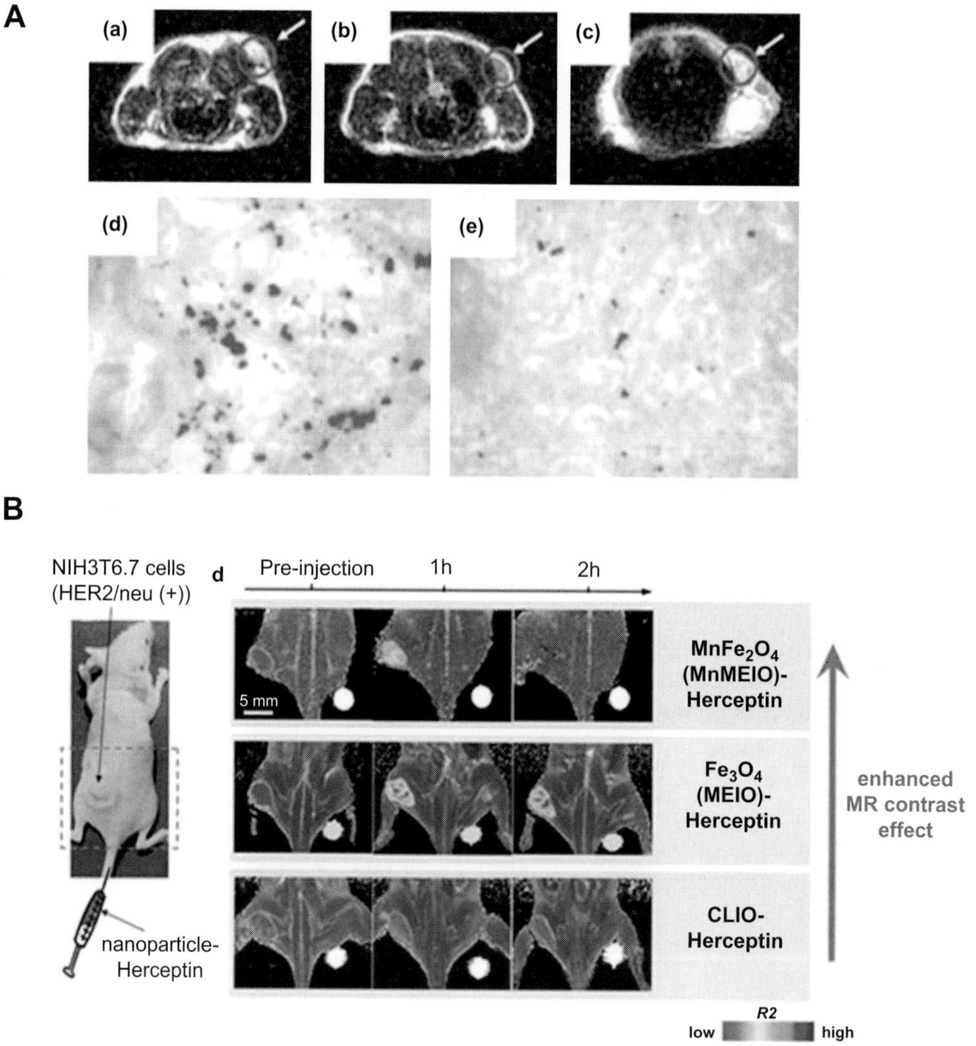

FIGURE 15.2 (A) MRI of U87MG tumor implanted in a mouse: (a) without NPs, (b) with c(RGDyK)-Fe_3O_4, and (c) blocking; and Prussian blue staining of U87MG tumors in the presence of (d) c(RGDyK)-Fe_3O_4 NPs and (e) blocking. (B) Highly-sensitive *in vivo* cancer detection by utilization of $MnFe_2O_4$ (MnMEIO)-Herceptin nanoparticle probes. For comparison, MEIO-Herceptin probes and CLIO-Herceptin probes were also tested. Color-mapped MR images of the mouse at different times following injection. See color plate section. *(Reprinted with permission from [16,18]. Copyright 2008, American Chemical Society. Copyright 2007, Nature Publishing Group.)*

NANOPARTICLE-BASED CT

Because of its unique nature, such as high-contrast resolution and unlimited penetration depth, x-ray CT is one of the most common imaging techniques in clinical use. The high-contrast images of anatomical structures are derived from the intrinsic differences in absorption of x-rays for body tissues such as bone, fat, air and water. However, its soft-tissue contrast is relatively poor. In order to detect tumors or structural abnormalities related to diseases, the adminstration of a large amount of contrast agent is often required. Current CT contrast agents in clinic are typically based on iodine- and gadolinium-containing molecules. A major concern in using these materials for CT is non-specific accumulation and rapid clearance. Compared to conventional contrast agents, nanoparticle-based CT contrast agents afford superior contrast enhancement and prolonged blood circulation time. Gold NPs have been recently developed for CT imaging, and they provide about 2.7 times greater contrast per unit weight than iodine, thereby reducing the interference from bone absorption [19]. However, a large dose of gold nanoparticles still needs to be adminstered to achieve high contrast images of tumors or other soft tissues.

Alternative nanoprobes for CT imaging have attracted more attention recently. The use of polymer-coated Bi_2S_3 NPs as a CT probe was reported by Weissleder and co-workers [20]. These NPs show a fivefold increase in x-ray absorption compared to iodine, very long circulation times *in vivo* and a favorable efficacy/safety profile. It is worth noting that though these NPs and their bioconjugates show potential for *in vivo* imaging of the vasculature, liver and lymph nodes, the intrinsic toxicity of Bi_2S_3 NPs may limit their clinical translation. Unlike Bi_2S_3 NPs, tantalum oxide NPs have relatively low toxicity and have been reported as nanoprobes suitable for high-performance CT imaging, and they exhibit remarkable potential for image-guided lymph node mapping [21].

Recently, several types of nanoparticles, including europium (Eu)-doped Gd_2O_2S or Eu-doped barium yttrium fluoride ($Ba_{0.55}Y_{0.3}F_2{:}Eu^{3+}$) phosphors, were found to emit near infrared (NIR) light under low dose x-ray excitation (cGy range) [22–25]. This discovery laid down the solid foundation for developing x-ray luminescence computed tomography (XLCT) techniques for molecular imaging.

XLCT is a new dual modality molecular imaging technique based on x-ray excitable phosphorescent nanoparticles. These nanoparticles, capable of converting x-ray radiation into NIR light, can be modified with biological molecules to target specific disease or biological processes. CT is performed by acquiring and reconstructing multiple projections. The XLCT technique merges anatomical imaging with functional assays of the disease, thereby, integrating these two functions. More importantly, XLCT may overcome the limitations associated with conventional x-ray, CT, or optical imaging alone, and find broad applications in cancer imaging and management.

NUCLEAR IMAGING

The major advantages of nuclear molecular imaging for cancer diagnosis are high sensitivity, accurate and reproducible quantitation ability, and unlimited tissue penetration [26]. Current nuclear imaging methods includes PET and SPECT. Basically, PET is one of the most common nuclear imaging techniques that detects pairs of photons emitted indirectly by a positron-emitting radionuclide (i.e. ^{11}C, ^{18}F, ^{64}Cu, ^{94m}Tc, ^{124}I, etc.) which could be introduced to the biomolecules or nanoparticles and administered to a live subject. Three-dimensional *in vivo* distribution or pharmacokinetics of PET probes can then be reconstructed by computer analysis. SPECT is another nuclear imaging technique using gamma rays emitted by radionuclides. Similar to conventional nuclear planar imaging, SPECT imaging is performed by using a gamma camera to acquire multiple two-dimensional (2D) projection images from multiple angles. After tomographic reconstruction of multiple 2D views, three-dimensional distribution of the nuclear probes *in vivo* can be freely reformatted as required. However, the detection efficiency of SPECT imaging is generally much lower than that of PET imaging because of the use of rotating detectors and collimators to reject many gamma photons. The spatial resolution of SPECT is also generally lower than that of PET because a PET scanner provides more localization information of radiation events than that of SPECT. The available gamma emitting radionuclides for SPECT are usually inexpensive, including ^{99m}Tc, ^{111}In, $^{123/125/131}I$, etc.

Nanoparticle-Based PET Imaging

Many radionuclides have been introduced into a variety of nanoparticles, including quantum dots, gold NPs, and carbon nanotubes [27–30]. The major advantages of nanoparticle-based PET imaging are to provide multiple functionalities and multiple modalities. The composition, structure, and surface coating of nanoparticles can be optimized to achieve controlled circulation *in vivo* and specific tumor targeting. By taking advantage of PET, radionuclide-labeled NPs have the ability to provide non-invasive imaging of nanoprobes, tissue distribution and pharmacokinetics.

Dai and co-workers investigated the biodistribution and tumor targeting ability of single wall carbon nanotubes (SWNT) coated non-covalently with polyethylene glycol (PEG), cyclic RGD peptides, and chelating agent for ^{64}Cu chelation [31]. The SWNTs of this study were uniform and short, typically having lengths of approximately 100 nm and diameters of 1–3 nm. PEG coating significantly prolonged the blood circulation times of NPs and reduced their non-specific uptake by RES. Efficient targeting of integrin-positive tumors was achieved with the resultant ^{64}Cu labelled SWNT-RGD.

Radionuclides can be incorporated into various NPs via appropriate chelating agents to form stable complexes [26]. However, the radiometal-chelator complexes attached to NPs could inevitably affect the physicochemical properties, biodistribution and pharmacokinetic of nanoprobes. Moreover, detached radiometals from NPs could lead to failure of reflecting the real *in vivo* behavior of whole nanoprobes. Therefore, it is necessary to develop a chelator-free strategy to validate the *in vivo* data. A wide selection of insoluble nano-sized metal salts has recently been screened for robust binding of various radionuclides [32]. Hydroxyapatite and aluminum hydroxide NPs showed particularly stable binding of ^{18}F in biological conditions. Although ^{18}F-labeled hydroxyapatite exhibited high stability in circulation, the gradual degradation was not negligible once they were trapped and accumulated in liver, resulting in release of ^{18}F which eventually accumulated in bone. In contrast, ^{18}F-labeled aluminum hydroxide showed much better stability than hydroxyapatite even though it was cleared through liver and spleen [32]. Another example is ^{64}Cu-doped CuS NPs. Li and co-workers demonstrated that, by integration of the positron emitter ^{64}Cu into the crystal structures, monodisperse CuS NPs with average size of 10 nm become an ideal nanoplatform for application of PET image-guided therapy [33]. These ^{64}Cu-doped CuS NPs exhibit relatively high uptakes in U87MG xenograft tumors as a result of EPR effect.

Nanoparticle-Based SPECT Imaging

Although the resolution and detection efficiency of SPECT imaging are lower than those of PET, the potential ability to achieve multiplex imaging by using different radionuclides makes SPECT very promising in molecular imaging [26]. However, no multiplexing SPECT imaging of radiolabeled NPs has been reported so far and such a possibility remains to be tested. Dependent on different applications, the NPs should be labeled with different radionuclides for SPECT because they have different decay time, energy and chemical properties. For instance, in order to study pharmacokinetics, tumor uptake and therapeutic efficacy of NPs within a week, ^{111}In-labeled chimeric L6 (ChL6) monoclonal antibody was conjugated with iron oxide NPs, which were administered to athymic mice bearing human breast cancer HBT 3477 tumors [34]. SPECT imaging was performed to monitor the tumor uptake of NPs and evaluate the tumor necrosis after treatment with alternating magnetic field (AMF) induced hyperthermia. The delay in tumor growth could be observed after hyperthermia treatment.

OPTICAL MOLECULAR IMAGING

One of the most inexpensive and rapid methods to image tumor at a molecular level is optical imaging [6,35,36]. Recent advances in design of both biologically compatible NIR nanoprobes and of highly sensitive photon-detection technologies have facilitated the application of optical molecular imaging modalities in cancer diagnosis. Fluorescence and luminescence imaging techniques are of particular interest to cancer diagnosis because of their low cost, high sensitivity and versatility. In fluorescence imaging, the excitation light illuminates the probes and the emission light at the shifted wavelength is collected. The probes detectable in the NIR region (700−900 nm) are highly desirable because of maximal tissue penetration and minimal autofluorescence. Bioluminescence imaging (BLI) is used to detect the luminescence released from a biochemical reaction between luciferin and luciferase [37]. The sensitivity of BLI is higher than that of fluorescence imaging because there is no inherent background signal. However, its potential application has been mainly limited to preclinical studies due to low tissue penetration of bioluminescence and complexity of transgenic techniques. Other optical imaging or optical-related imaging techniques such as *in vivo* Raman imaging and photoacoustic imaging (PAI), which are based on different physical principles and can generate images of different contrasts, have gained considerable interest mainly for their excellent tissue penetration ability and high spatial resolution.

NIR Fluorescent Nanoprobes

According to the excitation mechanism, NIR fluorescent nanoparticles can be classified into two categories: downconversion fluorescent NPs (DCNs) and upconversion fluorescent NPs (UCNs). Traditional fluorescent NPs such as fluorescent dye-loaded NPs, quantum dots, SWNTs and gold clusters are mainly based on downconversion fluorescence and generally emit long-wavelength light when excited by short-wavelength light. They often suffer from autofluorescence, photodamage or photobleaching. In contrast, UCNs convert low-energy light into high-energy fluorescence, thus making them another class of promising fluorescent labels for *in vivo* applications [29].

Fluorescent dye-loaded NPs are very common for biological applications. Generally, organic or metallorganic dye molecules are attached on or incorporated within the inorganic matrix which is optically transparent to excitation or emission light. Typical inorganic matrices include silica NPs and calcium phosphate NPs. Compared to NIR fluorophores alone, dye-loaded NPs show several desirable features for molecular imaging. First, the high payload of dye molecules per NP results in enhanced sensitivity. Secondly, chemically inert matrix could protect dyes from a harsh environment, thereby increasing photostability and biocompatibility of dye-loaded NPs. Thirdly, the matrix often provides versatile functionality for further biomodification with numerous targeting moieties such as antibodies or peptides.

The most widely studied NPs for optical imaging are quantum dots, which are nanometer-sized inorganic fluorescent semiconductor crystals with superior and unique photophysical properties, such as high quantum yield, broad excitation spectrum, narrow and symmetric emission spectra, size-dependent photoluminescent emission, and excellent photostability. In order to reduce further the autofluorescence and enhance tissue penetration, NIR-emitting QD, including cadmium (Cd) and non-cadmium based QDs, have been recently developed to overcome the limitations of dye-loaded NPs and meet the requirements of *in vivo* biological imaging applications [38–43]. Considering the potential acute and chronic toxicity and instability of some NIR QDs, it is necessary to coat them with a layer of less toxic ZnS or ZnSe, which can protect the NIR QDs against photo-oxidation and improve the quantum yield, such as CdTe/ZnS, $InAs_xP_{1-x}$/InP/AsSe [28,44]. Furthermore, the Cd metal in NIR QDs could be replaced by other more benign elements. The resultant NIR QDs, for example, $CuInS_2$/ZnS core/shell QDs, show great potential as biocompatible nanoprobes for molecular imaging. Gao and co-workers first demonstrated *in vivo* cancer targeting with QD-antibody conjugates using a small animal model [44]. Recently, Cheng and co-workers reported *in vivo* targeted imaging of tumor HER2 receptors using anti-HER Affibody protein-conjugated non-cadmium NIR QD (InAs/InP/ZnSe core/shell/shell), which exhibits specific tumor targeting ability and excellent tumor imaging contrast [39]. Like many other fluorescence probes, external light is required to illuminate the QDs to produce fluorescent signals. This limits the *in vivo* use of QDs due to strong background autofluorescence, poor tissue penetration of the excitation photons, as well as significant absorption and scattering of these photons in tissues. To circumvent these problems, self-illuminating QD conjugates that can emit red to NIR light without external excitation would be ideal. Cheng and co-workers developed a new strategy, and they successfully used luminescence produced by β-radionuclides as an internal light source simultaneously to excite multiple QDs with different sizes and emission wavelengths for *in vivo* multiplexing fluorescence imaging [45].

Upconversion NPs

The basic structure or composition of the upconversion NPs (UCNs) consists of two trivalent lanthanide ions (such as Tm^{3+}, Er^{3+} and Yb^{3+}) embedded in a matrix as dopants, one for absorption (e.g. Yb^{3+}) and the other for emission (e.g. Er^{3+}) [46]. The matrices or host materials include inorganic fluorescent nanocrystals such as fluorides (e.g. LaF_3, YF_3, and $NaYF_4$), oxides (e.g. Y_2O_3) and phosphates (e.g. $LaPO_4$). Like QDs, UCNs are extremely bright and photostable in biological conditions. Recently, UCNs have been demonstrated to be unique in their ability to image tissues at a great depth because UCNs can absorb and produce light in the NIR region with minimal autofluorescence background [46].

IN VIVO RAMAN SPECTROSCOPY AND IMAGING

Raman spectroscopy is a spectroscopic technique which gives vibrational information about molecules. The incident light interacts with molecular vibrations, phonons or other excitations in the system, resulting in the energy of the inelastically scattered light being shifted up or down, which reflects the information about the vibrational modes in the system. However, the spontaneous Raman scattering is typically very weak and the fluorescence sometimes overwhelms the bands in the Raman spectrum rendering the result useless. In order to enhance sensitivity and improve spatial resolution, several variations of Raman spectroscopy have been developed, including surface-enhanced Raman scattering (SERS) spectroscopy and Raman microscopy (Figure 15.3). SERS spectroscopy is now becoming one of the most promising non-invasive molecular imaging techniques because of its advantageous characteristics in terms of multiplexing ability, specificity, selectivity and sensitivity [47]. It provides structure-specific vibrational spectra of analytes with extremely high sensitivity, even much higher than fluorescence imaging.

A new class of colloidal metal NPs such as gold, silver NPs have been developed for enhancing the efficiency of SERS spectroscopy by exceptional enhancement factors in the order of 10^{14}–10^{15}, thereby making it possible for spectroscopic detection and identification of single molecules on the nanoparticle surface [48–50]. Such SERS NPs usually consist of metal cores, embedded reporter molecules and a silica or polymer encapsulation layer [51,52]. The phenomenon of SERS is generally explained by two different mechanisms: chemical enhancement and electromagnetic enhancement [53,54]. The former involves a new charge-transfer state between the reporter molecules and metal surface. The latter describes the surface electron movement in the metal substrate. The SERS NPs have the ability to provide very detailed spectroscopic information and are much brighter than QDs in the NIR region [55,56]. These properties raise new opportunities for multiplexed molecular diagnosis and *in vivo* molecular imaging.

The pioneering work of Nie and co-workers recently demonstrated *in vivo* tumor targeting and spectroscopic detection with SERS NPs [51]. They developed a class of biocompatible and non-toxic gold NPs coated with polyethyl glycol and embedded with many Raman reporters such as crystal violet, Nile blue, cresyl violet and diethylthiatricarbocyanine (DTTC). These PEGylated SERS NPs exhibit excellent stability under physiological conditions and provide brighter signals than wavelength-matched QD in the NIR region. After conjugation with tumor targeting molecules, such as single-chain variable fragment (ScFv) antibodies, the resultant SERS nanoprobes display the ability to target tumor cells overexpressing epidermal growth factor receptors (EGFR) in xenograft tumor models.

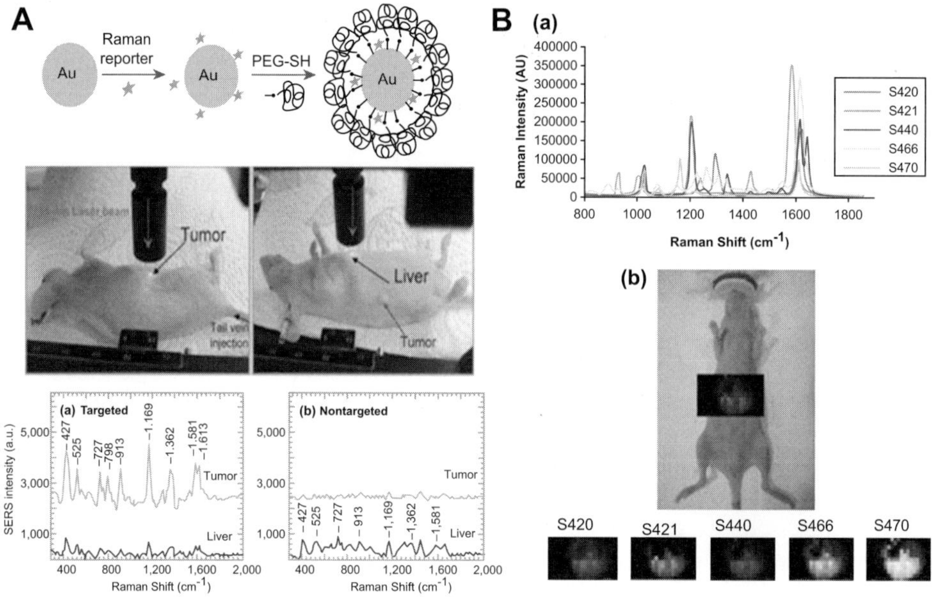

FIGURE 15.3 (A) Design and characterization of PEG-SH coated SERS (surface-enhanced Raman scattering) nanotags. *In vivo* cancer targeting and surface-enhanced Raman detection by using ScFv-antibody conjugated gold nanoparticles that recognize the tumor biomarker EGFR. Top: Photographs showing a laser beam focusing to the tumor site or to the anatomical location of liver. Lower: SERS spectra obtained from the tumor and the liver locations by using (a) targeted and (b) non-targeted nanoparticles. *In vivo* SERS spectra were obtained from the tumor site (red) and the liver site (blue) with 2-s signal integration and at 785 nm excitation. The Raman reporter molecule is malachite green, with distinct spectral signatures as labeled. (B) Demonstration of deep-tissue multiplexed imaging 24 h after i.v. injection of five unique SERS nanoparticle batches simultaneously. (a) Graph depicting five unique Raman spectra, each associated with its own SERS batch: S420 (red), S421 (green), S440 (blue), S466 (yellow), and S470 (orange). (b) Raman image of liver overlaid on digital photo of mouse, showing accumulation of all five SERS batches in the liver after 24 h post i.v. injection. Panels below depict separate channels associated with each of the injected SERS nanoparticle batches. See color plate section. *(Reprinted with permission from [51,57]. Copyright 2008, Nature Publishing Group. Copyright 2009, National Academy of Science.)*

SERS NPs have the ability to offer picomolar sensitivity and multiplexing capabilities for Raman molecular imaging. Gambhir and co-workers recently demonstrated multiplexed imaging of SERS NPs in live subjects [57,58]. Ten different types of SERS NPs injected subcutaneously in a living mouse could be separated spectrally. Furthermore, after intravenous injection of five most intense and spectrally unique SERS NPs, they all were successfully identified and spectrally separated in the liver using optimized non-invasive Raman imaging [28]. These SERS nanoparticles can also be modified with anti-EGFR Affibody protein (7 kDa) for specific Raman imaging of EGFR positive tumors *in vivo* [59]. SWNTs have also been explored as a nanoplatform for *in vivo* Raman imaging. These NPs were successfully used for *in vivo* targeted Raman imaging of tumor vasculature using RGD peptide-conjugated SWNTs, which exhibited specific targeting affinity of cell adhesion molecule integrin $\alpha_v\beta_3$ highly expressed in the tumor vasculature and tumor cells [57,60].

PHOTOACOUSTIC IMAGING

Photoacoustic imaging (PAI) is a newly emergent molecular imaging technique and has attracted significant research interest recently. It holds great promise for clinical translation and possesses several unique advantages over many other imaging modalities. As a promising non-ionizing, non-invasive and highly sensitive molecular imaging technique, PAI takes advantage of both optical and acoustic imaging, and thus overcomes the depth limitation of most traditional optical imaging techniques [61]. Basically, the optical energy is absorbed and partially converted into heat, resulting in transient thermoelastic expansion and thus ultrasonic emission, which is then detected by ultrasonic transducers to form images. It overcomes the depth limitation and can recognize small lesions at depths of up to 5 cm. Moreover, PAI provides functional and molecular information of tumors in real time with high imaging contrasts and ultrasonic spatial resolution through either endogenous or exogenous PAI contrasts. Due to available intrinsic contrast molecules, such as hemoglobin or melanin, PAI has been used to visualize blood vessel structure, thermal burns, and melanoma for tumor angiogenesis monitoring, blood oxygenation mapping, functional brain imaging, and skin melanoma detection. However, most diseases such as cancers do not show good PA contrast using endogenous PAI molecules, thereby it is necessary to develop new exogenous contrast agents which could provide sufficient PA signal even in low concentrations.

Recent advances in nanotechnology greatly benefit the development and design of high photoacoustic contrast agents. The amplitude of photoacoustic signal and contrasts are dependent on many factors such as optical absorption, optical-to-acoustic conversion efficiency, and heat transfer of the target medium [61]. Various light-absorbing NPs, such as Au-based nanostructures and carbon nanotubes, have been studied as PAI probes [62–75]. Although carbon nanotubes have shown promising properties as contrast agents for PAI, their imaging sensitivity is relatively low because of the low NIR light absorption coefficient. Moreover, their toxicity remains a concern [73]. The Au nanostructures such as Au-nanorods, nanocages and nanoshells exhibit superior optical properties, biocompatibility and safety, and thus they have been actively investigated in the fields of optical imaging and PAI. When Au nanostructures interact with incident light, they give rise to a strong extinction peak in the visible and NIR regions because of the localized surface plasmon resonance. Such extinction peaks are comprised of two components, scattering and absorption, which can be tuned by simply controlling the size, shape and other physical dimensions of the nanostructures [61]. Among various optical Au-NPs, ones with relatively dominant absorption over scattering are highly desirable for PAI. For example, Au-nanoshells with dielectric cores show strong optical resonance but relatively weak optical absorption, and therefore they are not ideal candidates for PAI [69,76,77]. By way of contrast, Au-nanorods and nanocages exhibit non-spherical symmetry and their optical absorption cross-sections (σ_a) in the NIR region could be maximized by manipulating their aspect ratios and thus are more suitable for PAI [61]. Although Au-nanorods and nanocages are good PAI enhancers, they generally have relatively large particle sizes ($\approx$50 nm) and unfavorable *in vivo* profiles [61,64,65,73]. In order to optimize the photoacoustic agents to achieve high imaging sensitivity and desired *in vivo* tumor targeting properties, novel Au-NPs with different shapes,

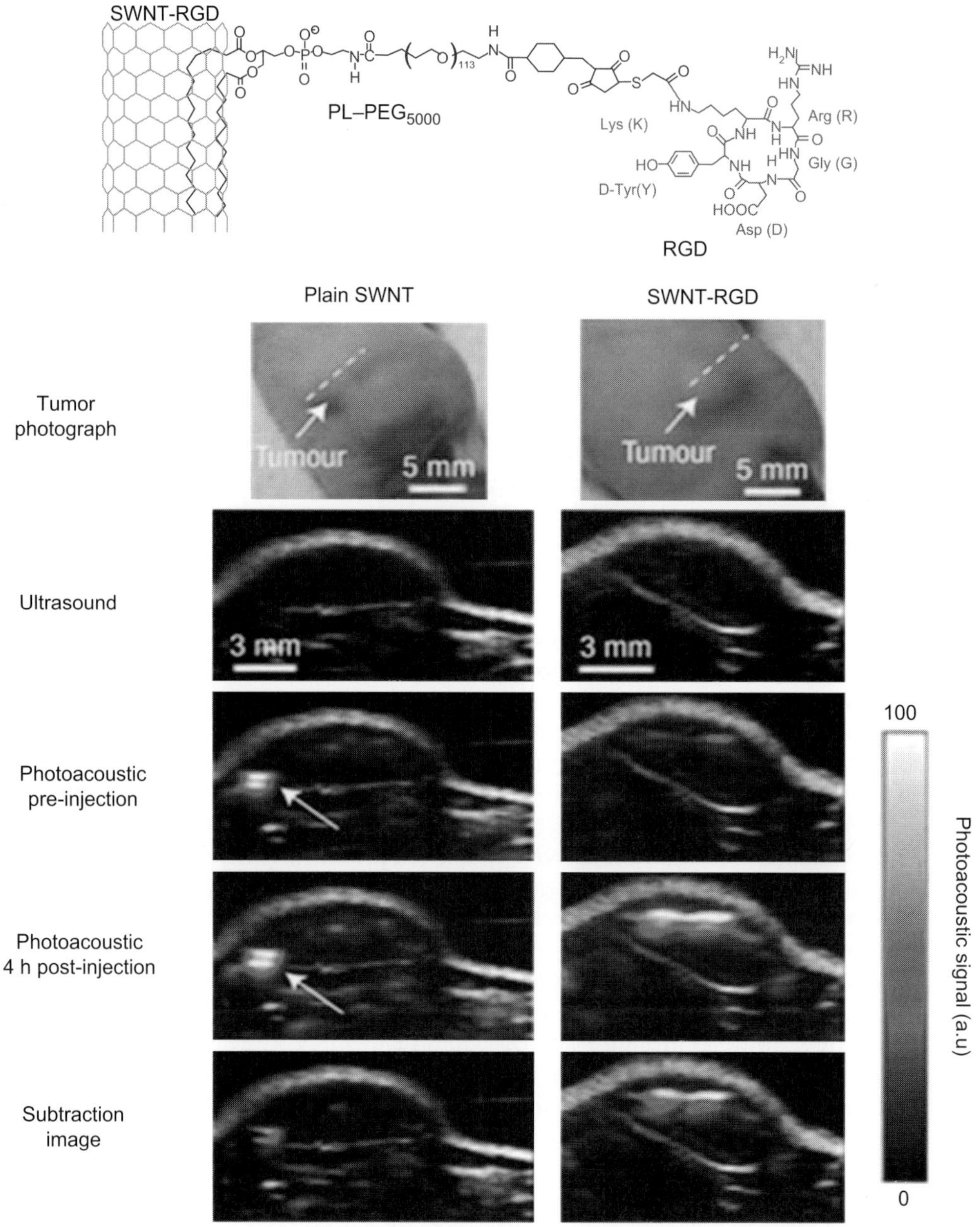

FIGURE 15.4 Single-walled carbon nanotube targets tumor in living mice. Ultrasound (gray) and photoacoustic (green) images of one vertical slice (white dotted line) through the tumor. The ultrasound images show the skin and tumor boundaries. Subtraction images were calculated as the 4 h post-injection image minus the pre-injection image. The high photoacoustic signal in the mouse injected with plain single-walled carbon nanotubes (indicated with a white arrow) is not seen in the subtraction image, suggesting that it is due to a large blood vessel and not single-walled carbon nanotubes. See color plate section. *(Reprinted with permission from [75]. Copyright 2010, American Chemical Society.)*

sizes and optical properties are under active investigation in the bionanotechnology and molecular imaging communities.

Gambhir and co-workers have recently reported that cyclic-RGD conjugated SWNTs could be used as PAI contrast agents to image $\alpha_v\beta_3$ integrins associated with tumor angiogenesis [66]. However, their minimal detectable concentration in living subjects was relatively high ($\approx$50 nM). This research group continuously optimized the detection limitation of SWNTs by attaching indocyanine green (ICG, NIR dye) to the surface of SWNT though π-π stacking interactions (Figure 15.4). The dye-enhanced SWNTs provided acceptable PAI contrast signals even at the subnanomolar level. These nanoprobes gave 300 times higher PA contrast in living subjects than bare SWNTs at the same concentration [75].

TAKE-HOME MESSAGES

Thanks to the rapid progress of bionanotechnology, numerous nanoparticle-based molecular probes have been developed, and diagnostic techniques have grown exponentially during the past decades. The successful development of diverse nanoprobes and their potential applications for cancer diagnosis have greatly inspired significant research interests in molecular imaging and personalized medicine. One of the most representative examples is iron oxide nanoparticles, which have been well-studied as contrast agents for MRI molecular imaging. The recent development of specific and sensitive NIR fluorescence-based NPs showed great potential for cancer molecular imaging. Combining the ultrasensitive detection and/or excellent spatial resolution of "state-of-the-art" imaging techniques (i.e. *in vivo* Raman spectroscopy or photoacoustic imaging) with the multiplexing capabilities and multifunctionalities of nanoparticles (gold-based NPs, carbon nanotubes) greatly facilitates the progress of non-invasive molecular imaging of tumors in live subjects. Due to functional diversity, composition and morphology of NPs, the multimodality molecular imaging such as PET/MRI, MRI/optical, MRI/PAI, etc. becomes possible and can offer synergistic advantages over each individual modality. The use of multifunctional nanoparticles with multimodality would have great advantages for monitoring the tumors with accurate location and anatomical structure and providing intraoperative surgical guidance for tumor tissue resection.

In order to achieve the objective for the clinical translation of nanoparticle-based molecular imaging, many concerns still need to be addressed, including biocompatibility, pharmacokinetics, targeting efficacy, and acute/chronic toxicity of engineered nanoparticles. Due to the enormous challenges, however, only a few nanoprobes have so far shown promising properties for translation into clinical use. Although there are successful examples for nanoparticle-based molecular imaging of tumors in small animal models, the unfavorable distribution due to non-specific uptake by RES, and undesirable toxicity inherently derived from the nanoparticle composition adversely influence the potential clinical use of nanoparticles. Therefore, there is an urgent need to exploit the rapid advances in nanomaterials by extensively examining their *in vivo* behavior, pharmacokinetics and toxicity.

Upon advances in synthesis and characterization of nanoparticles, remarkable progress in the molecular basis of cancers, and continuous improvement in imaging instrumentation, nanoparticle-based molecular imaging will eventually contribute in many important ways to the improved care of cancer patients in future.

References

[1] Jemal A, Siegel R, Xu JQ, Ward E. Cancer statistics, 2010. Ca-a Cancer J Clin 2010;60:277–300.

[2] Rudin M, Weissleder R. Molecular imaging in drug discovery and development. Nat Rev Drug Discov 2003;2:123–31.

[3] Ferrari M. Cancer nanotechnology: Opportunities and challenges. Nat Rev Cancer 2005;5:161–71.

[4] Weissleder R. Scaling down imaging: Molecular mapping of cancer in mice. Nat Rev Cancer 2002;2:11–8.

[5] Gambhir SS. Introduction. J Nucl Med 2008;49:1S–4.

[6] Willmann JK, van Bruggen N, Dinkelborg LM, Gambhir SS. Molecular imaging in drug development. Nat Rev Drug Discov 2008;7:591–607.

[7] Weissleder R. Molecular imaging in cancer. Science 2006;312:1168–71.

[8] Cai WB, Chen XY. Nanoplatforms for targeted molecular imaging in living subjects. Small 2007;3:1840–54.

[9] Frey NA, Peng S, Cheng K, Sun SH. Magnetic nanoparticles: synthesis, functionalization, and applications in bioimaging and magnetic energy storage. Chem Soc Rev 2009;38:2532–42.

[10] Na HB, Song IC, Hyeon T. Inorganic nanoparticles for MRI contrast agents. Adv Mater 2009;21:2133–48.

[11] Jun YW, Lee JH, Cheon J. Chemical design of nanoparticle probes for high-performance magnetic resonance imaging. Angew Chem Int Ed 2008;47:5122–35.

[12] Santini J, Formento JL, Francoual M, et al. Characterization, quantification, and potential clinical-value of the epidermal growth-factor receptor in head and neck squamous-cell carcinomas. Head Neck J Sci Specialt Head Neck 1991;13:132–9.

[13] Laurent S, Forge D, Port M, et al. Magnetic iron oxide nanoparticles: Synthesis, stabilization, vectorization, physicochemical characterizations, and biological applications. Chem Rev 2008;108:2064–110.

[14] Aboudpirak E, Hurwitz E, Bellot F, Schlessinger J, Sela M. Inhibition of human-tumor growth in nude-mice by a conjugate of doxorubicin with monoclonal-antibodies to epidermal growth-factor receptor. Proc Natl Acad Sci USA 1989;86:3778–81.

[15] Blum RH, Carter SK. Adriamycin—new anticancer drug with significant clinical activity. Ann Intern Med 1974;80:249–59.

[16] Xie J, Chen K, Lee HY, et al. Ultrasmall c(RGDyK)-coated Fe3O4 nanoparticles and their specific targeting to integrin alpha(v)beta(3)-rich tumor cells. J Am Chem Soc 2008;130:7542–3.

[17] Seo WS, Lee JH, Sun XM, et al. FeCo/graphitic-shell nanocrystals as advanced magnetic-resonance-imaging and near-infrared agents. Nat Mater 2006;5:971–6.

[18] Lee JH, Huh YM, Jun Y, et al. Artificially engineered magnetic nanoparticles for ultra-sensitive molecular imaging. Nat Med 2007;13:95–9.

[19] Popovtzer R, Agrawal A, Kotov NA, et al. Targeted gold nanoparticles enable molecular CT imaging of cancer. Nano Lett 2008;8:4593–6.

[20] Rabin O, Manuel Perez J, Grimm J, Wojtkiewicz G, Weissleder R. An X-ray computed tomography imaging agent based on long-circulating bismuth sulphide nanoparticles. Nat Mater 2006;5:118–22.

[21] Oh MH, Lee N, Kim H, et al. Large-scale synthesis of bioinert tantalum oxide nanoparticles for x-ray computed tomography imaging and bimodal image-guided sentinel lymph node mapping. J Am Chem Soc 2011;133: 5508–15.

[22] Carpenter CM, Sun C, Pratx G, Rao R, Xing L. Hybrid x-ray/optical luminescence imaging: Characterization of experimental conditions. Med Phys 2010;37:4011–8.

[23] Pratx G, Carpenter CM, Sun C, Rao RP, Xing L. Tomographic molecular imaging of x-ray-excitable nanoparticles. Opt Lett 2010;35:3345–7.

[24] Sun C, Pratx G, Carpenter CM, et al. Synthesis and radioluminescence of PEGylated Eu3+-doped nanophosphors as bioimaging probes. Adv Mater 2011;23:H195–9.

[25] Pratx G, Carpenter CM, Sun C, Lei X. X-ray luminescence computed tomography via selective excitation: a feasibility study. IEEE Transact Med Imaging 2010;29:1992–9.

[26] Welch MJ, Hawker CJ, Wooley KL. The advantages of nanoparticles for PET. J Nucl Med 2009;50:1743–6.

[27] Liu Z, Tabakman SM, Chen Z, Dai H. Preparation of carbon nanotube bioconjugates for biomedical applications. Nat Protocols 2009;4:1372–81.

[28] He X, Gao J, Gambhir SS, Cheng Z. Near-infrared fluorescent nanoprobes for cancer molecular imaging: Status and challenges. Trends Mol Med 2010;16:574–83.

[29] He X, Wang K, Cheng Z. *In vivo* near-infrared fluorescence imaging of cancer with nanoparticle-based probes. Wiley Interdiscip Rev Nanomed Nanobiotechnol 2010;2:349–66.
[30] Schipper ML, Cheng Z, Lee S-W, et al. MicroPET-based biodistribution of quantum dots in living mice. J Nucl Med 2007;48:1511–8.
[31] Kam NWS, O'Connell M, Wisdom JA, Dai HJ. Carbon nanotubes as multifunctional biological transporters and near-infrared agents for selective cancer cell destruction. Proc Natl Acad Sci USA 2005;102:11600–5.
[32] Jauregui-Osoro M, Williamson PA, Glaria A, et al. Biocompatible inorganic nanoparticles for [18F]-fluoride binding with applications in PET imaging. Dalton Transact 2011.
[33] Zhou M, Zhang R, Huang MA, et al. A chelator-free multifunctional [Cu-64]CuS nanoparticle platform for simultaneous micro-PET/CT imaging and photothermal ablation therapy. J Am Chem Soc 2010;132: 15351–8.
[34] DeNardo SJ, DeNardo GL, Natarajan A, et al. Thermal dosimetry predictive of efficacy of In-111-ChL6 nanoparticle AMF-induced thermoablative therapy for human breast cancer in mice. J Nucl Med 2007;48: 437–44.
[35] Santra S, Dutta D, Walter GA, Moudgil BM. Fluorescent nanoparticle probes for cancer imaging. Technol Cancer Res Treat 2005;4:593–602.
[36] Gao JH, Chen K, Xie RG, et al. Ultrasmall near-infrared non-cadmium quantum dots for *in vivo* tumor imaging. Small 2010;6:256–61.
[37] Bhaumik S, Gambhir SS. Optical imaging of Renilla luciferase reporter gene expression in living mice. Proc Natl Acad Sci USA 2002;99:377–82.
[38] Michalet X, Pinaud FF, Bentolila LA, et al. Quantum dots for live cells, *in vivo* imaging, and diagnostics. Science 2005;307:538–44.
[39] Gao J, Chen K, Miao Z, et al. Affibody-based nanoprobes for HER2-expressing cell and tumor imaging. Biomaterials 2011;32:2141–8.
[40] Smith BR, Cheng Z, De A, Rosenberg J, Gambhir SS. Dynamic visualization of RGD-quantum dot binding to tumor neovasculature and extravasation in multiple living mouse models using intravital microscopy. Small 2010;6:2222–9.
[41] Gao J, Chen K, Xie R, et al. *In vivo* tumor-targeted fluorescence imaging using near-infrared non-cadmium quantum dots. Bioconj Chem 2010;21:604–9.
[42] Smith BR, Cheng Z, De A, Koh AL, Sinclair R, Gambhir SS. Real-time intravital imaging of RGD–quantum dot binding to luminal endothelium in mouse tumor neovasculature. Nano Lett 2008;8:2599–606.
[43] Schipper ML, Iyer G, Koh AL, et al. Particle size, surface coating, and PEGylation influence the biodistribution of quantum dots in living mice. Small 2009;5:126–34.
[44] Gao JH, Chen XY, Cheng Z. Near-infrared quantum dots as optical probes for tumor imaging. Curr Topics Med l Chem 2010;10:1147–57.
[45] Liu H, Zhang X, Xing B, Han P, Gambhir SS, Cheng Z. Radiation-luminescence-excited quantum dots for *in vivo* multiplexed optical imaging. Small 2010;6:1087–91.
[46] Chatterjee DK, Gnanasammandhan MK, Zhang Y. Small upconverting fluorescent nanoparticles for biomedical applications. Small 2010;6:2781–95.
[47] Qian XM, Nie SM. Single-molecule and single-nanoparticle SERS: From fundamental mechanisms to biomedical applications. Chem Soc Rev 2008;37:912–20.
[48] Hutter E, Fendler JH. Exploitation of localized surface plasmon resonance. Adv Mater 2004;16:1685–706.
[49] Peng S, McMahon JM, Schatz GC, Gray SK, Sun Y. Reversing the size-dependence of surface plasmon resonances. Proc Natl Acad Sci USA 2010;107:14530–4.
[50] Willets KA, Van Duyne RP. Localized surface plasmon resonance spectroscopy and sensing. Annu Rev Phys l Chem 2007;58:267–97.
[51] Qian XM, Peng XH, Ansari DO, et al. *In vivo* tumor targeting and spectroscopic detection with surface-enhanced Raman nanoparticle tags. Nat Biotechnol 2008;26:83–90.
[52] Fernández-Lápez C, Mateo-Mateo C, Álvarez-Puebla R, et al. Highly controlled silica coating of PEG-capped metal nanoparticles and preparation of SERS-encoded particles. Langmuir 2009;25:13894–9.
[53] Nie S, Emory SR. Probing single molecules and single nanoparticles by surface-enhanced Raman scattering. Science 1997;275:1102–6.

[54] Le Ru EC, Blackie E, Meyer M, Etchegoin PG. Surface enhanced Raman scattering enhancement factors: A comprehensive study. J Phys Chem C 2007;111:13794–803.

[55] Blackie EJ, Ru ECL, Etchegoin PG. Single-molecule surface-enhanced Raman spectroscopy of nonresonant molecules. J Am Chem Soc 2009;131:14466–72.

[56] El-Sayed IH, Huang X, El-Sayed MA. Surface plasmon resonance scattering and absorption of anti-EGFR antibody conjugated gold nanoparticles in cancer diagnostics: Applications in oral cancer. Nano Lett 2005;5:829–34.

[57] Zavaleta CL, Smith BR, Walton I, et al. Multiplexed imaging of surface enhanced Raman scattering nanotags in living mice using noninvasive Raman spectroscopy. Proc Natl Acad Sci USA 2009;106:13511–6.

[58] Thakor AS, Luong R, Paulmurugan R, et al. The fate and toxicity of Raman-active silica–gold nanoparticles in mice. Sci Translat Med 2011;3:79ra33.

[59] Jokerst JV, Miao Z, Zavaleta C, Cheng Z, Gambhir SS. Affibody-functionalized gold–silica nanoparticles for Raman molecular imaging of the epidermal growth factor receptor. Small 2011;7:625–33.

[60] Zavaleta C, de la Zerda A, Liu Z, et al. Noninvasive Raman spectroscopy in living mice for evaluation of tumor targeting with carbon nanotubes. Nano Lett 2008;8:2800–5.

[61] Kim C, Favazza C, Wang LV. *In vivo* photoacoustic tomography of chemicals: High-resolution functional and molecular optical imaging at new depths. Chem Rev 2010;110:2756–82.

[62] Kim C, Cho EC, Chen J, et al. *In vivo* molecular photoacoustic tomography of melanomas targeted by bioconjugated gold nanocages. ACS Nano 2010;4:4559–64.

[63] Agarwal A, Huang SW, O'Donnell M, et al. Targeted gold nanorod contrast agent for prostate cancer detection by photoacoustic imaging. J Appl Phys 2007;102:064701.

[64] Mallidi S, Larson T, Tam J, et al. Multiwavelength photoacoustic imaging and plasmon resonance coupling of gold nanoparticles for selective detection of cancer. Nano Lett 2009;9:2825–31.

[65] Song KH, Kim C, Cobley CM, Xia Y, Wang LV. Near-infrared gold nanocages as a new class of tracers for photoacoustic sentinel lymph node mapping on a rat model. Nano Lett 2008;9:183–8.

[66] De La Zerda A, Zavaleta C, Keren S, et al. Carbon nanotubes as photoacoustic molecular imaging agents in living mice. Nat Nanotechnol 2008;3:557–62.

[67] Galanzha EI, Shashkov EV, Kelly T, Kim JW, Yang LL, Zharov VP. *In vivo* magnetic enrichment and multiplex photoacoustic detection of circulating tumour cells. Nat Nanotechnol 2009;4:855–60.

[68] Kim JW, Galanzha EI, Shashkov EV, Moon HM, Zharov VP. Golden carbon nanotubes as multimodal photoacoustic and photothermal high-contrast molecular agents. Nat Nanotechnol 2009;4:688–94.

[69] Jin YD, Jia CX, Huang SW, O'Donnell M, Gao XH. Multifunctional nanoparticles as coupled contrast agents. Nat Commun 2010;1:41.

[70] Nehl CL, Hafner JH. Shape-dependent plasmon resonances of gold nanoparticles. J Mater Chem 2008;18:2415–9.

[71] Bouchard LS, Anwar MS, Liu GL, et al. Picomolar sensitivity MRI and photoacoustic imaging of cobalt nanoparticles. Proc Natl Acad Sci USA 2009;106:4085–9.

[72] Wustholz KL, Henry AI, McMahon JM, et al. Structure–activity relationships in gold nanoparticle dimers and trimers for surface-enhanced Raman spectroscopy. J Am Chem Soc 2010;132:10903–10.

[73] Yang XM, Stein EW, Ashkenazi S, Wang LHV. Nanoparticles for photoacoustic imaging. Wiley Interdiscip Rev Nanomed Nanobiotechnol 2009;1:360–8.

[74] Moon GD, Choi S-W, Cai X, et al. A new theranostic system based on gold nanocages and phase-change materials with unique features for photoacoustic imaging and controlled release. J Am Chem Soc 2011;133:4762–5.

[75] Zerda Adl, Liu Z, Bodapati S, et al. Ultrahigh sensitivity carbon nanotube agents for photoacoustic molecular imaging in living mice. Nano Lett 2010;10:2168–72.

[76] Skrabalak SE, Chen JY, Sun YG, et al. Gold nanocages: Synthesis, properties, and applications. Acc Chem Res 2008;41:1587–95.

[77] Yavuz MS, Cheng YY, Chen JY, et al. Gold nanocages covered by smart polymers for controlled release with near-infrared light. Nat Mater 2009;8:935–9.

CHAPTER

16

Therapeutic Applications

Cyrill Bussy[1], *Christoph Alexiou*[2], *Robby A. Petros*[3], *Andreas M. Nyström*[4], *Laura Methven*[1], *Kostas Kostarelos*[1]

[1]Nanomedicine Laboratory, Center for Drug Delivery Research, The School of Pharmacy, University of London, London, UK

[2]Department of Oto-Rhino-Laryngology, Head and Neck Surgery, Section for Experimental Oncology and Nanomedicine, Erlangen, Germany

[3]Department of Chemistry, University of North Texas, Denton, TX, USA

[4]The Swedish Medical Nanoscience Center, and the Department of Neuroscience Karolinska Institutet, SE-171 77 Stockholm, Sweden

OUTLINE

Adverse Effects of Engineered Nanomaterials
DOI: 10.1016/B978-0-12-386940-1.00016-7

INTRODUCTION

Nanomedicine is the application of nanotechnology in medicine and is one of the most extensive and promising subdisciplines of research and development efforts at the nanoscale. Innovative engineered nanomaterials (ENM) constitute an essential component of this emerging field. Well-designed ENM have indeed the potential to overcome the limitations of current medicines (e.g. efficiency, specificity) and, thus, advance disease diagnosis and treatment.

In this chapter, we do not aim to offer an extensive survey of all the ENM developed for therapeutic applications so far, but hope to provide some pertinent examples of nanomaterials that are currently in use in the clinic or are in clinical trials, and others that have demonstrated promise for various biomedical applications but are still under preclinical development. The therapeutic potential and challenges offered by different types of nanomaterials, including nanoparticles, polymer-conjugates, polymerosomes, dendrimers and carbon nanotubes are therefore presented in the following sections.

One of the main and transversal applications of the selected ENM is their use as drug nanovectors in cancer intervention, with the aim to provide a more efficient and controllable delivery of chemotherapeutics compared to more common drug formulations. As knowledge and control of the different physical, chemical and biological properties of these nanoscale materials becomes more advanced, their promise as future therapeutic and diagnostic opportunities to fight cancer or other diseases becomes increasingly more realistic.

NANOMATERIALS FOR THERAPEUTIC APPLICATIONS

Nanoparticles

Nanoparticles display unique physical and chemical properties due to their size. Also within this range are antibodies, receptors, nucleic acids, proteins and other biological macromolecules. Biomimetic features as well as an extraordinary surface to volume ratio and the easy manipulation of properties are attractive tools for the field of medicine, particularly for imaging, diagnosis and therapy [1,2]. In the nano-sized range, agents which are only slightly soluble or even unstable can access cells by transportation [3]. Therefore, a multitude of different macromolecular agents, including antitumor drugs, radiotherapeutic nuclides, genetic material and antibodies, can use the nano-sized vehicle for focused delivery (Figure 16.1) [4]. Thus, substances with a high therapeutic potential can be utilized that

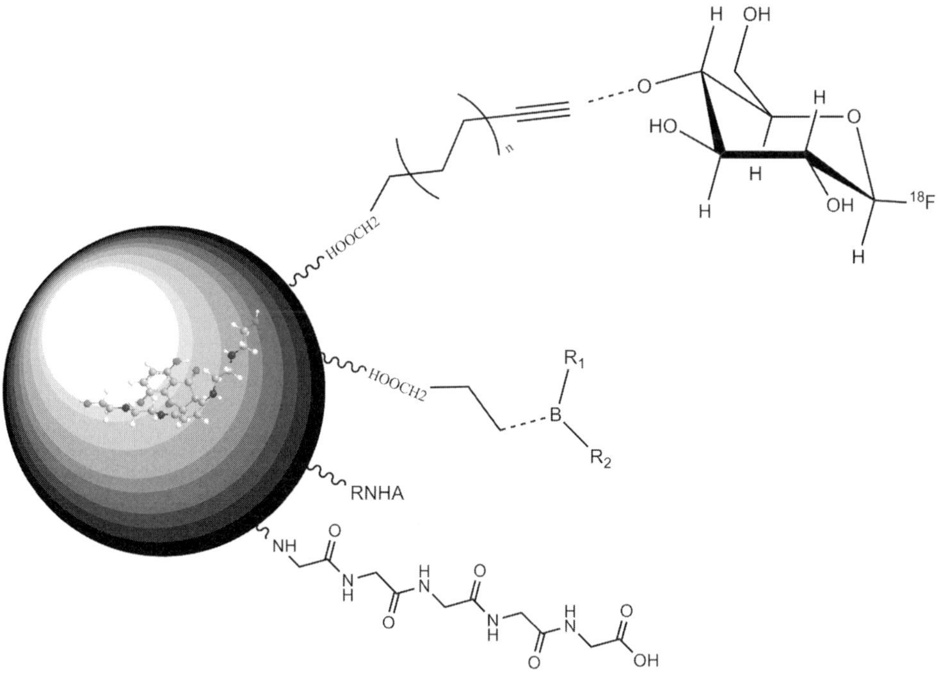

FIGURE 16.1 Schematic drawing of a therapeutic nanoparticle. Ideal model of therapeutic magnetic nanoparticles with multifunctional surface coating. As displayed there are imaging agents, substances for neutron capture therapy, antibodies and aptameres.

may have been disregarded previously because of their high systemic toxicity or because of metabolic barriers. For oncology purposes, this would mean a maximum level of antineoplastic agent reaching the tumor area with minimal levels in healthy tissue.

Nanoparticle Drug Delivery System Design

Nanoparticle systems were first used as pure vehicles for carrying drugs effectively over metabolic barriers. Current research focuses on nanoparticle systems able to release the agents in a controlled way in the targeted area. The release occurs by alteration of the physical and chemical parameters in the area of interest. Transportation and controlled release of therapeutic agents using nano- and micro-carrier systems in the target area of interest is a promising approach in current medicine [4,5]. For example, the specific attributes of diseased tissues can be exploited [6,7]: inflamed tissue can be differentiated from healthy tissue by the lower pH-value, increased temperature and a higher permeability of vessels [8]. These basic facts can be used for "targeting" with temperature- and pH-responsive nanoparticles [9,10].

Nanoparticle Drug Delivery

Most of the thermo-responsive systems are based on polymer-coated, drug-loaded nanoparticles and aggregates of them. Shen et al. developed temperature-sensitive polymer-conjugated albumine nanospheres, which were co-polymerized with a poly(N-isopropylacrylamid) moiety [11]. In this case, the release occurs by the increased temperature rupturing the interspaces of the polymer network on the outside of the nanoparticles. As a consequence, the nanospheres can be attacked and hydrolyzed easily by trypsin. Poly(N-isopropylacrylamid) is well known for facile alterations of conformation at a temperature level of about 32°C in aqueous media and is, therefore, often used as co-block polymer for thermolabile surfaces [12]. Defined profiles for release on demand can be achieved by connecting alternative polymers featuring different conditions for switching their conformation. Due to the multivalent properties of the surrounding polymer network temperature- and pH-responsive release systems are possible [10,13]. Nevertheless, the thermosensitive polymer system lacked sufficiently differentiated thresholds so the difference in the rate of drug release was only marginal. A different approach was undertaken by Stover et al. in which ceramides triggering apoptosis were used as effective anti-cancer agents [14]. Usually, ceramides barely penetrate the cell membrane and can be easily degraded. Nanoparticles with an incorporated thermoresponsive poly(N-isopropylacrylamid) moiety changed the whole polymer conformation above a threshold temperature. Due to the increased hydrophobicity, the particles could then cross the plasma membrane, become hydrolyzed and deliver the ceramide. Another concept under development is the use of light absorption to trigger drug release: the absorption of light induces heat, which leads to the bond between the drug and coating substance breaking [15]. This type of defined bond breaking enables an abrupt release of relatively large amounts of therapeutic substances and offers a novel way to progress drug release systems.

Theranostic Applications of Magnetic Nanoparticles

Magnetic iron oxide nanoparticles have already been introduced into medical fields. For diagnostic purposes, nanoparticles are used for magnetic cell separation whereby antibodies featuring a specific affinity are attached to particle surfaces. Hence, cells, DNA or bacteria can be separated magnetically. This technique is useful to verify the therapeutic effect after chemotherapeutic treatment [16]. Colloidal magnetic iron oxide particles (ferrofluid) are already used as contrast agents for magnetic resonance imaging (MRI) [17,18]. Commercially available contrast media are in use such as Resovist®, Endorem®, Sinerem® and Combidex®. Magnetite and maghemite particles are most commonly used in medicine and are generally well tolerated [19]. (Magnetic nanoparticles in clinical use and development for imaging, diagnosis and therapy are discussed also in Chapter 8 [see Table 8.1] and Chapter 15.) Their magnetic properties are activated during magnetic field application only.

An important parameter for *in vivo* application is the size of the particles. The magnetic force affecting the particles strongly depends on the radius ($\propto r^3$). Particles of a size of about 100 nm appear to be particularly useful since larger particles can be easily engulfed by MPS cells (mononuclear phagocyte system) and accumulate in the liver and spleen, especially following systemic (intravenous) application [20]. Furthermore, larger particles have a greater tendency to lead to vascular obliterations than smaller particles [21]. Alksne et al. reported one of the first clinical trials of magnetic particles in 1966 based on the use of ferromagnetic iron to close intracranial aneurysms [22]. Furthermore, magnetic albumine-microspheres have been used as a vehicle to carry the cytostatic drug doxorubicin to cause tumor remissions in rats [23]. Carbon-coated iron oxide particles carrying doxorubicin have also been developed for local chemotherapy. These particles (MTC-Dox, Ferex, USA) were applied intravascularly followed by the application of an external magnetic field in a phase I/II multicenter study on patients with hepatocellular carcinoma [24]. Previously, Lübbe et al. performed a novel human trial (phase I/II) with starch-coated magnetic nanoparticles whereby epirubicin was ionically bound to the particles. The formulation was applied intravenously and was well tolerated [25]. For magnetic drug targeting by intra-arterial administration, a sufficient particle size (80–150 nm) is required to attract the magnetic nanoparticles by an external magnetic field [26,27]. This type of application is advantageous as the particles accumulate in the desired area and by-pass the metabolic barriers of the MPS. Besides the particle size, the charge and chemical features are relevant. For *in vivo* applications, nanoparticles are usually coated for colloidal stability and to prevent segregation into particles and carrier medium [20,28]. A collapse of the colloidal stability would result in a thrombus formation. Figure 16.2 exhibits the principle of magnetic drug targeting (MDT).

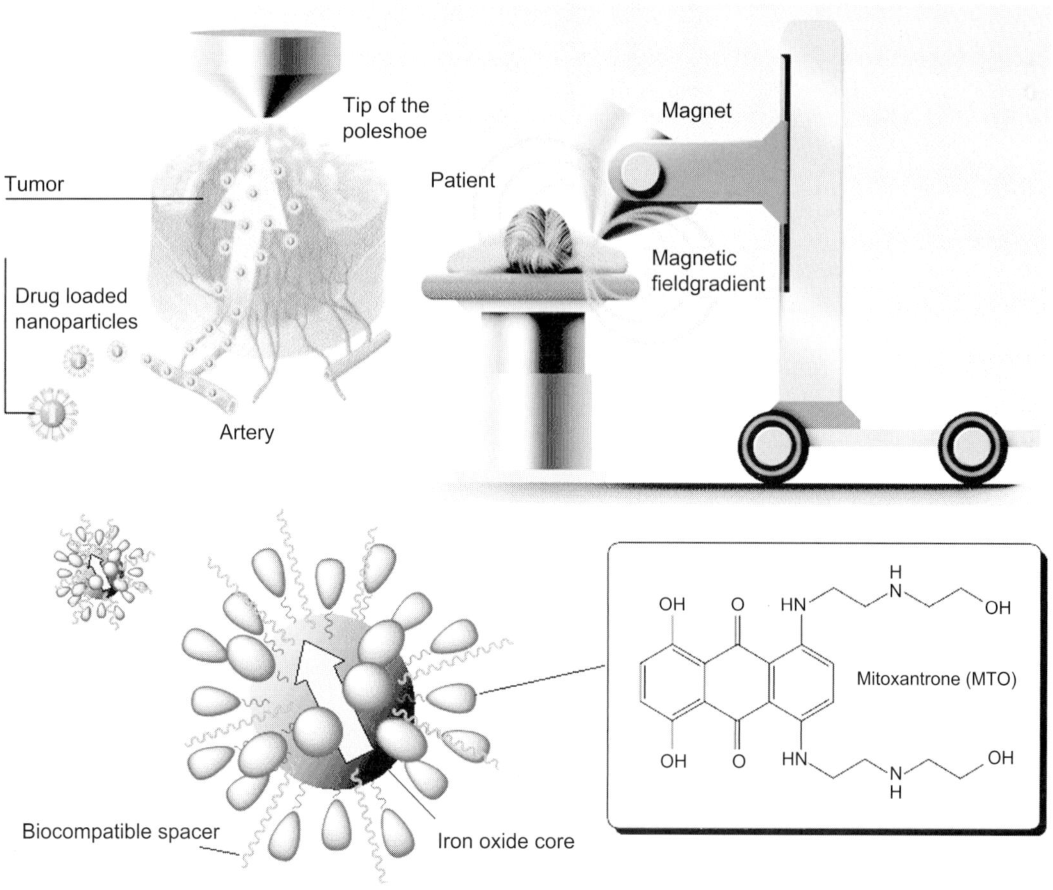

FIGURE 16.2 Principle of magnetic drug targeting. Magnetic particles that are functionalized with therapeutic agents are injected intra-arterially close to the tumor. An electron magnet positioned at the tumoral area directs and attracts the iron oxide particles in the tumor vascular system.

In animal studies, ferrofluids were injected intra-arterially into the tumor-supplying vessel (arteria femoralis) and a strong external magnetic field was applied simultaneously over a VX2-squamous cell carcinoma placed at the hind limb. An accumulation of ferrofluids could be demonstrated by histology (Figure 16.3), MRI and computed-micro-tomography (μCT)[29–32], without pathological alterations to the liver, kidneys, spleen, lung and brain. Metabolic decomposition of the ferrofluids proceeds in the liver and spleen in analogy to the physiological iron metabolism. In these organs, ferrofluids could be detected histologically for 3 months post-application and radiotracer studies using ^{59}Fe showed that iron arising from ferrofluids is embedded into the hemoglobin biosynthesis. Furthermore, these studies produced quantitative results showing a 114-fold higher enrichment of iron oxide nanoparticles per gram tissue due to the application of an external magnetic field (versus no external magnetic field) [26,33]. In a therapeutic study, the intra-arterial application into the arteria femoralis of rabbits transplanted with a VX2-squamous cell carcinoma was performed. Therefore, ferrofluids loaded with the chemotherapeutic agent mitoxantrone were used. It was shown that 20% of the systemic dose (regular systemic dose=10 mg/m^2 equates to 100%) was sufficient for complete tumor remission without any side effects and the cellular uptake of iron oxide particles was also shown [34,35]. Furthermore, electron microscopy

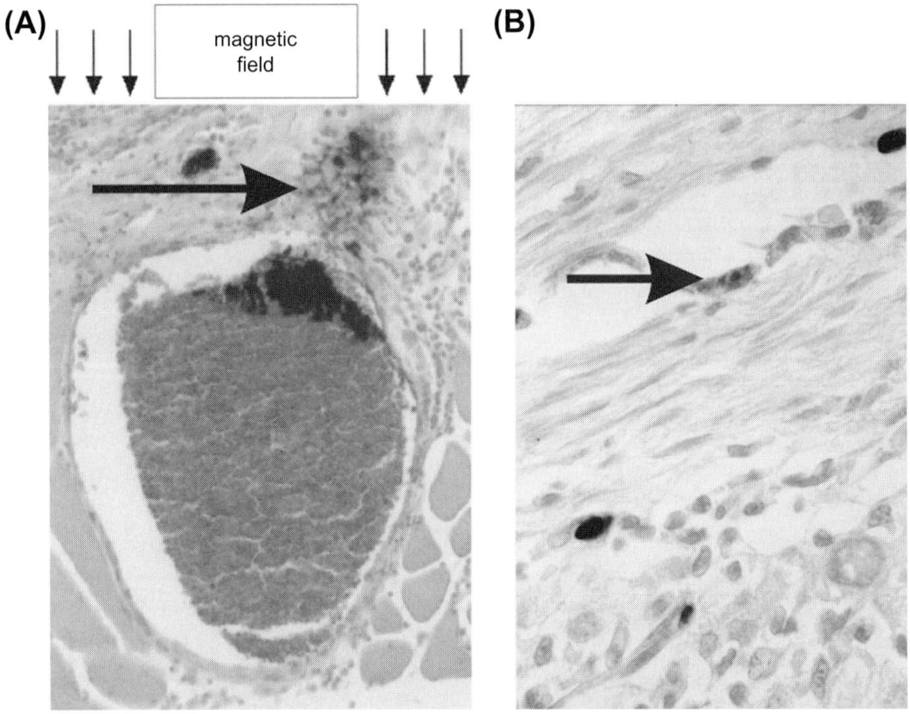

FIGURE 16.3 Particle concentration and infiltration into tumor tissue after MDT. (A) Tumor vessel and extravasation of the particles into the tissue (Magnification: 400×, Prussian blue staining), iron oxide appears dark (arrow). (B) Histological cross-sections of tumor tissue after MDT (Magnification: 1000×, Prussian blue staining), dark-colored iron oxide particle is labeled by an arrow [39].

verified by EDX analysis could visualize the iron oxide spots inside tumor cells, which is evident for the therapeutic effect [35].

Further Development of Nanoparticle Systems in Nanomedicine

By employing a combination of different approaches, different substances (classic drugs as well as non-classic agents like nucleic acids, antibodies etc.) and varying particle surfaces, multifunctional particles can be achieved. Thus, nanoparticles not only offer a platform for therapeutic agents (for a detailed discussion, see Chapter 15), but by binding a linker for functional imaging purposes (^{18}F-PET), multifunctional "theranostic" particles can be put on the stage [36–38]. Magnetic drug targeting and hyperthermia are emerging as promising and innovative applications in oncology [39] and, in future, could be utilized to achieve focused tumor therapy, increasing the success of a low drug dose and, therefore, reducing side effects.

Polymer Nanoparticles

Nanoparticle therapeutics are based on multifunctional particles, which are capable of extending circulation *in vivo*, targeting specific diseased cells, and releasing the therapeutic payload in a stimuli-responsive manner [40–43]. Each of these functions can often be engineered into the particle independently from other particle attributes. This section details some of the important design parameters in polymer nanoparticle engineering, progress made in both targeting specific cells and encapsulation/release of therapeutics, and demonstrates the promise of these materials in the fight against disease.

Design Parameters

One of the major advantages of polymeric nanoparticles in the targeted delivery of therapeutics is the ability to optimize particle size, modulus, shape, and *in vivo* stability independent of other particle attributes [40]. For example, a common misconception is that the ideal particle size for a drug delivery vector is $\approx$5–150 nm [40], but recent efforts in particle engineering clearly demonstrate that these defined limits apply only to highly rigid, spherical particles [44,45]. In one example, filamentous micelles (filomicelles) with single dimensions as long as 18 μm were shown to circulate for longer than 5 days [44]. In another example, particles designed to mimic red blood cells were fabricated where the particle modulus was varied over a physiologically relevant range [45]. The particles, which were discoid shapes 5.2–5.9 μm in diameter and 1.2–1.5 μm tall, showed widely differing biodistribution and circulation times as the particle modulus was varied from 7.8 to 63.9 kPa. As great as a 30-fold increase in elimination half-life was observed as the particle modulus was lowered and elimination half-life was extended to 93 h.

A variety of matrices can be used in particle fabrication with the most commonly employed being poly(lactic-co-glycolic acid) (PLGA) and biopolymers, such as chitosan and dextran. The choice of matrix determines much of the *in vivo* stability of the particle, as well as the degradation products that ultimately must be cleared from the body [46]. The polymers listed above are generally acknowledged to possess good biocompatibility/biodegradation profiles (although there is an ongoing debate over the definition of biocompatible [47]); however, a recent example of biodegradable acrylate-based particles illustrates the potential of acrylate-based materials in targeted drug delivery [48].

Targeted Drug Delivery

One of the most promising aspects of targeted drug delivery is the ability to target diseased cells specifically in order to reduce the exposure of toxic chemotherapeutics to healthy cells. In cancer applications, passive targeting can be achieved by exploiting the enhanced permeability and retention (EPR) effect [49,50]. Active targeting strategies are also routinely employed in targeted drug delivery and can be classified as either selective or non-selective. Selective targeting strategies rely on surface modification of the particle with ligands that recognize various biological receptors [51,52]. Non-selective targeting strategies target cell surface receptors that are expressed ubiquitously, but that are over-expressed on cancer cells. For example, cancer cells proliferate rapidly and as a result routinely express higher levels of the transferrin and folate receptors. As a result, many targeting strategies employ folate, transferrin, or antibodies to these receptors selectively to target rapidly proliferating cancer cells [53–56]. These receptors are also expressed to a lesser degree on many healthy cells, but selective targeting strategies utilize ligands capable of binding to receptors on targeted cells that are not expressed by other cell types [57].

Encapsulation and Release of Therapeutics

Several methods have been used to achieve encapsulation and release of therapeutics from drug delivery vectors. The majority of methods incorporate stimuli-responsive linkages designed to be cleaved once the delivery vector encounters a specified environment [58–60]. The stimuli-responsive linkages are designed either to destabilize the carrier, causing degradation and release of encapsulated therapeutic, or to cleave covalent bonds linking the therapeutic to the particle, facilitating release. Perhaps the most common approach exploits the pH gradient that occurs as a particle is internalized along an endosomal pathway where the pH drops from ≈ 7.4 to ≈ 5.5. A number of acetal and hydrazone linkages have been successfully employed in this regard in both nanoparticle therapeutics as well as polymer-drug conjugates. Some particle internalization pathways do not involve trafficking through an endosome, so other strategies must be developed for those situations. An approach that has received considerably less attention is the use of the reducing nature of the cytosol to break reductively labile bonds. The incorporation of disulfide cross-linkers has been particularly useful in this regard [61,62].

Targeted Delivery of Cisplatin

The potential benefits of utilizing targeted nanoparticles to deliver therapeutics is exemplified in a recent report detailing the performance of aptamer-targeted, cisplatin prodrug-loaded, poly(D,L-lactic-*co*-glycolic acid)-*b*-poly(ethylene glycol) nanoparticles (Pt-PLGA-*b*-PEG-Apt-NP) in a prostate cancer model [63]. Cisplatin was liberated from the prodrug via the intracellular reduction of the platinum complex (Figure 16.4). The particles were targeted by adding an aptamer targeting ligand (A10) to the surface, which recognizes the extracellular domain of PMSA.

The maximum tolerated dose (MTD) of un-targeted Pt-PLGA-*b*-PEG-NPs was tested in male Sprague Dawley rats and in Swiss Albino mice. In rats, the MTD of platinum contained in Pt-PLGA-*b*-PEG-NPs was twice that of cisplatin administered conventionally (MTDs of Pt-PLGA-*b*-PEG-NPs and cisplatin were 40 and 20 mg/kg, respectively). Similarly, the MTD of Pt contained in Pt-PLGA-*b*-PEG-NPs was higher than that for cisplatin in Swiss Albino mice.

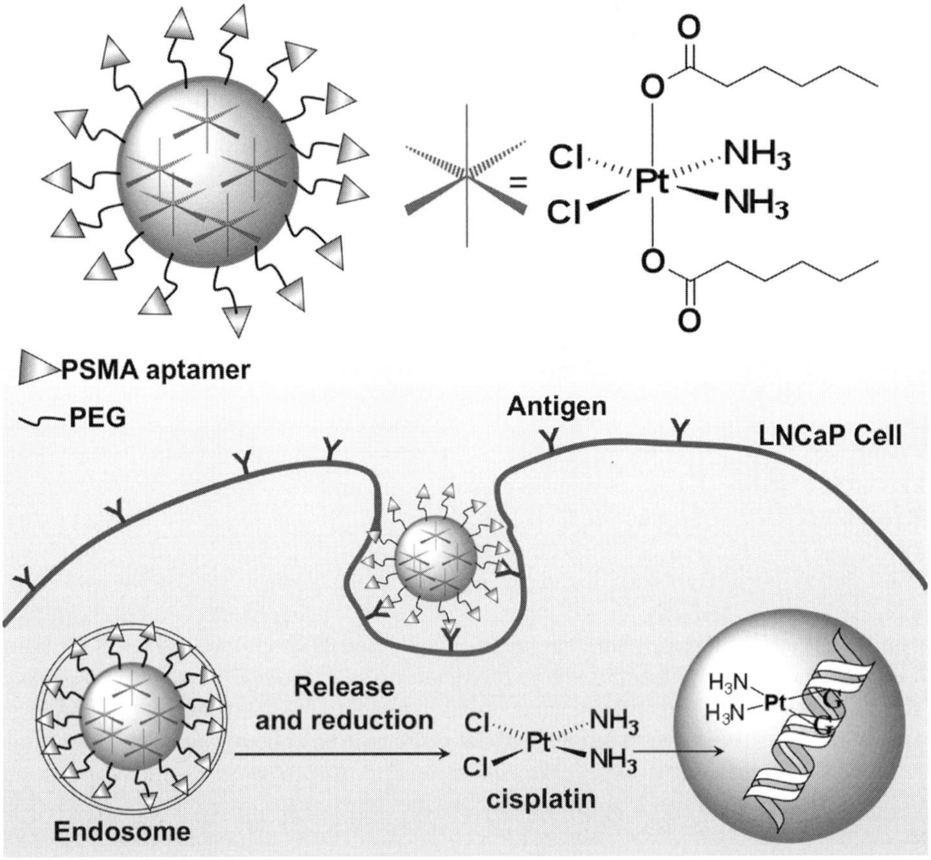

FIGURE 16.4 Schematic representation of the Pt-PLGA-b-PEG-Apt-NP construct. Chemical structure of the Pt(IV) prodrug 1 and intracellular reduction for the release of active cisplatin in PSMA expressing human prostate cancer LNCaP cells after receptor mediated endocytosis of Pt-PLGA-b-PEG-Apt-NP. *(Reproduced with permission from [63].)*

These combined results demonstrate a reduction in the toxicity of cisplatin when incorporated into a prodrug nanoparticle formulation, even in the absence of targeting ligands. It has also been shown that the circulation time of the prodrug is significantly enhanced: cisplatin was cleared rapidly from circulation with only 1.5% of the injected dose remaining after 1 h [64], whereas for the encapsulated cisplatin prodrug greater than 75% of the injected dose remained in the circulation at 1 h. Furthermore, total platinum excretion after 24 h was 17 times less for encapsulated cisplatin prodrug compared to dosing with free cisplatin. Thus, the increased circulation combined with reduced excretion leads to sustained exposure to the active therapeutic at a low dose.

Aptamer targeted Pt-PLGA-*b*-PEG-NPs (Pt-PLGA-*b*-PEG-Apt-NPs) were evaluated for efficacy and tolerability in a LNCaP subcutaneous xenograft mouse model of PCa. Pt-PLGA-*b*-PEG-Apt-NPs were more efficacious than cisplatin over the first 14 days but, at 28 days, Pt-PLGA-*b*-PEG-Apt-NPs and cisplatin tumor volumes were roughly equivalent. Even though Pt-PLGA-*b*-PEG-Apt-NPs and free cisplatin were comparable in efficacy after 28 days, the amount of cisplatin prodrug contained in Pt-PLGA-*b*-PEG-Apt-NPs was one third that of freely dosed cisplatin. Given the increased efficacy of Pt-PLGA-*b*-PEG-Apt-NPs compared to conventionally dosed cisplatin in a human prostate cancer model, these particles clearly demonstrate the potential of drug delivery vectors.

Hybrid Polymer Vectors—Targeted Nanogels

A recent study combining polymeric materials with liposomes and protein-based particles exploited the favorable properties of each of these vectors [65]. Targeted nanogels combine the ease of synthesis of liposomes and the molecular binding properties of proteins with the stability that can be achieved in polymer cross-linked nanoparticles to produce a stable drug delivery vector. These nanogels employ a cross-linked polymer/protein core for stable drug release with a lipid bilayer surface that can be functionalized with targeting ligands. A variety of therapeutics including natural products, kinase inhibitors, proteasome inhibitors, and chaperone inhibitors could be loaded. Human serum albumin and α1-acid glycoprotein were used in the polymer/protein core because of their known binding properties toward cancer therapeutics in the blood stream. Acrylate-based polymers were used to cross-link the protein core.

Synthesis of nanogels was achieved by adding the desired monomers, proteins, drug molecules, and photo initiator to a lipid film, followed by sonication and then extrusion (Figure 16.5). The protein core was then cross-linked via photo-polymerization of the encapsulated acrylate monomer. The nanogels were then targeted to $\alpha_V\beta3$ integrin by conjugating the cyclic peptide cRGDfK to the surface of the particle. The stability of the nanogels was investigated by measuring the release of the therapeutic in mouse plasma where over 60% of the encapsulated dose of docetaxel was retained in the cross-linked nanogel compared to $\approx$40% retention in the non-cross-linked analogue. *In vitro* experiments in M21 cells using targeted nanogels showed increased potency of each nanogel composition compared to dosing of the corresponding free drug. In fact, a 20-minute exposure to drug-loaded nanogels was comparable to a 72 h exposure to the respective free drug. The effect of targeting was also investigated *in vitro*, and a 13-fold enhancement in potency was reported for the targeted versus untargeted nanogels.

Taxane-loaded nanogel efficacy was later compared with the FDA-approved taxane formulation of Abraxane in an orthotopic breast cancer model. Abraxane at 1 mg/kg dose showed no tumor suppression at 30 days and was comparable to saline controls in terms of tumor volume (830 mm^3). Targeted nanogels showed significantly slowed tumor growth with a tumor volume of 190 mm^3 at 30 days. In a dose response experiment with Abraxane, a 1 mg/kg dosing of targeted nanogels induced the same level of tumor suppression as a 15 mg/kg dose of Abraxane. Similar improvements in efficacy of nanogels compared to Abraxane were reported in an orthotopic pancreatic cancer model. Furthermore, nanogels showed a 15-fold improvement in reducing metastasis to the hepatic hilar node compared to Abraxane. The efficacy and versatility of this drug delivery platform will undoubtedly lead to the application of nanogels in a variety of other cancer models.

The future is bright for polymer-based drug delivery vectors. Some of the benefits of targeted drug delivery, including increased efficacy and reduced non-specific toxicity, are beginning to be realized. Based on the success of using aptamer-targeted nanoparticles and targeted

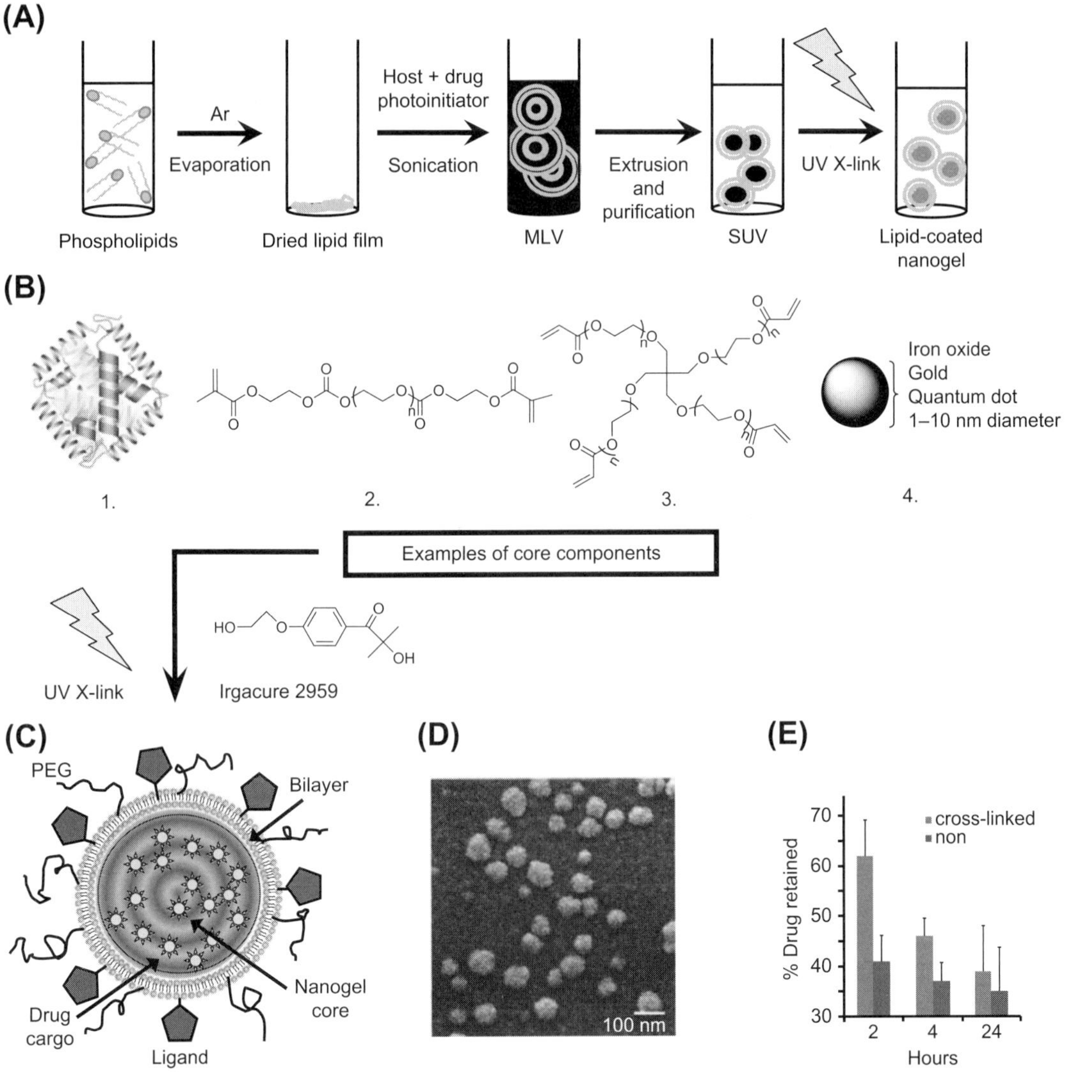

FIGURE 16.5 Formation of lipid-coated nanogels. (A) The method for controlled photo-cross-linking and formation of the nanogel core. (B) Examples of various monomeric inputs and imaging agents that can be used to form the nanogel core are shown. (C) A schematic representation of the final nanogel product with a lipid bilayer, presenting targeting ligands and polymeric coatings surrounding the gel core containing drug cargoes. (D) Scanning electron micrograph of a lipid-coated nanogel. (E) Nanogels with RGD peptide on the surface and docetaxel-loaded HSA in the cross-linked core were incubated in the presence of 20% plasma for the indicated time points. *(Adapted and reprinted by permission from the American Association for Cancer Research: Murphy, E.A., Majeti, B.K., Mukthavaram, R., Acevedo, L.M., Barnes, L., Cheresh, D.A. Targeted nanogels: a versatile platform for drug delivery to tumors,* Molecular Cancer Therapeutics, *2011, 10, 972–982.)*

nanogels loaded with therapeutics in preclinical studies, the potential impact of polymer-based vectors in the pharmaceutical industry is likely to be profound. One of the major obstacles in drug development has long been the poor pharmacokinetics of many small molecule therapeutics. The use of drug delivery vectors, which can themselves be designed to possess the appropriate biodistribution profile, facilitates the decoupling of therapeutic activity with pharmacokinetics. Thus, there will be a significant opportunity to re-examine a variety of potential drug candidates that have failed for reasons of poor pharmacokinetics. Additionally, if it is possible to guide a vector to the desired location *in vivo* and thereby concentrate its payload in target cells, the delivery of cargo that might otherwise be considered benign, such as essential transitions metals like Cu and Fe, may be capable of eliciting unexpected biological responses including targeted cell death via induced oxidative stress [66].

Dendrimers

Because of the precise structure and the extensive possibilities for designing dendrimers with ligands in exact positions in the structure, much work has been focused on applications in biomedical research [67,68]. Dendrimers offer specific advantages over many other classes of nanomaterials for therapeutic applications, such as controlled solubility, many end-groups, exact placement of functional groups, and control over size, shape, and molecular weight of the structure and, therefore, show great potential for the development of therapeutic applications.

Design Principles

Dendrimers represent a very special architectural type of polymeric materials. Dendrimers are synthesized from branched monomers of the type AB_x were x is >2 or higher. These monomer units are layered around a core molecule to build up an onion-like structure where each layer is called a generation. Typically, the synthesis is conducted either from the core and outwards (divergent method) or from the end groups and inwards (convergent method) (Figure 16.6). These growth steps are performed in a controlled manner, often utilizing organic protection/activation chemistry to yield spherically shaped dendrimers with almost perfect structure, and exact molecular weight and size. In this respect, dendrimers are similar to proteins, with controlled exact molecular weights (virtually no polydispersity [PDI] if produced correctly) and globular shapes. The size of the constructs is most often less than 5 nm in diameter, as it is difficult to control the synthesis of dendrimers of higher generations than 5 or 6. Because of the tedious synthetic preparation and purification necessary for generating high molecular weight dendrimers of high quality, they are typically only prepared in small quantities and the commercial prices are set thereafter. In addition to the dendrimer architecture, related cousins have been prepared that stem from the branched structure and the dendritic family now includes subclasses, i.e. dendrimers, dendrons, hyperbranched polymers, dendrigraft polymers and dendritic-linear polymers.

There are numerous examples of dendrimers synthesized in literature, made from virtually any AB_x monomer commercially available. To name a few, Poly(propyleneimine) dendrimers (PPI), Poly(amidoamine) dendrimers (PAMAM), Poly 2,2-bis(methylol)propionic acid (PBisMPA), Poly(benzyl ether) dendrimers (PBzE), poly(lysine) dendrimers (PLL), and polymelamine (triazine) dendrimers. Of these dendrimers the PAMAM, PPI, and

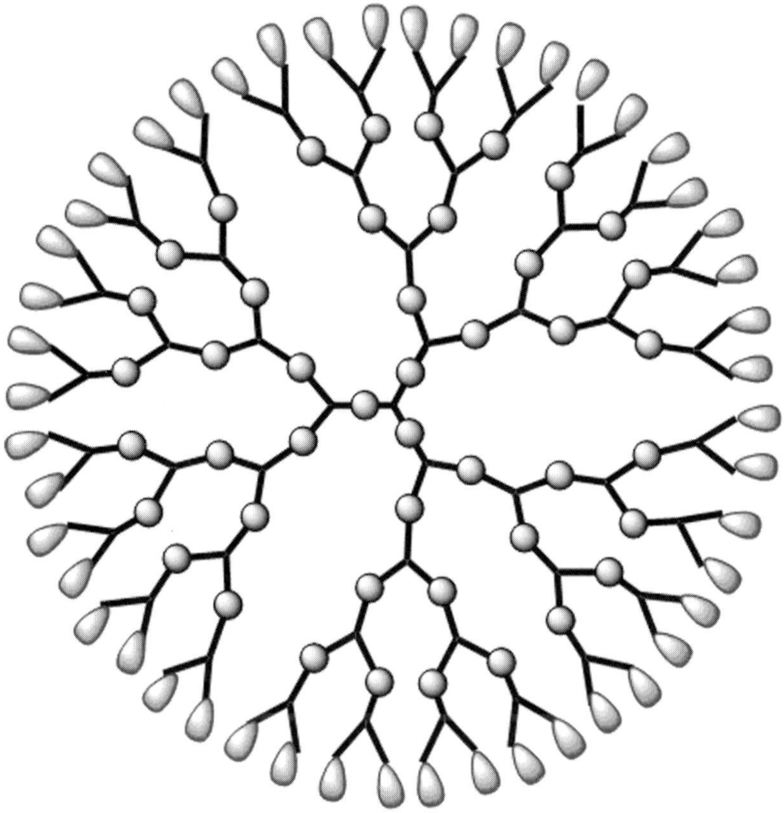

FIGURE 16.6 Schematic structure of a dendrimer.

PBisMPA dendrimers are readily available from large commercial vendors (some others are also produced on demand from other small companies). Here we will focus our attention on the most commonly used dendrimers in therapeutic applications, such as the PAMAM, PBisMPA, PPI and to some extent Poly(lysine) class, that have been evaluated in detail in animal experiments (Figure 16.7). Figure 16.7 depicts, from left a PAMAM dendrimer, PPI dendrimer and to the right a PBisMPA dendrimer.

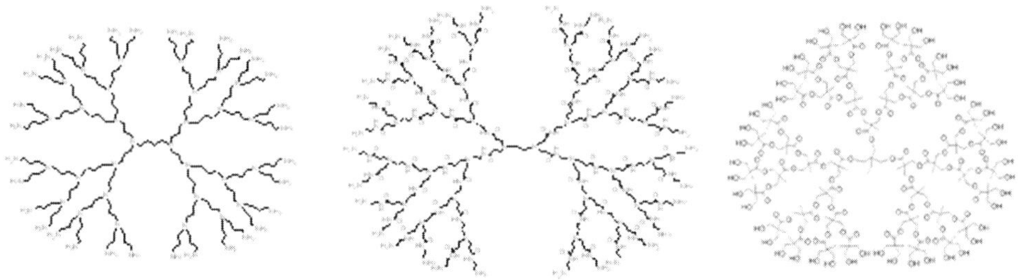

FIGURE 16.7 Commonly used dendrimers in biomedical applications.

Drug Delivery with Dendrimers

Up to now, most therapeutic applications of dendrimers have been focused on two topics, imaging and drug delivery. The examples in this chapter will focus on the drug delivery applications of dendrimers. Dendrimers are exact nano-sized molecules, therefore, they can be systematically controlled by size, surface and conformation, resulting in biodistribution and pharmacokinetic properties that can be precisely evaluated [69]. Similarly, the reproducibility and extremely low size distribution of these constructs in comparison to self-assembled systems, such as polymer micelles, can be a regulatory advantage [69]. Other factors that contribute to this interest is the high functional group density at the dendrimer surface that can be utilized for targeting applications and/or multivalent interactions, and the ability of dendrimer structures to be designed for controlled degradation. Moreover, unlike polymer micelles and liposomes that are assembled via hydrophobic interactions, dendrimers do not have a critical micelle concentration (CMC). This means that dendrimers used in systemic circulation applications cannot disassemble upon dilution. This avoids the burst release effect (very rapid release of drug cargo) often encountered with liposomes or polymer micelles which can be a major drawback [70–72].

However, dendrimers have other drawbacks such as their very small size. Dendrimers are typically less than 10 nm in diameter, and this small size often results in their very rapid renal excretion. This counteracts the effect of reaching a therapeutic concentration in blood, so, like most nanoparticles, dendrimers are usually equipped with polyethylene glycol chains to increase their size and therefore enhance their blood circulation time and reduce RES activation by providing a stealth surface. The other major drawback with dendrimers in therapeutic applications is related to the cost and scalability of manufacturing high generation dendrimers. So far, most studies have been conducted on a sufficiently small scale and this factor has not become an issue, but as the research progresses towards clinical testing this will be major challenge.

As with other nanoscale drug delivery agents, dendrimers have been explored in two major directions when it comes to drug loading: as passive drug delivery agents where the drug is encapsulated within the dendrimers framework via favorable drug–dendrimer interactions, or as covalent drug carriers with the pharmaceutical linked to the dendritic scaffold. Dendritic encapsulation or "dendritic box" was first demonstrated in 1994 [73], and since then a range of therapeutics including chemotherapeutics have been explored for dendrimer encapsulation. Most chemotherapeutics have hydrophobic structures and can, therefore, interact with the often hydrophobic interior of dendrimers, increasing the solubility of the pharmaceutical and using the dendritic exterior to shield the drug from its surroundings, which can reduce premature degradation. Further modifications, such as PEGylation of the dendrimers' end groups can extend and alter the pharmacokinetics of the drug. To generate successfully passive dendrimer-based drug carriers one must consider some issues that are general for passive drug delivery agents, mainly stability and premature release, concentration variations between batches, low drug-loading capability, which can result in high volumes of carrier, and rapid burst release under physiological conditions.

Dendrimer Encapsulated Drug Formulations

PAMAM and PPI dendrimers are commonly used for dendritic encapsulation studies based on their ability to be easily functionalized on the exterior, due to the presence of surface amine groups. Both classes of dendrimers are not degradable and are often found to evoke toxic or hemolytic responses in their original form [74,75]. Hence, it is necessary to shield the highly cationic amine exterior with less "provocative" groups, such as OH, COOH or PEG. In terms of chemotherapeutics, PAMAM have been explored for the formulation of several drugs including: Anastrozole [76], 6-mercaptopurine [77], 5-fluorouracil [78], Adriamycin [79], methotrexate [79] and camptothecin with some degree of success [80].

Polyester-based dendrimers, such as the class developed by Grinstaff et al., are constructed from glycerol and succinic acid and are degradable. They have been used to encapsulate topoisomerase inhibitors, such as 10-hydroxycamptothecin and 7-butyl-10-aminocamptothecin, with very promising results against a panel of breast, lung, colorectal and glioblastoma cell lines [81,82]. Similarly, poly-glycerol dendrimers have been shown to increase the solubility of paclitaxel by orders of magnitude compared to the traditional excipient Cremofor EL which can cause toxic side effects [81].

Covalent Dendrimer-Drug Carrier Formulations

Covalent drug carrier formulations can overcome some of the limitations that are often found for the sequestered passively loaded drug carriers. By covalent attachment via cleavable linkages that are activated by changes in such as pH, enzyme levels, reductive environment or temperature, the release of the pharmaceutical can be controlled circumventing burst release effects. In the case of the dendrimers, covalent attachment of drugs can also minimize batch-to-batch variations in drug loading, as the structure is monodisperse, and often higher drug concentrations can be accommodated by the carriers when the drug is covalently attached. Duncan et al. showed an early example of polymer-drug conjugation with a dendritic scaffold by utilizing PAMAM dendrimers carrying carboxylic acid end groups [83]. These COOH groups were used for attaching cisplatin in an effort to increase the effective solubility of the drug and minimize toxicity. The reaction of cisplatin caused intermolecular cross-linking of dendrimers generating larger dendrimer clusters and not discrete nanoparticles of one single dendritic molecule. However, as a model for drug delivery, this system demonstrated that the covalent attachment enabled 25 wt% of cisplatin to be conjugated with a 10-fold increase in solubility and significantly reduced toxicity as determined *in vitro*. *In vivo* investigations indicated that the blood clearance of these constructs was significantly reduced and that intratumoral concentration of platinum was enhanced compared to the free drug. This seminal work has since inspired a range of studies on both PAMAM dendrimers and many others. In terms of cancer therapy, Table 16.1 lists a representative number of dendrimers and the drugs investigated and below is a selected example.

Doxorubicin and PBisMPA Dendrimers

The use of polyester-based dendrimers for therapeutic applications has been well investigated by the group of Fréchet et al. These dendrimers are biocompatible with non-toxic and non-immunogenic properties which is highly desirable for pharmaceutical applications [98,99]. The research has evolved from simple dendrimer structures to more complex linear-dendritic constructs for chemotherapeutic delivery. In initial studies [37], the drug doxorubicin (DOX) was conjugated to the surface via acid-sensitive linkage that would allow

TABLE 16.1 Examples of Dendrimers and Drugs as Covalent Drug Carrier Constructs

Dendrimer	Drug	Reference(s)
PAMAM	Doxorubicin	[84–86]
PAMAM	Paclitaxel	[87–89]
PAMAM	Platinates	[83,90]
PAMAM	5-Fluorouracil	[91–96]
PBisMPA	Doxorubicin	[7–99]
Poly(melamine)	Paclitaxel	[100]
Poly(triazine)	Paclitaxel	[101]
Poly(triazine)	Camptothecin	[102]
PBzE	Methotrexate	[103]

drug release after cellular uptake. These structures depicted much higher IC50 values than the free drug, with no preferential organ uptake *in vivo* but a significantly increased half-life in blood.

The next generation of dendritic structures that was developed was coined "bow-tie" structures (Figure 16.8): dendrimers carrying long PEG chains on one side and reactive handles on the opposite side for drug attachment. These structures enable the number of PEG branches and drug density necessary for optimization of the PK and BioD characteristics to be tailored.

These carriers were shown to be highly effective drug carriers *in vivo*, with similar survivability as Doxil in BALB/c mice bearing subcutaneous C-26 tumors, far exceeding the effect

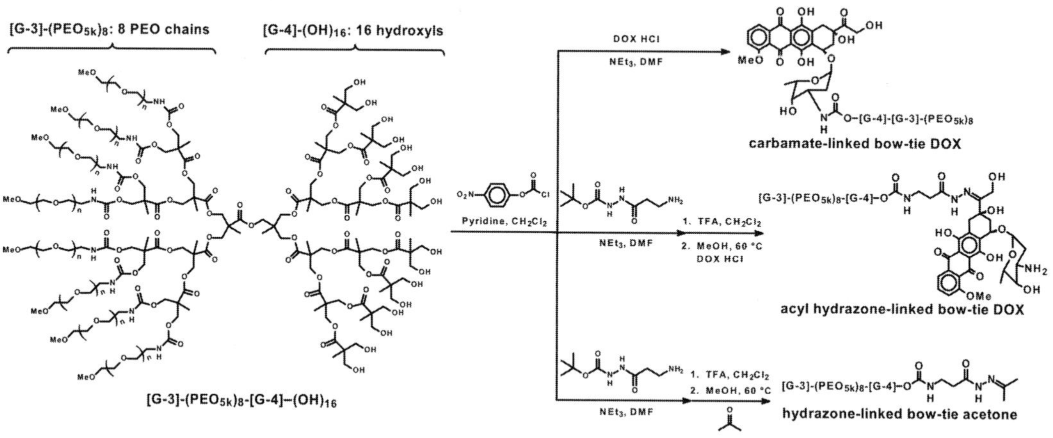

FIGURE 16.8 Bow-tie dendrimer structure. *(Reproduced with permission from [104]. Copyright 2006 Proceedings of the National Academy of Sciences of the United States of America.)*

of the free drug at the same concentrations. These carriers also have better stability and can potentially be developed into generic drug carriers for other types of chemotherapeutics as the half-life in blood can be varied from minutes to hours. The IC_{50} of the construct was determined to be 20 times higher than for the free drug, and the tumor uptake of the dendritic carrier was increased ninefold. The maximum tolerated dose was also increased on a DOX equivalent basis (Figure 16.9).

Perspectives Concerning Dendrimers

Research employing dendrimers in therapeutic applications is still in its infancy. Many opportunities have been demonstrated in the literature but several challenges exist. For these applications dendrimers must be designed for use in a biological system. Thus, dendrimers need to be non-toxic, degradable, equipped with non-charged groups and devoid of organ-specific accumulation. Currently, the number of dendrimers undergoing phase testing for therapeutic applications is limited, perhaps reflecting the complexity of using nanomedicine in reality or an effect of the complexity required to construct these structures.

Carbon Nanotubes (CNTs)

As for other ENM, the biomedical potential of CNTs is determined by their various physical, chemical and biological properties. CNTs are either single-walled (SWCNT) or multi-walled (MWCNT) depending on the number of concentric carbon cylinders they are made of. Their unique combination of intrinsic features (i.e. mechanical, electrical, thermal, optical and spectroscopic) along with a high surface tunability might indeed suit applications in the detection, imaging, monitoring and also the treatment of many diseases, spanning from cancer to neurodegenerative disorders. Furthermore, *in vitro* studies have demonstrated that CNTs present a key advantage due to their specific cylinder shape: they can be easily internalized by cells. They could, therefore, offer a unique solution for the intracellular delivery of small therapeutic molecules as well as macromolecules that do not easily cross the cell membrane barrier.

CNT Design

Pristine CNTs (as produced) are insoluble in most organic or aqueous solvents, and particularly in the high salt milieu that constitutes most biological environments. To overcome this issue and to advance their use for therapeutic purposes, modification of CNTs is required. Two main strategies have been developed, both of which are based on the modification of the nanotube surface, either by non-covalent physisorption of amphiphilic molecules or by covalent grafting of different kinds of chemical groups on the nanotube carbon backbone [105,106] (Figure 16.10). Both approaches result in improved solubility and dispersibility under physiological conditions and seem, according to many studies, to be less toxic when compared to their non-modified forms [107].

The hydrophobic and polyaromatic carbon surface of the nanotube offers a unique surface on which amphiphilic and aromatic molecules, such as polymers (e.g. glycodendrimer, pyrene-polyethylene glycol-lactose, and polysaccharide) or lipids, can easily bind by non-covalent interactions [108]. For instance, to improve the dispersibility of CNTs in aqueous solution PEGylated phospholipids have been used to functionalize SWCNTs which can be

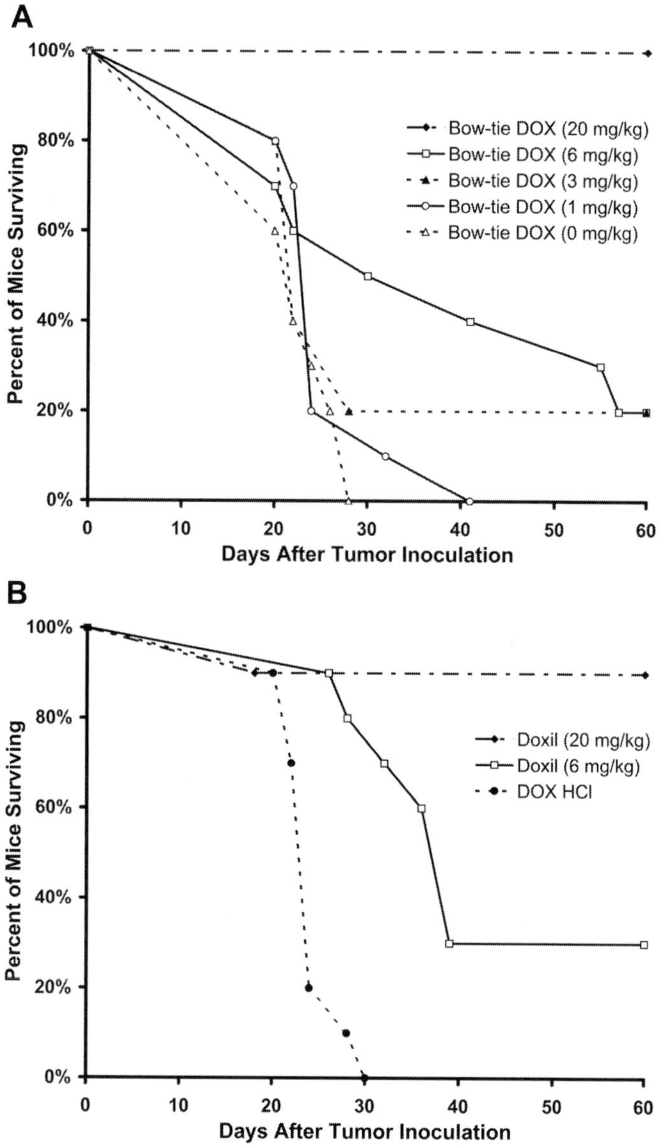

FIGURE 16.9 *In vivo* evaluation of PBisMPA bow-ties as doxorubicin carriers against c26 tumors. Survival versus time for BALB/c mice bearing s.c. c26 tumors. (A) Treatment consisted of a single i.v. dose of hydrazone-linked bow-tie DOX given 8 days after tumor implantation. (B) Treatment consisted of a single i.v. injection of either free DOX or Doxil given 8 days after tumor implantation. *(Reproduced with permission from [104]. Copyright 2006 Proceedings of the National Academy of Sciences of the United States of America.)*

then assessed for their therapeutic potential. PEGylated lipids offer multiple advantages. They improve CNT solubility and dispersibility while being more biocompatible than commonly used surfactants, such as sodium dodecylbenzene sulfonate or the triblock nonionic copolymer Pluronic F127. In addition, terminal amino groups present on the PEGylated moieties coating the CNT have also been exploited to load therapeutic molecules [109]. Another recently developed method to aid the dispersion of CNTs in physiological media is to wrap the CNT backbone with biocompatible DNA oligos. For compounds containing

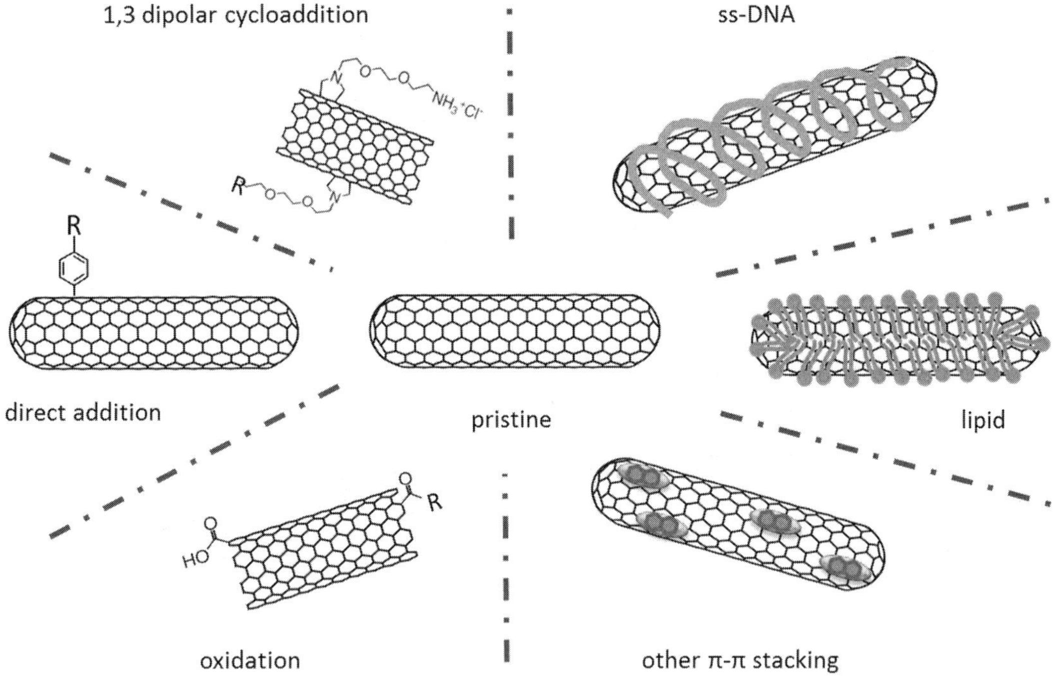

FIGURE 16.10 Different kinds of CNT surface modifications.

aromatic moieties, the supramolecular π-π stacking that is used to coat the surface of the nanotubes can also be exploited to load therapeutic molecules directly onto the CNT. For instance, following this procedure, doxorubicin has been successfully loaded onto CNT [110,111].

Alternatively, the covalent functionalization of CNT consists of the chemical modification of the carbon backbone. One of the most common methodologies is to heat under reflux a strong acid solution containing the CNTs. It is generally used as a first step to create defects on the nanotube sidewall (i.e. carboxyl groups) that can be utilized as mounting sites for further chemical modifications or to load the therapeutic molecule of interest. This oxidation process is also used for purification and shortening since impurities and length have been described as key parameters that may increase toxicological side effects. Alternatively, direct addition of chemical groups can be used for covalent functionalization. A popular strategy uses the 1,3-dipolar cycloaddition reaction of azomethine ylides by which terminal amino groups are used to conjugate different molecules either for imaging or therapeutic purposes or both [112,113].

Compared to other ENM, CNTs have a high surface area due to their specific cylinder shape. This characteristic can be exploited for a highly efficient loading of multiple molecules via different linkers (i.e. a multifunctional system) along the length of the CNT sidewalls. The greater external surface of MWCNT provides more possibilities for conjugation than SWCNT and, because of a larger diameter, encapsulation of therapeutic molecules is possible in the cavity of the internal cylinder.

Despite the early stage of the field in general, the range and adaptability of CNT properties have already enabled potential applications to be successfully developed for drug delivery systems and therapies for applications spanning from cancer to neurological disorders.

CNT Drug Delivery

ANTICANCER THERAPIES

One of the most studied biomedical applications for CNT is the treatment of cancer by chemical, immunological, or genetic therapies. Due to their ability to penetrate cells, functionalized CNTs have been used to deliver a variety of cytotoxic drugs, such as doxorubicin [110,111], methotrexate [114], paclitaxel [115] and cisplatin [116] into tumor cells. Similarly, an antitumor immune response can be induced by delivering tumor lysate proteins [117]. Finally, chemically functionalized CNTs have shown high efficiency for the delivery of DNA plasmid or siRNA (mainly *in vitro* for DNA [118] and *in vivo* for siRNA). It has even been demonstrated that CNT-siRNA conjugates are more effective in prolonging the survival of tumor-bearing animals than liposomal vector systems [119,120].

IMMUNOTHERAPY

The efficient intracellular delivery of molecules by CNTs has been also explored in the immunotherapeutic field either to stimulate the immune response by delivering antigenic proteins [120,121] or reduce the immune response in immune-related diseases by delivering siRNA into immune-competent cells [122].

Therapeutic Applications

CANCER

SWCNT have distinctive intrinsic optical properties with a strong optical absorption in the near-infrared (NIR) wavelength and can, therefore, also be utilized for photothermal therapy. The strategy is based on the uptake of CNTs by tumor cells, which will then be heated by NIR light and selectively induce the death of the cell that contains them [123].

NEUROLOGICAL DISORDERS

In the last few years, the potential of CNT utilization has been extended to brain diseases. In a rat stroke model, pretreatment with plasma aminated SWCNT has been demonstrated to reduce anatomic (i.e. neuronal loss) and functional (i.e. motor damage) sequelae after induced ischemic damage [124]. At the same time, non-ischemic animals injected with the same material show neither brain damage 3 weeks after injection, nor motor dysfunctions after 6 weeks. Compared to raw or carboxyl-functionalized SWCNT, these amine modified SWCNTs have been shown to be neuroprotective to the injured brains in the absence of drug-loading, possibly because of their surface properties (i.e. positive charges, surface energy) that provide a favorable environment for the neural tissue. It is one of the first *in vivo* studies to demonstrate that a chemically functionalized CNT (without any therapeutic agent) could be considered as a therapy on its own, and not only as a nanocarrier or nanodevice. In another study, using a rat stroke model, Al-Jamal et al. reported that localized administration of chemically functionalized MWCNTs carrying siRNA against caspase-3 can improve the behavioral performance of ischemia-induced animals [125]. More studies exploring the therapeutic capacity of CNTs against CNS diseases will surely appear.

Regenerative Medicine

Due to their size and shape, CNTs seem to influence cell adhesion behavior in such a manner that CNT-coated surfaces could be used to induce differentiation of stem cells [126,127] or to stimulate nerve [128], bone [129,130] or cartilage [131] regeneration. Moreover, interaction of CNT-based surface with neuronal tissue has been shown to favor neuronal electrical signaling [132]. Researchers are, therefore, now exploring the advantages of using CNT-coated electrodes for electric deep brain stimulation, which is one of the most efficient and spectacular treatments developed to reduce Parkinsonian symptoms (i.e. tremor, slow movement, rigidity, and gait disturbances) in an early stage of the disease.

Multimodal Applications

In addition to their intrinsic physical properties, different surface chemical modifications can be performed on the same CNT, leading to a platform that could carry multiple functionalities simultaneously, such as targeting, detecting and treating [133] (Figure 16.11). For instance, using carboxyl-modified CNTs to conjugate both cisplatin and epidermal growth factor, researchers have been able to target tumor cells overexpressing epidermal growth factor receptor specifically to kill them with the carried cytotoxic drug [134]. In another *in vitro* study, Wang et al. used an alternative strategy to kill neuroblastoma cells. They conjugated oxidized (carboxylated) SWCNTs with antibodies directed toward a tumor growth antigen (disialoganglioside, GD2) to target the tumor cells, and then used the intrinsic properties of the CNTs (optical absorbance in the near-infrared region) to generate heat and selectively kill the targeted cells by photothermal therapy [135]. Another *in vitro* work similarly used SWCNT coated with PEGylated phospholipids to target folate receptor-overexpressing cells. Once the SWCNT reached the target cells, their optical absorbance in the near-infrared region either to release intracellularly oligonucleotides (pulsed NIR irradiation), or to induce hyperthermia to kill the cells (continuous NIR irradiation) were demonstrated [123]. The same group also demonstrated that NIR irradiation could also be used for *in vivo* photoluminescence imaging of the PEGylated lipid-coated SWCNTs [136] and using another intrinsic property of the construct, the strong Raman signature of the SWCNT, they were also able to detect the presence of CNTs in physiological fluids (blood, urine, feces) [137]. In conclusion, clearly there is great potential to combine monitoring, imaging, targeting and multimodal treatments in the same CNT-based construct.

Health Monitoring

The high electrical conductivity and high electrochemical potential of the CNT surface have been exploited for biosensing. For instance, CNTs have been proposed as sensitive glucose detectors in the design of new *ex vivo* or implanted devices for monitoring glucose levels in diabetic patients [138]. Similarly, by attaching enzymes on drug-carrying CNT, it is expected that therapeutic biosensors will be generated that would be able not only to detect and measure biological molecules, but also subsequently to release their cargo to prevent or treat morbid processes in which these molecules take part (e.g. prevention of stroke in the case of detection of thrombin or other coagulation factors).

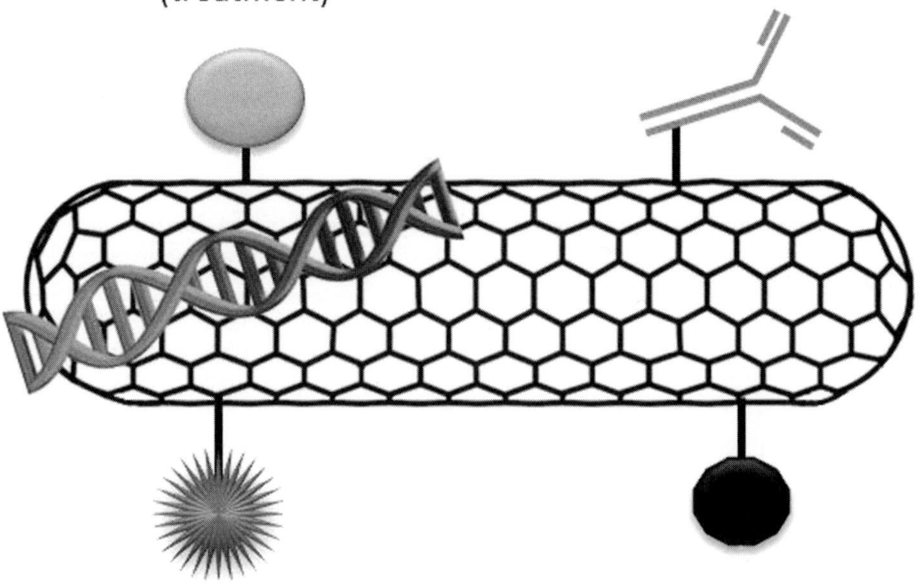

FIGURE 16.11 CNT-based multimodality platform.

Further Challenges for CNTs

Numerous questions remain unanswered before these promising nanomaterials reach the clinic [139]. In particular, three challenges must be overcome. First, CNT are not 100% pure "*nanotube*" material, but also contain different nanostructured impurities (amorphous carbon, multishell carbon balls, metal catalyst residues) that would affect their biological profile. Heterogeneity of the product (variation in length, diameter, number of walls, or graphitic structure) is similarly an important concern because it could impact both their toxicological profile and their behavior in biological systems (i.e. pharmacokinetics, biodistribution [140]). Control of CNT synthesis, purification and functionalization are thus imperatively needed to be overcome to allow clinical translation.

Another issue is the need to avoid the rapid elimination by macrophages and other phagocytic cells (i.e. cells of the reticuloendothelial system), because of the preferential uptake of CNTs by these sentinel cells. This type of cellular uptake is a problem for delivering the effective dose to a targeted tissue or cell population, and also for unwanted delivery of medicinal molecules to these immunocompetent cells. Further investigations that focus on escape from macrophage recognition, probably via improved surface functionalization or improved targeting, are therefore needed.

Finally, during the last few years there have been numerous reports regarding the potential toxicological profile of CNTs. Hence, many studies have demonstrated that pristine CNTs may exert various adverse biological effects leading to toxicity (see Chapter 7 for a detailed discussion). At the same time, there are various other studies demonstrating that chemical functionalization of the CNT backbone can significantly reduce or abolish such apparent immunological or toxicological responses [104,141,142]. Moreover, the long-term fate (i.e. biopersistence or clearance, biotransformation, and biodegradation) of CNT-based constructs for therapeutic applications have to be further documented. In the case of long-term biopersistence, CNTs could be either inert or induce reaction by the tissue(s) in which they reside. Also, if any biotransformation occurs within tissues, the products could potentially be more reactive and harmful than the original nanotube constructs. Even though biodegradation of CNTs can be seen as beneficial for their safety profile, partial degradation may also lead to questions about the inertness of such by-products. All of these questions have to be addressed by pharmacologists working in close collaboration with toxicologists to allow clinical translation of these promising materials.

TAKE-HOME MESSAGES

1. The field of ENM has the potential to fundamentally renovate pharmacological science and industry by proposing more efficient solutions for therapy and diagnosis.
2. The full potential and limits of ENM for different biomedical applications are still being explored by scientists working in nanomedicine and nanoscience.
3. Biomedical researchers face the grand challenge of translating ENM into a clinical reality.

Acknowledgments

CA would like to thank the Else Kröner-Fresenius-Stiftung for their support and the implementation of the Else Kröner-Fresenius-Stiftung-Professorship for Nanomedicine at the University Hospital Erlangen. Furthermore, the author would like to thank the DFG (PAK 151) and the Ministry of Science and Education (BMBF: FKZ: 01EX1012B) for their advancement. RAP acknowledges financial support from the University of North Texas, and the National Science Foundation (CHE-1004878, CHE-0840518). AMN acknowledges financial support from Knut and Alice Wallenbergs Foundation, Jeanssons Stiftelse, Karolinska Institutet, Åke Wibergs foundation, Eva and Axel Wallströms Foundation, Carl Bennet AB, The Swedish Research Council (VR) 2009-3259 (together with MM), The Swedish Medical Nanoscience Center and The Swedish Governmental Agency for Innovation Systems (Vinnova). LM, CB and KK wish to acknowledge the UK Engineering and Physical Sciences Research Council and the European Commission, FP7 Research Framework, for their financial support through the EPSRC Grand Challenge (Point of Care nanotechnology for early blood clot detection and characterization in disease screening, theranostic and self-monitoring applications), the EU Marie Curie Actions (Intra-European Fellowship, PIEF-GA-2010-276051), and the research program ANTICARB (Monoclonal ANTIbody-targeted CARBon Nanotubes against Cancer).

References

[1] Wagner V, Dullaart A, Bock AK, Zweck A. The emerging nanomedicine landscape. Nat Biotechnol 2006;24:1211–7.

[2] Peer D, Karp JM, Hong S, Farokhzad OC, Margalit R, Langer R. Nanocarriers as an emerging platform for cancer therapy. Nat Nanotechnol 2007;2:751–60.

[3] Perkins WR, Ahmad I, Li XG, et al. Novel therapeutic nano-particles (lipocores): Trapping poorly water soluble compounds. Int J Pharmaceut 2000;200:27–39.

[4] Pankhurst QA, Thanh NKT, Jones SK, Dobson J. Progress in applications of magnetic nanoparticles in biomedicine. J Phys D Appl Phys 2009;42:224001.

[5] Pankhurst QA, Connolly J, Jones SK, Dobson J. Applications of magnetic nanoparticles in biomedicine. J Phys D Appl Phys 2003;36:R167–81.

[6] Arap W, Haedicke W, Bernasconi M, et al. Targeting the prostate for destruction through a vascular address. Proc Natl Acad Sci USA 2002;99:1527–31.

[7] Torchilin VP. Drug targeting. Eur J Pharm Sci 2000;11:S81–91.

[8] Gerlowski LE, Jain RK. Microvascular permeability of normal and neoplastic tissues. Microvasc Res 1986;31:288–99.

[9] Shen Z, Wei W, Zhao YJ, et al. Thermosensitive polymer-conjugated albumin nanospheres as thermal targeting anti-cancer drug carrier. Eur J Pharm Sci 2008;35:271–82.

[10] Chuang CY, Don TM, Chiu WY. Synthesis of chitosan-based thermo- and pH-responsive porous nanoparticles by temperature-dependent self-assembly method and their application in drug release. J Polymer Sci a Polymer Chem 2009;47:5126–36.

[11] Shen ZY, Ma GH, Dobashi T, Maki Y, Su ZG. Preparation and characterization of thermo-responsive albumin nanospheres. Int J Pharm 2008;346:133–42.

[12] Lee CF, Wen CJ, Lin CL, Chiu WY. Morphology and temperature responsiveness-swelling relationship of poly(N-isopropylamide-chitosan) copolymers and their application to drug release. J Polymer Sci a Polymer Chem 2004;42:3029–37.

[13] Chuang CY, Don TM, Chiu WY. Synthesis and properties of chitosan-based thermo- and pH-responsive nanoparticles and application in drug release. J Polymer Sci a Polymer Chem 2009;47:2798–810.

[14] Stover TC, Kim YS, Lowe TL, Kester M. Thermoresponsive and biodegradable linear-dendritic nanoparticles for targeted and sustained release of a pro-apoptotic drug. Biomaterials 2008;29:359–69.

[15] Bakhtiari ABS, Hsiao D, Jin GX, Gates BD, Branda NR. An efficient method based on the photothermal effect for the release of molecules from metal nanoparticle surfaces. Angewandte Chem Int Ed 2009;48:4166–9.

[16] Bilkenroth U, Taubert H, Riemann D, Rebmann U, Heynemann H, Meye A. Detection and enrichment of disseminated renal carcinoma cells from peripheral blood by immunomagnetic cell separation. Int J Cancer 2001;92:577–82.

[17] Taupitz M, Wagner S, Hamm B, Dienemann D, Lawaczeck R, Wolf KJ. MR lymphography using iron oxide particles. Detection of lymph node metastases in the VX2 rabbit tumour model. Acta Radiol 1993;34:10–5.

[18] Harisinghani MG, Barentsz J, Hahn PF, et al. Noninvasive detection of clinically occult lymph-node metastases in prostate cancer. N Engl J Med 2003;348:2491–9.

[19] Gao Y. In: Weinheim KCS, editor. Biofunctionalization of nanomaterials. Wiley-VCH Verlag GmbH & Co. KGaA; 2005.

[20] Storm G, Belliot SO, Daemen T, Lasic DD. Surface modification of nanoparticles to oppose uptake by the mononuclear phagocyte system. Adv Drug Deliver Rev 1995;17:31–48.

[21] Muller RH, Jacobs C, Kayser O. Nanosuspensions as particulate drug formulations in therapy. Rationale for development and what we can expect for the future. Adv Drug Deliv Rev 2001;47:3–19.

[22] Alksne JF, Fingerhut A, Rand R. Magnetically controlled metallic thrombosis of intracranial aneurysms. Surgery 1966;60:212–8.

[23] Widder KJ, Morris RM, Howard DP, Poore GA, Senyei AE. Selective targeting of magnetic albumin microspheres to the yoshida sarcoma: Ultrastructural evalutation of microsphere distribution. Eur J Cancer Clin Oncol 1993;19:141–7.

[24] Wilson MW, Kerlan Jr RK, Fidelman NA, et al. Hepatocellular carcinoma: Regional therapy with a magnetic targeted carrier bound to doxorubicin in a dual MR imaging/conventional angiography suite—initial experience with four patients. Radiology 2004;230:287–93.

[25] Lübbe AS, Bergemann C, Riess H, et al. Clinical experiences with magnetic drag targeting: A phase I study with 4'-epidoxorubicin in 14 patients with advanced solid tumors. Cancer Res 1996;5:4686–93.
[26] Alexiou C, Jurgons R, Schmid RJ, et al. Magnetic drug targeting—biodistribution of the magnetic carrier and the chemotherapeutic agent mitoxantrone after locoregional cancer treatment. J Drug Target 2003;11:139–49.
[27] Alexiou C, Diehl D, Henninger P, et al. A high field gradient magnet for magnetic drug targeting. IEEE Transact Appl Superconduct 2006;16:1527–30.
[28] Brigger I, Morizet J, Laudani L, et al. Negative preclinical results with stealth((R)) nanospheres-encapsulated Doxorubicin in an orthotopic murine brain tumor model. J Contr Release 2004;100:29–40.
[29] Alexiou C, Jurgons R, Schmid R, et al. Magnetic drug targeting—a new approach in locoregional tumor-therapy with chemotherapeutic agents. Experimental animal studies. Hno 2005;53:618–22.
[30] Alexiou C, Arnold W, Hulin P, et al. Magnetic mitoxantrone nanoparticle detection by histology, X-ray and MRI after magnetic tumor targeting. J Magn Magn Mater 2001;225:187–93.
[31] Rahn H, Gomez-Morilla I, Jurgons R, Alexiou C, Odenbach S. Microcomputed tomography analysis of ferrofluids used for cancer treatment. J Phys-Condensed Matt 2008;20:204152–6.
[32] Tietze R, Rahn H, Lyer S, et al. Visualization of superparamagnetic nanoparticles in vascular tissue using X mu CT and histology. Histochem Cell Biol 2011;135:153–8.
[33] Alexiou Ch, Hulin P, Klein RJ, Bergemann Ch, Arnold W. Magnetic drug targeting: Biodistribution and dependency on magnetic field strength. J Magn Magn Mater 2002;252:363–6.
[34] Alexiou C, Arnold W, Klein RJ, et al. Locoregional cancer treatment with magnetic drug targeting. Cancer Res 2000;60:6641–8.
[35] Alexiou C, Schmid RJ, Jurgons R, et al. Targeting cancer cells: Magnetic nanoparticles as drug carriers. Eur Biophys J Biophys Lett 2006;35:446–50.
[36] Levy-Nissenbaum E, Radovic-Moreno AF, Wang AZ, Langer R, Farokhzad OC. Nanotechnology and aptamers: Applications in drug delivery. Trends Biotechnol 2008;26:442–9.
[37] Purushotham S, Chang PE, Rumpel H, et al. Thermoresponsive core-shell magnetic nanoparticles for combined modalities of cancer therapy. Nanotechnology 2009;20:305101–12.
[38] Welch MJ, Hawker CJ, Wooley KL. The advantages of nanoparticles for PET. J Nucl Med 2009;50:1743–6.
[39] Alexiou C, Tietze R, Schreiber E, Lyer S. Pharmakotherapie mittels nanomedizin—magnetische nanopartikel für drug delivery und hyperthermie-neue chancen für die krebstherapie. Bundesgesundheitsbl Gesundheitsforsch Gesundheitssch 2010;53:839–45.
[40] Petros RA, DeSimone JM. Strategies in the design of nanoparticles for therapeutic applications. Nat Rev Drug Discov 2010;9:615–27.
[41] Davis ME, Chen Z, Shin DM. Nanoparticle therapeutics: An emerging treatment modality for cancer. Nat Rev Drug Discov 2008;7:771–82.
[42] Zhang L, Gu FX, Chan JM, Wang AZ, Langer RS, Farokhzad OC. Nanoparticles in medicine: Therapeutic applications and developments. Clin Pharmacol Ther 2008;83:761–9.
[43] Shi J, Votruba AR, Farokhzad OC, Langer R. Nanotechnology in drug delivery and tissue engineering: From discovery to applications. Nano Lett 2010;10:3223–30.
[44] Geng Y, Dalhaimer P, Cai SS, et al. Shape effects of filaments versus spherical particles in flow and drug delivery. Nat Nanotechnol 2007;2:249–55.
[45] Merkel TJ, Jones SW, Herlihy KP, et al. Using mechanobiological mimicry of red blood cells to extend circulation times of hydrogel microparticles. Proc Natl Acad Sci USA 2011;108:586–91.
[46] Dobrovolskaia MA, Mcneil SE. Immunological properties of engineered nanomaterials. Nat Nanotechnol 2007;2:469–78.
[47] Kohane DS, Langer R. Biocompatibility and drug delivery systems. Chem Sci 2010;1:441–6.
[48] Parrott MC, Luft JC, Byrne JD, Fain JH, Napier ME, DeSimone J. Tunable bifunctional silyl ether cross-linkers for the design of acid-sensitive biomaterials. J Am Chem Soc 2010;132:17928–32.
[49] Maeda H, Greish K, Fang J. The EPR effect and polymeric drugs: A paradigm shift for cancer chemotherapy in the 21st century. Polymer Ther Ii: Polymers Drug Conjug Gene Deliv Sys 2006;193:103–21.
[50] Matsumura Y, Maeda H. A new concept for macromolecular therapeutics in cancer-chemotherapy—mechanism of tumoritropic accumulation of proteins and the antitumor agent Smancs. Cancer Res 1986;46:6387–92.
[51] Allen TM. Ligand-targeted therapeutics in anticancer therapy. Nat Rev Cancer 2002;2:750–63.

[52] Byrne JD, Betancourt T, Brannon-Peppas L. Active targeting schemes for nanoparticle systems in cancer therapeutics. Adv Drug Deliv Rev 2008;60:1615–26.
[53] Low PS, Henne WA, Doorneweerd DD. Discovery and development of folic-acid-based receptor targeting for imaging and therapy of cancer and inflammatory diseases. Acc Chem Res 2008;41:120–9.
[54] Gabizon A, Shmeeda H, Horowitz AT, Zalipsky S. Tumor cell targeting of liposome-entrapped drugs with phospholipid-anchored folic acid-PEG conjugates. Adv Drug Deliv Rev 2004;56:1177–92.
[55] Daniels TR, Delgado T, Rodriguez JA, Helguera G, Penichet ML. The transferrin receptor part I: Biology and targeting with cytotoxic antibodies for the treatment of cancer. Clin Immunol 2006;121:144–58.
[56] Daniels TR, Delgado T, Helguera G, Penichet ML. The transferrin receptor part II: Targeted delivery of therapeutic agents into cancer cells. Clin Immunol 2006;121:159–76.
[57] Patri AK, Myc A, Beals J, Thomas TP, Bander NH, Baker JR. Synthesis and *in vitro* testing of J591 antibody-dendrimer conjugates for targeted prostate cancer therapy. Bioconj Chem 2004;15:1174–81.
[58] Torchilin VP. Targeted pharmaceutical nanocarriers for cancer therapy and imaging. Aaps J 2007;9: E128–147.
[59] Oh KT, Yin HQ, Lee ES, Bae YH. Polymeric nanovehicles for anticancer drugs with triggering release mechanisms. J Mater Chem 2007;17:3987–4001.
[60] Ganta S, Devalapally H, Shahiwala A, Amiji M. A review of stimuli-responsive nanocarriers for drug and gene delivery. J Control Release 2008;126:187–204.
[61] Petros RA, Ropp PA, DeSimone JM. Reductively labile PRINT particles for the delivery of doxorubicin to HeLa cells. J Am Chem Soc 2008;130:5008–9.
[62] Oh JK, Siegwart DJ, Lee HI, et al. Biodegradable nanogels prepared by atom transfer radical polymerization as potential drug delivery carriers: Synthesis, biodegradation, *in vitro* release, and bioconjugation. J Am Chem Soc 2007;129:5939–45.
[63] Dhar S, Kolishetti N, Lippard SJ, Farokhzad OC. Targeted delivery of a cisplatin prodrug for safer and more effective prostate cancer therapy *in vivo*. Proc Natl Acad Sci USA 2011;108:1850–5.
[64] Avgoustakis K, Beletsi A, Panagi Z, Klepetsanis P, Karydas AG, Ithakissios DS. PLGA-mPEG nanoparticles of cisplatin: *In vitro* nanoparticle degradation, *in vitro* drug release and *in vivo* drug residence in blood properties. J Contr Release 2002;79:123–35.
[65] Murphy EA, Majeti BK, Mukthavaram R, Acevedo LM, Barnes L, Cheresh DA. Targeted nanogels: A versatile platform for drug delivery to tumors. Mol Cancer Ther 2011;10:972–82.
[66] Harris AN, Hinojosa BM, Chavious MD, Petros RA. Beyond platinum: Synthesis, characterization, and *in vitro* toxicity of Cu(II)-releasing polymer nanoparticles for potential use as a drug delivery vector. Nanoscale Res Lett 2011;6:445.
[67] Jang WD, Selim KMK, Lee CH, Kang IK. Bioinspired application of dendrimers: From bio-mimicry to biomedical applications. Prog Polymer Sci 2009;34:1–23.
[68] Nanjwade BK, Bechra HM, Derkar GK, Manvi FV, Nanjwade VK. Dendrimers: Emerging polymers for drug-delivery systems. Eur J Pharm Sci 2009;38:185–96.
[69] Lee CC, MacKay JA, Frechet JMJ, Szoka FC. Designing dendrimers for biological applications. Nat Biotechnol 2005;23:1517–26.
[70] D'Emanuele A, Attwood D. Dendrimer–drug interactions. Adv Drug Deliv Rev 2005;57:2147–62.
[71] Hawker CJ, Wooley KL, Frechet JMJ. Unimolecular micelles and globular amphiphiles—dendritic macromolecules as novel recyclable solubilization agents. J Chem Soc Perkin Transact 1993;1:1287–97.
[72] Moorefield CN, Newkome GR. Unimolecular micelles: Supramolecular use of dendritic constructs to create versatile molecular containers. Comptes Rend Chim 2003;6:715–24.
[73] Jansen JFGA, Debrabandervandenberg EMM, Meijer EW. Encapsulation of guest molecules into a dendritic box. Science 1994;266:1226–9.
[74] Margerum LD, Campion BK, Koo M, et al. Gadolinium(III) DO3A macrocycles and polyethylene glycol coupled to dendrimers—effect of molecular weight on physical and biological properties of macromolecular magnetic resonance imaging contrast agents. J Alloys Compounds 1997;249:185–90.
[75] Roberts JC, Bhalgat MK, Zera RT. Preliminary biological evaluation of polyamidoamine (PAMAM) Starburst(TM) dendrimers. J Biomed Mater Res 1996;30:53–65.
[76] Sarkar K, Yang H. Encapsulation and extended release of anti-cancer anastrozole by stealth nanoparticles. Drug Deliv 2008;15:343–6.

[77] Neerman MF. The efficiency of a PAMAM dendrimer toward the encapsulation of the antileukernic drug 6-mercaptopurine. Anti-Cancer Drugs 2007;18:839–42.
[78] Bhadra D, Bhadra S, Jain S, Jain NK. A PEGylated dendritic nanoparticulate carrier of fluorouracil. Int J Pharm 2003;257:111–24.
[79] Kojima C, Kono K, Maruyama K, Takagishi T. Synthesis of polyamidoamine dendrimers having poly(ethylene glycol) grafts and their ability to encapsulate anticancer drugs. Bioconj Chem 2000;11:910–7.
[80] Cheng YY, Li MH, Xu TW. Potential of poly(amidoamine) dendrimers as drug carriers of camptothecin based on encapsulation studies. Eur J Med Chem 2008;43:1791–5.
[81] Morgan MT, Carnahan MA, Immoos CE, et al. Dendritic molecular capsules for hydrophobic compounds. J Am Chem Soc 2003;125:15485–9.
[82] Morgan MT, Nakanishi Y, Kroll DJ, et al. Dendrimer-encapsulated camptothecins: Increased solubility, cellular uptake, and cellular retention affords enhanced anticancer activity *in vitro*. Cancer Res 2006;66: 11913–21.
[83] Malik N, Evagorou EG, Duncan R. Dendrimer-platinate: A novel approach to cancer chemotherapy. Anti-Cancer Drugs 1999;10:767–76.
[84] Choi SK, Thomas T, Li M-H, Kotlyar A, Desai A, Baker JR. Chem Commun 2010;46:2632–4.
[85] Kono K, Kojima C, Hayashi N, et al. Biomaterials 2008;29:1664–75.
[86] Zhu SJ, Hong MH, Zhang LH, Tang GT, Jiang YY, Pei YY. Pharmaceut Res 2010;27:161–74.
[87] Majoros IJ, Myc A, Thomas T, Mehta CB, Baker JR. Biomacromolecules 2006;7:572–9.
[88] Saad M, Garbuzenko OB, Ber E, et al. J Contr Release 2008;130:107–14.
[89] Khandare JJ, Jayant S, Singh A, et al. Bioconj Chem 2006;17:1464–72.
[90] Zhao XX, Loo SCJ, Lee PPF, Tan TTY, Chu CK. J Inorgan Biochem 2010;104:105–10.
[91] Zhuo RX, Du B, Lu ZR. J Contr Release 1999;57:249–57.
[92] Patri AK, Kukowska-Latallo JF, Baker JR. Adv Drug Deliv Rev 2005;57:2203–14.
[93] Thomas TP, Majoros IJ, Kotlyar A, et al. J Med Chem 2005;48:3729–35.
[94] Kukowska-Latallo JF, Candido KA, Cao ZY, et al. Cancer Res 2005;65:5317–24.
[95] Myc A, Kukowska-Latallo J, Cao P, et al. Anti-Cancer Drugs 2010;21:186–92.
[96] Majoros IJ, Thomas TP, Mehta CB, Baker JR. J Med Chem 2005;48:5892–9.
[97] Guillaudeu SJ, Fox ME, Haidar YM, Dy EE, Szoka FC, Fréchet JMJ. Bioconj Chem 2008;19:461–9.
[98] De Jesus OLP, Ihre HR, Gagne L, Fréchet JMJ, Szoka FC. Polyester dendritic systems for drug delivery applications: *In vitro* and *in vivo* evaluation. Bioconj Chem 2002;13:453–61.
[99] Ihre HR, De Jesus OLP, Szoka FC, Fréchet JMJ. Polyester dendritic systems for drug delivery applications: Design, synthesis, and characterization. Bioconj Chem 2002;13:443–52.
[100] Lee CC, Gillies ER, Fox ME, et al. Proc Natl Acad Sci USA 2006;103:16649–54.
[101] Lim JD, Simanek EE. Org Lett 2008;10:201–4.
[102] Lim J, Chouai A, Lo ST, Liu W, Sun XK, Simanek EE. Bioconj Chem 2009;20:2154–61.
[103] Venditto VJ, Allred K, Allred CD, Simanek EE. Intercepting the synthesis of triazine dendrimers with nucleophilic pharmacophores: A general strategy toward drug delivery vehicles. Chem Commun (Camb) 2009;3:5541–2.
[104] Dumortier H, Lacotte S, Pastorin G, et al. Functionalized carbon nanotubes are non-cytotoxic and preserve the functionality of primary immune cells. Nano Lett 2006;6:1522–8.
[105] Prato M, Kostarelos K, Bianco A. Functionalized carbon nanotubes in drug design and discovery. Acc Chem Res 2008;41:60–8.
[106] Liu Z, Tabakman SM, Chen Z, Dai HJ. Preparation of carbon nanotube bioconjugates for biomedical applications. Nat Protocols 2009;4:1372–82.
[107] Sayes CM, Liang F, Hudson JL, et al. Functionalization density dependence of single-walled carbon nanotubes cytotoxicity *in vitro*. Toxicol Lett 2006;161:135–42.
[108] Liu Z, Tabakman S, Welsher K, Dai H. Carbon nanotubes in biology and medicine: *In vitro* and *in vivo* detection, imaging and drug delivery. Nano Res 2009;2:85–120.
[109] Liu Z, Sun X, Nakayama-Ratchford N, Dai H. Supramolecular chemistry on water-soluble carbon nanotubes for drug loading and delivery. ACS Nano 2007;1:50–6.
[110] Liu Z, Fan AC, Rakhra K, et al. Supramolecular stacking of doxorubicin on carbon nanotubes for *in vivo* cancer therapy. Angew Chem Int Ed Engl 2009:7668–72.

[111] Ali-Boucetta H, Al-Jamal KT, McCarthy D, Prato M, Bianco A, Kostarelos K. Multiwalled carbon nanotube-doxorubicin supramolecular complexes for cancer therapeutics. Chem Commun (Camb.) 2008;4:459–61.
[112] Georgakilas V, Tagmatarchis N, Pantarotto D, Bianco A, Briand JP, Prato M. Amino acid functionalisation of water soluble carbon nanotubes. Chem Commun (Camb.) 2002;24:3050–1.
[113] Pantarotto D, Briand JP, Prato M, Bianco A. Translocation of bioactive peptides across cell membranes by carbon nanotubes. Chem Commun (Camb.) 2004;16–7.
[114] Samori C, Ali-Boucetta H, Sainz R, et al. Enhanced anticancer activity of multi-walled carbon nanotube-methotrexate conjugates using cleavable linkers. Chem Commun (Camb.) 2010;46:1494–6.
[115] Liu Z, Chen K, Davis C, et al. Drug delivery with carbon nanotubes for *in vivo* cancer treatment. Cancer Res 2008;68:6652–60.
[116] Bhirde AA, Patel S, Sousa AA, et al. Distribution and clearance of PEG-single-walled carbon nanotube cancer drug delivery vehicles in mice. Nanomedicine (Lond.) 2010;5:1535–46.
[117] Meng J, Duan J, Kong H, et al. Carbon nanotubes conjugated to tumor lysate protein enhance the efficacy of an antitumor immunotherapy. Small 2008;4:1364–70.
[118] Singh R, Pantarotto D, McCarthy D, et al. Binding and condensation of plasmid DNA onto functionalized carbon nanotubes: Toward the construction of nanotube-based gene delivery vectors. J Am Chem Soc 2005;127:4388–96.
[119] Podesta JE, Al-Jamal KT, Herrero MA, et al. Antitumor activity and prolonged survival by carbon-nanotube-mediated therapeutic siRNA silencing in a human lung xenograft model. Small 2009;5:1176–85.
[120] Pantarotto D, Partidos CD, Graff R, et al. Synthesis, structural characterization, and immunological properties of carbon nanotubes functionalized with peptides. J Am Chem Soc 2003;125:6160–4.
[121] Pantarotto D, Partidos CD, Hoebeke J, et al. Immunization with peptide-functionalized carbon nanotubes enhances virus-specific neutralizing antibody responses. Chem Biol 2003;10:961–6.
[122] Yang R, Yang X, Zhang Z, et al. Single-walled carbon nanotubes-mediated *in vivo* and *in vitro* delivery of siRNA into antigen-presenting cells. Gene Ther 2006;13:1714–23.
[123] Kam NW, O'Connell M, Wisdom JA, Dai H. Carbon nanotubes as multifunctional biological transporters and near-infrared agents for selective cancer cell destruction. Proc Natl Acad Sci USA 2005;102:11600–5.
[124] Lee HJ, Park J, Yoon OJ, et al. Amine-modified single-walled carbon nanotubes protect neurons from injury in a rat stroke model. Nat Nanotechnol 2011;6:121–5.
[125] Al-Jamal KT, et al. Functional motor recovery from brain ischemic insult by carbon nanotube-mediated siRNA silencing. Proc Natl Acad Sci USA 2011;108:10952–7.
[126] Nayak TR, Jian L, Phua LC, Ho HK, Ren Y, Pastorin G. Thin films of functionalized multiwalled carbon nanotubes as suitable scaffold materials for stem cells proliferation and bone formation. ACS Nano 2010;4: 7717–25.
[127] Chao TI, Xiang S, Lipstate JF, Wang C, Lu J. Poly(methacrylic acid)-grafted carbon nanotube scaffolds enhance differentiation of hESCs into neuronal cells. Adv Mater 2010;22:3542–7.
[128] Tosun Z, McFetridge PS. A composite SWNT-collagen matrix: characterization and preliminary assessment as a conductive peripheral nerve regeneration matrix. J Neural Eng 2010;7:066002.
[129] Usui Y, Aoki K, Narita N, et al. Carbon nanotubes with high bone-tissue compatibility and bone-formation acceleration effects. Small 2008;4:240–6.
[130] Khang D, Sato M, Price RL, Ribbe AE, Webster TJ. Selective adhesion and mineral deposition by osteoblasts on carbon nanofiber patterns. Int J Nanomed 2006;1:65–72.
[131] Khang D, Park GE, Webster TJ. Enhanced chondrocyte densities on carbon nanotube composites: The combined role of nanosurface roughness and electrical stimulation. J Biomed Mater Res A 2008;86:253–60.
[132] Cellot G, Cilia E, Cipollone S, et al. Carbon nanotubes might improve neuronal performance by favouring electrical shortcuts. Nat Nanotechnol 2009;4:126–33.
[133] Menard-Moyon C, Fabbro C, Prato M, Bianco A. One-pot triple functionalization of carbon nanotubes. Chem Eur J 2011;17:3222–7.
[134] Bhirde AA, Patel V, Gavard J, et al. Targeted killing of cancer cells *in vivo* and *in vitro* with EGF-directed carbon nanotube-based drug delivery. ACS Nano 2009;3:307–16.
[135] Wang CH, Huang YJ, Chang CW, Hsu WM, Peng CA. *In vitro* photothermal destruction of neuroblastoma cells using carbon nanotubes conjugated with GD2 monoclonal antibody. Nanotechnology 2009;20: 315101.

[136] Welsher K, Liu Z, Sherlock SP, et al. A route to brightly fluorescent carbon nanotubes for near-infrared imaging in mice. Nat Nanotechnol 2009;4:773–80.
[137] Liu Z, Davis C, Cai W, He L, Chen X, Dai H. Circulation and long-term fate of functionalized, biocompatible single-walled carbon nanotubes in mice probed by Raman spectroscopy. Proc Natl Acad Sci USA 2008;105:1410–5.
[138] Jacobs CB, Peairs MJ, Venton BJ. Review: Carbon nanotube based electrochemical sensors for biomolecules. Anal Chim Acta 2010;662:105–27.
[139] Kostarelos K, Bianco A, Prato M. Promises, facts and challenges for carbon nanotubes in imaging and therapeutics. Nat Nanotechnol 2009;4:627–33.
[140] Singh R, Pantarotto D, Lacerda L, et al. Tissue biodistribution and blood clearance rates of intravenously administered carbon nanotube radiotracers. Proc Natl Acad Sci USA 2006;103:3357–62.
[141] Gaillard C, Duval M, Dumortier H, Bianco A. Carbon nanotube-coupled cell adhesion peptides are non-immunogenic: A promising step toward new biomedical devices. J Pept Sci 2011;17:139–42.
[142] Heister E, Lamprecht C, Neves V, et al. Higher dispersion efficacy of functionalized carbon nanotubes in chemical and biological environments. ACS Nano 2010;4:2615–26.

APPENDIX

Annex: Synthesis of Engineered Nanomaterials

Muhammet S. Toprak[1], Gulaim A. Seisenbaeva[2], Vadim G. Kessler[2]

[1]KTH Royal Institute of Technology, Functional Materials Division, Kista-Stockholm, Sweden

[2]Department of Chemistry, SLU BioCenter, Uppsala, Sweden

OUTLINE

INTRODUCTION

The development of new approaches in medicine based on the application of nanomaterials sets special requirements on mastering the techniques of their preparation. It is also important to have a firm understanding of the specific properties and impurities that are associated with a particular applied synthetic process. The goal of this chapter is to provide an overview of chemical strategies employed for production of commonly used nanomaterials intended for biomedical applications.

CARBON NANOTUBES

Carbon nanotubes (CNTs) have recently attracted great interest in various applications due to their outstanding properties and special cylindrical structure, similar to those of tubular biological structures, such as those assembled from regular packing of proteins.

The characteristics of CNTs, namely, an ordered structure with a high aspect ratio, high mechanical strength, high electrical conductivity, and a large surface area, are promising for the application of CNTs to various fields, including nano-electronics, nanocomposites, energy storage, and biomedical fields [1–3].

Several methods have been devised to synthesize CNTs. They are generally based on gas-phase processes and can be classified into two main categories depending on whether they are running at high or at medium temperatures. High-temperature routes are based on the vaporization of a graphite target, like the arc-discharge or laser ablation process, whereas medium-temperature routes are based on vapor deposition processes, such as chemical vapor deposition (CVD) [4].

High-temperature methods involve sublimating (transforming from a solid state to gaseous one without turning into a liquid first) graphite in a reducing or inert gas atmosphere, bringing to temperatures above the sublimation temperature of graphite, and condensing the resulting vapor under a high temperature gradient [5].

In the *electric arc discharge* technique, an arc discharge is generated between two graphite electrodes under a partial pressure of an inert gas (helium or argon), which brings the temperature up to 6000°C. This is hot enough for the carbon contained in the graphite to sublime forming plasma. The ejected atoms are then transferred towards colder zones within the chamber, allowing a nanotube deposit to accumulate on the cathode. The type of nanotube that is formed depends crucially upon the presence of metal catalysts. If small amounts of transition metals, such as Fe, Co, Ni, are introduced in the target graphite, then single-walled carbon nanotubes (SWNT) are the dominant product. In the absence of such catalysts, the formation of multiwalled (MWNT) carbon nanotubes is favored.

The *laser ablation* technique also consists of sublimating graphite under a reducing or inert gas atmosphere. The process is named slightly differently based on the type of laser used; a high intensity pulsed laser [6,7] or a low intensity continuous laser [8]. The laser converts the composite solid material into small aggregates which recombine in a heated external furnace [9], creating a temperature gradient suitable for the formation and growth of long nanotubes [10]. The CNTs nucleate in the vapor phase, coalesce, get carried away by flowing argon and condense on the cooled collector. The material contains MWNTs or SWNTs depending on the experimental conditions. When the target was pure graphite the product was MWNTs, while when the target was a mixture of graphite and metallic catalysts (Ni-Co or Ni-Y) mixtures, SWNTs were obtained [8,11].

Sublimation of graphene was also achieved by using highly concentrated sunlight from a solar furnace (vaporization induced by a solar beam) for the formation of CNTs. Depending on the pressure and flow conditions, either MWNTs, SWNTs or a mixture of both were obtained [12,13].

Chemical vapor deposition growth relies on the capability to provide the atoms needed to form the solid deposit from a gas source that decomposes under the action of temperature. The *catalytic* CVD process makes possible the growth of carbon filaments of various sizes and shapes at low temperature (≤1000°C) from carbon-containing gaseous compounds, which decompose catalytically on transition-metal particles [14,15].

Autogenic reactions, based on the decomposition of one or more precursors at elevated temperatures with self-generated pressures (RAPET), can be used to prepare a wide range of materials including CNTs with interesting structural, morphological and technological properties [16].

A principal obstacle in development of a consistent view on the chemical activity and potential adverse effects of CNTs is the formation in all cases of complex mixtures of SWNTs and MWNTs. In addition, the MWNTs can exist in three different structural variants, possessing distinctly different physical and, most probably, chemical properties. This will seriously hinder both obtaining reliable data on toxicity that can vary from sample to sample in the same synthetic approach and complicate the certification of derived products for biomedical use.

Unfortunately, employing CNTs in biomedical systems with macromolecules is even questionable in the absence of any coating for different reasons: (1) the pronounced hydrophobicity of CNTs; (2) their aggregated nature derived from van der Waals interactions between the tubes; (3) the difficulty in controlling their size; and (4) the remaining impurities from their preparation, especially the transition metal content [1].

The encounter between CNTs and specific materials, including proteins, peptides, or DNA, can result in synergistic phenomena. Many of these materials are very effective in dispersing the hydrophobic CNTs in water and debundling them [17,18]. Principally, there are two types of method for conjugating other materials with CNTs: covalent and non-covalent methods. In the covalent method, through chemical modification, molecules or macromolecules are grafted onto the surface of CNTs through chemical linkages. There are several reviews on the functionalization of CNTs and a survey on the biomedical use of CNTs can be found in [1].

Heteroatomic nanotubes: for the synthesis of heteroatomic nanotubes (mainly BN and BCN), many methods were inspired by carbon nanotubes: arc discharge synthesis [19,20], laser ablation [21,22], CVD techniques, etc.

QUANTUM DOTS

Semiconductor nanocrystals exhibit remarkable size-dependent optical absorption and luminescence properties. These characteristics are due to so-called quantum confinement effects due to their nanometer size, and these nanoparticles are called quantum dots (QDs). QDs are commonly used in biological applications for labeling, among others. Figure A.1 [23] shows representative QD core materials scaled as a function of their emission wavelength superimposed over the electromagnetic spectrum.

The single most studied QD system is CdSe. QDs of CdSe can be synthesized with sizes ranging from 1 to 11 nm. Through quantum confinement of the photoexcited electrons, the optical properties of the nanocrystal can be tuned by size [24]. Growing defect-free nanocrystals is relatively easy because their length scale is too small to introduce major crystalline defects. The small length scale, however, makes controlling the size and surface of nanocrystals a tremendous challenge.

Synthesis of monodisperse CdSe nanocrystals was pioneered by the Bawendi group based on the thermolysis of organometallic precursors (see Figure A.1). They reported the synthesis of CdE (E=S, Se, Te) nanocrystals by a hot-injection method. Briefly, the process involves a rapid introduction of cold trioctylphosphine (TOP) solution containing the metal and chalcogenide precursors into hot trioctylphosphine oxide (TOPO) solution [25]. An immediate nucleation of the CdE nanocrystals is followed by growth at a lower temperature to avoid further nucleation, permitting a narrow size distribution of 5%. In this process, size control

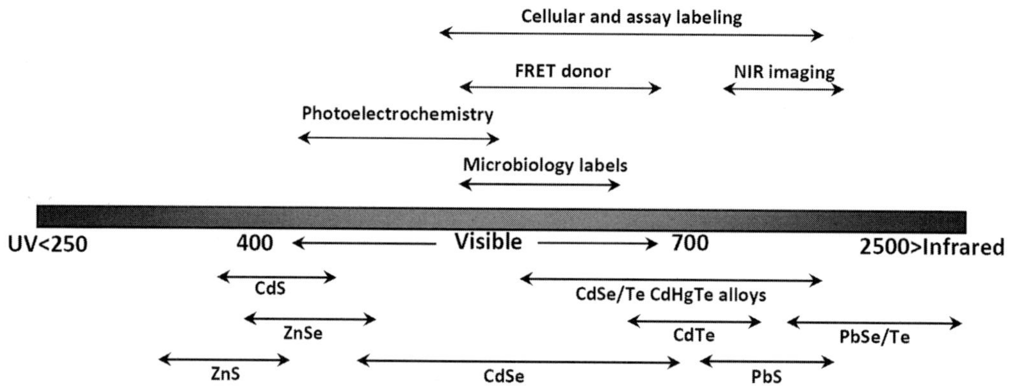

FIGURE A.1 Representative QD core materials scaled as a function of their emission wavelength superimposed over the spectrum. Representative areas of biological interest are also presented corresponding to the pertinent emission highlighting how most biological uses fall in the visible-near infrared (NIR) region. *(Adapted from [23].)*

is primarily achieved by kinetic control of the reaction conditions such as growth time, temperature and monomer concentrations. Weller et al. added hexadecylamine to the phosphine solutions to obtain a better particle size distribution. Here, the coordinating surfactant prevented agglomeration of the nanocrystals and controlled their growth rate to enhance uniformity [26]. Though both these approaches proved to be very popular, the organometallic cadmium precursor used in these processes was highly toxic dimethyl cadmium. Peng et al. used cadmium oxide and cadmium acetate as a cadmium precursor to decrease toxicity and improve the size control. Relatively higher thermal stability of these precursors resulted in a slower nucleation rate, minimizing the experimental uncertainty introduced by the rate of injection of the precursors [27]. This resulted in a further improvement in the size distribution and the repeatability of the obtained QDs. In order to avoid toxic phosphines in the synthesis, Mulvaney et al. used 1-octadecene as a solvent for Se [28].

Core-shell architecture QDs with a CdSe core and a CdS shell were synthesized through successive ion layer adsorption and reaction (SILAR), wherein the shell (CdS) was grown one monolayer at a time. This was achieved by alternating injections of CdO precursor solution and sulfur solution into a reaction mixture containing pre-synthesized CdSe QDs [29]. The presence of the surface layer of CdS reduced the surface defects that act as electron trapping centers, providing superior quantum yields in comparison to pure CdSe QDs. Subsequent improvements in SILAR technique allowed fabrication of shells of other materials, as well as a better crystalline matching between core and the shell, resulting in a very high quantum yield in the range of 80–90% [30,31].

In order to avoid direct exposure to Cd that may leach into biological fluids and cause acute toxicity, a more stable semiconductor, zinc sulfide (ZnS), is generally coated to form core-shell type particles. These particles also exhibit a higher luminescence due to the energetics of excitons.

Chirality is one of the most important factors of molecular recognition and therefore development of chiral luminescent nano-sized probes would provide very useful tools for both chemistry and biology. Moloney et al. [32] prepared the chiral CdS QDs by heating the naked

QDs with either left- or right-handed forms (enantiomers) of penicillamine. Despite the fact that the luminescence characteristics of QDs are composition and size dependent, these gave off a bright green-white light when excited with UV light. When a 50:50 (racemic) mixture of the two enantiomers was used, the QDs gave off a bright blue-white light instead [32]. These QDs could find important applications, including their use as fluorescent chemical and biochemical chirality sensors and molecular recognition nanodevices. It has to be noted, however, that the toxicity of QDs is very difficult to overcome and the search for more inherently biocompatible materials for bio-imaging is being intensively pursued worldwide (see also Chapter 15).

METAL NANOPARTICLES

Metal nanoparticles are highly attractive for both bio-imaging and therapeutic applications. They are derived normally from late transition metals and are highly toxic, which necessitates the simultaneous implementation of synthetic and protective strategies that are also described below.

Aqueous Synthesis of Metal Particles

Preparation of noble metal colloids (dispersions of nanoparticles in different media), especially those of gold, dates from antiquity, where the thermal dispersion of gold in glass was used for the creation of precious colored glass sorts (the deep red Cassius Purple, in particular). The aqueous dispersions of gold nanoparticles became available in the late Middle Ages, when the alchemists found a way to transform solid gold into a water-soluble precursor, tatrachloroaurate, $AuCl_4^-$, through its dissolution in aqua regia, a 3:1 volume mixture of concentrated hydrochloric and nitric acids. Historically the first reducing agents applied were starch and glucose, grape juice sugar. Already then size control was achieved by varying the concentration of the aurate solutions and heating time of the process, carried out as thermal reduction. (Using this technique today, the smallest particles obtained are 10 nm and give a red color, and by increasing the size, all colors of the rainbow can be produced finishing with blue for particles over 100 nm in size.) From this time dates the first reported attempt to use in medicine what we today identify as nano-gold, made by the alchemist Paracelsus, who described "potable gold". Its use as a potential panacea was apparently a result of superstition and not of any real medical insight.

Several new reducing agents for aurate have been proposed in modern times, including sodium citrate and ascorbate, the Turkevich method [33–35]. The produced particles, in this case, adsorb citrate and ascorbate anions on their surface and are stabilized by the produced negative surface charge. The size control is achieved by controlling concentrations of the reactants (the lower the concentration, the bigger the particles). Particles obtained using these reducing agents are usually spherical. Effective shape control, permitting the choice between spheres below 10 nm in size, nanorods with 4–5 aspect ratio and length up to 50 nm, and crystals with well-defined outer faces (octahedral and cubooctahedral), can be achieved using vitamin C as a reducing agent at different concentrations [36]. The produced particles in all these methods do not require any additional stabilizers in aqueous medium and are not accompanied by any non-biocompatible impurities.

The same chemical agents can be used to reduce thermally the salts of platinum, $PtCl_4^{2-}$ and $PtCl_6^{2-}$ anions [37], palladium, $PdCl_4^{2-}$ ions [38] and silver, Ag^+ cations, often applied in ammonia solution as $Ag(NH_3)^{2+}$ ions [39]. Stabilization in aqueous medium is produced just as for gold via capping with the anions of organic acids. An alternative way for enhancing the solution stability and controlling the shape of the particles is application of water-soluble polymers, for example, sodium polyacrylate [37]. To be able to control the nucleation process and, especially, the subsequent growth of metal particles in the thermal reduction processes, slow chemical evolution of reductive agents has been proposed, for example, hydrolysis of hexamethylene tetramine [40]. Facilitated nucleation of the particles of these metals on the glass walls of the reaction vessels can be a hindrance in the preparation of nanoparticle dispersions (the well-known silver mirror reaction on reduction of $Ag(NH_3)^{2+}$ ions by aldehydes, in particular by glucose). To reduce this process, a variety of irradiation techniques have been proposed instead of heating. In particular, laser irradiation and γ-radiation have been applied to prepare size-uniform silver nanoparticles [39].

A general approach to water-based colloids of both light (3*d*) late transition metals, such as Co, Ni and Cu [41], and heavy (noble) late transition metals [42] is reduction of salts with sodium borohydride, $NaBH_4$. It is important to note that this reaction leads to a poisonous cocktail of borate ions, in particular, because metal particles catalyze the decomposition of $NaBH_4$ in an aqueous medium producing hydrogen gas and sodium borate. This requires quick and efficient separation of the produced particles from the water solution. It is achieved through surface modification of the particles. Noble metal ones are often treated with alkanethiols [43]. Another possibility lies in the creation of a liquid double-phase system, an emulsion, where particles are immediately transferred into an organic phase (often toluene), using surfactants such as, for example, tetraoctyl ammonium bromide [42], making this approach intermediate between aqueous and non-aqueous synthetic approaches.

Light 3*d* transition metal particles are highly poisonous themselves, releasing appreciable concentrations of corresponding divalent cations, M^{2+}. To address this challenge two principal approaches have been tested: coating of metal seeds with a silica layer further modified covalently with targeting organic ligands [43] or using the magnetic 3*d* metal particles as seeds for the growth of noble metal shells, to provide, for example, Co-core-Pt-shell magnetic but chemically inert particles [44]. Application of such core-shell particles as contrast agents for MRI has been reported recently.

Both noble metal particles and 3*d* transition metal-plated ones are still appreciably toxic due to the possibility of their surface interaction with biomolecules. The common practice to overcome this obstacle is intentional surface modification with sulfur-containing biocompatible molecules such as functional mercaptanes or aminoacids like cysteine or methionine. Metal particles can be produced on a large scale with relatively low costs and their medical applications, especially for gold nanoparticles, are well on the way to broader practice.

Synthesis of Metal Particles in Non-Aqueous Media

The advantage of non-aqueous routes lies in the possibility of obtaining potentially smaller particles with high monodispersity. The smaller size is due to other stabilization mechanisms than surface charging, in the first case, the Van-der-Waals interactions that determine the size of the nuclei emerging in solution. Another important factor is that Ostwald ripening—the dissolution of some crystals with re-deposition of the material on the other ones—is most often principally hindered in organic media because of the lack of solution-stable species under the conditions of synthesis.

The non-aqueous approaches to gold nanoparticles are using actually the same starting reagent as aqueous ones, $HAuCl_4$, which is polar and requires polar solvents for dissolution, such as, for example, dimethyl formamide $(CH_3)_2NCHO$, acting as both solvent and reducing agent in combination with a surface-protecting polymer, often poly(vinyl)pyrrolidone [45]. An important advantage of non-aqueous solvents for production of core-shell magnetic metal particles covered by a gold shell is that the less oxidative media than water opens possibilities for production of metals that otherwise would not be possible to produce in nano form due to a facility for using much stronger reducing agents. For example, iron particles covered by gold have been synthesized by first reduction of $FeCl_3$ solution in alkyl pyrrolidone by sodium metal dissolved in the same medium, producing Fe metal seeds and then the addition of $HAuCl_4$ into the reaction medium giving the deposition of smaller Au nanoparticles (about 2 nm in size) onto the bigger iron ones (approximately 10 nm) [46].

For many light and heavy late transition metals there are commercially available metal-organic precursor compounds appreciably soluble in organic solvents that can easily be transformed into metal nanoparticles via thermal decomposition in solution—either simply on reflux or through solvothermal treatment. The ligand, often carboxylate (acetate) or β-diketonate (acetylacetonate), serves as the reducing agent (Figure A.2). If the solvent is a non-solvating one, such as, for example, xylene, the addition of surface-capping ligands

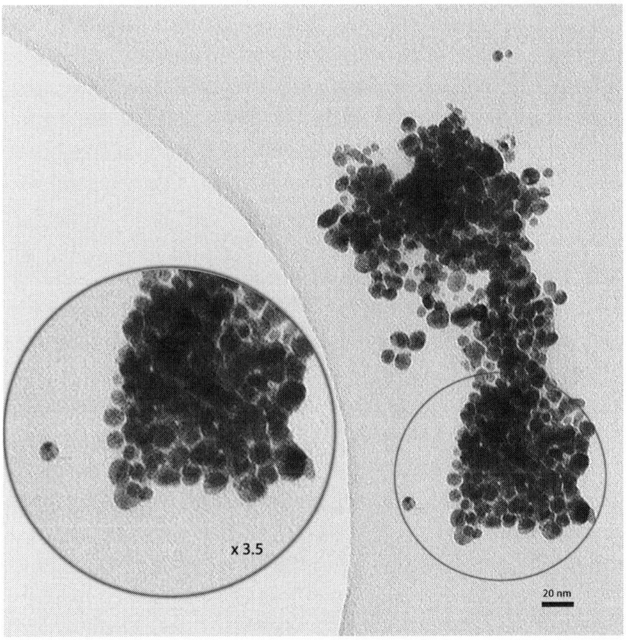

FIGURE A.2 TEM micrograph of Ni-Pd nanoparticles prepared by the solvothermal method.

stabilizing the particle dispersions via Van-der-Waals interactions is a common practice. The ligand in fact appreciably influences the crystallinity of produced particles. Thus, the stronger solvating ones result in poorer crystallinity, while those less efficiently interacting with the surface of the metal particles offer a bigger size of the crystalline domain within the particle and more complete crystallization. The most detailed study has been carried out with Pd nanoparticle synthesis in xylene, applying trioctyl phosphine as a stronger solvating ligand and dodecylamine as a less solvating ligand [47]. The synthesis in inert solvents like xylene provides hydrophobic particles, which have to be surface-modified further to become applicable for use in aqueous media [48]. It is important to note that some surfactants applied in the synthesis, such as phosphines, can be highly toxic and have to be replaced with biocompatible molecules for further practical use.

Hydrophilic metal particles can be produced by solvothermal synthesis in polar solvating media, such as high boiling point alcohols or carbonyl compounds (aldehydes or ketones). In particular, size-uniform hydrophilic particles were produced via thermolysis of metal acetylacetonates and alkoxides in fully biocompatible solvent acetophenone, $C_6H_5COCH_3$ [49]. The relative thermal stability of β-diketonate ligands makes them good candidates for carrying out size control via Ostwald ripening by duration of the solvothermal process.

Synthesis of Metal Particles in the Gas Phase

Gas phase reduction of precursors in a reductive atmosphere applied to both true gas phase and to spray pyrolyzed solutions is an established approach to synthesis of metal nanoparticles [50]. In fact, when metal-organic precursors are available, the decomposition itself produces a reductive medium. Maintaining an autogenic atmosphere at elevated temperatures offers an approach to uniform nanoparticles of late transition metals covered by a carbon shell with graphitic structure, having a typical thickness of 2–3 nm [51]. Carrying out of the process in a magnetic field causes rejection of the metal-doped graphite from the magnetic particles, and thus produces "naked" nanoparticles. It is important to note that gas-phase derived nanoparticles possess extremely high surface energy, which leads to their strong aggregation on transition to liquid media. Maintaining dispersion stability and a further use in nanomedicine relies on surface modification.

Oxide Mesoporous Nanostructured Submicroparticles

The discovery of ordered mesoporous structures, originally for silicon dioxide, SiO_2, by Kresge et al. [52], led very quickly to insight of the possibilities of their potential biomedical applications. The synthesis of these materials is based on polycondensation of silica in self-organized micelles of surfactants. In a liquid double phase oil–water system, high concentration of surfactant molecules results in their arrangement, dependent on the relative concentrations of components, into droplets, tubes, layered lamellae or, under proper conditions, into hexagonal or cubic porous structures that are then replicated by the growing inorganic polymer. The reaction mixture is comprised of water, alcohol or another less polar organic solvent, surfactant, tetraalkoxysilane, $Si(OR)_4$, and an acidic or basic catalyst used to enhance hydrolysis of the silicic acid ester. The structures are typically from 50 nm to several hundred nanometers in size with a pore diameter of 2 nm and higher. The originally applied surfactants were tetraalkyl ammonium salts, $R_4N^+X^-$, for example, n-hexadecyltrimethyl ammonium bromide, highly toxic and rather expensive compounds which, after synthesis, are retained in the pores, thus filling them. To obtain material with open pores and free from poisonous impurities, a thermal aftertreatment had to be applied (typically 6–8 h at temperatures over 500°C). This made the originally produced ordered mesoporous silicates, MCM-41 and alike, and aluminosilicates truly expensive.

A true revolution in the production of mesoporous silica was achieved with application of block-copolymers, for example, pluronics, combining hydrophilic blocks of polyethylene glycol with the hydrophobic ones of polypropylene glycol. Their use enabled minimization of the amounts of organic solvents and also offered the possibility to control the size of the pores via distribution and size of the blocks in the copolymer structure [53].

Very recently, a principally new approach using biocompatible surfactants self-assembled via hydrogen bonding, for example, folic acid, attractive itself as a targeting agent, has been proposed for the synthesis of ordered mesoporous silicas specifically aimed at medical applications [54]. Application of this approach enables avoidance of both the use of toxic chemicals in preparation of the material and also the prolonged thermal treatment for their removal.

The specific attractiveness of silica-based ordered mesoporous solids lies in their facile surface modification. In the case of traditional toxic surfactants, the functionalization is carried out post-synthesis, after the thermal treatment, while the application of biocompatible ligands would permit an *in situ* surface grafting during the synthesis procedure itself. The general approach to modification is explained in Figure A.3. Further development of the shell exploits the reactivity of the functional group X. It can be used to couple the targeting organic ligands to the produced silica particle via, for example, peptide bonding if $X=NH_2$: $-CH_2-NH_2+L-COOH \rightarrow -CH_2-NH-CO-L$, or metal centers/metal particles if X=SH. This chemistry is already really well developed, including fabrication of pH-sensitive gates, closed in the body fluids and opening only in cytosol to release medicines only inside the cells [55], molecular machines, working as tentacles providing propulsion within the cytosol, helping the particle to find the nucleus for release of biomedicines [56]. Recently, even a strategy for complete coating of the particle surface with phospholipids and grafted targeting biomolecules has been presented [57]. Application of mesoporous silica particles is well on the way into therapeutic practice.

In spite of their outstanding versatility on the side of covalent modification, the silica particles have a considerable drawback in their poor affinity to adsorbed non-grafted molecules of medicines, in particular, to carboxylic and phosphonic acids. To overcome this difficulty, the synthesis of mesoporous silica doped with zirconia has been applied rather successfully, as it

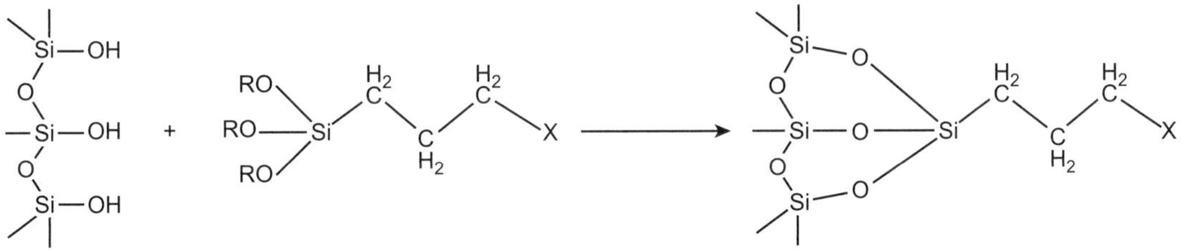

FIGURE A.3 General approach to functionalization of silica particles and surfaces.

avoided, for example, a typical burst-release of the diphosphonate bone growth factors, alendronate and zolendronate, otherwise identified as a problem in their delivery from mesoporous silica [58].

The principal alternative to silica in biodegradability and in specific affinity to biomolecules is offered by poorly soluble early transition metal oxides, in the first case, TiO_2, but even ZrO_2 and HfO_2 (Figure A.4). A major problem in the preparation of metal oxide mesoporous particles is that the produced materials are not polymers but phase-separated solids, which recrystallize via Ostwald ripening or aggregate, not following the self-assembly of introduced surfactant. It is possible to produce well-ordered films of mesoporous metal oxides with pores perpendicular to the substrate surface through evaporation-induced self-assembly (EISA) but, on attempts to obtain ordered particles as for SiO_2, the porosity usually collapses in the course of surfactant removal on thermal treatment [59]. Particles with semi-regular mesoporosity have been produced using pre-formed small particles of metal oxides aggregated under the action of pluronic surfactants in the presence of tungstate as binder material [60]. Potentially highly biocompatible hierarchically porous metal oxide particles have been obtained by microemulsion synthesis, applying metal alkoxide precursors, such as, for example, titanium ethoxide $Ti(OEt)_4$, modified by amphiphylic ligands like amino acids and small peptides [61]. In this approach, droplets of precursor solution in hydrocarbon solvent not miscible with water are sprayed into aqueous medium. They are then immediately covered by a semipermeable membrane of oxide nanoparticles that is transparent for water, diffusing inside by osmotic pressure caused by its affinity to amphiphylic ligands and creating a complex system of membrane channels within the solidified droplet. The obtained hierarchical particle retains a major part of the amino acid ligand, giving the possibility for its prolonged delivery. Subsequent thermal treatment eliminates the original structure-directing ligand and permits them to load the particle, via solution sorption, with a new one, for example, a phosphonate or bisphosphonate medicine. The applied TiO_2 is cheap and highly biocompatible, even being certified as a food additive E171. Some concerns about possible adverse effects of TiO_2 in sunscreens have been raised

FIGURE A.4 SEM micrograph of mesoporous nanostructured submicro TiO_2 particles.

based on the ability of fully crystalline titania (produced by gas phase synthesis especially) to generate reactive oxygen species (ROS) when exposed to sunlight (see Chapter 11). This property is, however, neutralized essentially by grafting of antioxidant ligands such as alkanolamines or amino acids [61].

Small Metal Oxide Nanoparticles

There are several metal oxide nanoparticles systems that have been used in the field of bionanotechnology. The most common type of these nanoparticles is magnetic nanoparticles for therapeutics and magnetic resonance imaging (MRI) applications. Several methods have been proposed to synthesize magnetic nanoparticles with controlled size, morphology and crystal structure, including chemical/physical vapor deposition [62], mechanical attrition [63] and chemical routes from solutions—including co-precipitation, sol–gel, hydrothermal, microemulsion and thermal decomposition techniques [64].

Chemical synthesis procedures are usually more reliable and give more flexibility on process parameters that lead to nanoparticles with tailored structures and magnetic properties to form stable colloidal suspensions in different solvents [65]. It is also most often cost-efficient, permitting up-scaling of the process necessary for broader applications. Several types of magnetic nanoparticles have been synthesized with a number of different compositions and phases, including iron oxides as Fe_3O_4 (magnetite) and γ-Fe_2O_3 (maghemite) [66], spinel-type ferromagnets as $MgFe_2O_4$, $MnFe_2O_4$, and $CoFe_2O_4$ [67], as well as alloys of $CoPt_3$ and FePt [68].

For biomedical applications, iron oxide nanoparticles are usually preferred due to their biocompatibility and absence of heavy metal ions that can cause toxicity and an immune response. These nanoparticles, based on Fe_3O_4 and Fe_2O_3, are currently used in clinical investigations, such as MRI contrast agents and they are predominantly synthesized by an aqueous co-precipitation process followed by surface modification with biocompatible coating layers [69,70] (see also Chapter 15). By changing the salt used as the source of ferrous and ferric ions as well as other reaction conditions, such as temperature and pH, the size, shape and composition of the magnetic nanoparticles can be varied [71]. Different organic additives and stabilizers can be used to obtain a stable colloidal suspension of the nanocrystals. The co-precipitation method in aqueous solutions allows the preparation of large quantities of nanoparticles in a single batch, however, with a broad size distribution [73].

Magnetic ferrite nanoparticles can also be prepared by a hydrothermal method requiring high temperature and pressure [72]. Reduction of metal ions under hydrothermal conditions has been used to prepare monodisperse nanoparticles, using metal salts, ethylene glycol, sodium acetate and polyethylene glycol [73]. Co-precipitation reactions to prepare magnetic nanoparticles can also be carried out in a confined environment, i.e. *microemulsion* droplets. A microemulsion is a thermodynamically stable isotropic dispersion of oil, water and surfactant, where the microdomains of liquids are stabilized by an interfacial film of surfactant molecules. Oil-in-water or water-in-oil (reverse) emulsion systems can be used to form micelles in which the synthesis of nanoparticles takes place via intermicellar exchange of reactants. In both cases, the dispersed phase consists of monodisperse droplets in the size range of 2–100 nm [74]. This dispersed phase provides a confined environment for the synthesis and formation of nanoparticles [74]. Several inorganic nanoparticle systems have been synthesized via an inverse microemulsion process [75,76]. Using biocompatible stabilizing agents, such as starch, it is possible to produce fully non-toxic iron oxide particles approved by the US Food and Drug Administration (FDA) in this form for MRI imaging applications on humans.

Non-Aqueous Synthesis of Small Metal Oxide Nanoparticles

The reactions leading to formation of small metal oxide nanoparticles in organic media can be subdivided into two major groups, involving hydrolytic and non-hydrolytic approaches respectively. The hydrolytic synthesis uses as starting reagents mostly metal chlorides or metal alkoxides. The water for hydrolysis is added on strong homogenization/stirring in small portions at a time. The nucleation of new solid phases in this case occurs via a series of coordination equilibria yielding oxometallate molecules as intermediates. When the size of an oxometallate aggregate exceeds approximately 2 nm its behavior changes due to non-negligible surface energy and it starts to resemble that of a phase separated solid particle. Hydrolysis, being a protolytic reaction, occurs rapidly and is kinetically unhindered in the view of high proton activity provided by water acting as rather strong acid in organic media (alcohol-based solution). This facility of reaction results in rapid nucleation with almost immediate formation of oxometallate particles having a core–shell structure: their core is well ordered and follows the construction principles for the corresponding metal oxide, while the shell is comprised of metal atoms bound to residual organic ligands and is amorphous [77]. The observed structural consequences are very much analogous to the observations made for formation of metal particles in organic media in [77]. The stability of produced nanoparticles against aggregation is then due to surface interactions of residual ligands with solvents. On removal of solvent, or simply in time, the produced particles often aggregate producing gels, which explains the common name of this approach referred to generally as sol–gel technology [78]. Nanoparticles with large size, ranging from tens to hundreds of nanometers with relatively narrow size distribution, can be obtained with this method [80]. A general approach to stabilization of the particles against aggregation and, at the same time, to improvement of their crystallinity and increase in size is solvo- or hydrothermal treatment in the presence of surfactants. Those most often applied for this purpose are non-ionic ones, dodecylamine and dodecanoic acid [81]. Typically, polycrystalline nanoparticles formed through coalescence of the original oxometallate cores with the size of 20–25 nm are produced after 12–24 h of hydrothermal treatment at 180–200°C in an autoclave with Teflon inset.

The non-hydrolytic approach is freer in the choice of precursors, using often commercially available chemicals relatively stable in ambient atmosphere such as carboxylates and β-diketonates. Two principal techniques can be applied, either thermolysis in hydrocarbon medium, e.g. xylene, occurring supposedly as radical decomposition, or the more specific ether elimination reaction, carried out in alcohols or carbonyl compounds (aldehydes or ketones). The synthesis of high-quality semiconductor nanocrystals and magnetic nanoparticles with an extremely narrow size distribution and high crystallinity was achieved via thermolysis [25,82]. In this process, organometallic precursors, such as metal acetylacetonates, metal cupferronates or carbonyls [83], are heated in high-boiling organic solvents containing stabilizing surfactants, e.g. fatty acids, oleic acid, oleylamine or hexadecylamine [84,85]. The ratio of the starting reagents, as well as reaction temperature and time are decisive parameters for the control of the size and morphology of magnetic nanoparticles [86]. The boiling point of xylene isomers is relatively low, about 140°C, and application of autoclaves with a process temperature over 200°C is preferred to obtain crystalline particles from more stable precursors. Simple injection of alkoxide precursor solution in a parent alcohol into boiling xylene turned out to be sufficient to produce well-crystallized maghemite, superparamagnetic γ-Fe_2O_3, nanoparticles from $Fe(O$-$C_2H_4OCH_3)_3$ [87]. In a similar method, iron oxide nanoparticles were prepared via decomposition of iron pentacarbonyl or through one-step decomposition of iron pentacarbonyl in the presence of a mild oxidant [88]. In order to avoid the use of highly toxic compounds (i.e. metal carbonyls), a two-step strategy has been proposed. The first step consists of the synthesis of iron oleate, stereate, or laureate precursors that decompose in a second step to form the desired nanocrystals. The size of the particles can be controlled by addition of thermally stable surfactants, e.g. dodecanoic acid (longer process duration then results in bigger particle size due to Ostwald ripening). The particles produced by this method are hydrophobic and this method allows large-scale preparation of nanoparticles with a very narrow size distribution, controlled morphology and high crystallinity [89].

Ether elimination from alkoxide precursors, the Bradley reaction [90], occurs for the least stable compounds already at room temperature, but is strongly accelerated by heating. It was developed for production of nanopowders of ferroelectrics in the early 1990s [91]. The facility of ether elimination increases with improved stability of carbo-cation R^+ generated in the crucial step of this process from the alkoxide group −OR and follows the series of R: $CH_3 < C_2H_5 < CH(CH_3)_2 << C(CH_3)_3 << CH_2C_6H_5 < C(C_6H_5)CH_3(OR)$ (semiacetal)$< CH_2$-$CH{=}CH_2$ [92]. Less solvating media, such as ketones, produce more crystalline powders compared with those derived from alcohols. High boiling point ketones, such as acetophenone, for example, with the boiling point of 202°C are especially attractive as the particles then are generated simply on reflux without any need for special equipment for autoclaving. The particles produced by the Bradley reaction are hydrophilic and display high affinity to carboxylate and phosphonate biomolecules making them attractive for biomedical applications [93].

Gas Phase Synthesis of Small Metal Oxide Nanoparticles

Development of spraying equipment in combination with the possibility to control thoroughly the combustion temperature and oxidative/reductive properties of the flame make spray pyrolysis a very attractive route to even small metal oxide nanoparticles [50]. In this case, the precursor role is not that crucial and solutions of different common metal salts, such as chlorides or nitrates (even hydrated ones do just fine), in, for example, alcohols, can be sprayed into an oxidative flame and transformed into well crystalline and often essentially monodisperse (as determined by the droplet size distribution) particles typically 20−50 nm in size, which is considerably bigger that the size typically available from solution techniques. The approach has found broad industrial application being commercialized by Degussa GMBH in Germany.

CONCLUSIONS

During the last 10 years, we have witnessed a truly amazing development in synthetic approaches to and characterization of engineered nanomaterials demonstrating their potential for medical applications. General approaches to improvement of biocompatibility and bioavailability and targeting have been proposed, showing highly probable paths for avoiding at least acute toxicity. Considerable effort is still required, however, to understand in detail the adverse effects of these materials and to ensure their safe application in the field of nanomedicine in the future.

References

[1] Lee Y, Geckeler KE. Carbon nanotubes in the biological interphase: The relevance of noncovalence. Adv Mater 2010;22:4076–83.
[2] Geckeler KE, Rosenberg E. Functional Nanomaterials. Valencia, CA: American Scientific Publishers; 2006.
[3] Geckeler KE, Nishide H. Advanced nanomaterials. Weinheim: Wiley-VCH; 2009.
[4] Loiseau A, Launois P, Petit P, Roche S, Salvetat J-P, editors. Understanding carbon nanotubes from basics to applications. Series: Lecture notes in physics, Vol. 677. Springer; 2006.
[5] Krätschmer W, Lamb LD, Fostiropoulos K, Huffman DR. Solid C_{60}: A new form of carbon. Nature 1990;347:354–8.
[6] Guo T, Nikolaev P, Rinzler AG, Tomanek D, Colbert DT, Smalley RE. Self-assembly of tubular fullerenes. J Phys Chem 1995;99:10694–7.
[7] Thess A, Lee R, Nikolaev P, et al. Crystalline ropes of metallic carbon nanotubes. Science 1996;273:483–7.
[8] Maser WK, Munoz E, Smalley. RE, et al. Production of high-density single-walled nanotube material by a simple laser-ablation method. Chem Phys Lett 1998;292:587–93.
[9] Bandow S, Asaka A, Saito Y, et al. Effect of the growth temperature on the diameter distribution and chirality of single-wall carbon nanotubes. Phys Rev Lett 1998;80:3779–82.
[10] Ishigaki T, Suzuki S, Kataura H, Krätschmer W, Achiba Y. Characterization of fullerenes and carbon nanoparticles generated with a laser-furnace technique. Appl Phys A 2000;70:121–4.
[11] Castignolles M, Foutel-Richard A, Mavel A, et al. In: Kuzmany H, Fink J, Mehring M, Roth S, editors. Electronic properties of novel materials science and technology of molecular nanostructures. American Institute of Physics; 2002.
[12] Laplaze D, Bernier P, Maser WK, Flamant G, Guillard T, Loiseau A. Carbon nanotubes: The solar approach. Carbon 1998;36:685–8.
[13] Guillard T, Cetout S, Alvarez L, et al. Production of carbon nanotubes by the solar route. Eur Phys J AP 1999;5:251–6.
[14] De Jong KP, Geus JW. Carbon nanofibers: Catalytic synthesis and applications. Catal Rev Sci Eng 2000;42: 481–510.
[15] Rodriguez NM. A review of catalytically grown carbon nanofibers. J Mater Res 1993;8:3233–50.
[16] Pol VG, Thackeray MM. Spherical carbon particles and carbon nanotubes prepared by autogenic reactions: Evaluation as anodes in lithium electrochemical cells. Energy Environ Sci 2011;4:1904–12.
[17] Mu Q, Broughton DL, Yan B. Endosomal leakage and nuclear translocation of multiwalled carbon nanotubes: Developing a model for cell uptake. Nano Lett 2009;9:4370–5.
[18] Kam NWS, Liu Z, Dai H. Carbon nanotubes as intracellular transporters for proteins and DNA: An investigation of the uptake mechanism and pathway. Angew Chem Int Ed 2006;45:577–81.
[19] Chopra NG, Luyken RJ, Cherrey K, et al. Boron nitride nanotubes. Science 1995;269:966–7.
[20] Loiseau A, Pascard H, Willaime F, Demoncy N, Hug G. Boron nitride nanotubes with reduced numbers of layers synthesized by arc discharge. Phys Rev Lett 1996;76:4737–40.
[21] Golberg D, Bando Y, Eremets M, Takemura K, Kurashima K, Yusa H. Nanotubes in boron nitride laser heated at high pressure. Appl Phys Lett 1996;69:2045–7.
[22] Zhou GW, Zhang Z, Bai ZG, Yu DP. Catalyst effects on formation of boron nitride nano-tubules synthesized by laser ablation. Solid State Commun 1999;109:555–9.
[23] Medintz IL, Uyeda HT, Goldman ER, Mattoussi H. Quantum dot bioconjugates for imaging, labelling and sensing. Nat Mater 2005;4:435–46.
[24] Brus LE. Electron–electron and electron–hole interactions in small semiconductor crystallites: The size dependence of the lowest excited electronic state. J Chem Phys 1984;80:4403–9.

[25] Murray CB, Norris DJ, Bawendi MG. Synthesis and characterization of nearly monodisperse CdE (E = sulfur, selenium, tellurium) semiconductor nanocrystallites. J Am Chem Soc 1993;115:8706–15.
[26] Talapin DV, Rogach AL, Kornowski A, Haase M, Weller HV. Highly luminescent monodisperse CdSe and CdSe/ZnS nanocrystals synthesized in a hexadecylamine–trioctylphosphine oxide–trioctylphospine mixture. Nano Lett 2001;1:207–11.
[27] Peng ZA, Peng X. Formation of high-quality CdTe, CdSe, and CdS nanocrystals using CdO as precursor. J Am Chem Soc 2001;123:183–4.
[28] Jasieniak J, Bullen C, van Embden J, Mulvaney P. Phosphine-free synthesis of CdSe nanocrystals. J Phys Chem B 2005;109:20665–8.
[29] Li JJ, Wang YA, Guo W, et al. Large-scale synthesis of nearly monodisperse CdSe/CdS core/shell nanocrystals using air-stable reagents via successive ion layer adsorption and reaction. J Am Chem Soc 2003;125:12567–75.
[30] Xie R, Kolb U, Li J, Basch T, Mews A. Synthesis and characterization of highly luminescent CdSe-core CdS/$Zn_{0.5}Cd_{0.5}S$/ZnS multishell nanocrystals. J Am Chem Soc 2005;127:7480–8.
[31] Reiss P, Bleuse J, Pron A. Highly luminescent CdSe/ZnSe core/shell nanocrystals of low size dispersion. Nano Lett 2002;2:781–4.
[32] Moloney MP, Gun'ko YK, Kelly JM. Chiral highly luminescent CdS quantum dots. Chem Commun 2007:3900–2.
[33] Turkevich J, Stevenson PC, Hillier J. A study of the nucleation and growth processes in the synthesis of colloidal gold. Discuss Faraday Soc 1951;11:55–75.
[34] Frens G. Controlled nucleation for the regulation of the particle size in monodisperse gold suspensions. Nat Phys Sci 1973;241:20–2.
[35] Kimmling J, Maier M, Okenve B, Kotaidis V, Ballot H, Plech A. Turkevich method for gold nanoparticle synthesis revisited. J Phys Chem B 2006;110:15700–7.
[36] Huang CJ, Chiu PH, Wang YH. Synthesis and characterization of gold nanodogbones by the seeded mediated growth method. Nanotechnology 2007;18:395603–6.
[37] Hikosaka K, Kim J, Kajita M, Kanayama A, Miyamoto Y. Platinum nanoparticles have an activity similar to mitochondrial NADH: Ubiquinone oxidoreductase. Colloids Surfaces B Biointerfaces 2008;66:195–200.
[38] Narayanan R, El-Sayed MA. Catalysis with transition metal nanoparticles in colloidal solution: Nanoparticle shape dependence and stability. J Phys Chem B 2005;109:12663–76.
[39] Henglein A, Giersig M. Formation of colloidal silver nanoparticles: Capping action of citrate. J Phys Chem B 1999;103:9533–9.
[40] Xu S, Yang Q. Well-dispersed water-soluble Pd nanocrystals: facile reducing synthesis and application in catalyzing organic reactions in aqueous media. J Phys Chem C 2008;112:13419–25.
[41] Tanori J, Duxin N, Petit C, Lisiecki I, Pileni MP. Synthesis of nanosize metallic and alloyed particles in ordered phases (Co, Cu, Fe-Cu). Colloid Polymr Sci 1995;273:886–92.
[42] Brust M, Walker M, Berthell D, Schiffrin DJ, Whyman R. Synthesis of thiol-derivatised gold nanoparticles in a two-phase liquid–liquid system. Chem Commun 1994:801–2.
[43] Martin MN, Basham JI, Chando P, Eah SK. Charged gold nanoparticles in non-polar solvents: 10-min synthesis and 2D self-assembly. Langmuir 2010;26:7410–7.
[44] Meng X, Seton HC, Lu LT, Prior IA, Thanh NTK, Song B. Magnetic CoPt nanoparticles as MRI contrast agent for transplanted neural stem cells detection. Nanoscale 2011;3:977–84.
[45] Han MY, Quek CH, Huang W, Chew CH, Gan LM. A simple and effective chemical route for the preparation of uniform nonaqueous gold colloids. Chem Mat 1999;11:1144–7.
[46] Ban Z, Bamakov YA, Li F, Golub VO, O'Connor CJ. The synthesis of core–shell iron and gold nanoparticles and their characterization. J Mater Chem 2005;15:4660–2.
[47] Liu Y, Wang C, Wei Y, et al. Surfactant-induced postsynthetic modulation of Pd nanoparticle crystallinity. Nano Lett 2011;11:1614–7.
[48] Grzelczak M, Vermant J, Furst EM, Liz-Marzan LM. Directed self-assembly of nanoparticles. ACS Nano 2010;4:3591–605.
[49] Wilkinson K, Palmberg L, Kupczyk M, et al. Solution engineered Pd-nanoparticles: Model for health effect studies of automotive particulate pollution. ACS Nano, revision 2011.
[50] Hahn H. Gas phase synthesis of nanocrystalline materials. Nanostruct Mat 1997;9:3–12.

[51] Pol VG, Pol SV, Gedanken A, et al. Applied magnetic field rejects the coating of ferromagnetic carbon from the surface of ferromagnetic cobalt: RAPET of $CoZr_2(acac)_2(O^iPr)_8$. J Phys Chem B 2005;109:6121–5.
[52] Kresge CT, Leonowicz ME, Roth WJ, Vartuli JC, Beck JS. Ordered mesoporous molecular sieves synthesized by a liquid-crystal template mechanism. Nature 1992;359:710–2.
[53] Zhao D, Feng J, Huo Q, et al. Triblock copolymer syntheses of mesoporous silica with periodic 50 to 300 angstrom pores. Science 1998;279(5350):548–52.
[54] Zhou CF, Garcia-Bennet AE. Release of folic acid in mesoporous NFM-1 silica. J Nanosci Nanotech 2010;10:7398–401.
[55] Chen HT, Trewyn BG, Wiench JW, Pruski M, Lin VSY. Urea and thiourea-functionalized mesoporous silica nanoparticle catalysts with enhanced catalytic activity for Diels–Alder reaction. Top Catal 2010;53:187–91.
[56] Thomas CR, Ferris DP, Lee JH, et al. Noninvasive remote-controlled release of drug molecules *in vitro* using magnetic actuation of mechanized nanoparticles. J Am Chem Soc 2010;132:10623–5.
[57] Ashley CE, Carnes EC, Phillips GK, et al. The targeted delivery of multicomponent cargos to cancer cells by nanoporous particle-supported lipid bilayers. Nat Mater 2011. DOI: 10.1038/NMAT2992.
[58] Colilla M, Manzano M, Izquierdo-Barba I, Valet-Regi M, Boissiere C, Sanchez C. Advanced drug delivery vectors with tailored surface properties made of mesoporous binary oxides submicronic spheres. Chem Mater 2010;22:1821–30.
[59] Soler-Illia GJ, de AA, Louis A, Sanchez C. Synthesis and characterization of mesostructured titania-based materials through evaporation-induced self-assembly. Chem Mater 2002;14:750–9.
[60] Wong MS, Jeng ES, Ying JY. Supramolecular templating of thermally stable crystalline mesoporous metal oxides using nanoparticulate precursors. Nano Lett 2001;1:637–42.
[61] Seisenbaeva GA, Moloney MP, Tekoriute R, et al. Biomimetic synthesis of hierarchically porous metal oxide microparticles—potential scaffolds for drug delivery and catalysis. Langmuir 2010;26:9809–17.
[62] Wirth S, von Monlnar S. Handbook of advanced magnetic materials: Fabrication and magnetic properties of nanometer-scale particle arrays, Vol. New York: I. Springer; 2006. p.294.
[63] Mishra SR, Dubenko I, Losby J, Ghosh K, Khan M, Ali N. Anomalous magnetic properties of mechanically milled cobalt oxide nanoparticles. J Nanosci Nanotech 2005;5:2076–81.
[64] Lu A-H, Salabas EL, Schüth F. Magnetic nanoparticles: Synthesis, protection, functionalization, and application. Angew Chem Int Ed 2007;46:1222–44.
[65] Lin XM, Samia ACS. Novel and efficient MR active aqueous colloidal Fe_3O_4 nanoassemblies. J Magn Magn Mater 2006;305:100–9.
[66] Sun S, Zeng H. Size-controlled synthesis of magnetite nanoparticles. J Am Chem Soc 2002;124:8204–5.
[67] Park J, An K, Hwang Y, et al. Ultra-large scale synthesis of monodisperse nanocrystals. Nat Mater 2004;3:891–5.
[68] Shevchenko EV, Talapin DV, Rogach AL, Kornowski A, Haase M, Weller H. Colloidal synthesis and self-assembly of $CoPt_3$ nanocrystals. J Am Chem Soc 2002;124:11480–5.
[69] Sun C, Lee JSH, Zhang M. Magnetic nanoparticles in MR imaging and drug delivery. Adv Drug Del Rev 2008;60:1252–65.
[70] Duguet E, Vasseur S, Mornet S, Devoisselle J-M. Magnetic nanoparticles and their applications in medicine. Nanomedicine 2006;1:157–68.
[71] Reiss G, Huetten A. Magnetic nanoparticles: Applications beyond data storage. Nat Mater 2005;4:725–6.
[72] Wang X, Zhuang J, Peng Q, Li Y. A general strategy for nanocrystal synthesis. Nature 2005;437:121–4.
[73] Deng H, Li X, Peng Q, Wang X, Chen J, Li Y. Monodisperse magnetic single-crystal ferrite microspheres. Angew Chem Int Ed 2005;44:2782–5.
[74] Capek I. Preparation of metal nanoparticles in water-in-oil (w/o) microemulsions. Adv Colloid Interface Sci 2004;110:49–74.
[75] Lopez-Quintela MA. Synthesis of nanomaterials in microemulsions: formation mechanisms and growth control. Curr Opin Colloid Interface Sci 2003;8:137–44.
[76] Uskokovic V, Drofenik M. Synthesis of materials within reverse micelles. Surf Rev Lett 2005;12:239–77.
[77] Kessler VG. The chemistry behind the sol–gel synthesis of complex oxide nanoparticles for bio-imaging applications. J Sol-Gel Sci Tech 2009;51:264–71.
[78] Kessler VG, Seisenbaeva GA, Håkansson S, Unell M. Chemically triggered biodelivery using metal-organic sol–gel synthesis. Angew Chem 2008;47:8506–9.

[79] Hench LL, West JK. The sol–gel process. Chem Rev 1990;90:33–72.
[80] Qin J. Nanoparticles for Multifunctional Drug Delivery Systems. Licentiate thesis. Stockholm: The Royal Institute of Technology; 2007.
[81] Seisenbaeva GA, Kessler VG, Pazik R, Strek W. Heteroleptic metal oxide "oxoclusters" as molecular models for the sol–gel synthesis of perovskite nanoparticles for bioimaging applications. Dalton Trans 2008:3412–21.
[82] Peng X, Wickham J, Alivisatos AP. Kinetics of II-VI and III-V colloidal semiconductor nanocrystal growth: "Focusing" of size distributions. J Am Chem Soc 1998;120:5343–4.
[83] Redl FX, Black CT, Papaefthymiou GC, et al. Magnetic, electronic, and structural characterization of non-stoichiometric iron oxides at the nanoscale. J Am Chem Soc 2004;126:14583–99.
[84] Li Y, Afzaal M, O'Brien P. The synthesis of amine-capped magnetic (Fe, Mn, Co, Ni) oxide nanocrystals and their surface modification for aqueous dispersibility. J Mater Chem 2006;16:2175–80.
[85] Huber DL. Synthesis, properties, and applications of iron nanoparticles. Small 2005;1:482–501.
[86] Dumestre F, Chaudret B, Amiens C, Renaud P, Fejes P. Superlattices of iron nanocubes synthesized from Fe $[N(SiMe_3)_2]_2$. Science 2004;303:821–3.
[87] Ahniyaz A, Seisenbaeva GA, Kessler VG, et al. Preparation of iron oxide nanocrystals by surfactant-free or oleic acid-assisted thermal decomposition of a Fe(III) alkoxide. J Magn Magn Mater 2008;320:781–7.
[88] Hyeon T, Lee CS, Park J, Chung Y, Na HB. Synthesis of highly crystalline and monodisperse maghemite nanocrystallites without a size-selection process. J Am Chem Soc 2001;123:12798–801.
[89] Jana NR, Chen Y, Peng X. Size- and shape-controlled magnetic (Cr, Mn, Fe, Co, Ni) oxide nanocrystals via a simple and general approach. Chem Mater 2004;16:3931–5.
[90] Bradley DC, Chakravarti BN, Chatterjee AK, Wardlaw W, Whitley A. Niobium and tantalum mixed alkoxides. J Chem Soc 1958:99–101.
[91] Yanovskaya MI, Turevskaya EP, Kessler VG, Obvintseva IE, Turova NYa. Application of metal alkoxides in the synthesis of oxides. Integr Ferroelec 1992;1:343–52.
[92] Pazik R, Tekoriute R, Håkansson S, et al. Precursor and solvent effects in the non-hydrolytic synthesis of complex oxide nanoparticles for bio-imaging applications by the ether elimination (Bradley) reaction. Chem Europ J 2009;15:6820–6.
[93] Pazik R, Seisenbaeva GA, Gohil S, et al. Simple and efficient synthesis of a Nd:$LaAlO_3$ NIR nanophosphor from rare earth alkoxo-monoaluminates $Ln_2Al_2(OiPr)_{12}(iPrOH)_2$ single source precursors by Bradley reaction. Inorg Chem 2010;49:2684–91.

Glossary

Adverse effect Change in morphology, physiology, growth, development or lifespan of an organism that results in impairment of functional capacity or impairment of capacity to compensate for additional stress or that results in an increase in susceptibility to the harmful effects of other environmental influences.

Aerodynamic diameter The notional dimension that is a function of the actual diameter, the shape and the density that determines where in the respiratory tract particles will deposit. ENM deposit in the distal tract of the respiratory system.

Agglomerates/agglomeration Collection of weakly bound nanoparticles. The forces holding an agglomerate together are weak forces, for example van der Waals forces, or simple physical entanglement.

Aggregates Particle comprising strongly bonded or fused particles where the resulting external surface area may be significantly smaller than the sum of calculated surface areas of the individual components. The forces holding an aggregate together are strong forces, for example covalent bonds, or those resulting from sintering or complex physical entanglement. Many ENM show a strong tendency to aggregation.

Airways (bronchi, bronchioles) The anatomical conduits that connect the outside air to the lung parenchyma. May be a site of particle deposition according to their aerodynamic diameter.

Alveolar lipoproteinosis A severe type of lung injury where the alveolar spaces are filled by lipoprotein. Observed in some animal studies after lung exposure to ENM (e.g. nickel nanoparticles).

Aneuploidy A state where a cell has gained or lost individual or multiple chromosomes leading to numerical changes in chromosome copy number. Observed in experimental studies after exposure to carbon nanotubes.

Antigen sampling Uptake and delivery of proteins and/or peptides to subepithelial lymphocytes to generate immune response. This process may have relevance for nanoparticles in certain biological compartments (e.g. the gastrointestinal tract).

Apoptosis Programmed cell death, i.e. an active, energy-consuming suicide program, which allows the removal of cells without damage to surrounding tissues, as opposed to necrosis, which is non-physiological, uncontrolled cell death. Nanoparticle-exposed or -loaded cells may undergo apoptosis sooner than normal, or be subjected to anomalous autophagic mechanisms. This is particularly relevant for the immune system.

Bioavailability The extent to which a compound (e.g. nanomaterial) reaches the systemic circulation.

Biological monitoring Continuous or repeated measurement of potentially toxic substances, their metabolites or their biochemical effects in tissues, secreta, excreta, expired air or any combination of these. Its purpose is to evaluate occupational or environmental exposure and health risk by comparison with appropriate reference values based on knowledge of the probable relationship between ambient exposure and resultant adverse health effects. This approach is in its early infancy for nanomaterials.

Biomarker Indicator signaling an event (changes, physiological adaption) or condition in a biological system and giving a measure of exposure, effect, or susceptibility. As related to biological monitoring, a biomarker is the presence of any substance, or a change in any biological structure or process that can be measured as a result of exposure.

Biopersistance The extent to which compounds (e.g. nanomaterials) remain in an organism/organ retaining their structure.

Blood–brain barrier Physical barrier between the central nervous system and the circulating blood, whose main components are tight junctions in the endothelial cells of the capillaries that restrict the diffusion of large molecules. It is presently not known if this barrier may be crossed by nanomaterials.

Brush-border Anatomical term for the surface coating of intestinal epithelial cells formed by densely packed microvilli. This structure greatly enlarges the surface of these cells, which are therefore highly exposed to ingested nanomaterials.

Carbon nanotubes Seamless tubular structures of graphene that can be either single-walled carbon nanotubes (SWCNT), comprising a single layer of carbon atoms, or multi-walled carbon nanotubes (MWCNT), comprising multiple concentric tubes. These materials are among the most important ENM.

Carcinogenicity Process of induction, promotion and progression of cancer. It is a concern for some nanomaterials, especially HARN (see definition).

Chemical vapor deposition (CVD) Method for the synthesis (growth) of e.g. carbon nanotubes that relies on the capability to provide the atoms needed to form the solid deposit from a gas source that decomposes under the action of temperature.

Clastogenicity Fragmentation or breakage of chromosomes. It has been observed in some experimental studies on titanium dioxide and zinc oxide nanoparticles (which are frequent components of e.g. sunscreens).

Clearance The processes whereby particles that deposit in biologic compartments (tissues) are removed and lost from the body. Relevant for nanoparticle accumulation in the lung.

Complement A system of at least 20 serum proteins which can interact with each other as well as with antibodies, pathogens and immune cells. Complement-mediated mechanisms are pro-inflammatory and play an important role for early and unspecific immune responses. A number of nanoparticle types have been reported to activate complement.

Condensation particle counter (CPC) Provides real-time number concentration measurements within their particle diameter detection limits. Without a nanoparticle pre-separator these counters are not specific to the nanometer size range. This instrument, used to assess nanoparticle concentration in an occupational setting, is generally sensitive to particles greater than 10–20 nm in diameter.

Corona Adsorbed material such as proteins, lipids, ions, organic matter, on a nanoparticle surface. It is believed that all nanoparticles entering the human body become coated with corona, which is therefore the biological interface between nanoparticles and the human body.

Dendritic cells A class of mononuclear leukocytes specialized in sensing non-self entities such as microorganisms and particles and in consequent antigen presentation. Resting dendritic cells (DCs) are located close to body surfaces, where they sample extracellular material within a large area through tentacle-like protuberances (dendrites). Activated DCs migrate to local lymph nodes where they interact with T cells and induce an immune reaction by presenting antigen.

Descriptor The way molecules are transformed into numbers, allowing some mathematical treatment of the chemical information contained in the molecule. The identification of "nanoparticle-specific" descriptors probably represents the most important research component of computational nanotoxicology

Developmental toxicity Any effect interfering with normal development both before and after birth, up to and including puberty. Recent experimental animal data suggest that some nanoparticles (e.g. carbon nanotubes) may cause developmental toxicity.

Diffusion chargers Provide estimates of external aerosol surface area when airborne particles are smaller than 100 nm in diameter. It is an instrument used to measure nanoparticle concentration in an occupational setting.

DNA adducts Formation of a complex between a DNA base and a chemical group through covalent bonding. Adducts are one specific type of DNA lesion, and some experimental studies report their induction by nanoparticles (e.g. silver nanoparticles).

DNA lesion Altered form of DNA or a base that forms following exposure to a genotoxin (i.e. an external agent with toxic effects on the genome). Some nanoparticles may cause DNA lesions (see DNA adducts).

Dose Total amount of a substance administered to, taken or absorbed by an organism. It is a crucial aspect in determining if and at what extent ENM may cause adverse effects in humans. Doses used in experimental studies are generally much higher than those expected in a human exposure scenario.

Dose–effect relationship Association between dose and the magnitude of a continuously graded effect, either in an individual or in a population or in experimental animals.

Dose–response relationship Association between dose and the incidence of a defined biological effect in an exposed population.

Drug delivery The process of (targeted) delivery of a pharmaceutical compound to a specific tissue or organ with the aim to achieve a therapeutic effect. Some ENM are very relevant in drug delivery, and may enable the selective delivery of toxic drugs to the diseased organ/tissue even at high therapeutic concentrations, sparing bystander cells and tissues.

Electrical low pressure impactor (ELPI) This device combines diffusion charging and a cascade impactor to provide aerosol size distributions by aerodynamic diameter as determined in real time by mass and number collected on a series of plates. It is an instrument used to measure nanoparticle concentration in an occupational setting.

Enhanced permeability and retention (EPR) effect This effect is based on specific attributes of diseased tissue (e.g. cancer tissue), which can be differentiated from healthy tissue by lower pH values, increased temperature and higher permeability of vessels. These attributes can be used for targeting with temperature- and pH-responsive nanoparticles.

Engineered nanomaterials (ENM) Nanomaterials intentionally engineered and produced with very specific properties or compositions (e.g. shape, size, surface properties, and chemistry) that distinguish them from other well-known man-made (welding fume, diesel exhaust) and/or environmental (e.g. volcanic ash) particulates.

Endocytosis Generic term for the uptake of material by cells by invagination of the plasma membrane and formation of membrane-bound compartments, the endosomes. This is one of the mechanisms by which nanoparticles may gain access into cells.

Eosinophils A type of leukocyte (white blood cell) that is involved in inflammatory and allergic type reactions. An increase in eosinophils in the lung has been detected after exposure to certain nanoparticles.

Fast mobility particle sizer (FMPS) This instrument is similar to the SMPS but has a much faster response time (approximately 1 second).

Fenton chemistry Oxidation of materials by hydrogen peroxide using an iron catalyst. This is one of the mechanisms by which exposure to ENM may cause an increased reactive oxygen species (ROS) production.

Fiber paradigm Structure–activity paradigm predicting adverse effects on the basis of the length ($>20\,\mu m$), diameter ($<3\,\mu m$) and biopersistence of a particle. Some carbon nanotubes and some other man-made materials, e.g. asbestos fibers, have these characteristics.

Frustrated phagocytosis The incomplete removal of fiber-like material (see fiber paradigm) by the (alveolar) macrophages. The process is linked with bio-accumulation.

Fullerene Closed-cage structure having more than 20 carbon atoms consisting entirely of three-coordinate carbon atoms. These materials do not seem to have a major role in ENM-induced toxicity.

Functionalization The purposeful insertion of chemical groups to the surface of ENM in order to modify their physicochemical properties. Generally performed for biomedical use of ENM, to increase biocompatibility, to enable targeted delivery, and so on.

Genotoxicity Toxicity related to genetic end-points, e.g. DNA damage, chromosome aberrations. There is experimental evidence that some nanoparticles may be genotoxic.

Graphene One-atom-thick planar sheet of carbon atoms organized in a honeycomb crystal structure. Carbon nanotubes (see above) derive from graphene.

Hazard Set of inherent properties of a substance, mixture of substances or a process involving substances that, under production, usage or disposal conditions, make it capable of causing adverse effects to organisms or to the environment, depending on the degree of exposure. Hazard evaluation of ENM exposure is one of the most relevant research areas in nanosafety.

High-aspect ratio nanoparticles (HARN) Engineered nanomaterials with one dimension that greatly differs from the other conferring to the nanomaterial a "fiber aspect" (see fiber paradigm).

In silico A process carried out by a computer. The development of reliable *in silico* models is one of the major challenges of computational nanotoxicology.

Inductively coupled plasma mass spectrometry (ICP-MS) Highly sensitive type of mass spectrometry. Inductively coupled plasma produces ions and the mass spectrometer separates and detects the ions. Can be used to determine tissue levels of a range of metal (and some non-metal) ENM down to ng/g.

Inflammasome An intracellular multiprotein complex which regulates the activation of the protease caspase-1. Activation of this protease leads to cleavage of the inactive pro-forms of the inflammatory cytokines interleukin (IL)-1β and IL-18, which results in release of their active forms and consequently in inflammation. This is one of the potential mechanisms of nanoparticle-induced damage.

Interstitium The anatomic compartment underlying the epithelial layer containing mesenchymal tissue and blood vessels. In experimental animal studies it has been shown that ENM reaching this compartment may cause fibrosis (e.g. pulmonary fibrosis).

Lung parenchyma The parts of the lung where gas exchange occurs. Since ENM deposit preferentially in this region and since pulmonary exposure is the most relevant way of exposure in humans, damage to parenchyma is a main concern for exposure in humans.

M cells Epithelial cells overlying lymphoid tissues which have high capacity for transporting material (including nanoparticles) as part of antigen-sampling (see above) function. Also known as microfold cells or membraneous cells, but neither of these names provides a universal description of their morphology so both terms have been usurped by the abbreviated name.

Macrophages Mononuclear leukocytes (white blood cells) with a major role in eliminating foreign and dangerous entities. Macrophages are scavengers, able to phagocytose and degrade foreign material, including microorganisms, effectors of inflammation, both in the destructive and in the repair phases, and also antigen-presenting cells. The term macrophage is usually adopted for tissue resident or stimulus-recruited cells, while the term monocyte designates the "immature" blood form of mononuclear leukocytes. At least one subclass of dendritic cells (see this term) is derived from blood monocytes. The interaction between ENM and macrophages is crucial for their potential to cause adverse effects.

Magnetic drug targeting (MDT) Is a therapeutic procedure by which magnetic particles (e.g. iron oxide nanoparticles) that are functionalized with therapeutic agents are injected intra-arterially close to the diseased area (e.g. tumor). An electron magnet positioned at the diseased area directs and attracts the iron oxide particles in the tumor vascular system.

Magnetic resonance imaging (MRI) A medical imaging technique used to image nuclei of atoms in the body through nuclear magnetic resonance. It allows to visualize detailed internal structures in the body. ENM may be used to improve the accuracy of this radiological method. Conversely, magnetic resonance imaging can track movement of nanoparticles around the body

Microvilli Hair-like projections on the surface of cells formed by membrane protrusions with cytoskeletal scaffold. In the intestine, these increase surface area for absorption of nutrients. Although nanoparticles are able to cross the intestinal epithelium with low efficiency, the huge increase in surface area caused by microvilli makes the total amount of absorbed nanoparticles potentially relevant.

Muco-ciliary escalator The cellular processes that waft mucus and any particle that are trapped in it (including nanoparticles) from the airways to the mouth for swallowing (normally).

Mucosal immune system A collective term for the cells and organs which interact to generate secretory and circulating antibody responses against luminal antigens and organisms. There is some evidence that nanoparticles may activate the mucosal immune system.

Mutations Permanent, structural, transmissible changes to the genetic material (usually cellular DNA or RNA). Mutations have been described to be induced in experimental studies by certain nanoparticles, e.g. silica nanoparticles.

Nanomaterial (NM) All materials designed to have at least one dimension less than 100 nm and exhibit unique qualities determined by this size (NNI 2011. National Nanotechnology Initiative 2011 Environmental, Health, and Safety Strategy). See also Chapter 1 for a detailed discussion on the definition of a nanomaterial.

Nanomedicine Covers the use of ENM for diagnostic and therapeutic purposes. The term may also be used, however, to include the adverse effects of ENM.

Nanoparticles (NP) Material which has all three dimensions in the nanoscale (of the order of 100 nm or less).

Nanotechnology Design, characterization, production and application of structures, devices and systems by controlling shape and size at the nanoscale.

Nanotoxicology Study of the toxicology of nanomaterials.

Nominal size The mean size of a given (nano)material.

Neutron activation Process whereby neutron radiation induces radioactivity in materials, occurs when atomic nuclei capture free neutrons. Methods using radioactive tracers can track movement of nanoparticles around the body.

Neural network A method for generating a quantitative structure—activity relationship (QSAR) relationship using an algorithm that "learns" the relationship in an analogous way to how the human brain learns. This is a key method in computational nanotoxicology.

Neuron The key cells of the nervous system that process and transmit information by electrical and chemical signaling. Experiments have shown damage to neurons which may be potentially relevant for human disease. However, it still uncertain if ENM may cross the blood—brain barrier (see above) *in vivo*.

OECD The Organisation for Economic Cooperation and Development (OECD) is an international economic organization of 34 countries with headquarters in Paris which aims to stimulate economic progress and world trade. The OECD established a Working Party on Manufactured Nanomaterials (WPMN) in 2006 in order to collaborate internationally on the safety research of nanomaterials (see Chapter 6).

Optical particle counter (OPC) Provides real-time number concentration measurements between their particle diameter detection limits. Particle size diameters begin at 300 nm and may go up to 10 000 nm. Although not measuring in the nano-range, this instrument may nevertheless be useful, associated with other instruments, in the screening of nanoparticle emissions in an occupational setting.

Overload An animal (rat)-specific response whereby normal particle clearance is overcome at high lung burden of low toxicity dust that culminates in fibrosis and cancer. Its relevance for human disease is questionable.

Oxidative stress An imbalance between the generation of harmful reactive oxygen species (ROS) and the cells' ability to neutralize these oxidants, resulting in damage to cellular components including proteins, lipids and DNA. It is considered by many researchers to be the main pathogenetic mechanism of ENM.

PEGylation Coating of particulates or proteins with polyethylene glycol or PEG. PEGylation may have several purposes in improving solubility, stability, masking or otherwise changing the physicochemical properties of pharmaceuticals. It is used in studies of biodistribution of nanomaterial and in nanomedicine (see this term).

Phagocytes Cells that engulf particles and microbes by the active process of phagocytosis. It is presumed that some ENM may impair the phagocytes' function.

Phagocytosis The cellular process of engulfing solid entities (microorganisms, particles, debris) by invagination of the cell membrane to form an internal phagosome (see also phagocytes).

Photoacoustic imaging (PAI) A newly emergent molecular imaging technique that takes advantage of both optical and acoustic imaging and thus overcomes the depth limitation of most traditional optical imaging techniques. The optical energy is absorbed and partially converted into heat, resulting in transient thermoelastic expansion and ultrasonic emission, which is then detected by ultrasonic transducers to form images. Some ENM like gold nanoparticles and carbon nanotubes are being actively investigated as potential agents enhancing the imaging power of this technique.

Placenta The organ that develops during pregnancy of most mammalian species on the interface between the pregnant organism and the fetus. It functions as an exchange surface for gases and nutrients between the mother and the fetus, without the mixing of maternal and fetal blood. At birth, the placenta is delivered soon after the baby. Recent experimental data show that some ENM may cross the placenta, with the potential to induce direct damage to the embryo, and some others may induce morpho-functional damage to the placenta itself, thus causing indirect damage to the embryo.

Pleura The layer that covers the lungs (visceral pleura) and inside of the chest wall (parietal pleura) such that the two pleura are opposed to each other during normal breathing. Some fiber-like ENM may be unable to cross the stomata (see this term) and thus become trapped between visceral and parietal placenta. The main concern linked to this event is the potential for the development of mesothelioma, a malignant tumor of the pleura associated to other fiber-like particles such as asbestos.

Point mutation Localized DNA alterations involving only one or a few bases; for instance, a deletion of base(s) from a sequence, inclusion of bases in a given sequence, or replacement with a different base (known as a base substitution) may occur. In some experiments point mutations have been shown to be induced by ENM (e.g carbon nanotubes).

Polymorphonuclear leukocytes (PMN) Short-lived cells equipped with an arsenal of oxidative and proteolytic processes for the killing of microorganisms that usually die rapidly upon encounter with the offending agent (indeed, their death is a key element in pathogen destruction). PMN activation has been described in some animal studies after exposure to ENM.

Positron emission tomography (PET) Nuclear medicine imaging technique used to obtain computer-produced 3D images of functional processes in the body through the detection of pairs of photons emitted indirectly by a positron-emitting radionuclide (e.g. ^{11}C, ^{18}F, ^{64}Cu, ^{94m}Tc, ^{124}I, etc) which could be introduced to biomolecules or nanoparticles.

Proximal alveolar region The proximal parts of the terminal airways and their surrounding parenchyma. This is a potential target for ENM-induced pulmonary damage.

QSAR Quantitative structure–activity relationship, a statistical or computational modeling method that finds relationships between the intrinsic properties of molecules of nanoparticles and their physical or biological properties.

Quantum dots Crystalline nanoparticle that exhibits size-dependent properties due to quantum confinement effects on the electronic states. May be used in imaging techniques, but their use is currently limited by their potential toxicity.

Raman spectroscopy Technique used to study vibrational, rotational or other low frequency modes in a system. Relies on inelastic (or Raman) scattering of monochromatic light (usually from a laser) through interaction of the laser light with the vibrations or other excitations in the system. This technique is being used for tracking of some ENM (e.g. carbon nanotubes) in biological systems.

Reference material A material which is used for calibration, method validation, the establishment of metrological traceability, method development, and other quality control purposes. The use of appropriate reference materials is very important for a correct evaluation of experimental studies on ENM.

Reticuloendothelial system (RES) Consists of the phagocytic cells located in reticular connective tissue. The cells primarily include the Kupffer cells of the liver, tissue histiocytes, monocytes and macrophages in lymph nodes and spleen. It has the ability to take up ENM more avidly than other cells. Thus, the tissues more rich in RES cells are preferential sites of accumulation (and potential toxic effects) of ENM.

Risk Possibility or expected frequency of occurrence of a harmful event (death, injury or loss) arising from exposure to a chemical or physical agent. The evaluation of the risk posed by ENM is a crucial issue of current nanosafety research.

Risk assessment Identification and quantification of the hazard resulting from a specific use or occurrence of a chemical or physical agent. Risk assessment considers any possible harmful effects on individual people or on society of using the chemical or physical agent in the amount and manner proposed and via all possible routes of exposure. Quantification ideally requires the establishment of dose–effect and dose–response relationships in likely target individuals and populations. Risk assessment is generally performed based on the chemical composition of the material in question. In the case of nanomaterials, however, this may not be sufficient as chemicals generally considered non-toxic in macro-scale ranges may become toxic in the nano-scale range. This thus constitutes one of the major challenges today.

REACH Registration, Evaluation, Authorisation and Restriction of Chemicals (REACH) is a European Union Regulation that addresses the production and use of chemical substances, and their potential impacts on both human health and the environment. REACH is the strictest law to date regulating chemical substances and will affect industries throughout the world. REACH entered into force on 1 June 2007.

Reactive oxygen species (ROS) Oxygen-based ions or radicals that are very chemically reactive. They are formed during normal aerobic cellular metabolism and play an important role in several cellular functions (e.g. immune defenses). Their excess may cause cellular damage, so a balance is maintained by antioxidants that scavenge or neutralize ROS (see also oxidative stress).

Scanning electron microscopy (SEM) An imaging technique that provides highly magnified images of surface morphology (10–50 nm resolution). Useful for ENM identification.

Scanning mobility particle sizer (SMPS) This instrument measures particle diameters from 2.5 to 1000 nm and can display data as a size and number distribution using up to 167 size channels. It is being used in the occupational setting for assessing the presence of nanomaterials.

Shape Physical shape or morphology of particles including surface structure at the nanoscale. It is considered a key element in the biological activity of nanomaterials (i.e. shape is an independent factor in determining ENM toxicity).

Single-photon-emission computed tomograpy (SPECT) A nuclear imaging technique using gamma rays emitted by radionuclides. Similar to conventional nuclear planar imaging, SPECT imaging is performed by using a gamma camera to acquire multiple two-dimensional (2D) projection images from multiple angles. Its performance may be improved by ENM and, conversely, may be useful for ENM tracking *in vivo*.

Size selective sampler (SSS) Cascade impactors are currently the only devices giving a direct estimation of mass concentration with a cut-off point around 100 nm. These are being used to evaluate the presence of nanomaterials in occupational settings.

Stomata Small holes in the parietal pleura involved in clearance of particles and fibers (see pleura).

Superparamagnetic iron oxide nanoparticles (SPIONs) Nanoparticles which may improve the performance of MRI imaging (see above).

Surface area Ideally, the surface area of particles involved in biological interactions (biologically relevant particle surface area). This is perhaps the best metric associated with the biological activity of ENM.

Surface plasmon field A property of certain metal nanoparticles such as gold to generate oscillating electrons around the metal surface with the application of light. This property is very useful to improve the performance of some diagnostic imaging techniques such as MRI (see above).

Surfactant (surface active agent) Compound that lowers the surface tension of a liquid, allowing dispersion of insoluble compounds in solution. Frequently used in experiments involving ENM, thereby posing the problem of interpretation of the data, due to the potential intrinsic toxicity of some surfactants.

Surfactant (pulmonary) Phospholipid and protein layers located in the lungs at the air–liquid interface and made up of 85–90% phospholipids and about 10% surfactant proteins. Surfactant is very important in the biological evaluation of ENM in the lung, since it causes a phospholipid coat (corona—see this term) on the surface of ENM, which modifies its biological interactions.

Tapered element oscillating microbalance (TEOM) This method measures nanoaerosol mass concentration on-line with a suitable size selective inlet. It is being used in the occupational setting.

Teratogen Chemicals and environmental agents that are able to interfere with embryonic or fetal development. Formerly used of agents inducing structural malformations, but at present also refererring to chemicals that induce other kinds of maldevelopment, e.g. developmental neurotoxicity. Recent experimental data suggest that some ENM (e.g. carbon nanotubes) may be teratogenic.

Theranostic Describes a nanoparticle that can be used both for diagnostic and therapeutic applications.

Tight junctions Areas of close membrane contact between adjacent epithelial cells, underpinned by protein complexes, that form a barrier to the movement of material between cells and separate the distinct apical and basolateral membrane domains. They may nevertheless be crossed by ENM.

Toxicokinetics and toxicodynamics The branch of toxicology that studies how toxins are absorbed by, distributed in, metabolized by, and eliminated from the living systems. The pathogenetic mechanisms by which the toxin causes toxicological effects on the body are called toxicodynamics. The terms particokinetics and particodynamics may be used when studying these processes for ENM.

Transmission electron microscopy (TEM) Transmission electron microscopy is an imaging technique that uses a beam of highly energetic electrons to examine objects very closely, on a fine scale (it has a resolution <10 nm). Very useful for the detection of ENM in biological samples.

Ultrafine particles (UFP) Particles with a nominal diameter (such as geometric, aerodynamic, mobility, projected-area or otherwise) of 100 nm or less. The term is often used in the context of particles produced as a by-product of a process such as fuel combustion, i.e. incidental nanoparticles.

Villus epithelium The surface layer of cells covering the villi, finger-like protrusions within the small intestine. The cells lining the villi have microvilli at their apical surface (see above).

X-ray luminescence computed tomography (XLCT) XLCT is a new dual modality molecular imaging technique based on x-ray excitable phosphorescent nanoparticles.

Zeta potential The potential difference between the dispersion medium and the stationary layer of fluid attached to the dispersed particle. The zeta potential is an ENM characteristic linked to their biological (toxicological) effects.

Index

K

L

M

N

Color Plates

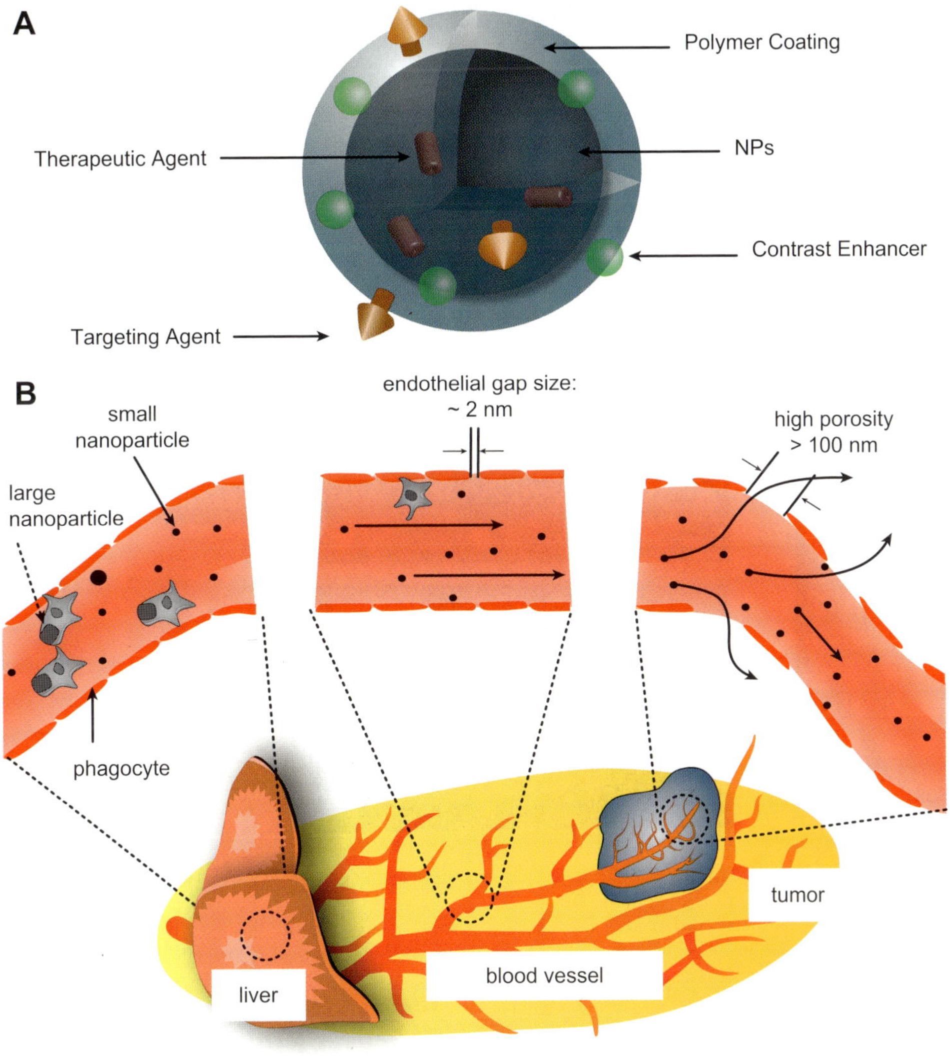

FIGURE 15.1 (A) Multifunctional nanoparticles (NP). Coupled with biological targeting agents and drug molecules, such multifunctional NPs have shown great promise as target-specific delivery vehicles with enhanced imaging sensitivity and therapeutic efficacy. (B) Behavior of nanoparticles in blood vessels in the liver, normal tissue, and tumor tissue regions. The enhanced permeability and retention (EPR) effect of the tailored multifunctional NPs is greatest at cancer cells. *(Reprinted with permission from [3]. Copyright 2005, Nature Publishing Group.)*

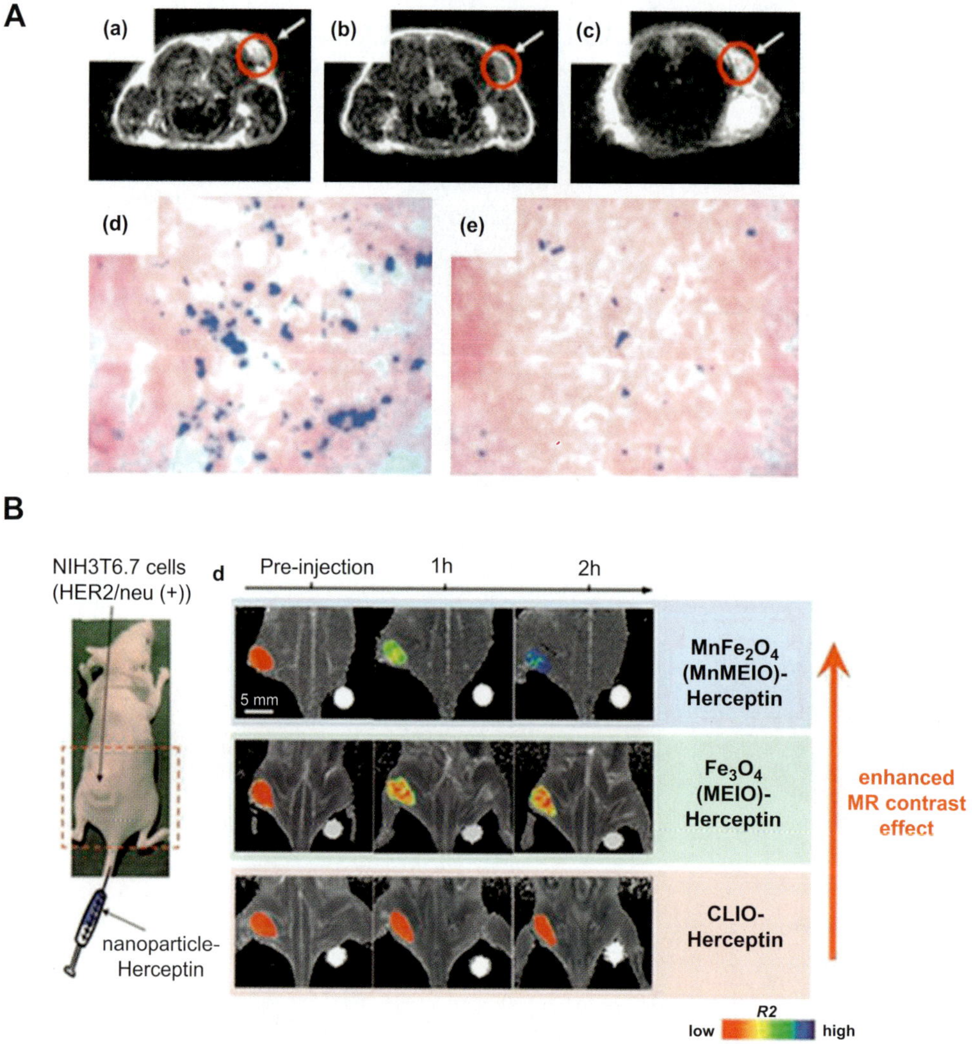

FIGURE 15.2 (A) MRI of U87MG tumor implanted in a mouse: (a) without NPs, (b) with c(RGDyK)-Fe_3O_4, and (c) blocking; and Prussian blue staining of U87MG tumors in the presence of (d) c(RGDyK))-Fe_3O_4 NPs and (e) blocking. (B) Highly-sensitive *in vivo* cancer detection by utilization of $MnFe_2O_4$ (MnMEIO)-Herceptin nanoparticle probes. For comparison, MEIO-Herceptin probes and CLIO-Herceptin probes were also tested. Color-mapped MR images of the mouse at different times following injection. *(Reprinted with permission from [16,18]. Copyright 2008, American Chemical Society. Copyright 2007, Nature Publishing Group.)*

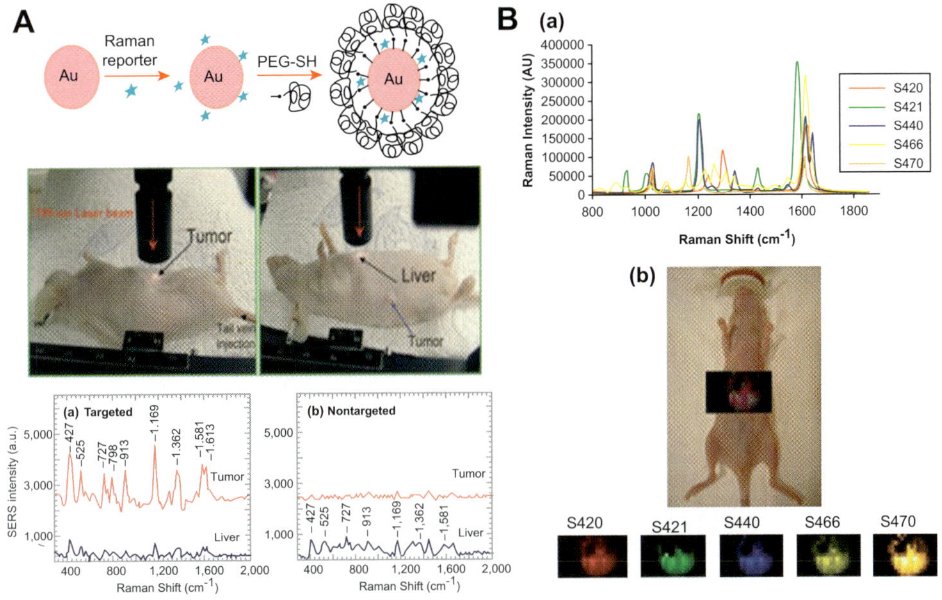

FIGURE 15.3 (A) Design and characterization of PEG-SH coated SERS (surface-enhanced Raman scattering) nanotags. *In vivo* cancer targeting and surface-enhanced Raman detection by using ScFv-antibody conjugated gold nanoparticles that recognize the tumor biomarker EGFR. Top: Photographs showing a laser beam focusing to the tumor site or to the anatomical location of liver. Lower: SERS spectra obtained from the tumor and the liver locations by using (a) targeted and (b) non-targeted nanoparticles. *In vivo* SERS spectra were obtained from the tumor site (red) and the liver site (blue) with 2-s signal integration and at 785 nm excitation. The Raman reporter molecule is malachite green, with distinct spectral signatures as labeled. (B) Demonstration of deep-tissue multiplexed imaging 24 h after i.v. injection of five unique SERS nanoparticle batches simultaneously. (a) Graph depicting five unique Raman spectra, each associated with its own SERS batch: S420 (red), S421 (green), S440 (blue), S466 (yellow), and S470 (orange). (b) Raman image of liver overlaid on digital photo of mouse, showing accumulation of all five SERS batches in the liver after 24 h post i.v. injection. Panels below depict separate channels associated with each of the injected SERS nanoparticle batches. *(Reprinted with permission from [51,57]. Copyright 2008, Nature Publishing Group. Copyright 2009, National Academy of Science.)*

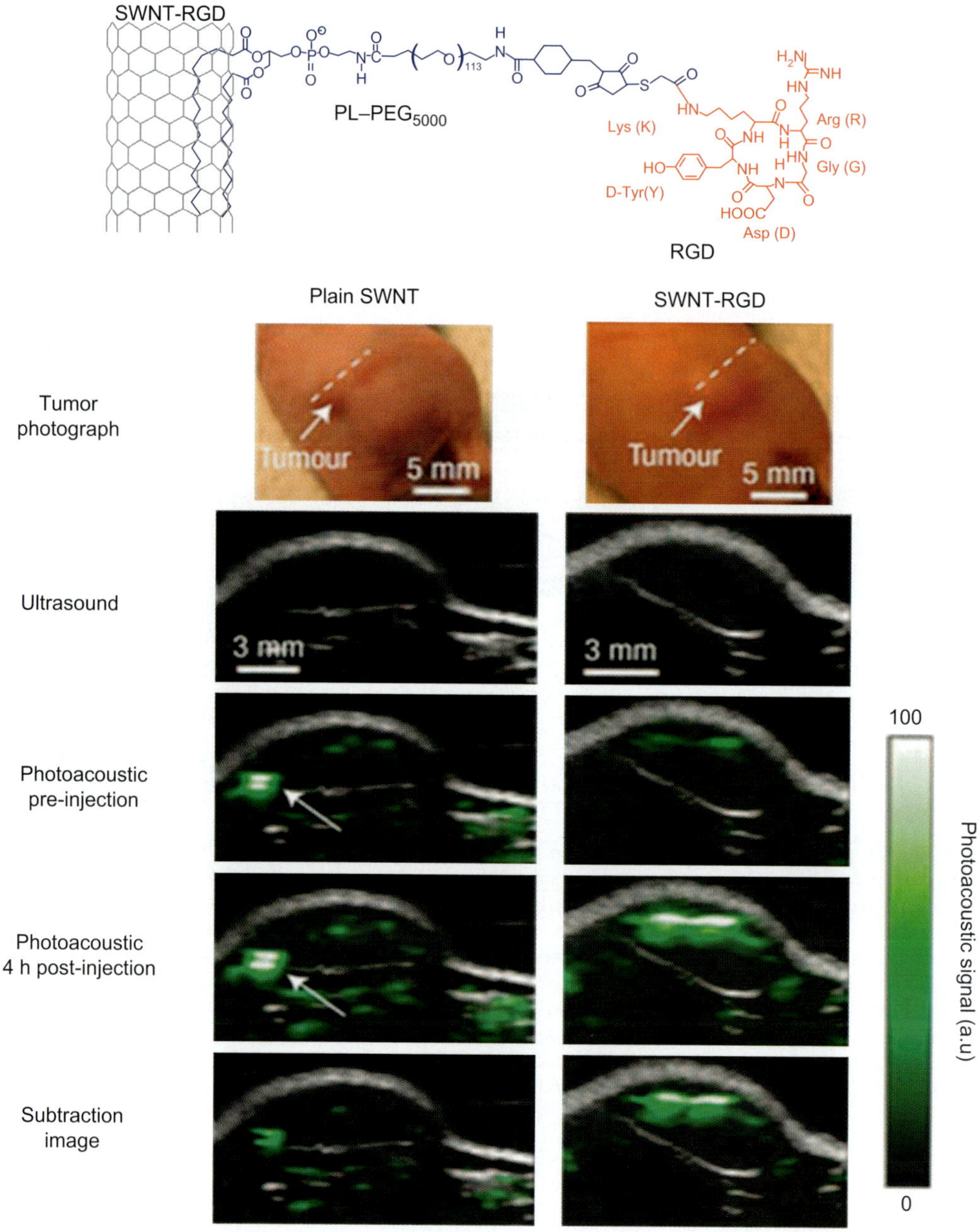

FIGURE 15.4 Single-walled carbon nanotube targets tumor in living mice. Ultrasound (gray) and photoacoustic (green) images of one vertical slice (white dotted line) through the tumor. The ultrasound images show the skin and tumor boundaries. Subtraction images were calculated as the 4 h post-injection image minus the pre-injection image. The high photoacoustic signal in the mouse injected with plain single-walled carbon nanotubes (indicated with a white arrow) is not seen in the subtraction image, suggesting that it is due to a large blood vessel and not single-walled carbon nanotubes. *(Reprinted with permission from [75]. Copyright 2010, American Chemical Society.)*